Thomas Wunder

Essentials of Strategic Management

Effective Formulation and Execution of Strategy

2016
Schäffer-Poeschel Verlag Stuttgart

Dr. Thomas Wunder is Professor of Strategic Management at the Neu-Ulm University of Applied Sciences in Germany. He has been teaching strategy at universities in Switzerland and the USA. Previously he was president of Péter Horváth & Partners Inc. in Boston und Atlanta as well as Executive Relationship Manager at Highland Worldwide. He received his doctorate from the European Business School (EBS). His research interest areas are strategy execution, international strategy alignment, and sustainable strategic management.

Dozenten finden weitere Lehrmaterialien unter
www.sp-dozenten.de

FSC
www.fsc.org
MIX
Papier aus verantwortungsvollen Quellen
FSC® C103849

Gedruckt auf chlorfrei gebleichtem, säurefreiem und alterungsbeständigem Papier

Bibliografische Information der Deutschen Nationalbibliothek
Die Deutsche Nationalbibliothek verzeichnet diese Publikation in der Deutschen Nationalbibliografie; detaillierte bibliografische Daten sind im Internet über http://dnb.d-nb.de abrufbar.

Print ISBN 978-3-7910-3285-6 Bestell-Nr. 20635-0001
EPDF ISBN 978-3-7992-6812-7 Bestell-Nr. 20635-0150

www.schaeffer-poeschel.de
info@schaeffer-poeschel.de

Lektorat: Traudl Kupfer, Berlin
Einbandgestaltung: Umschlagentwurf: Goldener Westen
Satz: Claudia Wild, Konstanz
Umschlaggestaltung: Kienle gestaltet, Stuttgart (Abbildung: Shutterstock)
Druck und Bindung: Schätzl Druck & Medien GmbH & Co. KG, Donauwörth

Printed in Germany
Januar 2016

Schäffer-Poeschel Verlag Stuttgart
Ein Tochterunternehmen der Haufe Gruppe

Preface

Regardless of their position, career aspiration, role or department, most practitioners have to deal with aspects of strategic management either in making or in supporting strategic decisions. This book provides the essentials to master this task successfully. It is written for students who are on their way to becoming practitioners as well as for practitioners who want to look—again or for the first time—at the most essential and practical aspects of strategic management.

A general trend for textbooks on strategic management is that authors try to address all theories and research areas and skip nothing which is covered by other textbooks on this subject. Furthermore, there is a general tendency in mainstream strategic management textbooks to focus more on strategic analysis and on formulating strategies but less on executing them. This book takes a different approach in two ways:

- First, the guiding principle in deciding what to include and what to leave out is to answer two key questions. Is it actionable and does it have a significant impact? Instead of providing an overview over all theories and aspects of the academic field, the book provides a selection of scientifically grounded approaches, methods and ideas that promise to be practical and to make a difference in a company.
- Second, knowing that a great portion of strategies fail due to bad execution, great emphasis is put on this area. In practical strategy work as well as strategy literature, strategy execution typically gets relatively little attention—sometimes even viewed as "graveyard of strategy"—although it is one of the biggest strategic management challenges of today.

This book is intended to provide students and practitioners with a selection of practically applicable methods at a level of detail that is necessary to make an impact in their future or current job situations. With this focus, the book should be particularly relevant for *students of applied sciences*. For practitioners the book provides actionable approaches to deal with their strategic situations in a systematic and comprehensive, yet pragmatic way. I focus on the essentials of strategic management drawing from state-of-the-art literature and own experience aided by the fortune of spending the past 15 years supporting leaders of mainly large and medium size multinational companies to refine and execute strategies. Having spent more than half of my professional career in the United States of America, I had the chance to not only consult on the topic but to practice it myself at the helm of a small U. S. subsidiary where I had to develop and execute my own business strategy. For six years I was a direct report to the CEO of Horváth & Partners Management Consultants, from whom I learned many essentials about strategy and leadership.

During this time I also had the opportunity to learn from two inspiring U. S. consulting firms we were affiliated with. Working two years in Lincoln, Massachusetts, surrounded by the excellent consultants of the Balanced Scorecard Collaborative (now "Palladium Group") helped me to understand how to innovate continuously and make intellectually highly sophisticated ideas actionable. Key management concepts for successful strategy execution originally developed by the two founders, Harvard Business School Professor Robert S. Kaplan and David P. Norton, and further advanced by Horváth & Partners with some bestselling publications in Europe, are drawn on in the second part of this book. During another four years as an affiliate to the North Highland Company in Atlanta, Georgia, I was fortunate to learn what true customer dedication, sales excellence and effective business development are like. I was fortunate to gain a lot both professionally and personally being exposed to their unique company culture, their business spirit and the outstanding strategic leadership skills of their CEO.

How this Book Works

The book is designed to be actionable and user-friendly. One of the most important principles that I learned during my career is the power of simplicity. When dealing with complex organizations and facing complex strategic challenges the worst thing to do is adding complex strategic management processes, tools and methods. I tried to focus on the essentials and to illustrate many ideas with about 200 easy to understand charts. Throughout the book there are "Strategy Practice" boxes presenting real-life examples from globally renowned companies such as BAYER, Beiersdorf, BMW, Bosch, DB Group, Henkel, Hewlett-Packard, Hilti, Nestlé, Procter & Gamble, PUMA, Quiagen, Recaro, SAP, Siemens, Toyota, Unilever, Volkswagen, and ZEISS. I am grateful that many of those companies gave permission to present their examples in its original company format. This makes strategy approaches and methodologies tangible for readers and underlines their relevance and way of application in company practice today. Although many examples relate to large multinational firms, methods are generally applicable to businesses or non-profit organizations of any size and industry. Historical, anecdotal, more academic or other adjacent elaborations that I consider helpful will be briefly explained in "Strategic Snapshot" boxes. If you are a student, these boxes are detailed enough to contain valuable information for a better comprehension of the topic or for encouraging further readings. If you are a time-lacked practitioner and want to focus fast on key methods that you need to consider for your strategy work, you might skip those boxes.

The book is based on the idea that people and organizations that engage in the essentials of a systematic strategic management process are generally more likely to outperform those that do not. The foundations of strategic management are laid in chapter one. Key approaches for strategic analysis and for formulating strategies including the more recent approach of business modeling are presented in the following two chapters. The fourth chapter of this book is dedicated to one of the biggest challenges in today's strategic management: strategy execution. It is intended to help you understand why even the best strategies might not come true and how to get it right. Detailed guidelines will demonstrate how to build an effective strategy execution system. Throughout the book there are a number of management approaches and examples which relate to the important link of strategic management and sustainability. They illustrate practical ways of how to integrate economic, ecological and social factors. At the end of this book there is an exercise designed to facilitate an application of the essentials of strategic management explained in the various chapters from analysis through strategy formulation to execution and control: the Strategy Workout. With this exercise, readers will be able to take what they have learned in each step of the strategy process and apply it to their own or any other company in a seamless and integrated way.

In order to capture the key components and relationships of the strategic management approach presented in this book I developed a simple application-oriented strategy framework which I called "Wheel of Strategy" (WOS). It is meant to provide coherent and illustrative guidance of how to bring strategic management with its essential methods and ideas to life in company practice. It is a bird's eye perspective on the set of integrated elements required to develop strategies and make them happen. Review questions for each chapter are summarized in the appendix at the end of this book, which provides students in particular an opportunity to reflect what they have learned and prepare for an exam.

Strategy can be fun, so, enjoy yourself, work hard and use your imagination.

Origins of this Book and Acknowledgements

The initial roots of this book were laid in 1995 at the University of Birmingham (UK) where I was fortunate to visit a Strategic Management class taught by an inspiring strategy practitioner. His hands-on ideas on how to devise and implement strategy and what can go wrong particularly related to, what he called, "the abnormal psychology of senior leaders" inspired me to further pursue this topic. Many ideas in this book are drawn from my experience gathered during the 15 years as a management consultant, doctoral student, business leader, and finally full professor of strategic management since then. The practical foundations were laid at Horváth & Partners Management Consultants where I had the chance to work strongly with their global strategy practice. Multinational corporations facing the challenge of executing corporate strategy and aligning their international subsidiaries, strategic business units and functional areas around common strategic themes gave me the opportunity to conduct strategy sessions in more than 15 countries and learn from strategy discussions in highly intercultural management teams. During my doctorate in strategic management at the European Business School (EBS), I was fortunate to build on that experience and work on the question of how to implement transnational strategies effectively. Coming from a highly planning-based strategy approach from management consulting, it was particularly inspiring to discuss this topic with academics that fundamentally questioned the controllability of companies and approached strategic management in a completely different way, that is complexity and evolutionary management. Some of those thoughts are addressed in this book although, for practical reasons, most of it is definitely more on the planning side.

I would like to thank some special people without whom this book would not have been possible. First and foremost, I sincerely appreciate the sacrifice writing this textbook has meant for my wife and our two little children. Over the last few years, I spent more hours on conducting research, writing texts and developing charts than when I was a consultant or graduate student. Therefore, I dedicate this book to my family. I was also fortunate to work with Schäffer-Poeschel publishing house that approached me a few years ago with the idea to write an application-oriented, concise and innovative strategic management textbook in English leveraging my practical experience and academic background. In addition to the entire Schäffer-Poeschel team, I would like to particularly thank my superb editor, Martin Bergmann, for his appreciated ideas, flexibility and great support to make this project come true. I would also like to thank Traudl Kupfer and Claudia Dreiseitel for professionally proofreading my manuscript and for lots of useful feedback. The Neu-Ulm University of Applied Sciences (HNU) provided a conductive environment and outstanding institutional support to make this project possible. I am grateful to the University Executive Board and the Dean of the Department of Business and Economics for providing their appreciated support. I would also like to thank my students at HNU and other universities at master and undergraduate levels as well as executive MBA programs. They provided lots of useful insights with their intellectual curiosity, perspectives, opinions, and questions. Last but not least, I wish to thank the reviewers both in Germany and the United States who shared their expertise with me. Knowing that any list of acknowledgements will almost always be incomplete, I am grateful to all people who supported this project and helped to make it happen. I feel truly gifted by their sophisticated feedback and support.

Thomas Wunder

Table of Contents

PART I:
Developing Strategies: Strategic Analysis, Ideation, and Choice

Part II:
Bringing it into Reality: From Strategy to Action and Control

Notes to the Reader

The content structure of this textbook is illustrated by the reader-friendly layout that facilitates learning and working with the content in many different ways.

Marginalia: Marginalia next to the text briefly guide readers through the essential content of each chapter. They serve for orientation, underline the structure of the text and summarize the passage. Additionally, marginalia facilitate finding certain keywords and sections quickly. Furthermore, students can use the additional space in the margin for their own notes.

Learning objectives: Each chapter aims to achieve several „learning objectives", which are mentioned right in the beginning of each chapter. These learning objectives introduce the following topics and indicate the knowledge and skills that readers are supposed to acquire.

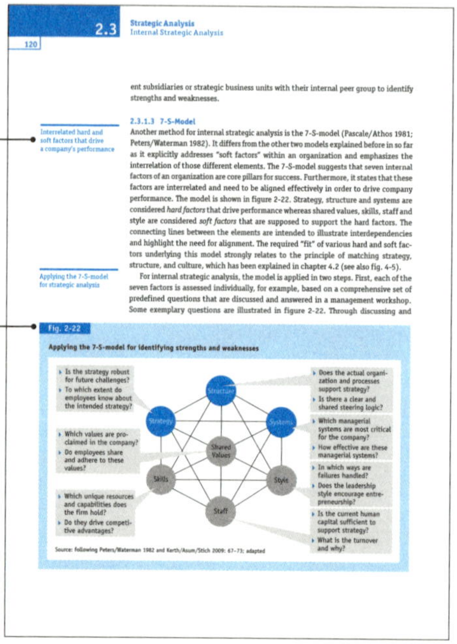

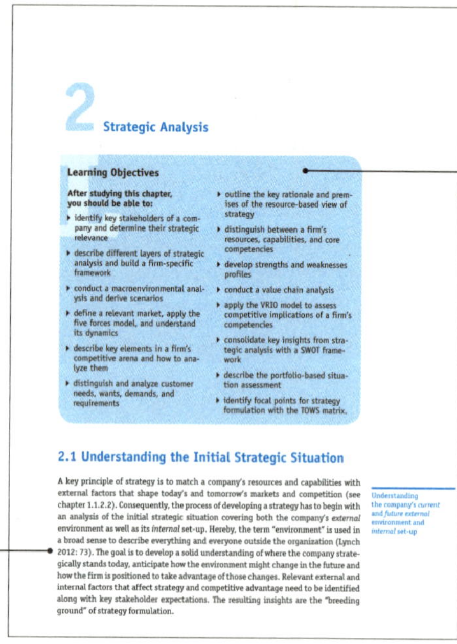

Figures: Numerous figures visualize and amplify the ideas that are described in the text. The text makes reference to and refers to the figures where relevant.

References: References are mentioned directly within the text, namely including the name of the authors as well as the year of publication. At the end of this book all reference literature and indication of sources are listed completely.

Strategic Snapshots: Valuable additional aspects such as historical, anecdotal, more academic or other explanatory elaborations are provided in highlighted boxes. They are suitable for a better comprehension of the topic and for encouraging further readings.

Strategy Practice Examples: Comprehensive additional examples, mostly from corporate practice, illustrate the theory. They are also pointed out separately.

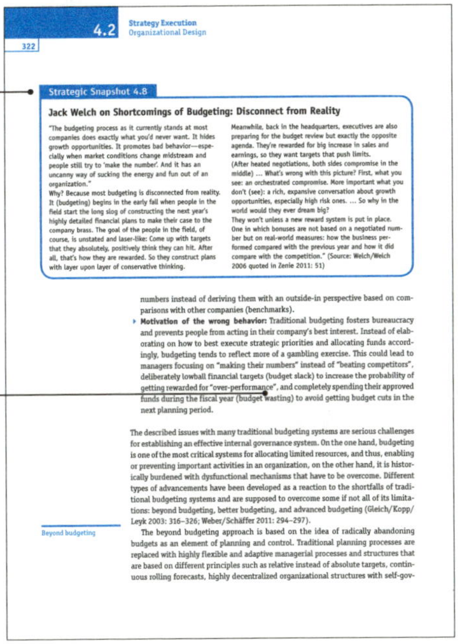

4.2 Strategy Execution
Organizational Design

322

Strategic Snapshot 4.8

Jack Welch on Shortcomings of Budgeting: Disconnect from Reality

"The budgeting process as it currently stands at most companies does exactly what you'd never want. It hides growth opportunities. It promotes bad behavior—especially when market conditions change midstream and people still try to 'make the number'. And it has an uncanny way of sucking the energy and fun out of an organization."
Why? Because most budgeting is disconnected from reality. It (budgeting) begins in the early fall when people in the field start the long slog of constructing the next year's highly detailed financial plans to make their case to the company brass. The goal of the people in the field, of course, is unstated and laser-like: Come up with targets that they absolutely, positively think they can hit. After all, that's how they are rewarded. So they construct plans with layer upon layer of conservative thinking.
Meanwhile, back in the headquarters, executives are also preparing for the budget review but exactly the opposite agenda. They're rewarded for big increase in sales and earnings, so they want targets that push limits.
(After heated negotiations, both sides compromise in the middle) ... What's wrong with this picture? First, what you see: an orchestrated compromise. More important what you don't (see): a rich, expansive conversation about growth opportunities, especially high risk ones. ... So why in the world would they ever dream big?
They won't unless a new reward system is put in place. One in which bonuses are not based on a negotiated number but on real-world measures: how the business performed compared with the previous year and how it did compare with the competition." (Source: Welch/Welch 2006 quoted in Zenie 2011: 51)

numbers instead of deriving them with an outside-in perspective based on comparisons with other companies (benchmarks).

- **Motivation of the wrong behavior:** Traditional budgeting fosters bureaucracy and prevents people from acting in their company's best interest. Instead of elaborating on how to best execute strategic priorities and allocating funds accordingly, budgeting tends to reflect more of a gambling exercise. This could lead to managers focusing on "making their numbers" instead of "beating competitors", deliberately lowball financial targets (budget slack) to increase the probability of getting rewarded for "over-performance", and completely spending their approved funds during the fiscal year (budget wasting) to avoid getting budget cuts in the next planning period.

The described issues with many traditional budgeting systems are serious challenges for establishing an effective internal governance system. On the one hand, budgeting is one of the most critical systems for allocating limited resources, and thus, enabling or preventing important activities in an organization, on the other hand, it is historically burdened with dysfunctional mechanisms that have to be overcome. Different types of advancements have been developed as a reaction to the shortfalls of traditional budgeting systems and are supposed to overcome some if not all of its limitations: beyond budgeting, better budgeting, and advanced budgeting (Gleich/Kopp/Leyk 2003: 316–326; Weber/Schäffer 2011: 294–297).

Beyond budgeting

The beyond budgeting approach is based on the idea of radically abandoning budgets as an element of planning and control. Traditional planning processes are replaced with highly flexible and adaptive managerial processes and structures that are based on different principles such as relative instead of absolute targets, continuous rolling forecasts, highly decentralized organizational structures with self-gov-

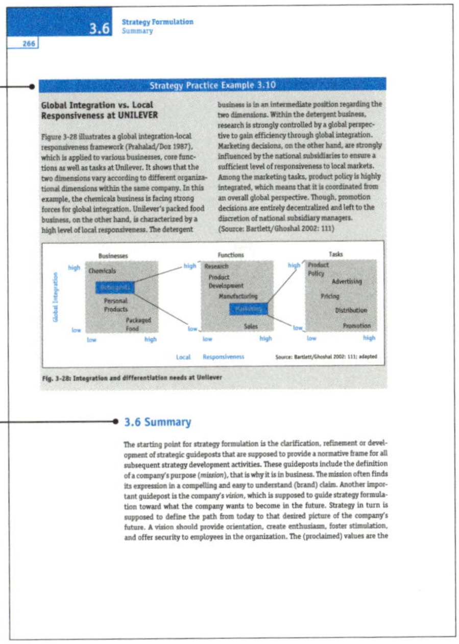

3.6 Strategy Formulation
Summary

266

Strategy Practice Example 3.10

Global Integration vs. Local Responsiveness at UNILEVER

Figure 3-28 illustrates a global integration-local responsiveness framework (Prahalad/Doz 1987), which is applied to various businesses, core functions as well as tasks at Unilever. It shows that the two dimensions vary according to different organizational dimensions within the same company. In this example, the chemicals business is facing strong forces for global integration. Unilever's packed food business, on the other hand, is characterized by a high level of local responsiveness. The detergent business is in an intermediate position regarding the two dimensions. Within the detergent business, research is strongly controlled by a global perspective to gain efficiency through global integration. Marketing decisions, on the other hand, are strongly influenced by the national subsidiaries to ensure a sufficient level of responsiveness to local markets. Among the marketing tasks, product policy is highly integrated, which means that it is coordinated from an overall global perspective. Though, promotion decisions are entirely decentralized and left to the discretion of national subsidiary managers.
(Source: Bartlett/Ghoshal 2002: 111)

Fig. 3-28: Integration and differentiation needs at Unilever

3.6 Summary

The starting point for strategy formulation is the clarification, refinement or development of strategic guideposts that are supposed to provide a normative frame for all subsequent strategy development activities. These guideposts include the definition of a company's purpose (*mission*), that is why it is in business. The mission often finds its expression in a compelling and easy to understand (brand) claim. Another important guidepost is the company's *vision*, which is supposed to guide strategy formulation toward what the company wants to become in the future. Strategy in turn is supposed to define the path from today to that desired picture of the company's future. A vision should provide orientation, create enthusiasm, foster stimulation, and offer security to employees in the organization. The (proclaimed) values are the

Index: An index at the end of this book provides a fast and easy access to terms, definitions, tools and methods.

Summaries: At the end of each chapter readers will find consolidated repetitions of the key subjects of these sections. Summaries can be used well as part of the exam preparation or for a quick repetition of chapters that have been worked through some time ago.

Strategy Workout and Review Questions: The appendix provides a strategy workout and review questions that test students' knowledge and understanding of the central elements and coherencies of all chapters and should be transferred to cases from practice. If students identify knowledge gaps or uncertainties, we recommend to repeat the equivalent sections thoroughly once again.

1 Strategic Management Foundations

Learning Objectives

After studying this chapter, you should be able to

- explain triggering events for companies to rethink their strategic orientation.
- understand the roots of strategy and their relevance for management today.
- describe different business views of strategy and how they can be linked.
- explain the six principles of strategy and their relevance for strategic management.
- explain ways for measuring competitive advantage and company performance.
- outline foundations of decision making and apply them to strategic situations.
- understand the impact of cognitive biases and recommend how to deal with them.
- explain the evolution, schools of thought and paradigms of strategic management.
- describe the wheel of strategy framework and explain its interrelated components.
- explain how to make strategic management effective and describe its resulting benefits.

1.1 What Is Strategy?

Outperforming competitors and long lasting survival

How does a company outperform its competitors and win in the market? How can it perform well for a prolonged period of time and ensure long-lasting company survival? These are definitely not easy questions to answer. Only few companies have remained in the market and been successful forever and a day. An example is the Bavarian state brewery Weihenstephan, which celebrated its 970th birthday in 2010 and is the oldest company in Germany. Another long-term performer is the Merck Group. Founded in 1668, it is not only the oldest pharmaceutical and chemical company in the world but also ranks among the world's leading firms in this industry today. Other multinational companies like Bosch, Daimler, Henkel or Siemens have been writing their success stories for more than 100 years. On the other hand, firms such as AEG, Chrysler, Nixdorf Computers, Kodak, and Agfa that used to be famous for their innovative products in times past have gone out of business, were swallowed up by others or had to sell major divisions.

Strategic Snapshot 1.1

"Few large corporations live even half as long as a person"

"Few large corporations live even half as long as a person. In 1983, a Royal Dutch/Shell survey found that one third of the firms in the Fortune '500' in 1970 had vanished (de Geus 1988). Shell estimated that the average lifetime of the largest industrial enterprises is less than forty years, roughly half the lifetime of a human being! The chances are fifty-fifty that readers of this book will see their present firm disappear during their working career. In most companies that fail, there is abundant evidence in advance that the firm is in trouble. This evidence goes unheeded, however, even when individual managers are aware of it. The organization as a whole cannot recognize impending threats, understand the implications of those threats, or come up with alternatives."
(Source: Senge 1990: 17)

Average life span of companies

Looking at the average life span of major companies actually reveals a sobering picture. For example, less than 15% of the largest companies in the US listed in 1917 in Forbes Magazine still exist today. The rest vanished from the business landscape because they either have been taken over by other companies or gone out of business. Looking at even so-called world-class companies identified by Peters and Waterman or by Collins and Porras in their bestselling books "In Search of Excellence" or "Built to Last" reveals a similar picture. Only a minor portion of all companies perform well over several decades (Watson 2012: 287; Foster/Kaplan 2001: 7 f.). The average life span of companies in Europe is about 12 years, 28 years if they are publicly listed on a stock exchange, and 48 years if they are large corporations with more than 10,000 employees or $5 billion market capitalization (Stadler/Wältermann 2012: 10). Most companies that were high performers at a certain time are not able to remain superior in the long run, for example, for 10 years or more (Wiggins/Ruefli 2005 and 2002). Why did so many prominent companies vanish from the business landscape or lose the superior performance level they had at a certain time? They have failed to adapt their *strategy* to the changing environment and had to learn that past performance is no guarantee of future success.

Triggering events to rethink strategy

When do companies typically rethink their strategic orientation and engage in a (re-)definition of their strategy? Some triggering events that may act as stimuli for changes in strategy are (Wheelen/Hunger 2010: 24):

- **Performance gap (unmet performance expectations):** Many companies with revenues or profits that are no longer increasing, falling behind major competitors or even decreasing, rethink their strategic orientation to get back on course in the long term. This may also be the case for anticipated future performance issues caused by, for example, advancing competitors or disruptive technologies.
- **Changes in ownership:** New shareholders may alter financial and strategic expectations of the company and require a new strategic orientation. For example, family-owned companies differ significantly from private equity firms when it comes to risk aversion, short- and long-term profit orientation or the desire for appreciation in the community.
- **New anticipated trends:** New trends such as new technologies or changes in customer preferences may make existing strategies ineffective and require new strategic approaches. For example, Nokia underestimated the trend from cell phones

Strategy Practice Example 1.1

CEO Eras and Strategy at Daimler

Diversification was one of the key corporate strategic thrusts of Edzard Reuter in his era of being CEO of Daimler from 1987 to 1995. He wanted to transform the company into an "integrated technology corporation" and diversified Daimler with acquisitions in the aerospace and electrical industries such as MBB, MTU, Dornier, and AEG. His successor, Jürgen Schrempp, changed this diversification strategy and refocused the group onto the automobile business during his 10 years as CEO. He wanted to refocus the corporation and create a world auto giant with a strong shareholder value orientation. Under his leadership, the Daimler-Benz AG merged with the US corporation Chrysler in 1998. Due to unsatisfactory results the era DaimlerChrysler was finished in 2007 by Dieter Zetsche who has become CEO in January 2006. The company was renamed to its current name Daimler AG.

to smart phones and was forced to rethink its strategic position once companies like Apple introduced their new innovations such as the iPhone.

- **New CEO or executive leadership team:** New executives at the helm of companies usually bring in new strategic ideas and approaches (see Strategy Practice Example 1.1). This is similar to what frequently happens in sports when a new coach or manager comes in and fundamentally changes the strategic thrust.
- **Intervention from other external stakeholders:** Interventions from other external stakeholders than customers and competitors such as governments, non-governmental organizations, banks, etc. may have a significant impact on the company. For example, recent decisions of the German government to exit nuclear energy until 2022 and increase the share of renewable energy sources for supply in Germany to 80% by 2050 triggered significant changes in corporate strategy at some German energy corporations such as E.On.
- **More unstable environment:** More and more companies are facing unstable environments with a high degree for volatility. For example, some firms are required to advance their strategic agility and adapt their strategies more frequently than they used to. Other companies such as ZEISS (see 3.2.2.2) may foster diversification to get a balanced risk profile of their businesses portfolio with regard to turbulence in the global economy (see chapter 3.2.2.2).

1.1.1 Origins and Views of Strategy

Multiple applications of the strategy term

Nowadays, the term "strategy" is used in all kind of areas ranging from computer gaming, gardening, Poker as well as food to sports, dating or even housecleaning. This variety of applications makes it hard to develop a common and shared understanding. The same is true in a business context. The range of strategy interpretations reaches from anecdotal statements like "strategy is what makes money"—a perspective of the CEO of a Fortune 500 company in a strategy workshop—to more philosophical understandings like "strategy is revolution; everything else is tactics" (Hamel 1996: 70). Furthermore, the strategy term is frequently related to its origin in military history.

How relevant this might be for tackling today's management challenges will be discussed first followed by a brief overview of some actual views of strategy that can be applied to company practice today.

1.1.1.1 Roots of Strategy

"Stratos" and "agein"

The etymological roots of the term "strategy" go back to ancient Greek (6th/5th century BC) where—based on the terms stratos (army, military force) and agein (to lead)—stratégos was an expression for an army leader or military general. Later on in history, various military leaders referred to militarily motivated strategies in their renowned publications, such as:

- "The Art of War" written by the Chinese military strategist and mathematician Sun Tzu (544–496 BC). Nowadays it is considered the first strategy book in history. One of his important strategy principles is the idea of victory without fight.
- "The Prince" and "About the Art of War" written by the Italian Renaissance political philosopher and historian Niccoló Machiavelli (1496–1527).
- "The Book of Five Rings" written by Miyamoto Musashi (1584–1645), a Japanese swordsman and samurai. One of the important strategy principles he refers to in his famous publication is the importance of situation assessment from a bird's eye perspective.
- "On War" written by the German-Prussian General and military theorist Carl von Clausewitz (1780–1831). He defined strategy as the use of combat for the purpose of war.

Relevance of military strategy principles for business

Figure 1-1 provides some fundamental principles of historic military strategies that seem to be timeless and relevant also for business management today (Kotler/Berger/Bickhoff 2010: 7): Concentration of resources strongly relates to one of the key challenges in management, that is the effective allocation of limited resources. The element of surprise is leveraged by companies in a variety of ways such as establishing a first mover advantage with their products in certain markets or by unexpected mergers, acquisitions, etc. Furthermore, companies elaborate very carefully on which competitive arenas they are playing in, based on their strengths and core competencies. Communication between the top leadership team and the employee base is a key pillar of any modern strategy execution system. The precise coordination of strategic objectives and resources relates to the management approach of strategy and organizational alignment which is vital for making strategies happen effectively. Finally, how companies can gain a substantial advantage through innovation can be seen over and over again by innovative companies launching new technologies, products or business models.

Limitations of military principles for business management

Anecdotally speaking, strategy "arranges strengths, means, time, space and methods in a guiding principle of action. It is, therefore, nothing else but an efficient success plan, whose fundamental elements *What do I want?—What can I do?* and *What do I do?* have not been changed since Seneca's "Want—Can—Dare" *(Note: Roman philosopher, 4 BC–65 AD)*" (Werle 2005: 197). Historical insights from warlords may provide simple and highly generic inspirations for staying on target and capturing new perspectives—that might be useful in, for example, dynamic situations or when

Fig. 1-1

Applicability of military strategy principles for business management

Source: Kotler/Berger/Bickhoff 2010: 7

competing in multiple markets. However, based on a more critical evaluation, those principles provide hardly any answers for solving strategic challenges today. Although many ideas are timeless and valid, they lack uniqueness. The insights are revolutionary for its time but self-evident for most managers today. Most historic strategy ideas are on a high level. They particularly lack precise recommendations for one of the biggest challenges in strategic management, that is execution. Furthermore, "big names" are frequently used for marketing purposes (see Strategic Snapshot 1.2).

Strategic Snapshot 1.2

Return of the "Warlords"

Although the relevance of military principles for solving today's business challenges can be questioned, there are a number of popularized business publications of the last 20 years leveraging historical strategy perspectives. Examples are:

- "Miyamoto Musashis Book of Five Rings. 52 Brilliant Ideas for Your Business" (2012; published in German)
- "Sun Tzu and The Art of Business. Six strategic principles for Managers" (2011)
- "Sun Tzu. The Art of War for Managers. 50 Strategic Rules Updated for Today's Business" (2010)
- "Clausewitz on Strategy. Inspiration and Insight from a Master Strategist" (2001), a publication of the Strategy Institute of the Boston Consulting Group
- "The New Machiavelli. The Art of Politics in Business" (1999).

Practitioners might be inclined to use the supposed credibility of historic military leaders for reducing complexity in strategic decision-making situations. One of the key limitations of military strategies for business is its traditional focus on opponents which is equivalent to focusing only on competitors in markets. Not only does this neglect cooperative strategies for the most part, but also the customer as most important stakeholder for business is not addressed at all. And finally, translating certain rules of war into business is inappropriate today if the recommendations are not compliant with law or ethical principles (Werle 2005).

1.1.1.2 Business Views of Strategy: the 5 P's

To better understand what strategy means in a business context, it is helpful to recognize different interrelated views of strategy that can be found in management practice and science alike (Mintzberg 2000: 23–29, 1987 and 1978). Understanding and integrating those different views is supposed to reduce some of the confusion that comes with the strategy term and establish a foundational terminological framework that will be applied in this book:

- Most practitioners are likely to define strategy as a **plan** that specifies what the company intends to do and when. It is made purposely in advance of the actions to which it applies.
- Others may understand strategy more as a specific competitive move or **ploy** to preempt an opponent's response in a head-on competitive situation. In this understanding it is also a plan but more in a sense of outmaneuvering an opponent in a 1:1 business setting.
- Another understanding of strategy is specifically related to the **position** in the environment that allows an organization to generate sufficient "rent". In practical terms this might be a particular industry or the financially most promising product-market combination within the competitive arena a company focuses its resources on.
- Whereas the position is outside the organization, the understanding of strategy can also be based on a more internal view. Here, strategy is seen as a collective **perspective** in people's minds. It is a kind of shared mental model that builds the strategic orientation of the company.
- A final understanding of strategy as a **pattern** is not related to the intention of people but to the resulting behavior of an organization. Here the key is consistency in behavior, whether intended or not. Successfully realized strategies are not always planned in advance and planned strategies are not always realized (see also chapter 1.1.1.5).

Integrating the 5 P's of strategy

Figure 1-2 provides an attempt to integrate the different views of strategy in a hierarchical order following the strategic planning logic of this book. Strategic guideposts are established by a company's vision, mission and values. They provide a high-level normative direction and strategic context. This is a view of strategy as *perspective*. Framed by the strategic guideposts the company has to decide where and how it wants to compete based on industry and market attractiveness and dynamics as well as its own resources and capabilities. This is the core of formulating corporate and

Fig. 1-2

5 P's of strategy and strategic planning logic

Source: following Simons 18; adapted

business strategies and reflects a view of strategy as *position*. Once certain strategy options have been decided on and strategy is set, it is refined with strategic goals, quantified with metrics and targets, and translated into strategic action programs as well as corresponding budgets and incentives. This is related to the view of strategy as *plan*. Some of those goals or actions may capture certain moves to outwit rivals or fight competitive threats, that is viewing strategy as *ploy*. Finally, strategy is what really happens. The strategic behavior of an organization may have been planned according to the process described or it may emerge unplanned through learning and trial and error. In any case, it is a consistent stream of actions and decisions. This is strategy as *pattern*. By describing how companies like Ikea, Starbucks, Apple or others present themselves in the market, key strategic elements can be clearly identified as they reflect consistency in behavior of those companies. Cornerstones of their strategies are visible for everybody without knowing whether they were ever planned in formal strategy sessions or just emerged over time.

1.1.2 Understanding Strategy with Six Principles

Strategy is a multifaceted phenomenon that can hardly be described with a single definition. The 5 P's described before are a practical and a frequently used consolidation of various views of strategy. Different authors emphasize different elements

Strategy is a multifaceted phenomenon about how to compete.

when they provide their understanding of strategy. A common theme that can be found in most strategy definitions is the idea of competition. Consequently, strategy can be seen as an approach of how to compete. To shed additional light on what strategy is, selected aspects from various understandings of strategy are pointed out. According to certain authors, strategy is about

- the creation of competitive advantage (Ohmae 1982: 36)
- the determination and pursuit of basic long-term goals (Chandler 1962: 13)
- the idea of being different and choosing what to do and what not (Porter 1996: 70)
- the description of a "path" from a current to a targeted future state (Kirsch 1991: 301)
- the integration of an organization and its environment based on consistent patterns of organizational decisions over time (Mintzberg 1978)
- the definition of businesses in which to compete on a corporate level and how to compete on a business level (Andrews 1980: 18 f.).

Instead of trying to formally define what strategy is, these elements are used to derive the six principles shown in figure 1-3. They are supposed to sharpen the strategy concept that is relevant for moving forward and will be explained next.

Fig. 1-3

Six principles of strategy

Strategy is about gaining, sustaining and renewing competitive advantage

1.1.2.1 Quest for Competitive Advantage

Most people will agree that strategy makes a major difference between winning and losing in any competitive situation whether it is in business, sports, politics, or others. In business, it helps a company to establish some sort of advantage relative to its competitors that is crucial for outperforming them. As illustrated in figure 1-4, the strategy a company pursues is supposed to lead to competitive advantage, and thus, ultimately to superior performance in a given competitive arena. Strategy is about gaining, sustaining and renewing competitive advantage as a base for superior per-

Fig. 1-4

Relation of strategy, competitive advantage and performance

formance. The question what "superior" performance means is discussed more in detail in chapter 1.1.3.

Gaining competitive advantage—the strategic triangle

Firms that are capable of providing their customers goods and services that are better, cheaper or delivered faster than those of their competitors are generally more likely to outperform their rivals and win in the market. Companies take strategies to achieve such a competitive advantage, and thus, ultimately superior performance compared to their competitors in the same industry or to the industry average. The competitive advantage a company wants to realize always needs to be assessed relative to other companies competing in the same—clearly defined—competitive arena or market and with respect to the needs and wants of the customer. Simply put, competitive advantage can be created when companies are able to utilize their resources and capabilities for meeting customer needs and delivering customer value in a way their competitors cannot, given the specific context in which they compete. This can also be referred to as the "strategic sweet spot" (Collis/Rukstad 2008: 89). A key principle of any business strategy is to consider these three main players—the so-called "strategic three C's" or "strategic triangle" (Ohmae 1982: 91 f.)—and identify the strategic sweet spot as illustrated in figure 1-5.

Sustaining competitive advantage

Companies typically try to keep their advantage relative to their rivals over a prolonged period of time. In those cases, the companies would have a **sustainable competitive advantage** (Porter 1985: 11). This can be pursued, for example, by making the advantage difficult to understand and hard to imitate by rivals through, for example, a unique business model (see chapter 3.4.4) or by protecting it from imitation with patents, for example. However, a company can typically sustain its competitive advantage only for a certain period of time. As rivals work hard to imitate and neutralize it, protection might only last for a while and then expire. Furthermore, changes and discontinuities in the environment such as technological leaps might erode the advantage. For example, Leica Camera has been renowned for its high quality cameras over a long time. The firm ignored the trend of digital photography technology in the 1990s and tried to compete with high-class analog cameras. This strategy did not work out and lead to a competitive disadvantage. The company—meanwhile recovered—faced a severe crisis, had to look for investors and realized losses of more than €15 million in 2005.

Fig. 1-5

The strategic sweet spot

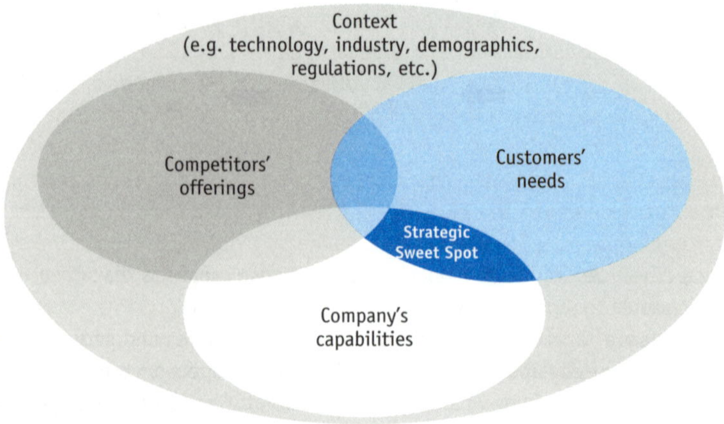

Source: Collis/Rukstad 2008: 89; adapted

Renewing competitive advantage

Given today's highly volatile and dynamic environment, companies might not be able to or do not strive to sustain competitive advantage for a long time. They focus on continuously building **temporary competitive advantage**—that is a competitive advantage that lasts only for a very short period of time—based on speed, organizational agility and innovation. For example, in the food industry only a few companies focus on radical (breakthrough) innovation based on new to the world products that are proprietary and become blockbusters. Most companies have to work hard to continuously establish temporary advantage based on incremental innovation or so-called renovations, that is slight changes in taste, packaging, or size, for example (Wunder/Bausch 2014a). Also, some companies seek new products or technologies through acquisition strategies as they lack own capabilities to develop true breakthroughs. Firms may even purposely avoid long-lasting competitive advantage or alter their competitive advantage after a while. They do not strive to sustain competitive advantage as it makes them predictable and vulnerable from the perspective of their aggressive competitors. Instead there are pursuing strategies directed toward continuously renewing their competitive advantage (D'Aveni 1994; see also Strategic Snapshot 1.3).

Outside-in versus inside-out

1.1.2.2 Fit of Markets and Resources

In the quest for competitive advantage, a company has two key options for elaborating on strategies. First, it can take an outside-in perspective and try to identify sources for competitive advantage in the industry or market. Hereby, the market a firm decides to compete in and its strategic positioning in this competitive arena are

Strategic Snapshot 1.3

Competitive Advantage—Illustrated With Soccer

In the 2012/13 European soccer season, FC Bayern Munich became the first German team that won the so-called "Triple", that is the National Championship, the National Cup and the European UEFA Champions League. This means that—in this season—Bayern Munich gained a **competitive advantage** over its rivals based on different sources such as individual players, the total team composition, the coach, as well as effective alignment of what the coach wanted and what the team delivered. The club was able to execute a strategy that led to this competitive advantage, and thus, to superior performance. However, one can also say that the club has been able to realize **sustainable competitive advantage** as they won the National Championship 24 times in 51 years since the foundation of the "Bundesliga" in 1963. The second best teams in this ranking won only five times. Although the players and coaches of Bayern Munich have been continuously changing over the past 51 years, the club was able to outperform its competitors for such a prolonged period of time. In the 2013/14 German soccer season, FC Bayern Munich—now with a new coach named Josep ("Pep") Guardiola—won the German Championship at the earliest time ever in German Bundesliga history, that is with 77 points at the 27th of 34 games. The new coach did not try to merely sustain the competitive advantage the team gained in the previous season. He changed the strategic system fundamentally in a way that there seem to be no fixed positions for the players anymore. In and between matches, players that used to be excellent in certain positions and play well in "traditional" formations have now been changed continuously. Based on this new agility of the team and an outstanding squad, the coach continuously creates **temporary competitive advantage** that makes it very hard for competitors to anticipate their ploys and come up with an effective game plan.

considered key determinants for success. Second, the company can focus on its resources and develop strategies following an inside-out perspective. Hereby the key drivers of success are seen in the company's resources, capabilities and competencies. These two perspectives are expressed in the market-based view of strategy and the resource-based view of strategy. They are traditionally used for explaining differences in competitive advantage, and thus, the ultimate performance of companies.

Market-based view of strategy

The market-based view of strategy has its origins in the 1980s and is strongly linked to the work of Michael Porter (Porter 1985 and 1980; see also Porter 2008). It is theoretically grounded in industrial economics. Following the so-called structure-conduct-performance paradigm (SCP-paradigm), the success of a company (performance) is primarily determined or limited by the characteristics of an industry (structure), which strongly determine the company's behavior (conduct). According to the market-based view of strategy, the firms operating environment is considered the most important factor for achieving competitive advantage, and thus, superior performance. Consequently, the company is well advised to identify and focus its business activities on the most attractive industries and markets. Furthermore, it needs to have a strong position in those industries to achieve above normal profits. This position can be analyzed and measured relative to competitive forces that are used to characterize the industry: bargaining power of buyers, bargaining power of suppliers, threat of new entrants, threat of substitutes, and the rivalry among established firms (competitors). The application of an enhanced framework of those competitive forces to analyze an industry structure is explained more in detail in chapter 2.2.2.2. Based on their evaluation, the company decides to remain in or enter an industry. If the company is part of the industry, it leverages competitive strategies to establish an optimal shield against those competitive forces as a foundation for realizing competitive advantage. Some companies may be in a position to even influence and shape the industry structure in a way that provides them with new sources of competitive advantage. Cost leadership, differentiation or focusing on market niches are examples of generic strategies for positioning a firm in a specific industry or market and realizing competitive advantage (see chapter 3.4.2).

Although the market-based view provides a wealth of insights and guidelines for strategy formulation, it also has its limitations (Chandler/Werther 2014: 52):

- **Narrow view on only three stakeholders:** Only the firm's customers (buyers), suppliers and competitors are focused on. However, there are many other stakeholders that may influence the competitive environment and can squeeze the performance of a company. Thus, the requirements and relative power of other stakeholders such as governments, nongovernmental organizations (NGOs), local communities, unions, creditors, etc. need to be thoroughly considered as well.
- **Confrontational perspective on stakeholder relationships:** The general relationship with particularly buyers and suppliers is considered primarily combative. In order to gain competitive advantage and survive, it is suggested that a firm needs to analyze the power of buyers and suppliers and ultimately "beat" its stakeholders. More cooperative strategic approaches such as collaboration between business competitors (coopetition) or creating shared value with suppliers are not considered.
- **Neglecting internal organizational characteristics of companies:** Internal resources and capabilities of a company that are likely to predict how a firm is able to compete are neglected. Competitive advantage, and thus, performance is primarily related to the external industry structure. Strengths and weaknesses of a company and its corresponding strategic behavior are considered only of minor relevance.

Resource-based view of strategy

Based on criticism of the strong outside-in success logic in the market-based view, a different perspective on strategy emphasizes the role of an individual firm's resources, capabilities and competencies for gaining and sustaining competitive advantage (Barney 2001 and 1991; Hamel/Prahalad 1994; Wernerfeldt 1984; Peteraf 1993). Anecdotally speaking, whereas the market-based view is suggesting to establish a market position, and thus, answers the question of "where to be", the resource-based perspective is suggesting to leverage resources and capabilities, and thus, answers the question of "what to be". This does not mean that the importance of understanding the industry is neglected in the resource-based perspective. However, companies within an industry are advised to exploit their individual resources and capabilities—or core competencies (Prahalad/Hamel 1990)—to make a difference in competition when compared with other firms. According to this view, it is not the general availability of resources and capabilities that provide competitive advantage but their combination in a way that is a particular strength or competence relative to rivals. To provide such a competitive edge in the long run, a firm's competence (i. e. the combination of resources and capabilities) needs to fulfill the following criteria (Barney 1991: 105–114; Barney/Hesterly 2010: 68–83; see also chapter 2.3.2):

- **Value:** It must enable a firm to exploit an opportunity or neutralize a threat in its environment.
- **Rarity:** It must be available for and controlled by only a small number of competitors.
- **Imitability:** It must be expensive and a cost disadvantage for another firm that lacks and wants to obtain it.
- **Substitutability:** There must be no strategic equivalent available, that is no substitute competence that allows implementing the same strategies.
- **Organization:** The firm's organization must be designed in a way that enables exploiting the resources and capabilities for competitive advantage. For example, even when firms are competing with similar product and service offerings, one may have a competitive advantage because it is able to put together marketing and sales programs, technology such as billing systems, incentives, and training in a way other firms struggle with.

According to the resource-based view of strategy, a company can expect to enjoy a sustainable competitive advantage when it has particular resources and capabilities that are valuable, rare, in-imitable and non-substitutable, and when it is organized in a way to exploit these resources. A more detailed description of what resources, capabilities and competencies are and how to apply the corresponding framework (VRIN) is provided in chapter 2.3. When applying a resource-based view of strategy in company practice, it needs to be considered that there are also some fundamental limitations of this approach:

- **Neglecting a firm's operating environment:** Competitive advantage is primarily related to a unique combination of resources and capabilities and the resulting customer and economic value. Changing industry conditions and market dynamics are mainly ignored although they are likely to have a strong impact on a company's ability to develop and exploit competencies for competitive advantage (Chandler/Werther 2014: 48f.). A more situational perspective emphasizes the role of a specific company context on performance and suggests a basic "fit" between the firm's internal configuration and its external environment.
- **Lack of core competences:** In company practice, combinations of resources and capabilities that fulfill all the criteria for being a source of sustainable competitive advantage are hard to find. For many firms it is simply not possible to develop competencies that are valuable, rare, not imitable and non-substitutable. They are not able to combine their resources and capabilities in such a way that is difficult to replicate. There are a number of practitioners left disillusioned after a failed attempt to find core competencies in their companies based on those criteria.

Integrating market and resource-based view

Conceptually, the market-based and resource-based views of strategy are competing. In company practice, though, they are more a question of sequence in the strategic planning process than a question of which approach to follow and which not. Both views provide a valuable perspective on strategy for company practice, and thus, need to be considered within the strategic management process. This can be practically achieved by leveraging the SWOT framework (see chapter 2.4.1). Hereby, the company's external environment is analyzed to identify **o**pportunities and **t**hreats, which can be related to a market-based view of strategy. The results are combined with an internal perspective that represents a resource-based view in which **s**trengths and **w**eaknesses are analyzed. To expand the strategic ideation process beyond the boundaries of today's operating environment (e. g., industries and product/market combinations), it is suggested to supplement the identification of strengths and weaknesses with a particular analysis of (core) competencies along with the criteria provided by the resource-based view of strategy (see chapter 2.3.2. Tools for elaborating on alternative strategy options based on the identified opportunities, threats, strengths, and weaknesses such as the TOWS-matrix (Weihrich 1982; see chapter 2.4.3) are essentially based on the idea of integrating a market-based with a resource-based perspective on strategy.

Dynamic perspective on markets and resources for competitive survival

An established company will typically start its strategic analysis by looking at macroenvironmental trends and the industry, markets, customers, etc. they are currently operating in or serving. Due to increasing market dynamics and volatility, though, product-/market-combinations that have been successfully exploited by a company leveraging its competences might lose its strategic relevance in the future. To survive and grow, organizations have to execute in the present and adapt to the future (Beinhocker 2006 and 1999). This may well require an alteration or advancement of the firm's resources and capabilities. As explained and illustrated at the beginning of this book, past performance is no guarantee for future success. Existing competencies may become less relevant in the future if market requirements are changing and firms may need to develop new competencies to survive and grow in the long run. In other words, the integration or "fit" between the company's internal configuration and its environment has to be managed in a dynamic and evolutionary way to ensure competitive survival (Teece/Pisano/Shuen 1997). Driven by the quest for growth particularly public companies will seek new business areas when their traditional markets lose attractiveness. In company practice, this is not an easy undertaking as successful organizations will usually find it difficult to change and adapt to new conditions in their environment.

In addition to the market- and resource-based views of strategy there are a variety of other perspectives such as the (dynamic) capability-based view (Teece/Pisano/Shuen 1997) or knowledge-based view (Grant 1997), which are both related to the resource-based perspective. Furthermore, there is the value-based view, the society- or sustainability-based view (see chapter 1.1.3.4) and others which are not further elaborated on here.

1.1.2.3 Being Different and Making Choices

Doing things right—doing the right things

Probably one of the most quoted all-time principles for managers is to distinguish between *doing the things right*, that is efficiency, and *doing the right things*, that is effectiveness (Drucker 1966; see also Strategic Snapshot 1.4). This basic rationale can also be applied for sharpening the understanding of what strategy is. Excellent efficiency means performing similar activities better than rivals whereas strategy means performing different activities than rivals. This requires clear strategic positioning

Strategic Snapshot 1.4

Efficiency Versus Effectiveness According to Peter Drucker

"In the ongoing business markets, technologies, products, and services exist. Facilities and equipment are in place. Capital has been invested and has to be serviced. People are employed and are in specific jobs, and so on. The administrative job of the manager is *to optimize the yield from these resources*. This means *efficiency*, that is, doing better what is already being done. It means focus on costs. But the optimizing approach should focus on *effectiveness*. It focuses on opportunities to produce revenue, to create markets, and to change the economic characteristics of existing products and markets." It does not ask: "How do we do this or that better?" It asks: "Which of the products really produce extraordinary economic results or are capable of producing them? Which of the markets and/or uses are capable of producing extraordinary results?" It then asks: "To what results should, therefore, the resources and efforts of the business be allocated so as to produce extraordinary results rather than the 'ordinary' ones, which is all efficiency can possibly produce? Of course efficiency is important. Even the healthiest business, the business with the greatest effectiveness, can die of poor efficiency. But even the most efficient business cannot survive, let alone succeed, if it is efficient in doing the wrong things, that is, if it lacks effectiveness." (Source: Drucker 2008: 31 f.)

and choice. Strategy is about being different and making choices about what to do and what not to do. This will be explained more in detail next, drawing primarily on the ideas of Michael Porter (Porter 1996):

Operational excellence

Being asked on how to outperform their competitors many business practitioners, particularly in functional areas, will refer to improving efficiency in their operational activities. They continually optimize processes to reduce failure rates and customer complaints, increase productivity and cut down non-value add activities, reduce lead times and working capital, eliminate waste, etc. Many companies are adapting best-in-class practices in various areas of their supply chain to outperform rivals and gain competitive advantage particularly when it comes to cost and pricing. Corresponding performance improvement tools for eliminating inefficiencies can be derived from famous management concepts such as Six Sigma, Lean Management, Total Quality Management, Business Process Optimization, Benchmarking, etc. (Shingo 1986 and 1985; Womack/Jones 2003; Wunder/Bausch 2015). Firms leveraging those techniques are ultimately trying to improve efficiency which, generally speaking, relates to the ratio of output to input. They want to gain competitive advantage by performing the same or similar value chain activities *better* than their rivals. All those approaches can be summarized under the umbrella of "operational excellence" also referred to as "operational effectiveness" (Porter 1996: 61 f.).

Productivity frontier

Companies achieving operational excellence, that is best-in-class operational processes, are likely to benefit from their relative cost position as a source for competitive advantage and profitability at a certain time. However, there is a downside to this approach. Knowledge about best practices and techniques for eliminating inefficiencies are widely available and can be adopted by all competitors in an industry leading to competitive convergence in operational excellence. The more companies are embracing and implementing benchmarking, Six Sigma or lean principles, etc., the more indistinguishable they become from one another regarding their operational processes. Figure 1-6 illustrates this idea with a productivity frontier. Assuming a tradeoff between cost and quality, the productivity frontier represents the maximum value a company is capable to deliver at a given cost leveraging all available best practices in the operational processes. It can be applied to certain individual operational activities, a group of activities or the entire operational framework. The more companies in an industry applying best practices and striving for excellence, the more they move toward the productivity frontier. This maximum state of best practice shifts outwards with improvements in technology, overall skill levels or management practices that allow companies in an industry to improve the value and lower cost at the same time. Firms are getting better at the same time as their competitors. With an outward shift of the frontier, competitive pressure on operational excellence for individual companies increases, typically to the benefit of customers.

Strategic positioning

Instead of focusing on outperforming competitors based on operational excellence, the core of strategy is about establishing a unique position within a competitive arena through performing different activities than rivals do. In this understanding, strategy is about deliberately being *different* rather than only being *better* as a source for a more sustained competitive advantage. Hereby, the aspired strategic position comes from unique customer value the company is able to create and capture with its products, services or—most importantly—its business model (see chapter 3.4.4). Competitive advantage is based on creating and preserving what is distinctive about a company, that is its distinct strategic position in the competitive arena. This requires choice. "(…) strategy is an integrated set of choices that uniquely positions the firm in its industry so as to create sustainable advantage and superior value relative to the competition." (Lafley/Martin 2013: 5) Key principles underlying strategic positioning are (Porter 1996):

- **Creating a unique set of integrated activities that is different from rivals.** Hereby, the distinction may come from the activities themselves or from the way

Fig. 1-6

The productivity frontier

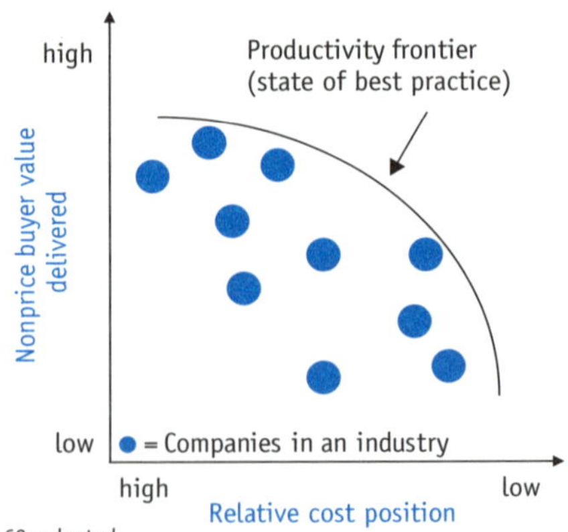

Source: Porter 1996: 62; adapted

they interact and reinforce one another ("fit"). The principle of strategic positioning based on a unique activity system is strongly related to the idea of a business model (see chapter 3.4.4). The company may position itself by serving a few needs and wants of many customers differently than rivals, by serving broad needs and wants of few customers, or by satisfying broad needs and wants of many customers but in a narrow market.

- **Choosing what to do strategically and what not to do.** "Strategy is making trade-offs in competing. The essence of strategy is choosing what *not* to do. Without trade-offs, there would be no need for choice and thus no need for strategy." (Porter 1996: 70) For some executives, making choice instead of working with very broad and generic strategic directions might be frightening as it comes with the risk of being blamed for a bad choice. Furthermore, strategic choice with tradeoffs may be perceived as a constraint to capture growth opportunities (see Strategic Snapshot 1.5). Lafley and Martin wrote: "It is natural to want to keep options open as long as possible, rather than closing off possibilities by making explicit choices. But it is only through making and acting on choices that you can win. Yes, clear, tough choices force your hand and confine you to a path. But they also free you to focus on what matters." (Lafley/Martin: 2013: 5) Companies with a lack of clear strategic choice end up allocating little but insufficient resources to everything, and thus, set the firm up for failure instead of focusing their resources on what really matters.

Superior performance requires both strategy and operational excellence.

Superior performance essentially requires a hybrid approach of both strategy and operational excellence, although the two elements work in different ways. First, competitive advantage may well come from best-in-class operational processes. Looking at annual savings that companies regularly report as percentage of their revenues once they finished efficiency improvement programs clearly reveals that operational excellence contributes to superior performance. Furthermore, those cost savings provide additional financial resources that can be invested in strategic themes such as innova-

Strategic Snapshot 1.5

Strategic Positioning and the "Growth Trap"

"Among all other influences, the desire to grow has perhaps the most perverse effect on strategy. Trade-offs and limits appear to constrain growth. Serving one group of customers and excluding others, for instance, place a real or imagined limit on revenue growth. Broadly targeted strategies emphasizing low price result in lost sales with customers sensitive to features or service. Differentiators lose sales to price-sensitive customers. Managers are constantly tempted to take incremental steps that surpass those limits but blur a company's strategic position. Eventually, pressures to grow or apparent saturation of the target market lead managers to broaden the position by extending product lines, adding new features, imitating competitors' popular services, matching processes, and even making acquisitions. (...) Compromises and inconsistencies in the pursuit of growth will erode the competitive advantage a company had with its original varieties or target customers. Attempts to compete in several ways at once create confusion and undermine organizational motivation and focus. Profits fall, but more revenue is seen as the answer. Managers are unable to make choices, so the company embarks on a new round of broadening and compromises. (...) Too often, efforts to grow blur uniqueness, create compromises, reduce fit, and ultimately undermine competitive advantage."
(Source: Porter 1996: 75–77)

tion, international growth or others. To keep pace with competition, operational excellence tends to be particularly relevant in industries where strategies of the market players do not differ significantly. As a matter of fact, unique strategies are desirable but rare. Being asked about the uniqueness of their strategies, the majority of companies actually state that they pursue similar strategies to their competitors (Horváth & Partner GmbH 2008: 9; Wunder/Bausch 2014b: 58). This means both a need for having efficient operations and an opportunity for strategic differentiation.

It also means that in addition to the skill of crafting an effective strategy, companies can gain competitive advantage by means of excellent execution. This is particularly the case where the strategies of rivals tend to be similar (Greiner/Wolf 2010: 5; Horváth & Partner GmbH 2014: 4). Note that this understanding does not necessarily relate to operational excellence but to the effectiveness of how well a defined strategy is executed (see chapter 4).

Operational excellence does not substitute strategy.

In markets where scale is important for achieving competitive advantage, operational excellence can be particularly effective. However, although it may well support the effort to establish a unique strategic positioning, it is not a substitute for strategy. Moreover, competing on operational excellence only may be a risky undertaking as it leads to a rather destructive path of efficiency-oriented competitive races of companies tempted to copying competitors rather than differentiating from them. A company can outperform its rivals in the long run only when it is able to establish a difference that it can preserve as long as industry conditions allow and establish new strategic positions in case there are major changes in the environment. Companies trying to be everything to everybody may end up being nothing to nobody. However, companies relying solely on their strategy might become vulnerable regarding operational excellence and may be outperformed in terms of flexibility, speed or cost. As described below and illustrated in figure 1-7, clearly distinguishing a management agenda of operational excellence from strategy but combining both elements seems to be the most promising recipe for sustained superior performance (see also Strategy Practice Example 1.2):

- **Management agenda 1: Operational excellence**
  Driving a continuous effort to achieve best practices, be better than rivals and possibly even shift the productivity frontier when it comes to operational processes.
- **Management agenda 2: Strategy**
  Establishing a unique strategic position with clear choices about what to do and what tradeoffs to apply, resulting in a unique integrated activity system or business model.

1.1.2.4 Path to a Destination

Strategy is a means to an end.

A company usually has an explicit or implicit aspirational idea about where it wants to be in the long run. This desired picture of the company's future is often articulated in terms of a vision (see chapter 3.1). The aspiration may be derived from various elements such as shareholder expectations regarding profit and growth, the company's long-term objectives regarding its stakeholders such as customers, employees or the society, a personal aspiration of the CEO to transform the organization or stretch it toward a certain performance level, an entrepreneur's "dream" about the future, or a desire to ensure long-term survival of the company and pass it on from one generation to another in the case of a family-owned firm. A strategy can be perceived as a

Fig. 1-7

Superior performance based on both strategy and operational excellence

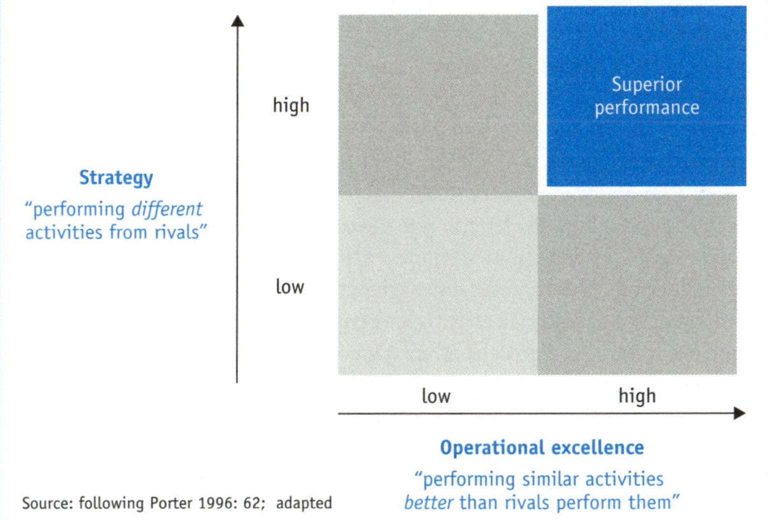

Source: following Porter 1996: 62; adapted

Strategy Practice Example 1.2

Outperformers Leverage Both Strategy and Operational Excellence

In a 2012/13 management survey of more than 100 companies in the food industry, 70 % of the participants stated that they pursue strategies that are similar to those of their competitors. Only 12 companies emerged that were able to grow both revenues as well as profit stronger than their competitors for three years in a row. All of those 12 outperformers reported that they are well positioned in their corresponding competitive arena regarding both strategy as well as efficiency. The hybrid approach of combining strategy and operational excellence seems to pay off in terms of profitable growth. Among the rest of participating food producing companies more than 50 % stated that their strategic positioning is rather bad or very bad and roundabout 40 % gave this answer related to their relative positioning in efficiency. What did the outperformers do differently in their strategic planning process than their peers? The survey found four success factors:

- Better leverage of a systematic and formal strategy process
- Application of more solid tools for external strategic analysis
- Stronger engagement in business models
- More effective management of strategy execution

Source: Wunder/Bausch 2014b

path from today to a long-term destination (see fig. 1-8). It represents a means to an end, that is a course of action into the future.

Vision guides strategy formulation.

At the beginning of each entrepreneurial activity stands an idea of what the company wants to achieve in the future, that is a vision, even though it might not be explicitly articulated (Hinterhuber 2004: 75). Based on this understanding, a desired future state or vision has to be defined *first* before a strategy can be formulated (see fig. 1-8). "The vision creates the picture of the destination. The strategy defines the logic of how this vision will be achieved. Vision and strategy are essential complements." (Kaplan/Norton 2001: 74) Together with the company's mission and values the vision represents a normative strategic guidepost that is supposed to frame strategy formulation. Whereas the vision states what the company ultimately wants to accomplish in the long run, the mission articulates why the company exists (purpose) and what it actually does, and the values represent ethical norms and standards that govern behavior of individuals. More detailed elaboration on the strategic guideposts is provided in chapter 3.1.

Long-term aspiration versus strategic goals

Practically speaking, strategy is *not* the articulation of a long-term aspiration regarding market leadership, achieving certain growth or profitability thresholds or other foundational long-term objectives but the fundamental approach or combination of approaches how to make them come true. This must not be confused with the development of strategic goals that are supposed to refine and describe a strategy after it has been developed to focus management attention and resources on the most important strategic priorities (Schreyögg 1984: 87). Strategic goals along with performance metrics and targets as well as strategic action programs are elements of strategy execution. They are in place to describe and operationalize strategy to make it happen and not to formulate strategy (see chapter 4.3).

Fig. 1-8

Strategy as path from today to the vision

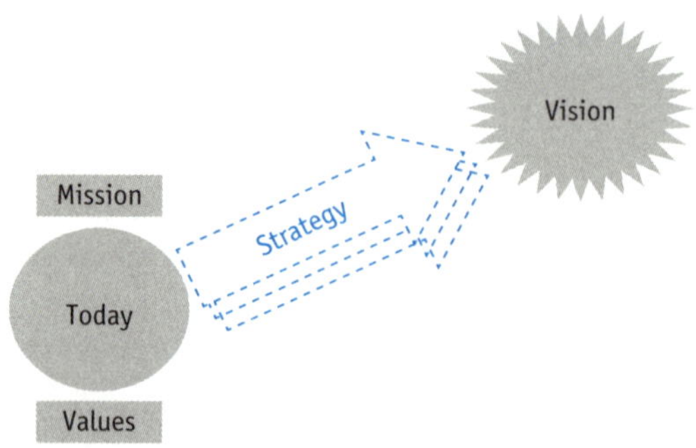

Source: Wunder 2004: 26; adapted

1.1.2.5 Consistency in Behavior

An understanding of what strategy is can be further sharpened by asking the question of what a strategy actually constitutes: intention or realization? On the one hand, strategy can be perceived as an intention or plan, which means it constitutes prior to the action (ex ante). On the other hand, strategy can also be viewed as something that constitutes once it happens (ex post). To better reflect and understand the reality of strategy, it is helpful to distinguish between different types of strategy making, which are illustrated in figure 1-9 (Mintzberg 2000: 23–25; Mintzberg 1987 and 1978):

Strategic intention versus what actually happens in reality

A strategic plan (see chapter 1.1.1) reflects an **intended strategy.** It is typically the outcome of a rational and structured development process based on external and internal analysis and top-level executive decisions about the course of action to be pursued by the organization. Business schools and strategic management literature have a general preference of teaching such a rational planning model (see also chapter 1.2). In case strategic intentions or plans are fully realized it can be called **deliberate strategy.** In reality, though, this does not happen too often. In business practice, only a small portion of effectively planned strategies are also effectively realized due to several reasons, such as unexpected events that have significant implications for the company and make strategic intentions obsolete or ineffective strategy execution (see chapter 4.1.1). Strategic intentions that do not come true can be called **unrealized strategy.**

Intended strategy are either realized or unrealized.

By looking back to the strategic behavior of a company that shows some sort of consistency of pattern over the course of, for example, the last five years, a **realized strategy** can be identified. When asking business people to describe what strategy they have realized over the past five years, for example, and whether the strategy was intended, they generally follow three different lines of reasoning:

Realized strategies are a combination of intention and emergence.

Fig. 1-9

Realized strategy as result of deliberate and emergent strategies

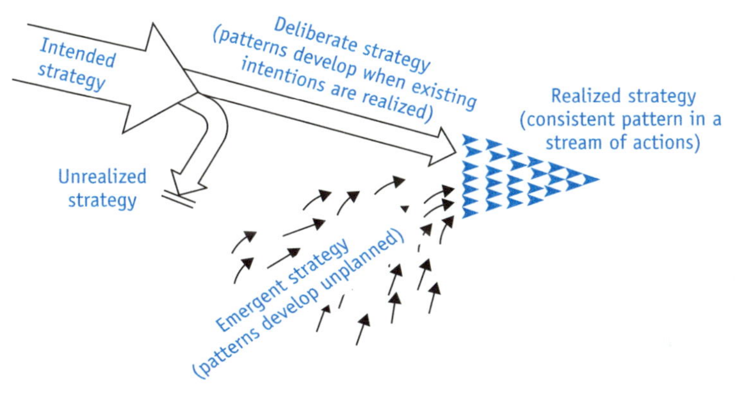

Source: Mintzberg 1978: 945; adapted

- Some may claim that what they strategically planned five years earlier was realized perfectly.
- Others may claim that their realized strategies in the past five years had nothing to do with the strategic plan they developed five years ago.
- Most managers will probably give an answer that falls between these two extremes, that is some strategic intentions have been fully realized, portions of the strategic plan may have been deliberately adapted to reflect changes in the environment and some actions with strategic magnitude just evolved without previously expressed intention.

A realized strategy pattern or consistent strategic behavior that was not expressly intended is called **emergent strategy**. It evolves from autonomous unplanned strategic initiatives undertaken by employees of their own desire and choice that converge in time in some sort of consistent strategic behavior of the organization. For a sound understanding of strategy it is critical to accept that a realized strategy is typically based on elements from both intended strategies and emergent strategies. "The real world inevitably involves some thinking ahead of time as well as some adaption en route." (Mintzberg 2000: 24) Hereby, neither intended nor emergent strategies can be labeled good or bad only because of the type of their formation. For example, strategies emerging bottom-up in the organization may provide innovative strategic ideas for creating competitive advantage. However, a lack of strategy alignment (see chapter 4.3.3) can also result in emergent strategies that may be in conflict with planned strategies, and thus, breed friction and failure in an organization.

Strategy as planned emergence

What is the practical consequence of the described distinction of strategies? Managers should be aware that a planning approach—typically top-down—is only one way of trying to control a company's strategic behavior. They also have to be sensitive to emerging strategic initiatives—typically bottom-up—that must be identified, evaluated and, if promising, integrated into upcoming strategic planning cycles. Through their strategic leadership, executives can create an environment that is fruitful for emergent strategies, for example, by encouraging entrepreneurship and strategic ideation on multiple management levels and providing strategic context and basic direction with room for self-control mechanisms instead of detailed strategic plans that have to be executed rigorously. This approach of deliberately combining some sort of top-down strategic planning with room for bottom-up strategies to emerge can be called *planned emergence* (Rothaermel 2013: 46). Some companies pursue "what may be called *umbrella* strategies: the broad outlines are deliberate while the details are allowed to emerge within them. (...) effective strategies mix these characteristics in ways that reflect the conditions at hand, notably the ability to predict as well as the need to react to unexpected events." (Mintzberg 2000: 25)

1.1.2.6 Multiple Level and Theme Alignment

Strategies at different firm levels address different themes.

There is not *one* strategy in an organization. Whereas a corporation's top executive leadership team has to formulate corporate (or divisional) strategy, the CEO of a strategic business unit (SBU) works with competitive strategies and business models, leaders of functional areas develop their corresponding functional strategies to sup-

Strategic Snapshot 1.6

Planned Versus Emergent Strategy—Illustrated With Soccer

At the forefront of important sports team events such as the soccer world cup or the super bowl, finalist teams typically analyze their opponents very thoroughly leveraging a wealth of data. They conduct a number of planning cycles and develop a sophisticated game plan (**intended strategy**) including certain contingencies. The game plan is then communicated to the entire team and a disciplined execution is seen as a key success factor.

Imagine we are watching the final of the soccer world cup and some unforeseen circumstances are happening early in the game: A key player becomes injured and has to be replaced, another one is sent off and the opponent team is playing in a way which was not anticipated at all. In such a case, the previously defined game plan might not be valid anymore and might be abolished by the playing team (**unrealized strategy**). However, the team we watch will have no possibility for a new planning cycle to develop a new game plan that they can follow to deal with the new situation (**deliberate strategy**), at least not until the half-time break. There is also no "time-out" in soccer. A certain unplanned behavioral pattern is likely to evolve on the field based on a consistent stream of individual decisions and actions as well as the resulting team dynamics (**emergent strategy**). Independent of whether it was previously intended (planned) or not, as spectators we will be able to see and describe the team's strategy once it has happened (**realized strategy**).

The question of whether the evolving strategic team behavior is successful or not is not driven anymore primarily by thorough analysis and planning combined with disciplined execution of the game plan. Winning or losing now strongly depends on other success factors such as trust between the players, individual excellence, strong will and motivation to perform, key players taking initiative and carrying the others along, well trained plays and automatic moves, etc. A successful **realized strategy** on the field is likely to be a combination of previously defined strategic intentions as well as emerging behavioral patterns.

port corporate and the businesses. A regional management team is driven by their regional strategy and general managers of international subsidiaries typically follow country strategies. Finally, product managers have to come up with product or service strategies that are typically based on some kind of marketing mix. Depending on the organizational area, each type of strategy has to deal with different questions and usually addresses different strategic themes. Here are some examples (Wheelen/Hunger 2010; see also chapter 3):

- **Corporate strategy** is primarily about choosing the general direction of a company as a whole. This includes the choice of industries and businesses in which the company wants to compete (division and business portfolio) and the related question of focus versus diversification. In multiple-business companies, corporate headquarters is classically managing various strategic business units in a way that they are collectively delivering more value under the corporation's stewardship than if they were acting alone. Corporate strategy also includes decisions regarding the value creation activities the company should perform in those businesses (integration or de-integration). Furthermore, in multinational enterprises it is concerned with decisions about local integration and global responsiveness and the corresponding strategy postures such as a global, international, multinational or transnational strategies. For example, companies following a transnational strategy simultaneously respond to local needs, global demands as well as cross-border learning opportunities and consequently manage their international subsidiaries as an integrated network. In large corporations, divisions on the second management level are often huge and consist of multiple businesses, functions, etc. themselves. Therefore general strategic thoughts of corporate strategy do also apply to divisional strategy.

- **Business strategy** is concerned with how to position a company's (strategic) business units within their specific competitive arena (e. g., industry or market segment). Whereas corporate strategy asks what businesses the company should be in and with what intensity regarding the company's resources and management capacity, business strategy asks how a business should compete. It is about choosing generic competitive strategies such as cost leadership, differentiation or outpacing and may also include cooperative approaches such as strategic alliances with other firms in an industry to gain competitive advantage relative to rivals. At the core of business strategy is a business model which represents the anticipated mechanisms how the business is supposed to be successful in its competitive arena. It can be considered as refinement of a generic strategy on a business level by describing the fundamental principle of how a product and service offering for certain target customers (value proposition) is created and delivered (value creation and delivery) resulting in economic benefits for the firm's business (value capture). (Wunder 2013)
- **Functional strategy** is the strategic approach of a functional area in the organization to support achievement of corporate and business strategies with due consideration of resource efficiency or productivity. This includes strategies of core functional areas like R&D, Marketing, Supply Chain Management or Operations as well as support functions such as HR, Finance or IT, etc. In each of those areas there is a variety of strategy options to choose from in order to support the strategic thrust of one or several business units. In general, the resources and capabilities of functional areas are a crucial foundation for the company and its businesses to create and sustain competitive advantage. Hereby, the strategic orientation of a functional area strongly depends on the corporate parenting approach such as leveraging functional resources and capabilities across different businesses to create synergies or for entering new businesses based on the firm's core or distinct competencies. This may lead to the foundation of centers of excellence where functions in certain subsidiaries get a global strategic mandate based on local best practices. Furthermore, local content regulations in certain countries may lead to local production strategies in order to get access to regional markets targeted by corporate or business strategy.
- **Regional strategy and country strategy** deals with how competitive business strategies or functional strategies vary from one region or country to another. From a sales perspective, competitive arenas among regions and countries may differ significantly based on criteria such as level of market maturity (emerging versus mature regional markets), consumer preferences and buying power, local competitive environment, government regulations, local infrastructure, etc. As a result, companies may need to adjust their competitive strategies from one country to another. From a functional perspective, value chain activities may need to vary from region to region to gain competitive advantage. For example, some corporations put their R&D activities in certain countries to exploit local resources and

infrastructure, other concentrate global production facilities in certain countries to utilize factor cost differences and achieve economies of scale. Marketing has to think about establishing global (power) brands or local brands or both. Various international subsidiaries may fulfill different strategic roles for the implementation of the corporate or business strategies considering their local resource strength as well as the importance of the local market. (Bartlett/Goshal 2002: 121–128)

- **Product strategy or service strategy** is primarily concerned with the design and structure of the marketing mix. This includes questions about the product or service concept, breadth and depth of the product range, radical and incremental product innovation, branding strategies as well as strategic pricing, communication and sales decisions. In case of services, the classical marketing mix is supplemented with the three additional components *personnel decisions* (e.g., competence, motivation, loyalty, and customer orientation of employees), *physical environment decisions* (visible factors of the service infrastructure) as well as service *process decisions* to ensure high quality service levels. (Homburg/Kuester/Krohmer 2013: 82 and 375 f.)

Strategy alignment

Converting the illustrated strategic variety that comes from multiple organizational areas and multiple strategic themes into some sort of consistent strategic behavior (see 1.1.2.5) of the entire organization requires strategy alignment. Depending on the company's steering logic, the different strategy types build a hierarchy of strate-

Fig. 1-10

Strategy as a result of multi-level and multi-theme alignment

Level	Strategy type	Sample themes
Corporate (or division)	Corporate/divisional strategy	▸ Focus/diversification ▸ Integration/de-integration ▸ Global, multinational …
Business	SBU strategy/competitive strategy/ business models	▸ Differentiation/cost leadership/outpacing ▸ Business model patterns
Function	Functional strategy (core and support functions)	▸ R&D-, Marketing-, Supply Chain-, Operations Strategy ▸ HR-, Finance-, IT Strategy
Region	Regional/country strategy	▸ Market penetration/ market development ▸ Subsidiary strategic roles
Product/ service	Product/service strategy	▸ Product/service concept and product range ▸ Marketing mix (4 Ps/7 Ps)

ALIGNMENT

gic priorities that have to be aligned for being effective. Ultimately, all the different strategies have to complement and support one another as far as possible and desired by the overall strategic approach. (Wunder 2014b; see also chapter 4.3.3)

This concludes the elaboration on principles for sharpening an understanding of what strategy is. It should now be clear which strategy concept is considered foundational for the upcoming elaborations. An important question has not yet been addressed. How does a company know whether its strategy pays off? The next chapter deals with this question and provides some approaches of how to define and measure what a strategy is actually supposed to deliver in terms of results.

1.1.3 Measuring Competitive Advantage and Company Performance

Superior performance

How can a company measure whether its strategy works out and whether it has a competitive advantage? As explained earlier, competitive advantage is reflected in superior firm performance. Whereas it is relatively easy to define the "winner" or "outperformer" in a sporting event (see Strategic Snapshot 1.3), it becomes a bit trickier in business. For a company, it needs to be defined how its performance is measured, what is considered as "superior", and whether it is relative to its competitors or other stakeholders. A company's performance can be assessed by looking at different types of performance data, which leads to different ways for measuring competitive advantage (Barney/Hesterly 2010: 13–29; Rothaermel 2013: 115–128; Schaltegger/Wagner 2006):

- **Economic value:** A firm has a competitive advantage when it creates and captures more economic value—derived from the customer value it generates—than its rivals.
- **Accounting performance:** A firm has a competitive advantage when selected profitability ratios from accounting are greater than the industry average.
- **Shareholder value:** Companies that earn above their cost of capital are realizing above normal performance and are generally said to have a competitive advantage.
- **Corporate sustainability performance:** A firm has a competitive advantage when its integrated economic, ecological and social performance ("triple bottom line") is greater than the one of its rivals or the industry average.

1.1.3.1 Economic Value

Capturing more economic value than rivals

The question of how a company is performing based on its competitive advantage can be perceived from the perspective of an individual product or service. On the most basic level, profitability of a company depends on the value, price and cost as illustrated in figure 1-11 (Rothaermel 2013: 115–117 and Hill/Jones 2010: 77–81). The **value** (V) customers place on the company's product or service reflects the satisfaction or happiness the customer gets from it and is thus also called **utility**. It captures how much a customer is willing to pay for it at the maximum. This utility is the individually perceived benefit gained by the customer that purchases the firm's products or services. Given a generic product definition, this benefit can be created by different attributes such as (Homburg/Kuester/Krohmer 2013: 108 f.):

- Core product features (e. g., number of airbags in a car) and additional features generating a benefit (e. g., a car's sound system)
- The packaging regarding its function or design (e. g., 10 individually packed 10 g chocolate pieces instead of one 100 g bar) or the tangible environment—that is "packaging"—in case of a service (e. g., branch office of a bank or waiting and treatment room at a dentist)
- Basic services (e. g., a company website where manuals of current and past products can be downloaded) and value added services (e. g., 0 % financing option when purchasing a car)
- The brand (e. g., Polo Ralph Lauren reflecting exclusivity, PUMA telling a story about hipness and belonging, or Fiji water one of the bestselling brands of bottled water in the USA reflecting purity).

If this value or utility is greater than the **price** (P) charged, customers are likely to buy it. Typically, the **cost** (C) to produce the product or service does not matter to the customer but greatly to the company producing and selling it. The difference between the utility (or value) in the eyes of customers and the company's cost to produce it is the **economic value** created, which is split between these two parties. Some of this value is captured by the customer in the form of a **consumer surplus**. The other portion—what economists call the **producer surplus** and managers call the profit—is captured by the company, which usually tries to get as much of it as possible. If a company is able to create and capture more economic value than its rivals it has a competitive advantage.

Utility versus market price

Why do companies not charge the price that represents the customer's maximum willingness to pay (value or utility)? Customers are able to capture a consumer surplus because companies are in competition with each other regarding their products and services, and thus, have to charge a lower price than in a pure monopolistic situation. Furthermore, it is practically impossible for companies to charge each customer a price that reflects the unique individual assessment of utility as the perceived value changes based on income, time and various other factors including cognitive or neuropsychological processes (Scheier/Held 2012). Even by looking at purchasing habits or analyzing customer preferences on an individual level with statistical techniques, like the conjoint analysis, will not fully solve this problem.

Improving producer surplus by increasing customer value or reducing cost

If the goal is to drive bottom-line growth—that is increasing the profit—a company can improve its cost situation through, for example, operational effectiveness techniques such as Lean Management or Six Sigma or changes in its cost structure and lower C relative to P, which is typically the initial response of engineers being asked that question. Another option—that might be the first answer of product managers or marketers—is to strengthen the utility provided by its products and services which gives the company a variety of pricing options. Typically, this will initially raise (production) cost because the company has to spend money on one or more value increasing mechanisms like quality improvements, value added services or emotionalizing the brand as listed above. Due to the increased utility, one pricing option is to

Fig. 1-11

Economic value creation per unit and competitive advantage

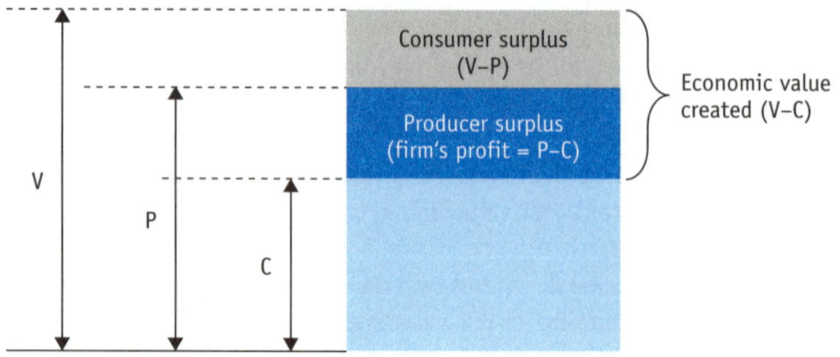

V = (perceived) customer **v**alue (consumer's maximum willingness-to-pay)*
P = market **p**rice
C = company's **c**ost

Source: following Rothaermel 2013: 116 and Hill/Jones 2010: 78

raise prices more than the cost increase, which will result in a higher profit per unit. Another option is based on a very different rationale. Despite the increased utility, the company is lowering the price so that customers clearly recognize that they are getting a great bargain. Based on the increased demand, unit sales volume expands allowing economies of scale. As a result, the average unit cost of production fall and the profit per unit increases.

Limited applicability on a firm-level

Knowing the principles of economic value creation and the fundamental dynamics among its components is a good starting point not only for measuring but for understanding competitive advantage. It provides managers a conceptual framework for strategic dialog and ultimately supports decision making. However, trying to operationalize competitive advantage this way on a firm level would require estimating and calculating the different components for all products and services of the company. For precise numbers on the performance of a company, people usually rely on accounting data.

1.1.3.2 Accounting Performance (Profitability)

Top- and bottom-line performance

Competitive advantage can also be assessed by looking at information from a company's published profit and loss (P&L) and balance sheet statements. As these statements are typically created based on widely accepted accounting standards such as US-GAAP (generally accepted accounting principles) and audited by certified public accountants they are widely used to calculate performance metrics. Data is usually available as financial statements of publicly traded companies are publicly available and many privately owned firms typically release some information about their financial results. Of particular relevance is the firm's top- and bottom-line performance. This relates to the achieved revenue and profit numbers in a certain time

Strategy Practice Example 1.3

Comparing Toyota and General Motors in Creating Value per Unit

"According to a 2008 study by Oliver Wyman, in 2007 Toyota made $922 in profit on every vehicle it manufactured in North America. GM, in contrast, lost $ 729 on every vehicle it made. What accounts for the difference? First, Toyota has the best reputation for quality in the industry. According to annual surveys issued by J. D. Power and Associates, Toyota consistently tops the list in terms of quality, while GM cars are at best in the middle of the pack. The higher quality translates into a higher utility and allows Toyota to charge 5 % to 10 % higher prices than GM for equivalent cars. Second, Toyota has a lower cost per vehicle than GM, in part because of its superior labor productivity. For example, in Toyota's North American plants, it took an average of 30.37 employee hours to build a car, compared to 32.29 at GM plants in North America. That 1.94-hour productivity advantage translates into lower labor costs for Toyota; hence, a lower overall cost structure. Therefore, as summarized in the illustration below, Toyota's advantage over GM derives from greater utility (U), which has allowed the company to charge a higher price (P) for its cars, and from a lower cost structure (C), which taken together implies significantly greater profitability per vehicle (P-C)."

(Source: Hill/Jones 2010: 80)

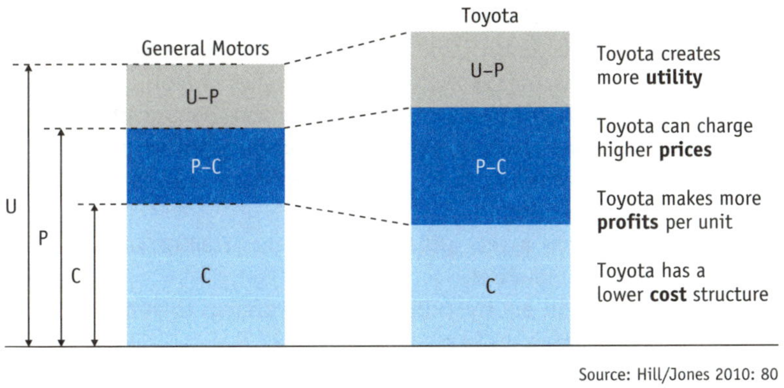

Fig. 1-12: Comparing Toyota and GM

frame, usually one fiscal year. Those numbers can be found when looking at the top line (revenue) and bottom line (profit) of a firm's profit and loss (P&L) statement. Correspondingly, managers speak about "top-line growth" when they refer to sales (revenue) increase and "bottom-line growth" when they refer to profit increase. A common understanding of competitive advantage is based on the idea of above normal returns, and thus, relates to profits that are excess of revenues over expenses. Common metrics for profit are EBIT (earnings before interest and taxes), EBITDA (earnings before interest, taxes, depreciation and amortization) or the Net Income. However, the problem of using siloed profit numbers for performance comparisons to understand competitive advantage is that they are only absolute values and do not say anything about the assigned capital or the revenue required to realize the profit, for example. Directly comparing financial statements of two companies for assessing their performance is almost impossible because of differences in size (see Strategy Practice Example 1.4).

Performance ratios

One way to overcome the problems involved when comparing companies of different sizes is to calculate and compare financial ratios. With a ratio analysis the relationship between different pieces of financial information is investigated. Ratios using data from accounting can be grouped into the following categories (Ross/Westerfield/Jordan 2010: 54–63):

- Ratios focusing on the firm's ability to meet its short-term financial obligations (liquidity or short-term solvency ratios)
- Ratios addressing the company's long-term ability to meet its obligations or its financial leverage (leverage or long-term solvency ratios)
- Ratios describing how intensely a firm uses its assets to generate sales (asset management, turnover or activity ratios)
- Ratios focusing on the bottom-line and measure how efficiently a firm manages its business, operations and assets to generate income (profitability ratios).

Achieving greater profitability than the industry average

Profitability metrics are probably the best known and most widely used financial ratios, particularly when it comes to measuring competitive advantage. These figures represent the profit (EBIT, EBITDA, etc.) proportionately to a measure like assets, equity, capital, or revenues. Some examples of profitability metrics that are most commonly used for direct performance comparisons of different firms are:

- **Profit margin, return on sales (ROS)** or **return on revenue (ROR)** measure the profit earned per unit (Euro, US-Dollar, etc.) of revenue (see Strategy Practice Example 1.4)
- **Return on assets (ROA)** measures the profit earned per unit (Euro, US-Dollar, etc.) of total asset
- **Return on equity (ROE)** measures earnings to owners or how the shareholders fared during the year
- **Return on investment (ROI)** or **return on invested capital (ROIC)** measure the profit in relation to the investment required to obtain that profit. It shows how effectively the company uses the (owned or borrowed) capital invested in its operations. Given the same risk, the investor wants the maximum return for any given amount of resources.

Without comparison, the listed ratios say very little about the performance of a company. They need to be compared with some standard such as the measures of main competitors (see Strategy Practice Example 1.4), benchmarks from within or outside the industry, or with the average of other firms within the industry. Using accounting information, a firm is said to have a competitive advantage when selected profitability ratios are greater than the industry average. When comparing those numbers, it is important to keep in mind that the listed profitability ratios are rarely computed in exactly the same way by different companies. Profitability does not mean the same thing to all people. One example is the question whether it is based on net income, EBIT or EBITDA. When numbers from different sources or people are compared it

Strategy Practice Example 1.4

Comparing Automakers by Using Difference Types of Performance Indicators

No other company sold more automobiles in 2012 than Toyota (9748 thousand cars). General Motors (9286) and Volkswagen (9075) ranked 2 and 3. The next biggest companies in terms of numbers sold are Ford with 5668 thousand cars and Nissan (4940). This huge difference in sales numbers shows the outstanding size of the top three as compared to the rest of global automakers. Only a slightly different picture is revealed when looking at total revenue in 2012. Toyota is ranked number 1 with € 206,894 million followed by Volkswagen (€ 192,676 million), General Motors (€ 115,479 million), Daimler (€ 114,297 million), and Ford (€ 101,823 million). Regarding the profit of those automakers, Toyota is again at the top of the ranking. With € 14,213 million in 2012, Toyota was the automaker with the highest profit (EBIT) followed by Volkswagen (€ 11,510 million), Daimler (€ 8,615 million), BMW (€ 8,300 million) and Hyundai (€ 5,791 million).
Those absolute performance indicators do not tell us much about competitive advantage. For understanding competitive advantage we need to look at performance ratios and eliminate the impact of size. For example, when using the Return on Sales as profitability ratio for measuring performance, the ranking changes significantly with BMW, Hyundai, Daimler, and Kia at the top, followed by Toyota at rank 5 (see fig. 1-13). It shows that BMW had a significant competitive advantage over its rivals in the automotive industry in 2012, not only based on a comparison with the industry average but also directly with main competitors. The average EBIT margin considering all automakers listed was 5.4 %. Hereby, German automakers achieved an average Return on Revenue of 7.4 %, Southern European companies 0.2 %, Japanese automakers 4.5 %, and US car manufacturers 3.8 %. The numbers cover the entire business results of the listed automakers including financial services from January to December 2012, independent of their fiscal year, based on different sources and an own calculation of Ernst &Young.

(Source: Ernst & Young GmbH 2013)

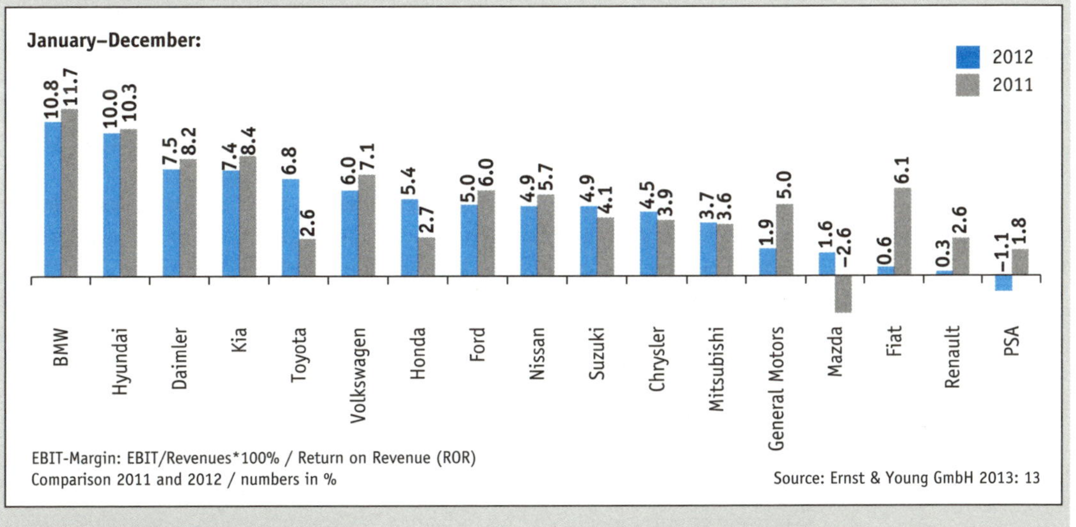

Fig. 1-13: EBIT-margin of automakers

needs to be ensured that they are computed the same way—according to the same standards and principles—or, if not, they need to be made comparable.

Limitation of profitability ratios

Using profitability ratios from accounting for measuring competitive advantage offers the great advantage that information is widely available and that they are relatively easy to compute. However, it also has some significant limitations. For example, it is important to note that profitability ratios such as ROA, ROE and ROI/ROIC are calculated based on the actual book value. This means that a sinking remaining book value of investments corresponds with increasing profitability although the performance of the company may not have changed at all. For example, a public company buying back stocks reduces both its assets (cash) as well as its equity because the number of outstanding shares is reduced. This improves ROA and ROE ratios even if profits are stable. Furthermore, accounting data is backward-looking. This means that the ratios reflect the results of past decisions and that they are no indicator for how competitive advantage might be affected in the future. Another major pitfall of using accounting data for measuring competitive advantage is the omission of the cost of capital that a firm employs to produce and sell its products in most accounting ratios.

1.1.3.3 Economic Performance and Shareholder Value

Providing a firm's shareholders higher "utility" than rivals

For some schools of thought, accounting measures understate the importance of a company's suppliers of capital. If the firm's equity and debt holders do not receive an expected return as a compensation for their risk, value will be lost and they will withdraw their capital in search for better investment options. Companies do not only want to avoid that their shareholders shift the investments to higher yielding companies but, moreover, attract even more capital to grow and expand their business. With this philosophy the focus shifts from traditional figures like revenue, profit and accounting performance ratios to shareholder value. The ultimate measure for a company's success is the capability to enrich its shareholders (shareholder value added).

Comparing profit with weighted average cost of capital

To quantify its shareholder value creating performance, a company needs to look at the cost of capital that a firm needs to operate. This is the rate of return suppliers of capital—for example, banks and bondholders (debt) or individual and commercial stockholders (equity)—expect for their investments given a certain risk level. Whereas the cost of dept equals the interest a company must pay for borrowing money from the debt holder, the cost of equity is the rate of return equity holders expect in order to invest in the company considering firm- and market-specific risk factors. Both elements are combined in the so-called weighted average cost of capital (WACC), which is the weighted average of the cost of equity and the after-tax cost of debt. Companies that measure competitive advantage this way need to compare their profit to its cost of capital instead of comparing profitability with main competitors or the average profitability of firms in the industry. If the profit is higher than the WACC, the company creates value, if it is equal it retains value, and if the profit is lower than the WACC, it decreases value. Increasing value means that the company is able to realize above normal economic performance. It has a competitive advantage based on which it can expect to get access to cheap capital for further expanding its business. Companies that earn their cost of capital just fulfill the expectations of their equity and debt holders. Based on their competitive parity they get access to the capital they need to survive but have only limited growth opportunities.

Stock market performance: total return to shareholders

Another way to measure competitive advantage from a shareholder perspective is by comparing a firm's total return to shareholders with the industry average or a broader market index. A firm has a competitive advantage if it is able to provide its shareholders or owners a higher "utility" than the competitors or market. Hereby, the total return to shareholders is the return shareholders receive for their risk capital over a specific period. It includes stock price appreciation plus dividends received. On the one hand, measuring performance through total return to shareholders is considered as an objective external performance indicator representing a stock market view that takes into account all available information about the company's past, present, and anticipated future performance with a special emphasis on growth. On the other hand, this type of measurement has severe limitations due to the psychological mood of investors reflecting irrational expectations, the impact of macroeconomical factors on stock prices rather than the company's strategy, and the increased volatility in the firm's environment that makes particularly short-term performance assessments based on total return to shareholders difficult. (Rothaermel 2013: 121 f.)

Shareholder value—relevance lost?

The visual impact of climate change, ecological and social calamities as well as human injustices have lead to an increased criticism of company practices that are solely driven by economic performance considerations. Although maximizing shareholder value might be a prime objective of strategy development in some companies, more and more firms embrace the idea of not only improving their profitability or economic performance but also their social and ecological performance (see Strategic Snapshot 1.7). This can be captured by looking at the corporate sustainability performance which will be explained next.

1.1.3.4 Corporate Sustainability Performance

Sustainability and the "triple bottom line"

Today's environmental and social challenges call for a more balanced definition of *superior performance* based on the principle of "sustainability". According to the World Commission on Environment and Development "Sustainable Development is development that meets the needs of the present without compromising the ability of future generations to meet their own needs." (WCED 1987: 43) Applied to a business context, "corporate sustainability can accordingly be defined as meeting the needs of a firm's direct and indirect stakeholders (such as shareholders, employees, clients, pressure groups, communities, etc), without compromising its ability to meet the needs of future stakeholders as well." (Dyllick/Hockerts 2002: 131) In this sense, it relates to the stakeholder approach and the shared value idea (see Strategic Snapshot 1.7). Foundational for the principle of corporate sustainability performance is the so-called triple bottom line framework (Elkington 1997), which is applied by many companies today. In this approach, economic, ecological and social achievements are simultaneously considered and integrated to one holistic perspective on performance (see fig. 1-15). The term is based on the English and Anglo-American profit and loss statement (P&L) where the "bottom line" traditionally reflects the profit or net income. Following a triple bottom line perspective, companies should assess their overall performance not only by looking at their financial results—that is

Strategic Snapshot 1.7

Shareholder Value, Stakeholder Value, Shared Value

Companies exist to create value. The key question, however, is "value" for whom? Shareholder Value Management or Value Based Management is an approach that puts the creation of **shareholder value** at the focal point of all economic activities within a company. According to this philosophy, shareholders do not only provide the risk capital necessary to run the company's operations but are also the legal owners of public companies, and thus, deserve a special emphasis. Originated in the mid 1980s and 1990s and rooted in Anglo-American management philosophy (e.g., Rappaport 1986, Stewart 1991), the overall strategic intent of a company and the corresponding management methods and processes are directed on increasing or even maximizing shareholder value. Strategic management decisions regarding R&D, production, sales, marketing and other areas are very much assessed with respect to key drivers of shareholder value, the so-called "value drivers". Critics of that approach argue that a strong shareholder value orientation leads to short-termism in a way that, for example, quarterly financial results for shareholders are overrated and investments or innovations that may reduce short-term profits but drive success in the long term are neglected. Also, critics state that focusing solely on shareholder value exaggerates the importance of the company's equity and debt holders, often to the disadvantage or at the expense of other stakeholders such as employees or the society in general. The global financial crises 2008–2010 as well as global social and ecological problems lead to an increased demand for a stronger consideration of other stakeholders.

"Strategy strives to provide the business with a source of sustainable competitive advantage. For any competitive advantage to be sustainable, however, the strategy must be acceptable to the wider environment in which the firm competes." (Chandler/Werther 2014: 11) The Stakeholder Management approach—conceptually originating prior to Shareholder Value Management—(e.g., Freeman 1984) puts the management and integration of interests of different stakeholder groups in the center of strategic management. "A stakeholder is any group or individual who can affect, or is affected by, the achievement of a corporation's purpose. Stakeholders include employees, customers, suppliers, stockholders, banks, environmentalists, government and other groups who can help or hurt the corporation". (Freeman 2010: vi). The **stakeholder value**

Creating Shared Value
across Nestlé's Value Chain

Nestlé
Creating Shared Value
Nutrition | Water | Rural Development

	Agriculture and rural development	Manufacturing and environment	Products and consumers
Activity	▸ Training farmers ▸ Water management ▸ Direct procurement	▸ Local investments ▸ Water & energy efficiency ▸ Worldwide standards	▸ Nutritional optimisation ▸ PPPs ▸ Education progr.
Value for society	▸ Generating higher income for farmers ▸ Conservation of natural ressources	▸ Value creation in local communities ▸ Natural ressources ▸ Knowledge transfer	▸ Public health ▸ Local investments ▸ Economic growth
Value for Nestlé	▸ Qualtity and availability of raw materials ▸ Efficient/sustainable supply chain	▸ Cost savings ▸ Risk management ▸ People development	▸ Trust in Nestlé brands ▸ Market shares ▸ Profitable growth

Source: Nestlé 2009. Reprinted with permission of Nestlé Deutschland GmbH.

Fig. 1-14: Creating shared value at Nestlé

Continued on next page

Continued from previous page

added can be seen as a consolidation of value added of the different stakeholder groups, such as shareholder value added, employee value added, creditor value added, customer value added, supplier value added, etc. (Figge/Schaltegger 2000: 29). The strategic task of management is to satisfy those critical groups who have a stake in the business in a way that ensures long-term success of the company. This perspective does not automatically mean that there is a conflict with the basic idea of the shareholder management approach. Actively managing all critical stakeholders such as customers, employees, suppliers, or the communities in which the firm operates can be seen as a key driver for increasing profits and shareholder value over the long term, as all of them have an impact on the success and failure of the company. Neglecting important stakeholders might destroy shareholder value and threaten the survival of the firm (Figge/Schaltegger 2000; Chandler/Werther 2014: 53–83).

In recent years, reconciliation between the shareholder value creation and the stakeholder orientation has been under way, partly driven by the increased role of corporate social responsibility. One example related to society as a key stakeholder is the recommendation that companies should not focus on short-term financial performance but on **shared value** that combines value creation for shareholders and for society: "The solution lies in the principle of shared value, which involves creating economic value in a way that also creates value for society by addressing its needs and challenges. Business must reconnect company success with social progress. Shared value is not social responsibility, philanthropy, or even sustainability, but a new way to achieve success." (Porter/Kramer 2011: 64) An example of how the shared value philosophy is applied across a high-level value chain at Nestlé is illustrated in figure 1-14:

their P&L's bottom line or derived profitability measures—but also consider their ecological performance (e. g., carbon footprint, use of renewable materials and energies, waste, etc.) and their social performance (e. g., physical and mental well-being of employees, fair trade, equal opportunities, etc.).

Measuring corporate sustainability performance

There are various approaches that consider multiple bottom lines such as TIMM–Total Impact Measurement and Management (PWC 2015) or SROI–Social Return on Investment (Social Value UK 2015). However, such an undertaking is incredibly complex given all the possible metrics in the various dimensions that need to be captured in an objective way allowing comparisons across firms and industries. Currently there is no standardized way of measuring corporate sustainability performance by integrating all dimensions in one consolidated performance indicator. A common way of dealing with this issue is the application of standardized sustainability or corporate social responsibility (CSR) reporting guidelines such as the Global Reporting Initia-

Fig. 1-15

Integration of economic, environmental and social performance

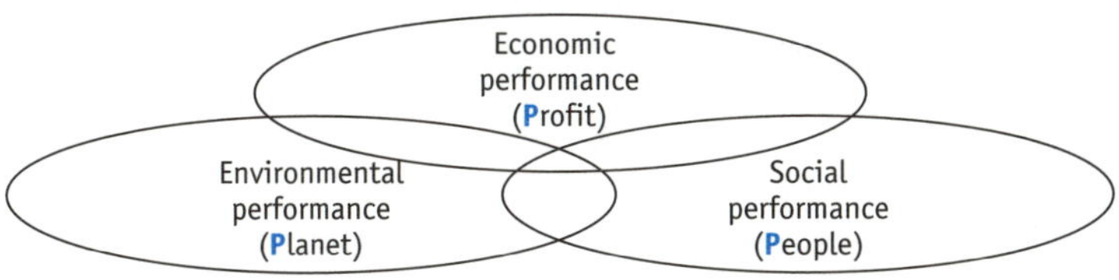

Source: following Elkington 1997

tive (GRI G4 2015), the Commission on ESG Environmental, Social and Governance Issues (EFFAS CESG 2015), the International Organization for Standardization ISO 26000:2010 guidance on social responsibility (ISO 2015) and others with their corresponding key performance indicators measuring different sustainability dimensions (Kleinfeld/Martens 2014). As of today, a number of companies have integrated their financial and sustainability reporting into one report on performance (Eccles/Krzus 2010; Frank 2014; see also Strategy Practice Example 1.5). Additionally, there are examples of companies trying to capture their sustainability performance. An example is the sports and lifestyle company Puma SE and its parent company Kering, which were the first to globally publish an Environmental Profit and Loss Account (E P&L). PUMA accounted for € 145 million in externalized environmental damage in their fiscal year 2010, which was about 5 % of revenues and about 50 % of the accounting profit (EBIT) at this time (Hengstmann/Seidel 2014). A corresponding but more challenging approach for measuring the social dimension of performance is the Social Profit and Loss Account (S P&L). Imagining a requirement for all companies to pay for their externalized environmental and social damage and also getting rewarded for positive externalities—both based on a standardized and objective way to measure this—current profitability rankings and assessments of competitive advantage would change significantly (Elkington/Zeitz 2014: 71–85).

Triple bottom line and competitiveness

For an increased number of managers today, thinking about strategy in terms of the triple bottom line has become a critical element when analyzing competitive advantage and survival. They manage their performance not only in terms of financial profit or shareholder value but also with regard to their ecological and social conduct. For example, nowadays all 30 companies of the German Stock Index (DAX)—the so-called "Blue Chips"—commit and communicate a clear confession to their social responsibility (Blanke/Godemann/Herzig 2007; see also Strategy Practice Example 1.5). Moreover, the shareholder value philosophy of putting special emphasis on equity and debt holders because they provide the risk capital necessary to run the company's operations (see Strategic Snapshot 1.7) has been challenged recently. Studies have shown that companies performing high not only in economic but also in environmental and social categories have lower cost of capital regarding both debt and equity, meaning that they get better access to risk capital than firms with a low sustainability performance (Deutsche Bank Group 2012; Haßler/Häßler 2014; Löbbert 2014). As a result, integrating economic, ecological and social aspects of performance is not only a question of ethics and compliance providing boundaries for strategy but can be seen as levers of competitive advantage and key pillars for competitive survival.

Strategic relevance of sustainability

Considering non-economical factors in strategic management is not a new trend but has been an integral element of many strategic management models. For example, in one of the first integrated models of strategic management—the LCGA scheme of the Harvard Business School (see also Chapter 1.3.1)—the "acknowledgement of noneconomic responsibility to society" is considered one of four key components of strategy (Christensen et al. 1982: 99). Corporate social responsibility or sustainability has been moving beyond a primarily ethical-driven or compliance-oriented conduct. It has become a topic of high strategic relevance for the long-term survival of a company. Here are some of the reasons (Wunder 2014a: 66):

Strategy Practice Example 1.5

Performance and the Triple-Bottom-Line Approach at SAP

Toward a Sustainable Future

"Our commitment to a sustainable world is strong, as we seek to minimize our own environmental footprint and positively impact the communities in which we work. In 2012, for example, 60 % of the electricity consumed by SAP came from renewable sources. We reduced our greenhouse gas emissions slightly despite strong business growth. And, our employees delivered more than 130,000 hours of volunteering in the communities in which we work. Our commitment to achieving sustainable success is further demonstrated in this report, where we integrate our sustainability and financial reporting."

(Source: McDermott/Snabe 2013)

- In many industries, critical stakeholders such as customers, employees or public institutions are increasingly asking for sustainable business practices and present new requirements for companies.
- Companies are increasingly identifying sustainability as a source for gaining a competitive advantage over their rivals (e. g., brand positioning, company image, product differentiation, cost advantages, etc.).
- Being only reactive can lead to a competitive disadvantage if competitors are taking sustainability-related market opportunities faster.
- Implementing sustainability has an impact on all areas in the company and requires new ways of thinking and new competencies across the entire organization.
- The topic is not a short-lived management fashion but reflects the huge and serious global problems of our society that need to be solved urgently.

Business case for sustainability

Instead of a single-minded focus on short run economic results, the economic, ecological and social performance of a company is integrated in a "triple bottom line" perspective (see fig. 1-15). Corporate sustainability performance is achieved when environmental and social issues are managed in line with increased economic performance. Hereby, the substantial contribution of a company's activities to solve environmental and social problems of our society is considered a strategic success factor for ensuring long-term competitiveness and survival. In this understanding, competitive advantage is consequently not only related to a market-driven idea of creating greater customer value than competitors and capturing the corresponding economic value for the primary sake of shareholders but to creating and distributing stakeholder value in a broader sense (Loew/Clausen 2010; Dyllick/Hockerts 2002; Schaltegger/Wagner 2006). Hereby, a firm's environmental and social contributions are actively managed to create a business case, that is to "realize economic success through (not just with) an intelligent design of voluntary environmental and social activities" (Schaltegger/Lüdeke-Freund/Hansen 2011: 7 f.) Such a business case typically requires the incorporation of sustainability within the strategic management process (Chandler/Werther 2014: 65; Wunder 2014a). There are a variety of examples of companies that embrace and leverage sustainability as a principle for their business strategies. Many of those firms were able to gain competitive advantage coming from corporate sustainability (see chapter 3.4.4.5).

1.2 Strategic Decision Making

Strategy decision situation are not primarily analytical problems.

Strategy is about making choices. To accomplish this task, strategic management seminars, university classes as well as mainstream literature on strategic management are typically presenting methods and tools that essentially follow rational decision making models. Hereby, logical norms and a mathematical approach is seen as the basis for determining whether judgments are rational or not. While corresponding approaches are widely applied in business practice and seem to provide considerable benefits for tackling strategic management issues, they do not always consider how people actually make decisions. Strategic decision situations are often seen as primarily analytical problems that can be best solved with a sophisticated rational and mechanistic approach. The importance of intuition or "gut feeling" is often ignored although it is a powerful mechanism that can be highly effective, particularly when dealing with complex problems and uncertain or unpredictable environments (Gigerenzer 2007). Underlying "rules of thumb" (heuristics), however, can easily become biases to good decision making when improperly applied.

Managers are largely unaware of how their minds function.

Most managers remain largely unaware of how their minds help them to make strategic choices. Without basic knowledge about how the brain works, managers cannot anticipate when their cognitive processes that usually serve them well are likely to lead to severe and systematic errors in strategic decision making. The following provides a brief introduction to the foundations of decision making with the goal to create awareness for basic judgmental mechanisms and particularly for cognitive biases. It is intended to sensitize to the limitations of rather analytical methods and tools presented later on in this book and provide some guidelines of how to combine analytical approaches and intuition to ensure effective decision making in the context of strategic management.

1.2.1 Foundations of Decision Making

Decision making requires alternatives. If there are no alternatives to chose from, there are no decisions to make. Choosing among alternatives is not a simple act but a comprehensive process of decision making. There are various models to describe this process which will be briefly explained next.

1.2.1.1 Rational Model

Objective and logical decisions for optimal results

The rational model of decision making assumes that people make fully objective and logical decisions that lead to an optimal result. When solving a defined problem, "optimal" means it can be proven that there is no better solution to this problem. A decision maker who was perfectly rational would optimally and consciously follow the six steps illustrated in figure 1-16 (Bazerman/Moore 2013: 2 f.): Based on a clear problem statement, criteria for evaluating various decision options are identified and weighed according to their relative importance. Each generated alternative is rated for each criterion. After multiplying rate and weigh the alternative with the highest score is chosen as it is the one with the highest value or utility. The model is prescriptive in a

Fig. 1-16

Steps in the rational model of decision making

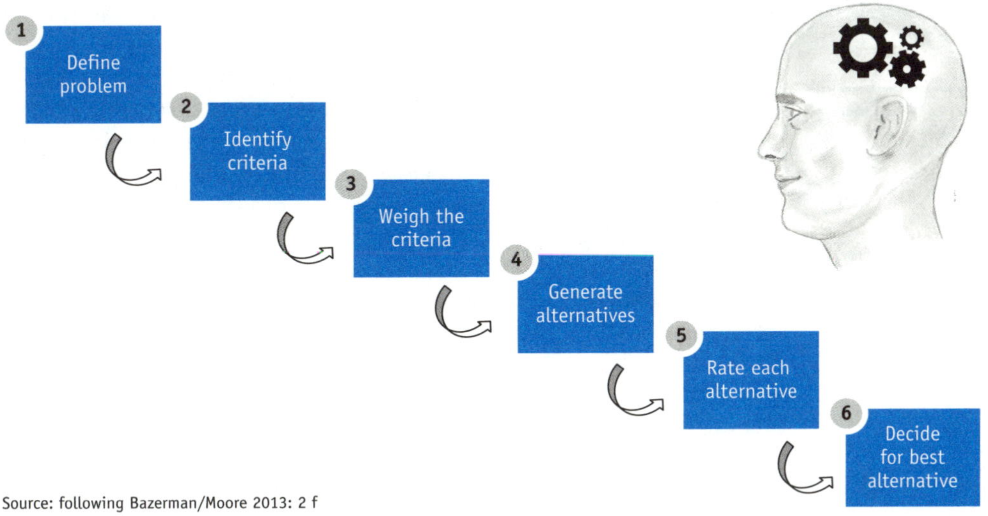

Source: following Bazerman/Moore 2013: 2 f

way that it defines how decisions *should* be made rather than describing how they *are* actually made. In a business context, the rational model can be seen as consistently deciding for the alternative that maximizes a certain goal or expected *utility* like company performance (see chapter 1.1.3) within specified constraints such as available resources as well as the decision maker's values and risk preferences.

Assumptions of rational decision making

Deciding rationally has a number of assumptions. It assumes that managers know the utility of all available options and are able to compute what the optimal decision is. Furthermore, it relies on assumptions such as a clear and unambiguous problem statement, a precisely defined single goal to be achieved, complete information about all available alternative options that are identified in an unbiased manner as well as clarity about the decision maker's preferences that are constant and stable. (Robbins/DeCenzo 2008: 104; Robbins/Judge 2013: 209)

1.2.1.2 Bounded Rationality

Deciding rationally for satisficing ("good enough") results

It is unlikely that people actually decide completely rationally based on the rather optimistic assumptions of perfect rationality explained before. Nobel Prize winner Herbert A. Simon coined the term of "bounded rationality" to explain the fundamental incapability of a human being to make an optimal decision based on all the assumptions explained before (Simon 1957; March/Simon 1958). According to this model individuals are still consciously attempting to make rational decisions. However, a number of limitations prevent them from accurately computing the optimal decision such as:

- Time and cost constraints limit the quantity and quality of information to define the problem, identify criteria, generate all available alternatives and their utility, etc.
- Cognitive limitations of the decision makers such as the amount of available information in their usable memory, intellectual limitations, perceptual errors, etc.

Managers are satisficers not maximizers.

As a consequence of bounded rationality, individuals look for a solution that is "good enough" to solve the problem rather than an optimal one. This solution represents a so-called *satisficing* choice, that means it is satisfactory and sufficient. In the real business world, managers tend to behave within the confines of bounded rationality. They respond to a complex problem by reducing it to a level they can understand and construct simplified models derived from essential problem features. A few easy to find alternative solutions reflecting familiar criteria are identified. Ideally, those solutions have been tried and "proven" to work before. The first solution that meets an acceptable level of performance is selected and ends the search. (Robbins/Judge 2013: 210)

The concept of bounded rationality has been supplemented with a variety of other identified bounds to decision making such as bounded willpower, bounded self-interest, bounded awareness, and bounded ethicality (Bazerman/Moore 2013; Thaler 2000). Those concepts are not further elaborated on here. However, related common biases and errors in decision making and how to avoid them will be explained later on in this chapter. Before this attempt is meaningful, a third category of decision making needs to be introduced: intuition.

1.2.1.3 Intuition

Deciding unconsciously based on experience, associations, and affection

Intuition is often seen as the least rational way of making decisions. Truly intuitive decisions are made in an unconscious process which is based on the decision maker's experience, associations and affection (Robbins/Judge 2013: 211). In other words, intuition is not some sort of paranormal sixth sense but a result of past experience, knowledge, skills, individual perception, and feelings of the decision maker. This is why experts tend to make good intuitive decisions more frequently than others (Rausch 2013: 17). Intuitions come to mind quickly, are strong enough to trigger a decision, individuals are not completely aware where their intuitions come from, and they are based on two key principles (Gigerenzer 2007: 16–19 and 47 f.):

- **Heuristics**, that is simple rules of thumb or judgmental shortcuts that leverage the most important information and ignore the rest, and thus, enable fast decisions and actions. They are used by human beings to compensate their limited cognitive capabilities to handle with the complex environment.
- **Evolved capacities** of the brain, that is capacities such as cognitive abilities and social instincts, the ability to trust and to experience emotions such as love that evolved over millennia. They enable the heuristics.

Intelligence of the unconscious

Management decisions that are based on heuristics are neither necessarily false nor worse. In a positive sense, these decision-making strategies can be considered as efficient techniques to solve problems. Without them it would be impossible to find a solution facing complex and unstructured situations (Beck, 2009, 122). An example is the *availability heuristic,* which can be very useful for managerial decision making. It is a tendency for people to base their judgment on information that is readily "available" in their memory. Information regarding events of greater frequency is recalled more easily than rare events, which is often an indicator for their relevance in decision making. Another example is the *representativeness heuristic,* which fre-

Strategic Snapshot 1.8

System 1 and System 2 Thinking

Two systems or "modes of thinking" in the brain can been seen as the cognitive foundations for any kind of decision making (see fig. 1-17). Although these two modes have received many labels, the terms System 1 thinking and System 2 thinking are widely used. System 1 refers to intuition and operates automatically. Its functioning can be further described as unconscious with no sense of voluntary control, quick, implicit, emotional, and effortless. Examples of automatic activities related to System 1 are the interpretation of verbal language or visual information, orienting to a loud unexpected sound, brushing teeth, or driving a car on an empty road. By contrast, System 2 refers to reflection and operates controlled and conscious. It refers to reasoning that is slower, more explicit and logical, as well as more effortful. The rational decision making model illustrated in figure 1-16 can be seen as prototype of System 2 thinking. Mental activity of System 2 literally requires "paying attention" and deliberately disposing a portion of the limited attention capacity to corresponding mental operations. Examples are computing the product of 17 x 24, counting how often a letter appears on a page, learning to drive, or filling out a tax form. Most decisions in life are made using System 1 thinking.

Time-pressures and busy managers that have many issues on their minds are likely to rely more on System 1 thinking. Not only that more logical System 2 operation is not required for any managerial issue, it may even provide worse decisions than the intuitive System 1 thinking depending on the specific situation. Many managers have a great deal of trust in their intuitions. Sometimes, however, System 1 thinking can lead to severe and systematic errors in decision making caused by cognitive biases. The main reason is that people are usually unaware that they rely on heuristics. Left unchecked, these subconscious cognitive biases can undermine important strategic decision making. A key challenge for managers is to be aware of those pitfalls and identify situations in which they should move from the intuitive System 1 thinking to the more logical System 2 thinking.

(Source: Stanovich/West 2000; see also Bazerman/Moore 2013: 3 f.; Kahneman 2011: 20–24 and Kahneman/Lovallo/Sibony 2011)

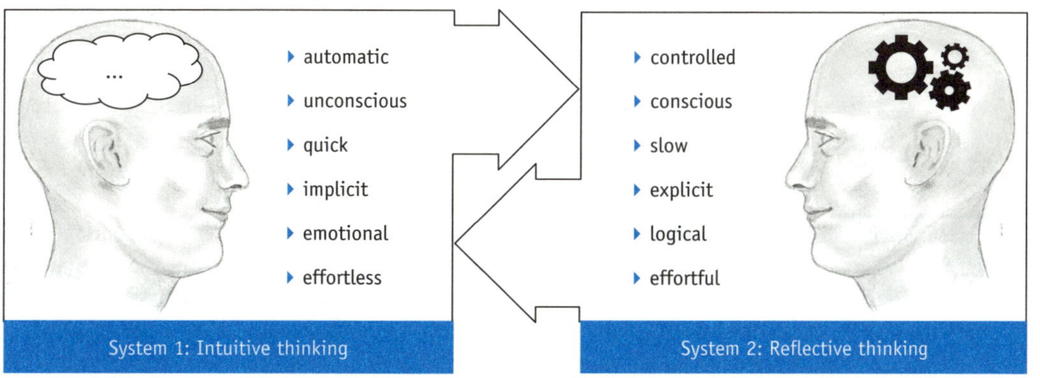

Fig. 1-17: Two modes of thinking

quently offers a good first-cut approximation. Managers who use it judge the likelihood of an occurrence (e. g., performance of an individual, event, etc.) based on how strongly its traits fit to something they are already familiar with (e. g., previously formed stereotypes they have in mind). Both heuristics provide time-pressured managers useful ways of dealing with a complex world. However, along with other heuristics they can also lead to serious errors in decision making which are outlined in table 1-1. (Bazerman/Moore 2013: 8 f.; Robbins/DeCenzo 2008: 108)

Intelligence of the unconscious

"A gut feeling is not good or bad, rational or irrational per se. Its value depends on the context in which the rule of thumb is used. (...) Gut feelings may appear simplistic, but their underlying intelligence lies in selecting the right rule of thumb for the right situation." (Gigerenzer 2007: 48 f.) Intuition is a highly complex form of decision making that does not substitute the conscious rational approach but rather complement it. The situational context determines how well a heuristics and our evolved capacities are working and how important intuition is for good decision making. Eventually, the quality of decision making does not only have to be a matter of elaborating on pros and cons or cost-utility analyses. What this means for typical strategic decision situations will be addressed next.

1.2.2 Strategic Decision Situations

Judgmental cognitive processes are important in strategic management.

Traditionally there has been an emphasis on rational analysis and decision making in strategic management. The more sophisticated analytical tools and techniques for preparing decision making are, according to the general assumption, the better will the strategic decisions be. Nowadays, trend analysis and future prediction industries are continuously developing more and more sophisticated tools that are essentially based on this type of approach. However, it is sometimes neglected that strategic management is not a purely technical process. Even the "best" strategic management processes and tools sometimes fail (Wunder/Stemmermann 2013). The role of intuition and judgmental cognitive processes in strategic management has gained tremendous attention in recent years (Lovallo/Sibony 2010; Powell/Lovallo/Fox 2011; Roxburgh 2003; Schrager/Madansky 2013). It is increasingly recognized that rational analysis has been overemphasized and that relying on intuition can actually improve decision making in certain instances (Gigerenzer 2007). But what is so "special" about strategic decision situations?

1.2.2.1 Characteristics of Strategic Decision Situations

Strategic management differs from operational management.

Typical strategic decision situations tend to differ fundamentally from operational decisions but both need to be aligned (see fig. 1-18). Strategic decisions are directional with long-term impact. Instead of dealing with a partial problem in a subarea of the organization, strategic decisions are often company-wide and more holistic. The decision base relates more to a political than a technical environment with problem solving that is often novel or rare instead of routine or a replication of precedent approaches. Whereas operational management has to deal with how to utilize resources in terms of productivity, strategic management requires decisions of how to

Fig. 1-18

Characteristics of strategic and operational decisions

	Strategic decisions (strategic management)	Operational decisions (operational management)
Focus:	direction ("What?")	efficiency ("How?")
Time horizon:	long-term	short-term
Impact:	company-wide; holistic	subarea; partial
Decision base:	political	technical
Problem solving:	novel/rare	routine/precedents to follow
Resources:	allocate/commit resources	utilize resources
Situation:	ambiguous	definite/clear

best allocate and commit limited resources in the organization. One of the biggest challenges of strategic management is the ambiguity of decision situations. In operational management, decisions tend to relate to rather definite and clear situations (e.g., the breakdown of a production line because a machine has stopped working, customer complaint, etc.). In strategic management the situation is often vague and uncertain (e.g., capturing a new market trend and adapting the business model). Whereas operational decision making may call for a rational process that is more related to System 2 thinking, this approach might not be sufficient for dealing with strategic decisions.

The essential characteristics of those ambiguous strategic decision situations can be described with complexity, dynamics, volatility, intransparency, and lack of "objective" knowledge about reality (Dörner 1992: 58–66 see also Strategic Snapshot 1.9):

Strategic Snapshot 1.9

Strategic Decision Making Illustrated With Chess

The characteristics of strategic decision making can be illustrated by using the analogy of a chess game where the players have to use a chess set with many chess pieces that are interconnected with rubber threads, so they cannot move one piece without moving the others (*complexity*). Additionally, the own pieces and the ones from the opponent are moving on their own based on rules the players do not know exactly or have false assumptions about (*dynamics and volatility*). On top of that, some of the own and the opponent's pieces are in the fog and not or only barely visible (*lack of transparency*). (Source: Dörner 1992: 66)

Complexity, dynamics and volatility

Strategic situations tend to be rather complex. This means that there are many interdependent elements inside and outside the organization that ideally need to be considered. Furthermore, if one element is altered it might impact the others. For example, if a competitor launches a new product it might have an impact on the behavior of our customers and the perception of our own portfolio. Launching a new product by our own might impose competitor response that we did not know before. Also, if the government releases new laws, it strongly affects the market. The recent turnaround in the German energy policy toward renewable energy, for example, changed the competitive landscape in the energy sector dramatically. Furthermore, strategic decision makers are facing a high level of dynamics as development occurs also when they do not act. This means that even the most comprehensive analysis of today's state is not sufficient as the competitive landscape is continuously changing. Moreover, companies have to deal with an increasing level of volatility. This means that the frequency and magnitude of change between growth and decline in the economy is increasing, which has widespread impact due to globally interconnected and fast moving markets.

Lack of transparency and "objective" knowledge about reality

Another characteristics is a lack of transparency, that is that neither the planner nor the decision maker has access to "all" information that would be necessary to consider. A decision needs to be made even though the current situation can never be fully understood, not to speak about the future. Furthermore, there is no "objective" knowledge about reality. People decide based on their individual explicit or implicit world view or assumptions based on which the individual brain is constructing its own reality. Most people strive for certainty which prevents them from dealing with the possibility that their assumptions might be wrong or their view might be incomplete.

1.2.2.2 Dealing with Strategic Decision Situations

Complexity and uncertainty call for good intuition

The challenge of strategic situations is that decisions need to be made in a highly complex environment today based on an anticipation of the hardly predictable future. Complex problems require complex solutions, we are frequently told. Finding an "optimal" solution for a strategic decision situation based on the characteristics explained, managers may want to use as much information as possible and feed it into the most sophisticated analytical decision-making process. Management leadership teams, for example, that are recruited from operational experts with technical backgrounds shaped by reasoning of natural sciences rationality, tend to approach strategic decision situations in a more "technical" manner and rely primarily on analytical tools to "compute" an optimal solution. Finding an "optimal" solution in such a situation, however, requires two preconditions:

- First, there actually is an "optimal" solution
- Second, there is a rational way to find it.

Given the complexity, dynamics, volatility, intransparency and flaws in perception of such strategic situations, it is very unlikely that the two preconditions apply. In such highly complex and uncertain situations, simplicity and good intuition becomes critical for success. Depending on the situation they may provide good or even better results than time consuming rational problem solving and decision-making processes. (Gigerenzer 2007: 81–92; Gigerenzer/Brighton 2009; Gigerenzer/Todd/ABC Group 1999)

In a strategic decision situation and its underlying conditions of complexity and uncertainty, executives "are not handed nicely distilled comprehensive summaries of the situations they face. Instead, the 'facts' that confront executives—if they can be called facts—are typically ambiguous, contradictory, and far-flung, and they emanate from various parties who have their own motives. As a result, the situations that executives face are not knowable; they are only interpretable. (...) the individual filters the facts through a web of personal qualities—including what he or she values, and how his or her mind works. As such, the person's actions are much more a reflection of the person than of the situation." (Finkelstein/Hambrick/Cannella 2009: 3 f.) Strategic decisions are not only driven by factors like customers, competitors, technology or the company's resources, capabilities and competencies but by the people making strategic decisions at different levels in the organization and their biases and dispositions. They are also affected by the decision making environment at a particular company, such as whether open discourse is encouraged with dissenting views or whether senior leaders reject input and seek affirmation of their own viewpoints (see also chapter 1.3.3.3).

Strategic decisions not only reflect the situation but also the person and environment.

1.2.2.3 Cognitive Biases and How to Counter Them

A "simple" and intuitive approach to strategic decision making has pitfalls. Relying too heavily on experience, gut feelings and convenient rules of thumb can also distort rationality. Heuristics sometimes lead to severe and systematic errors in decision making. The main reason is that people are usually unaware of the heuristics they base their decisions on. Besides emotional and motivational influences, this can lead to errors in intuitive judgment due to subconscious cognitive biases (Kahneman/Lovallo/Sibony 2011; Tversky/Kahneman 1974). Those cognitive biases breed failure and interfere with the idealistic assumption of rational decision making underlying most of the strategy tools. Left unchecked, this cannot only affect but undermine important strategic decision making resulting in a negative effect on the company. Helping organizations and management teams to become aware of the potential adverse impact of using heuristics that cause the corresponding cognitive biases are likely to improve their decision-making processes, and thus, lead to more beneficial outcomes. Organizations that worked on reducing such bias-based flaws in their strategic decision-making processes significantly improved their return on investment by up to seven percentage points (Lovallo/Sibony 2010: 6).

Relying too strongly on heuristics has pitfalls.

Cognitive biases are errors in intuitive judgment. They have to be distinguished from logical fallacies which are errors in logical argumentation. Cognitive biases are caused by System 1 thinking (see Strategic Snapshot 1.8). They unconsciously undermine the effectiveness of decisions that are based on heuristics. "Because System One is so good at making up contextual stories and we're not aware of its operations, it can lead us astray. The stories it creates are generally accurate, but there are exceptions. Cognitive biases are one major, well-documented example. An insidious feature of cognitive failures is that we have no way of knowing that they're happening: We almost never catch ourselves in the act of making intuitive errors. Experience doesn't

Cognitive biases unconsciously undermine decision making.

Tab. 1-1

Common biases in decision making

Availability bias	Reliance upon the knowledge that is readily available in the mind rather than examining additional sources of information.
Anchoring effect	Giving disproportionate weight and fixation to the initial information the mind receives without adequate adjustment for subsequent information.
Confirmation bias	Only seeking information that confirms pre-existing views or past choices and ignoring information that supports alternative judgments.
Egocentric bias	Decisions are influenced by the goal to maximize own gain. Also, own contributions are seen as overly responsible for outcomes of groups.
Escalation of commitment	Having increased commitment to and staying with a previous decision even when there is clear evidence that the decision is wrong.
Conjunction fallacy	Overestimating the likelihood of a scenario with more detail than with less. More specific conditions are seen more probable than general ones.
False consensus bias	Overestimating the extent to which own thoughts, opinions, beliefs, and experiences are common and shared by others.
Framing effect	Decisions are influenced by the way or expression of how logically equivalent information is presented (e.g., as a loss or as a gain).
Groupthink	A group consensus for a first remotely plausible option overrides the realistic appraisal of alternatives and engagement in critical thinking.
Halo effect	Drawing overall impressions about something (e.g., person, product, situation, etc.) based on only a single characteristic.
Hindsight bias	Believing that a known situation could have been accurately predicted and dismissing the possibility that it could have turned out differently.
Inappropriate attachment	Distorting judgments by emotional attachment of decision makers to people, projects, organizations, or other things.
Illusion of control	Overestimating the ability and power to control events. Illusionary believe to have influence on things that are beyond one's control.
Overconfidence bias	Placing exaggerated confidence in one's own knowledge or abilities to solve a certain situation other than it should be given by objective parameters.
Optimism bias	Believing oneself is less at risk of experiencing a negative event or more likely to be successful compared to others neglecting actual probability.
Positive outcome bias	Overestimating the probability of good things happening ("wishful thinking") and underestimating the probability of undesirable outcomes.
Reactance bias	Overrating something that is limited or taken away (e.g., decision option) as a reaction (resistance) to perceived constraints of freedom.
Self-serving bias	Attributing success as one's own personal contribution but failure to the result of external factors and rejecting the validity of negative feedback.
Status quo bias	Avoiding any changes from the current baseline (status quo), and thus, judging and deciding in a way that things remain the same.
Sunk cost fallacy	Judging investment decisions not only on their own merits but considering costs that have already been incurred and cannot be recovered.

Source: Bazerman/Moore 2013: 14–59; Kahneman 2011; Klatt/Möller 2011; Weber/Schäffer 2011, 263–268; Dobelli 2011; Montier 2007, 19; Riesenhuber 2006, 69–134; Pohl 2012; Robbins/Judge 2013: 211–216; Roxburgh 2003; Baron 2008 help us recognize them. (By contrast, if we tackle a difficult problem using System Two thinking and fail to solve it, we're uncomfortably aware of the fact.)" (Kahneman/Lovallo/Sibony 2011: 4) In strategic management, cognitive biases are likely to affect the most important strategic decisions made by even the smartest executives in top companies. Their decisions may be biased in ways that seriously compromise their quality. Managers can hardly eliminate their own biases on an individual level even if they know they have them. However, creating awareness of biases and incorporating those insights into the strategic decision making processes of management leadership teams helps to "debias" those collective strategic decisions in organizations (Kahneman/Lovallo/Sibony 2011). There is a very wide spectrum of the behavioral phenomena related to cognitive biases. Table 1-1 provides an overview and brief explanation of 20 common biases in decision making in alphabetical order.

Insights about the modes of strategic decision making and how to best deal with dangerous errors and biases have to be build into the construction logic of a strategic management process. It may be considered when selecting tools and methods for strategic analysis, strategy formulation and execution, defining standards of how strategic decision proposals should look like, forming strategic planning teams or an office of strategic management (see Strategic Snapshot 1.10 in chapter 1.3.3.1), building management leadership teams for making strategic decisions or defining norms for reaching consensus. The chapter on strategic decision making concludes with some suggestions for how to improve strategic decisions and avoid falling in errors caused by cognitive biases.

No logical reasoning for intuitive decisions

First, management teams making strategic decisions are well advised to take intuitive decisions of their members very seriously and complement conscious rational strategic decision making approaches they may follow with those intuitive judgments. Hereby, it makes little sense to ask a member who decided intuitively to explain the reasons for his judgment because those decisions are typically made in an unconscious process. Some management leadership teams give equal weight to purely intuitive decisions. Particularly when making important strategic decisions, they accept an answer like "I don't know why but my gut feeling tells me it is the wrong (or right) decision" as veto for their management team conclusion without asking for a more logical reasoning (Gigerenzer 2010).

"Debias" strategic decisions

Second, deliberately looking or asking for information that disconfirms one's beliefs or a strategic recommendation can help to counteract a variety of biases such as overconfidence, optimism or hindsight (Robbins 2004: 164–168). Asking questions like "What are major things that can go wrong? How likely are they?" may help to see how thoroughly a strategic recommendation has been investigated and may reveal potential biases of the recommending or deciding team. For example, those questions may help to reduce the liability risk of supervisory boards when providing their legally required approval for certain strategic decisions of management boards. Figure 1-19 provides a checklist of 12 questions for reducing cognitive biases and errors when making strategic decisions. Most questions can be related to both the team who prepared a strategic decision and makes recommendations based on deep analysis as well as to the management leadership team who has to make the decision with less time for its judgment.

Fig. 1-19

12 questions for reducing biases and errors in collective strategic decision making

1 Are there any self-interests in the team?	7 Where did the numbers come from?
2 Did the team fall in love with one proposal?	8 Could there be a HALO effect?
3 Were there dissenting opinions in the team?	9 Are we overly attached to past decisions?
4 Are there wrong or salient analogies used?	10 Is the base case overly optimistic?
5 Were credible alternatives considered?	11 Is the worst case bad enough?
6 How would we decide in a year's time?	12 Is the team overly cautious?

Source: following Kahneman/Lovallo/Sibony 2011, adapted

1.3 What is Strategic Management?

General management perspective on how to manage a firm strategically

Strategic management is an integrated management field that deals with decisions and actions that determine the long-run performance of a company. It explicitly applies a general management and cross-functional point of view that elaborates on decisions and strategies in the light of the entire company. Managing an organization strategically requires integrating knowledge and processes of different business disciplines such as finance and managerial control, marketing, operations, human resource management, organizational and human behavior, leadership, and others. Strategic management includes strategic analysis, strategy formulation at different levels in the organization including business modeling, as well as strategy execution along with control, testing and adaption of strategy. The wheel of strategy framework introduced in chapter 1.3.3 embodies this view with all relevant aspects. It is an illustrative holistic framework of strategic management that can be used as a reference point. Many concepts and techniques in strategic management have been developed and used by or for rather large multinational corporations. Foundational thought models and approaches, however, can be applied to business and non-profit organizations of any size.

1.3.1 Evolution of Strategic Management

From financial planning to strategic planning

Few if any companies today are operating only based on *basic financial planning* focusing solely on the following year's budgets. A firm is likely to engage in any kind of strategy process where it deals with the question of how to secure its future beyond next year based on some sort of analysis. Companies may conduct this more in the sense of *long-range planning* which could still be focused on financials but with a time-horizon of 3–5 years. Hereby, decisions are based on internal and available (ad hoc) external information and corresponding extrapolated trends. It typically relates to projects with durations longer than one year. Long-range planning activities can be found primarily at the middle to lower management levels. Traditional *strategic planning*, on the other hand, is based on a more systematic and formalized internal and external analysis. The goal is to increase the organizations responsiveness to changing competition with more comprehensive planning processes. Top management is the primary planning level and is supported by market and competitive intelligence units or strategic planning groups. Implementation is left to lower management levels.

Strategic management goes beyond long-range and strategic planning

Like strategic planning, *strategic management* is also based on a systematic external and internal analysis or even a continuous monitoring of the business environment in the sense of strategic foresight. The time horizon is 3–5 years and beyond. Some companies elaborate on potential pictures or scenarios of the future with a 10, 20 or even longer time perspective, derive contingency strategies and make decisions about their desired position in an anticipated competitive arena of the future. Corporations such as Siemens with its technology oriented "Pictures of the Future" or Nestlé with its magazine "GOOD" regularly publish trends that will shape our lives over the next 10 to 20 years, draw on scenarios of the future and explain their corresponding strategic thrusts. Contrary to long-range or strategic planning, strategic management is not too focused on a certain planning level but tries to involve people at all levels in an interactive way in ongoing strategic thinking and ideation. Strategy execution as well as strategic agility and adaptability are considered as particular important elements given today's dynamic and volatile environment (see chapter 4.3.7.3). Also, behavioral aspects such as intuition or heuristics and the related biases are more and more elaborated on in strategic management to improve the strategic decision making process (see chapter 1.2). In company practice, the terms strategic planning and strategic management are sometimes used synonymously, that is companies officially engaging in "strategic planning" may actually practice strategic management and vice versa.

Strategy as own academic discipline

The origins of strategy as an own academic discipline can be found at the Business Schools at universities in the USA. One of the forerunners was the Harvard Business School where a course titled "Business Policy" in the Senior Management Training was based on solving company case studies dealing with selected general management issues including the strategic orientation of a company. The initial focus was not on scientific research or developing theories related to strategy but on solving real company problems. Dealing with strategy as an academic discipline advanced in the 1960s and 1970s. One example is Alfred Chandler (1918–2007) with his book "Strategy and Structure" in 1962, where he writes about the growth process of four companies (Sears, General Motors, DuPont and Standard Oil). Based on his insights, he formulated his famous strategic management principle "Structure follows Strategy" that is still frequently quoted by executives today. Another pioneer of today's strategy practice and one of the founders of strategic planning was H. Igor Ansoff (1918–2002) of Massachusetts Institute of Technology with his work "Corporate Strategy" in 1965. He described in depths the foundations of strategic planning and developed a detailed flow chart of analysis, gap identification and decision making. This was mapped out in the Ansoff Model for Strategic Planning which is shown in figure 1-20 (Ansoff 1965: 202 f.). It is an illustrative example for the level of planning detail and

Fig. 1-20

The Ansoff model for strategic planning

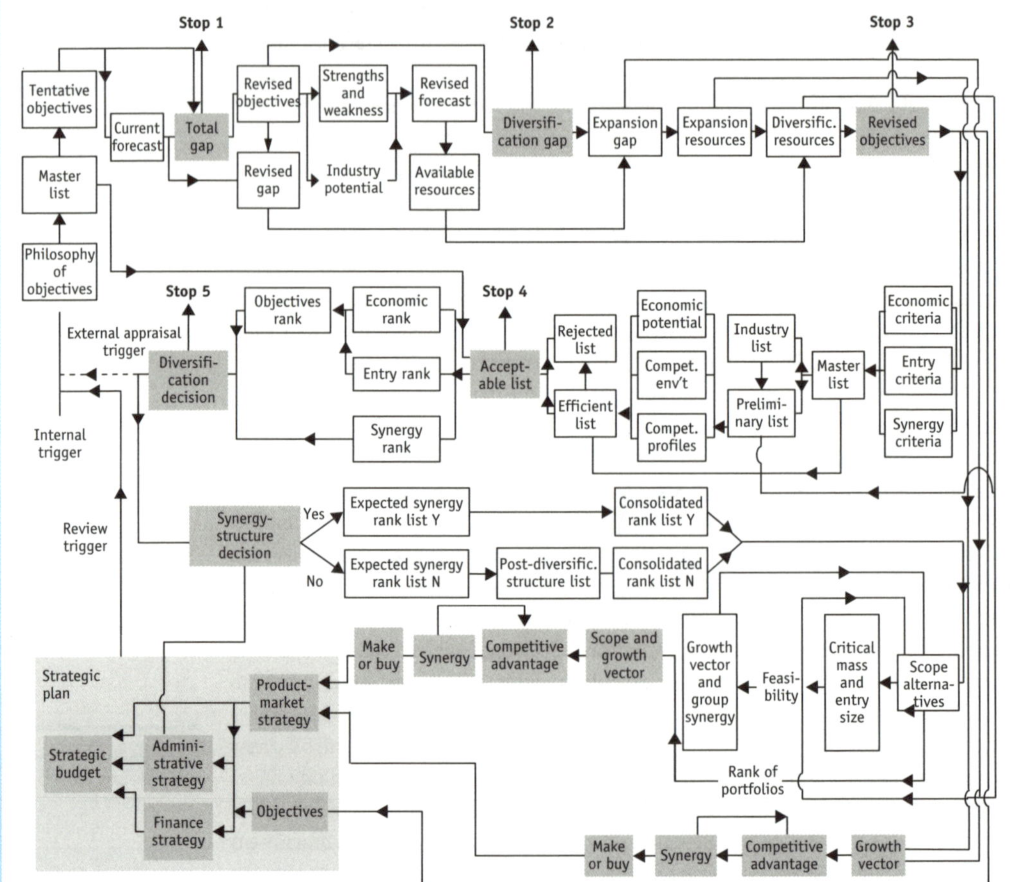

Ansoff 1965: 202 f.

granularity some of those early models of strategic planning where based on. One of Ansoff's most famous concepts is the product-market decision framework for growth strategies that is often referred to as "Ansoff Matrix" and has been used by many managers until today. He advocated the SWOT analysis—a "tool" which was also derived from research conducted at the Stanford Research Institute in the 1960s (Humphrey 2005: 7)—to get a comprehensive understanding for creating a "fit" of the business environment and the organization.

LCAG-model: controlled and conscious strategy formulation implementation

Also in 1965, Edmund Learned et al. published the renowned text-book "Business Policy: Text and Cases". Here, the strategy process is separated into a strategy formulation phase that is followed by a strategy implementation phase. The corresponding strategic management framework is often referred to as the Harvard Business School's "LCAG-model". Figure 1-21 shows its basic logic that evolved over time and is still the foundation of many mainstream strategic planning and management models in company practice today (see Strategy Practice Example 1.6). It is based on the belief that strategies can be "designed" in a controlled and conscious thought process by applying basic principles principles such as matching factors from the company's external environment (i. e. opportunities and threats) with its internal organizational factors (i. e. strengths and weaknesses) to strategically position the firm in its environment. This simple and easy to apply logic is supplemented by elements of *social responsibility* and *managerial values* at the interface of strategy creation and

Fig. 1-21

Basic logic of strategy formation in the LCAG-model

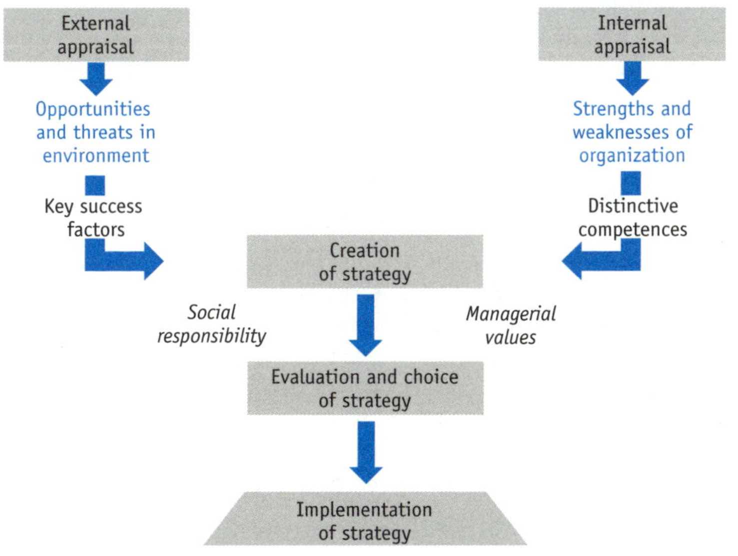

Source: Mintzberg 2000: 37 following Christensen/Andrews/Bower/Hamermesh/Porter: 1982; adapted

evaluation and choice. In today's era of sustainability those two elements seem to be relevant more than ever. One of the co-authors of this early book, Kenneth Andrews (1916–2005), elaborates further on this strategic management approach in his publication "The Concept of Corporate Strategy" where he clearly differentiates between corporate strategy and business strategy. He also elaborated on matching external opportunities with corporate capabilities at an acceptable level of risk—an approach that has been widely applied in companies. Many of the strategic management models today are still based on the fundamental concepts and methods developed in the 1960s and 1970s by those renowned pioneers of this discipline.

1.3.2 Schools of Thought and Paradigms in Strategic Management

Different views on how strategies form

Over the years a variety of perspectives on strategic management evolved that differ particularly regarding the question of how strategies form. Each of those "schools" appeared at different stages in the development of the field, can be found in fairly clearly delineated types of literature and have significantly influenced business practice and vise versa. Figure 1-22 provides an overview of different schools of thought in strategic management leveraging one single adjective that seems to capture the view best and lists some exemplary thinkers who significantly contributed to the field. The various schools of thought can be grouped in prescriptive and descriptive schools and are briefly explained in the following (Mintzberg/Ahlstrand/Lampel 1998).

Prescriptive schools of strategy

Suggestions of how strategies *should* be developed

The first three schools are *prescriptive* in nature, that is they primarily deal with the question of how strategies *should* be developed and are less concerned with how strategies are actually form. The first (*design school*) suggest strategy formation being a process of more or less formal design or conception (see fig. 1-22). Strongly related is the *planning school,* which is emphasizing a highly systematic and formal process for strategy making (see fig. 1-20). Compared to the first two perspectives, the *positioning school* is focusing less on the process of strategy making and more on strategic content, for example, on selecting the "right" strategic positions in the competitive arena for achieving superior performance. Prescriptive strategy models are widely applied in company practice, academic textbooks and college lectures, practical management guides as well as executive trainings. Generally speaking, prescriptive strategic management models suggest starting a strategy development process by getting a comprehensive understanding of where a company currently stands and where it will be in the future if no strategic changes are made. They provide guidance for the identification and selection of strategic options to move forward and for the successful execution of strategic initiatives.

A practical example of such a process is provided in Strategy Practice Example 1.6 and figure 1-23. Some companies follow prescriptive strategic management or planning processes in a more formal, systematic and analytical way, others prefer a less formal approach. Although the value of strategic planning for firm performance can be seen critically (Mintzberg 2000), research in this field also emphasizes the importance of formal strategic planning or strategic management processes for improving company performance (e. g., Dye/Sibony 2007; Becker/Freeman 2006; for further sources see Wheelen/Hunger 2010: 6 f.). For example, a recent survey of more than 100 companies in the German food industry found that 90 % of the participating executives perceived a formal strategy process as a key success factor for their company. The best performing food companies—that were able to grow revenues as well as profit three years in a row stronger than their competitors—clearly engaged in a more systematic and formal strategic management process (Wunder/Bausch 2014b). In these companies, strategic management methods are applied to identify developments in the company's environment at an early stage. Companies systematically elaborate on strategic responses with, for example, adjusted or new business models. Furthermore, they systematically design an effective execution process to make their strategies happen.

Descriptive schools of strategy

Descriptions how strategies are actually get formed

The six schools of strategy labeled *descriptive* are less suggesting how to ideally make a strategy but rather *describing* how strategies are, in fact, *get* formed. Two schools are trying to understand strategy formation focusing on the individual. Hereby, strategy is seen as the result from the vision of great leaders (*entrepreneurial school*) or from the complex cognitive processes operating in a person's head (*cognitive*

Fig. 1-22

Schools of thought in strategic management

School of thought ...	views strategy formation as a ...	Exemplary thinkers
Prescriptive (SHOULD)		
Design school	... process of *conception*	Selznick, Chandler, Andrews, Learned
Planning school	... *formal* process	Ansoff, Steiner, Lorange, Campbell
Positioning school	... *analytical* process	Porter
Descriptive (AS IS)		
Entrepreneurial school	... *visionary* process	Schumpeter, Bennis/Nanus
Cognitive school	... *mental* process	Simon, March, Tversky, Kahneman
Learning school	... *emergent* process	Weick, Lindblom, Quinn, Burgelman
Power school	... process of *negotiation*	Macmillan, Sarrazin, Pettigrew, Doz
Cultural school	... *collective* process	Rhenman, Normann, Brunsson, Barney
Environmental school	... *reactive* process	Hannan, Freeman
Integrative		
Configuration school	... process of *transformation*	Chandler, Miller, Miles, Snow

Source: following Mintzberg/Ahlstrand/Lampel 1998, adapted

school). The latter has enjoyed great attention in recent years driven by research in fields like behavioral strategy or behavioral controlling (see also chapter 1.2). Four other schools open the view on strategy formation beyond the individual and include other actors. According to the *learning school* strategies cannot be developed solely based on plans or visions because the world is too complex. Strategies rather emerge and evolve in small incremental steps as the organization adapts to its environment and "learns" along the process of adaption (see also chapter 4.3.7.3). The *power school* views strategy as a result of using politics and power for negotiating strategies favorable to particular interests that might be beyond purely economic considerations. It relates to corresponding processes within the organization or by organizations themselves. For another group of scholars—the *cultural school*—strategy is essentially the result of a collective and cooperative process that is deeply rooted in the culture of an organization (see also chapter 4.2.1). The *environmental school* is driven by the contingency theory considerations. Strategy is seen as strongly related to the context of a firm and results from the organization's response to primarily external factors. If leadership does not read the environment properly and the company does not properly react to those external forces, it sooner or later will be "selected out". The last school (*configuration school*) tries to integrate different views of strategy formation. It views strategy as a transformational process where concepts from prescriptive schools of strategy as well as insights from descriptive schools are integrated to manage strategic change in practice.

Two different paradigms in strategic management

Even though a number of prescriptive strategic management models seem to be widely used and effective in business practice, it is important to note that they rely on a specific paradigm. According to the philosopher of science Thomas S. Kuhn a paradigm is a conceptual view of the world. It is a way of thinking or common intellectual framework that members of a scientific community share at a certain time. It constitutes what scientific solutions are acceptable and what scientific questions can be asked (Kuhn 1970). In the strategic management field, two key paradigms—classic-rational versus comprehension-oriented—can be distinguished that will be briefly explained next (Ungericht 2012).

Classic-rational paradigm of strategic management

In the classic-rational paradigm it is assumed that reality is something objective and the human perception is a 1:1 copy of it. People decide and act purely rational and consciously to fulfill their economic self-interest. Internal as well as external influencing factors can be entirely controlled (cause-and-effect logic). The complexity and uncertainty of a strategic situation can be reduced through planning. The classic-rational paradigm is based on a mechanistic-analytical world-view, that is:

- Priority of planning: Strategies are developed and executed based on conscious and rational decisions derived from solid analysis (first think than act than check).
- Strategic management is primarily a technical challenge. With the right techniques, concepts and tools we are able to find and implement the right strategies.
- The key terms of this paradigm are: planning, analysis and control.

Strategy Practice Example 1.6

Strategy Process at the RECARO Group

RECARO is one of the leading seat brands in the world. The RECARO Group encompasses the two main divisions Aircraft Seating and Child Safety as well as the RECARO Holding. Figure 1-23 shows the strategy process of the group that is essentially following the classic strategic management phases of strategic analysis, strategy formulation and strategy implementation and review. It is a particularly thorough and systematic process that operates on an annual cycle. Strategic planning at RECARO starts with a bottom-up analysis of the current situation regarding the company's product areas, regions and customer groups in all strategic business units. A variety of tools are used to analyze the competitive arena and market trends and derive so-called "strategic challenges". Based on this, strategic options are developed, evaluated and chosen. They are the foundation for deriving strategic objectives, corresponding metrics and targets as well as strategic programs to make the selected options happen. Progress in strategy implementation is regularly reported leveraging a strategy cockpit. (Kellinghusen 2012: 121 f.)

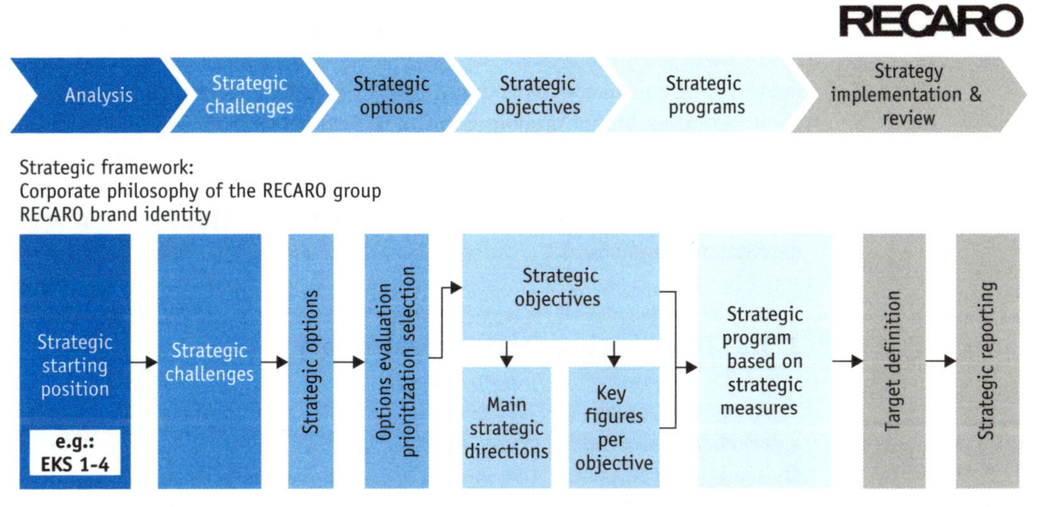

Source: Kellinghusen 2012: 121. Reprinted with permission of RECARO Holding GmbH.

Fig. 1-23: RECARO Group strategy process

Comprehension-oriented paradigm of strategic management

In the comprehension-oriented paradigm it is assumed that reality is a construct in the human brain based on subjective perceptions and factors. People decide and act influenced by unconscious cognitive processes and different motives. Internal and external influencing factors cannot be controlled (company = social/political system). Complexity and uncertainty cannot be reduced. We can only learn how to deal with it. The comprehension-oriented paradigm is based on a critical-realistic world view, that is:

- Classical mechanistic "wisdoms" of strategic management are fundamentally questioned.

- Organizations are seen as complex social/political systems and not as rationally designed "machines".
- Priority of acting (acting = realizing/learning) instead of planning: Strategies emerge based on learning, experimenting, negotiating as well as intuitive processes.
- The key terms of this paradigm are: process, adaption, innovation, learning.

Mainstream versus system-theoretical or evolutionary approaches

Many mainstream strategic planning and management models as well as a wealth of current textbooks, practical management guides, industry applications, and consulting firm approaches are essentially based on a classic-rational paradigm. Approaches that are more based on a comprehension-oriented paradigm include strategic management models based on system-theory or evolutionary strategic management models as well as the research field strategy as practice (e. g., Bleicher 2011; Grimm 1996; Kirsch 1997; Malik 2008; Ulrich 2001). It is hard to say which of the different approaches and paradigms of strategic management is right or wrong because they are based on very different assumptions. On the one hand, it is an illusion to believe that social processes within organizations can be entirely planned. On the other hand, there are good reasons for applying the "rationality" of analytical strategic management models and tools to systematically elaborate on the future of the company and corresponding strategy options. It is important, though, to understand the limitations of such methods and to be aware of behavioral and other aspects that impact strategic decisions and their execution. Those do not only need to be considered but ideally built into the construction logic of more rational approaches (Wunder/Stemmermann 2013).

1.3.3 The Wheel of Strategy Framework

Essentials for managing organizations strategically

The wheel of strategy (WOS) framework introduced in this book is intended to create a discipline of strategic thinking and strategic practice within an organization. It is an analytical and prescriptive model for strategic management that captures the essential aspects to be considered when managing organizations strategically. It is supposed to help companies formulate and execute strategies effectively, and thus, ultimately improve performance. The framework consists of four basic components that can be illustrated with reference to a wheel with bearing and axle (see fig. 1-24). The wheel of strategy is intended to be a simple to use, yet comprehensive model that frames and structures the thought process from strategy to action. It is based upon 15 years of practical strategy work in organizations and grounded in state-of-the-art strategy theory and research. The framework captures the components strategy process, structure, culture, and strategic leadership with corresponding methods and approaches for each component. Its company-specific refinement and application is typically facilitated by functional areas such as the Office of Strategic Management (see Strategic Snapshot 1.10) or Strategic Planning, Corporate Development and Strategic Controlling departments.

Fig. 1-24

The wheel of strategy (WOS) framework

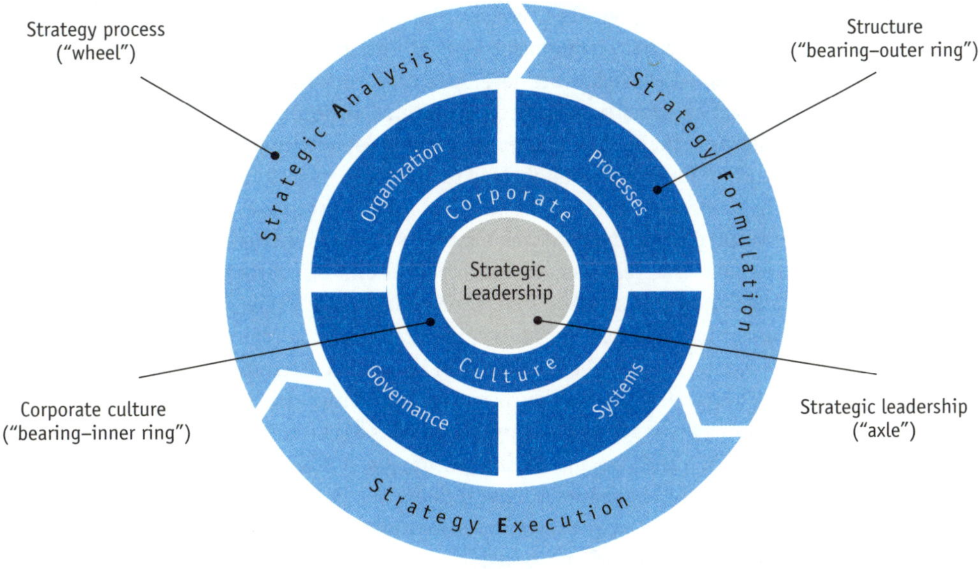

Metaphorically speaking, in some companies the wheel is spinning very fast to reflect dynamics in the firm's competitive arena and corresponding ongoing strategy adjustments. The speed is so high that the interfaces between analysis, formulation and execution diminish and are only barely perceptible. In other industries, a slower movement of the wheel may be sufficient. Furthermore, even the best wheel will not rotate well if the bearing lacks lubrication. In business terms this relates to the structure within a company—its organization, business processes, systems and governance—and its corporate culture. If sand or grit gets into the bearing and is mixed up with grease, it will hinder mobility and lead to frictional losses just as it is with companies whose organizational design does not match strategy. For effective strategic management, strategy, structure and culture must fit together (see also chapter 4.2). Strategic leadership is considered the axle of the wheel of strategy. This is where everything else is built upon. It needs to be strong to properly stabilize the bearing, and thus, enable the wheel to move. In business practice, strategic leaders and leadership teams do significantly matter for a strategy process to be effective. They essentially affect which strategies are pursued and how those strategies are translated into action and ultimately firm performance.

A spinning wheel needs a lubricated bearing and strong axle.

Integrated tasks: strategic analysis, strategy formulation, strategy execution

1.3.3.1 Strategy Process

The first WOS component—the outer layer or *wheel*—is where the "rubber meets the road". It describes the strategy process (or strategic management process) with its three integrated management tasks strategic analysis, strategy formulation and strategy execution. In this closed-loop and repetitive strategy process, the organization's external environment and internal resources and capabilities are analyzed, strategic alternatives are developed, the best options to move forward are decided on and their execution is managed. Interfaces between these three tasks are typically not clear cut but highly interdependent. Key activities in strategy formulation or execution, for example, might lead to new focal areas that require further investigations in strategic analysis. Cascading and aligning strategic priorities of different organizational units and management teams during strategy execution might lead to new insights that require a refinement of the strategic direction, and thus, further activities related to strategy formulation. Creating a shared understanding and commitment for strategy across the entire organization—typically a key activity in strategy execution—might be facilitated by involving corresponding people in charge of execution already in strategy formulation activities, and thus, making them "owners" of strategy. In each of the three broad tasks of the strategy process, managers focus on certain questions and utilize specific management concepts, methods and tools to answer those questions. Typical questions in the three areas are listed next. Corresponding management concepts, methods and tools are addressed in the chapters to come, as indicated in brackets.

Strategic analysis: What is happening in the environment and where do we stand?

- How do we capture the initial strategic situation of the company? (2.1)
- What key opportunities and threats are evolving outside the firm? (2.2)
- What are major strengths and weaknesses within the company? (2.3)
- What are the strategic key insights we need to consider for strategy formulation? (2.4)

Strategy formulation: What and where do we want to be and how will we succeed?

- What are the prevailing strategic guideposts of the company? (3.1)
- Where and how should we compete from a corporate perspective? (3.2)
- Where and how should we compete from an international perspective? (3.3)
- Where and how should we compete on a business level? (3.4)
- What are the best strategic options to move forward? (3.5)

Strategy execution: How do we get it done?

- How do we perform on key drivers for execution excellence (4.1)?
- How can we design our organization to put our strategies into practice? (4.2)
- What strategy execution system do we need to make our strategies happen? (4.3)

Strategic Snapshot 1.10

Office of Strategic Management—Governing the Strategy Process

An office of strategic management (OSM) governs the company's strategic management activities. It is defining and applying consistent definitions, standards, methods and tools that are applied in the strategy process. It acts as a process owner, performance driver, integrator and advisor to management. "The OSM integrates and coordinates activities across functions and business units to align strategy with operations. The OSM can be viewed as the designer of an intricate watch, keeping all the various planning, execution, and control processes synchronized despite their running at various frequencies: daily and weekly for dashboards and operational control meetings; monthly for updates of strategy maps, scorecards, strategic initiatives, and strategy review meetings; and quarterly or annually for scanning the external environment and performing the analytic studies that inform strategy testing and adapting meetings. The OSM can also be viewed as an orchestra leader. It is not the creator or the producer of the music being played. Instead, it keeps all the diverse organizational players—executive team, business units, regional units, support units (finance, human resources, information technology), theme teams, departments, and ultimately the employees—aligned so that they can create beautiful music together, executing the enterprise's strategy in unison, with each component playing its distinctive part." (Kaplan/Norton 2008: 281–303)

The OSM tasks can be fulfilled by an own department that would typically report directly to the CEO to ensure unbiased corporate performance management and the ability to orchestrate different areas in the organization that are vital for making strategy happen. Some companies supplement their OSM team with dedicated people from the area of operational excellence (OE) to establish a strong link and integration of strategy and operations. In recent years, more and more companies have been following such an approach by involving their operational project management experts also in the strategic management process (Hobbs 2007: 25 f.).

In these cases, OSM does not only manage the portfolio of strategic action programs but support execution of critical strategic actions through effective analysis or project management support by, for example, people trained in Six Sigma techniques such as (master) black or green belts (Mikel/Schroeder 2005: 188–213). As an alternative to a separate department, OSM activities may also be covered by functional areas such as Strategic Planning, Corporate Development or Strategic Controlling (Horváth 2011: 221–231). Furthermore, companies who want to more tightly integrate their sustainability management with strategic management may leverage the OSM approach and create an *Office of Strategy and Sustainability* (Wunder 2014a: 77).

1.3.3.2 Structure and Corporate Culture

Organizational design

The second WOS component—the middle layer or *bearing* with its two rings (structure and corporate culture)—supports an effective strategy process. Both structure and corporate culture are elements of organizational design. Structure summarizes the organizational structure, its business processes and managerial systems as well as the governance mechanisms for the corporation including the strategy process itself. It all needs to be properly aligned for effective strategic management. The more appropriate the organization is designed and the more sophisticated it is governed, the more effectively managers can fulfill their corresponding tasks of analyzing, formulating and executing strategy as described above. Key considerations of *organizational design* are addressed in chapter 4.

Corporate governance and strategy process governance

In the WOS framework, governance is explicitly addressed as a structural element (see fig. 1-24). It includes corporate governance—the mechanisms to direct and control a company both strategically and legally— as well as the governance mechanisms of the strategy process itself, which are considered a key success factor of the effective formulation and execution of strategies. While governance will not be further elaborated on in the scope of this book, some important questions that have to be clarified particularly regarding the desired strategy process governance are (Müller-Stewens/Lechner 2011: 60–86):

Team

- Who should drive and facilitate the strategy process? (see Strategic Snapshot 1.10)
- Who should provide information and insights for strategic analysis?
- Who should be the core management teams elaborating on strategic options?
- What is the right level of involvement of the organization in the strategy process?

Timing

- What triggers the strategy process activities (e. g., calendar, major events or incidents)?
- How frequently is progress in strategy execution reported and reviewed?
- How often does the strategy need to be revised comprehensively?
- How frequently should certain activities in the strategy process be conducted?

Tools and techniques

- What is the standard of management tools that are used for tasks in the strategy process?
- What methods or procedures are followed for the different tasks in the strategy process?
- What templates should be used to document and communicate results?
- How is the organizational change and transformation process managed?
- What type of IT support is applied for the different tasks of the strategy process?

Strategic Snapshot 1.11

Tools versus Techniques (Methods)

For a strategy process to be effective, the differentiation between management tools and management techniques (methods) is important. A tool never works on its own. It needs a skilled person or team to apply the tool and there are a variety of possibilities of how to apply it. A technique, on the other hand, is based on knowledge gained from past experience and on the task. It is a thought-through and planned method of how to conduct a sequence of steps to achieve a certain result. The same tool can be leveraged in different ways by various methods or techniques for different purposes. For example, the balanced scorecard (BSC) as a tool provides a frame of how to think about and structure strategic priorities and measurements in particular (see chapter 4). The real benefit, however, comes from the process of how the tool is applied to solve a certain management problem. In which way will a balanced scorecard be developed? How will the strategic dialog be facilitated? How will strategic ideas be developed, prioritized and decided on? For effective strategy execution, the elaboration and management processes which are based on applied techniques are usually more important than the final results (Wunder 2004a: 113–157). An entire strategic management framework as well as its corresponding tools can only be effective, when they are adjusted and applied according to the specific company, its environment and the particular problem to be solved.

1.3.3.3 Strategic Leadership

Executive behavior, styles and characteristics affect strategic management.

Strategic leadership is the third WOS component—the core or *axle*—and builds the foundation of the framework. It summarizes the behavior, styles and characteristics of executives when it comes to managing an organization strategically. The way how leaders in a company behave with regard to formulating and executing strategies are considered as an important factor for determining the effectiveness of a strategic management process (Finkelstein/Hambrick/Cannella 2009: 3–42; Wunder/Stemmermann 2013). Strategic leaders—their aspirations, experiences and values—influence, for example, in which competitive arena the company plays, which business models it applies and which elements it emphasizes (e. g., products, brands, customer groups, channels, etc.), how the firm deals with competitors, how fast it wants to grow and which role sustainability plays for the strategic agenda of the company. Their current decisions are constrained by past decisions and their future decisions by their current ones. Hereby, human factors like biases, egos, experiences, aptitudes of key executives greatly affect what happens in companies (Dammann 2007; Kets de Vries 2004). Strategic leadership becomes visible in executive decisions and behavior such as

- what strategic matters leaders pay attention to and at which level of detail
- what they measure and control regularly
- how they react to critical incidents or unmet strategic expectations
- how strategically focused or ambiguous they are
- whether they are egocentric or inspire others to entrepreneurship and strategic ideation
- what they get upset about
- whether they encourage open discourse with dissenting views or reject input and only seek affirmation of their own viewpoints
- how they reward and punish
- how open and flexible they are for strategic change
- how they emphasize analytics over intuition or vice versa
- whether they "walk the talk".

Interaction of decision-making levels as strategic "breeding ground"

As strategy happens at different levels and areas in an organization (see chapter 1.1.2.6), a variety of different parties are subject to strategic leadership: first, *individual* executives at different management levels, such as the CEO and other CxOs at the top level, as well as general managers leading divisions, strategic business units, regions or countries, and functional areas. Second, executive or management leadership *teams* at different levels in the organization. And third, other *corporate governance bodies* such as the board of directors, supervisory boards or owners. Strategic leadership is also reflected in the interaction among and between those different decision making levels in the organization. This interaction can be seen as the "breeding ground" for strategies and the decision-making levels as its "layers". Companies that want to manage their activities more strategically need to consider this interaction and identify and solve potential deficits (see fig. 1-25). Here are some typical strategic leadership issues that can be found in company practice (Wunder/Stemmermann 2013: 111–118):

Fig. 1-25

Issues of strategic leadership interaction between different levels

Source: Wunder/Stemmermann 2013: 110; adapted

- **No grip on reality and high distance:** Almost classical is the allegation against decision-making levels to keep a long distance from each other. The leadership team appears as being disconnected from the strategic issues they need to decide on. Their grip on the day to day business reality of subordinate levels is questioned. Without sufficient and unbiased preparation of strategic decisions, executives as well as other bodies such as supervisory boards may not only take wrong judgments but also face liability risk if something goes wrong.
- **Missing reliability and credibility:** An indicator for reliability is the way how executives react to information that strategic aspirations could not be accomplished. If such messages that have been communicated bottom-up from lower management levels are denied or strongly disqualified at the top without sufficient reasoning, it undermines reliability and credibility of strategic leaders. This may result in a climate of mistrust and an interaction that is driven more by tactical constraints than strategic considerations.
- **Frequent breaks in strategic continuity:** Leaders that are frequently changing their strategic direction undermine the stabilization of processes and business models within their organization. If this is triggered by discontinuities in the firm's competitive environment it is not an issue but rather an advantage if the company is agile and capable of strategy adaption. However, if the continuity breaks are caused by overly optimistic leaders that are repeatedly caught up by business reality it becomes an issue. In such situations, the organization may perceive strategic changes as a result of actionism and rush in their leadership teams and lose faith in new strategic pursues.
- **Lack of clarity and enforcement:** On a highly generic strategy articulation level, it is easy to be in agreement. For strategic leaders such situations have the supposed advantage that changes in the strategic direction are less in need of expla-

nations. However, vague and ambiguous strategy statements provide room for interpretations and undermine accountability. This deficit in strategic leadership becomes a particular issue at subordinate management levels that are supposed to make strategy happen but may not fully understand the strategic priorities. The lack of shared understanding and commitment for strategy results in inefficient resource allocation. Furthermore, a lack of enforcement when things become unaligned also undermines the strategy. People may agree to a strategic direction in a group but follow their personal strategic priorities once left to their own devices. Effective strategic leadership drives alignment and accountability.

- **Exaggerated strategic aspirations:** Shareholder expectations or personal aspirations of top executives may lead to revenue growth, profitability or other targets that are not achievable if evaluated realistically. Why do strategic leaders produce such "stretched" targets? Some may want to unleash additional performance potential in the organization and make their managers think "out of the box". However, it is a thin line between inspiring or ambitious goals and aspirations that are too excessive and frustrating. Hereby, rather moderate growth targets in the first year may be followed by increasing growth rates in the following years, which is also referred to as "hockey stick effects". To achieve the ambitious goals, capital expenditures are signed off in the first years. If it becomes clear that the growth targets will not be achieved, they are not revised but moved to the next planning period because expenditures already incurred. This "sunk cost effect" can lead to recurring and "self-producing" growth patterns—also referred to as "wandering dune"—which is disconnected from economic reality and may become a burden on the organization (see also chapter 1.2.2.3).
- **Micromanagement:** Frequently, senior executives practicing strategic leadership have progressed to their positions due to excellent track records from operational areas. These leaders may feel uncomfortable with aggregated strategic information and delegation of decision power and may prefer to observe and control every management task in the organization closely. There is a risk that strategic leaders try to cope with the ambiguity of strategic decision situations by a scrupulously precise planning and management approach that provides an illusion of control. This may result in extensive strategic performance reports and leaders giving direct orders to lower management levels without following their chain of command and direct reporting lines. As explained earlier, dealing with strategic situation requires a different approach than solving operational issues (see chapter 1.2.2). This needs to be captured and addressed in corresponding leadership development programs.

Even the best strategy processes with the most sophisticated tools and techniques might fail if they do not support and reflect the strategic leadership that is practiced by key executives, management teams and other decision-making bodies in an organization. A strategic management framework like the suggested wheel of strategy can only be as effective as the strategic leaders that apply it. For example, a strategy process with a high level of involvement of managers at different levels in the organization that is very structured and leverages a wealth of analytical tools might be appro-

Strategic leadership must fit the applied strategic management framework.

priate for one company but considered as overly time consuming, bureaucratic and complex in another. There are a great number of leadership texts describing the roles and "ideal" attributes of strategic leaders. Those topics will not be further elaborated on in this book. It is very unlikely that there is only one best way for strategic leadership based on certain "ideal" characteristics. Effective strategic leadership will more likely depend on the specific strategic circumstance of a company and its internal configuration as well as the strategic skills and intuition of managers preparing or making strategic decisions.

1.3.4 Making Strategic Management Effective

After having explained essential foundations of strategic management, the final chapter deals with the question of how to make strategic management effective. This includes some general guidelines to consider for applying strategic management, corresponding benefits companies can expect from this approach and also reasons or situations when engaging in formal strategic management might not be appropriate (Wheelen/Hunger 2010: 6 f.):

Generic guidelines for practicing strategic management

Strategic management in general and corresponding frameworks such as the wheel of strategy in particular suggests what a company should do in its strategy process. It assumes that companies that work more systematically, disciplined, diligently and continuously on analyzing the changing strategic situation in which they operate and on formulating and executing corresponding strategies are more likely to outperform those that do not. To make strategic management effective the following guidelines are suggested:

- **People process:** Strategies should not be developed by "strategists" and then handed over to a leadership team or line managers for execution. The strategy process is about strategic dialog, interaction, understanding, ideation, commitment, and continuous strategic learning and adaption, not about developing a nicely bound strategy document. It has to be a people process more than a paper process.
- **Continuous strategic learning:** There might be no sense of urgency for strategic management because things are fine as they stand. Company examples at the beginning of this book have illustrated that firms might fail to adapt their strategies to the changing environment even or particularly when they have been successful so far. Past performance is no guarantee for future success and might even provide the impression to live in a fool's paradise. With this in mind, the application of the suggested strategic management framework needs to facilitate a continuous strategic learning process rather than a onetime or once-every-five-year event.
- **Power of simplicity:** As with many management concepts developed for practical application, the strategic management process and its methods and tools need to be simple and pragmatic, yet comprehensive enough. Companies and managers are acting in such a complex and dynamic environment that adding another shovel of complexity with processes, management methods and tools is typically doomed

to failure in corporate life. This becomes a particular issue in case people who are supposed to get value from tools and methods do not fully understand how they work. Manager's impression that the applied strategy framework becomes an overly formal and bureaucratic mechanism and more of a burden than a benefit is a major pitfall that needs to be avoided.

- **Analytical reflection and intuition:** No strategic management approach must be considered as a ready-to-use prescription for success. The wheel of strategy framework, for example, is supposed to frame vital questions and support managers in finding the answers by integrating rational or reflective as well as intuitive elements of strategic decision making. However, the framework—like any other strategic management model—does not provide the strategic answers itself. This is generally the mandate of management leadership teams.
- **Top management support:** A top management that is not actively supporting the strategy process for any reason is another pitfall particularly in the area of strategic management where you would assume a high level of top management commitment. Active support could, for example, be reflected in leaders who welcome impartiality and insightfulness as well as a spirit of exploring, scrutinizing, challenging, suggesting, and forward-looking from everybody involved in the strategy process. They drive people to envision what could be and to identify or create strategic opportunities. They also ensure that the strategy process is not misused for political purposes, for example, to gain control over resources.

Benefits of strategic management

If those guidelines are followed and corresponding pitfalls are avoided, companies can expect a variety of benefits from applying a strategic management approach as suggested. Managers will have a clearer sense of the company's current strategic situation and its changing environment, and thus, will make more informed decisions. The company is more proactive than reactive in identifying opportunities and threats, and for shaping its future. Also, managers will explore strategic options of how to continuously advance the company's success logic more systematically, and are thus less at risk to miss critical developments in the firm's environment. Further benefits that are primarily related to effective strategy execution are (see chapter 4.3):

- Top and line management will get a shared understanding and buy-in for the most critical strategic priorities of the company.
- Functional and regional strategic perspectives are better aligned with corporate and business strategies to optimize the overall results.
- The management leadership team gets a focused, balanced, consistent, and reliable view on the strategic performance of the organization.
- Resources and incentives are allocated against the highest value strategic projects to improve the overall business impact.
- The organizational structure, business processes and systems are aligned with strategy.
- People in the entire organization are motivated because they have line of sight from their role to the ultimate strategy of the company.

Reasons for not engaging in formal strategic management

Given the potential benefits of strategic management explained before, what are reasons why companies do not want to engage in a formal strategic management process and do not apply a corresponding framework? Companies that are in a *current crisis situation* need to focus more on firefighting activities with short-term impact. They have to ensure survival before they can direct their resources and management attention to long-term strategic decisions. For example, many companies in the 2008–2010 global economic crises were facing decreasing order income, revenues and liquidity. Here, the focus was clearly on measures to ensure short-term profitability and liquidity, such as cost saving programs as well as improved working capital management through optimizing the supply chain and improved order-to-cash and purchase-to-pay processes. A crisis, however, also offers a unique opportunity to outpace competition. After successfully navigating through the decline, some companies leveraged their strategic management process to be better prepared for the economic upturn than competitors who still had to recover from the impact of the crisis (Wunder/Müller 2009). Another reason why companies might not want to initiate a formal strategy process is *resistance to potential strategic change*, particularly in situations where they did major investments in assets and resources that might not be well-suited for newly emerging trends in the business environment (sunk cost effect). Also, *highly confident leaders* that have been successful based on their entrepreneurial intuition tend to rely more on gut feeling than on formal strategic management processes. Other reasons might be a *lack of knowledge or experience* with formal strategy work and a corresponding fear of failure or loss of control. Sometimes, a more formal strategic management process is *perceived as too expensive and effortful*, and is thus abandoned due to budget considerations. This is particularly true in small companies or when managers involved had prior bad experience with overly formal, bureaucratic and time consuming approaches.

1.4 Summary

Companies are on an ongoing quest to win in their competitive arena and perform well for a prolonged period of time. They pursue strategies to outperform competitors and ensure long-lasting survival. A variety of reasons may trigger companies to rethink their strategic orientation. They range from economic expectations and emerging new trends in the company's environment to personal aspirations of owners or senior leaders and intervention from other stakeholders such as governments, non-governmental organizations or the society in general. Rooted in military history, a business "strategy" can be viewed as a plan, ploy, position, perspective or pattern. It is a multifaceted phenomenon that can hardly be described with a single definition. According to six key characteristics, a strategy is about

- gaining, sustaining and renewing competitive advantage
- creating a fit between a company's resources and its environment
- being different and making choices, that is choosing what to do and what *not* to do
- defining a path to a destination (e.g. vision)

- creating consistency in behavior capturing both intended and emergent strategic initiatives
- aligning multiple strategic themes and perspectives at different levels in an organization.

To see whether a strategy is effective, companies can measure competitive advantage and performance by looking at their accomplished economic value, accounting performance (profitability), shareholder value, or corporate sustainability performance. As strategy is about making choices, their effectiveness is influenced by the way decisions are made, that is rationality, bounded rationality and intuition. Hereby, strategic decision situations are different from operational ones in a way that they are highly ambiguous, that is complex, dynamic and volatile as well as non-transparent. This calls for more intuitive decision making based on heuristics, which comes with an increased risk of cognitive biases that need to be dealt with in the different tasks of a strategy process.

Strategic management can be seen as another stage in the evolution from financial to long-term planning and strategic planning. Contrary to descriptive schools of strategic management that focus on how strategies form in reality, prescriptive approaches suggest how strategies should be made and essentially follow a classic-rational paradigm of strategic management. The wheel of strategy framework is such a model that captures essential aspects to be considered when managing organizations strategically. It covers the four interrelated components *strategy process* with its tasks strategic analysis, strategy formulation and strategy execution, *structure* capturing the organizational structure, business processes, managerial systems and governance, *corporate culture* and *strategic leadership*. For making strategic management effective it needs to be a people process that drives continual strategic learning in a simple way leveraging analytical reflections and intuition and receives a great deal of top management support. If corresponding pitfalls are avoided, companies may enjoy a number of benefits from strategic management, which ultimately result in a more proactive and systematic approach to shape their future and ensure long-lasting survival.

PART I

Developing Strategies: Strategic Analysis, Ideation, and Choice

2 Strategic Analysis

Learning Objectives

After studying this chapter, you should be able to:

- identify key stakeholders of a company and determine their strategic relevance
- describe different layers of strategic analysis and build a firm-specific framework
- conduct a macroenvironmental analysis and derive scenarios
- define a relevant market, apply the five forces model, and understand its dynamics
- describe key elements in a firm's competitive arena and how to analyze them
- distinguish and analyze customer needs, wants, demands, and requirements
- outline the key rationale and premises of the resource-based view of strategy
- distinguish between a firm's resources, capabilities, and core competencies
- develop strengths and weaknesses profiles
- conduct a value chain analysis
- apply the VRIO model to assess competitive implications of a firm's competencies
- consolidate key insights from strategic analysis with a SWOT framework
- describe the portfolio-based situation assessment
- identify focal points for strategy formulation with the TOWS matrix.

2.1 Understanding the Initial Strategic Situation

Understanding the company's *current* and *future external* environment and *internal* set-up

A key principle of strategy is to match a company's resources and capabilities with external factors that shape today's and tomorrow's markets and competition (see chapter 1.1.2.2). Consequently, the process of developing a strategy has to begin with an analysis of the initial strategic situation covering both the company's *external* environment as well as its *internal* set-up. Hereby, the term "environment" is used in a broad sense to describe everything and everyone outside the organization (Lynch 2012: 73). The goal is to develop a solid understanding of where the company strategically stands today, anticipate how the environment might change in the future and how the firm is positioned to take advantage of those changes. Relevant external and internal factors that affect strategy and competitive advantage need to be identified along with key stakeholder expectations. The resulting insights are the "breeding ground" of strategy formulation.

Strategic business fields and strategic business units

The "environment" of a multi-business corporation can be characterized by various external *strategic business fields* (or *strategic business areas*) which may be further decomposed into submarkets or market segments (see chapter 2.2.3.1). Factors affecting a company's strategy and competitive advantage may differ significantly in those different fields. Many companies are internally managing their activities regarding the different external fields they are operating in with specifically formed organizational units— *strategic business units (SBU)*. Generally speaking, a strategic business unit is the internal analog to an external strategic business field. Whereas a company can be seen as a combination of different strategic business units, its environment can be seen as a combination of different strategic business fields (Müller-Stewens/Lechner 2011: 143–150). Companies operating in those multifaceted environments are likely to face diverse conditions for their business activities, and thus, need to apply strategic analysis in a differentiated manner. In line with separating corporate strategy from business strategy for the purpose of strategy formulation (see chapter 3.2 and 3.4), those firms may need to distinguish between strategic analysis on a corporate and on a strategic business unit level (Hungenberg 2011). The following elements and methods for strategic analysis can be applied to both the environment of a company in total as well as to an individual strategic business area. The goal is to understand the initial strategic situation of various businesses as well as the company in total. Even if the company is operating in only one business area, strategic analysis should not be narrowed down to this area as the firm may need to change its business focus and diversify to ensure a successful future development given changes in its environment.

2.1.1 The Company and Its Stakeholders

Foundation of strategic analysis: organizational, economic and societal stakeholders

Foundational for strategic analysis is an understanding of the company as a system of *stakeholders*. Stakeholders are individuals, groups or institutions that have an interest, claim or stake in the company and can affect or are affected by its decisions and activities (Freeman 2010). *Organizational stakeholders* are internal to the firm and include employees, managers, board members, and stockholders. External stakeholders are composed of *economic stakeholders* (e. g., customers, competitors, creditors, suppliers, unions) as well as *societal stakeholders* such as communities, governments and other regulators at a local, state, federal or supranational level (e. g., the European Union), media, and nonprofit or non-governmental organizations (Chandler/Werther 2014: 53–56). Selected key stakeholder groups of a company are illustrated in figure 2-1. Stakeholders pursuing their own interests wish to influence aspects of a company's strategy. Some of them such as managers can directly control a firm's resources whereas other stakeholders such as non-governmental institutions influence a firm's activities more indirectly. If managed well, stakeholders can strongly foster the performance and continued survival of a firm. However, powerful stakeholders exerting their influence can also be a major pitfall for a company's strategy in case their expectations are neglected. When conducting strategic analysis, a company has to recognize the manifold relationships with key stakeholders and

Fig. 2-1

Selected key stakeholders of the company

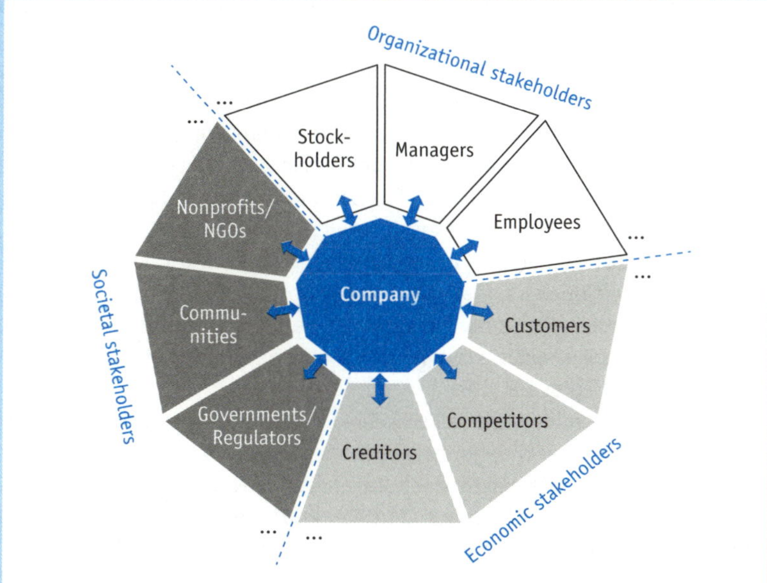

understand the potential opportunities and constraints they may impose. Ultimately, the likelihood of a strategy to be successful strongly depends on how well it captures expectation of relevant stakeholder groups. Therefore, it does not only have to be considered for situational assessment but also for evaluating strategy options as part of strategy formulation (see chapter 3.5.1.5).

Stakeholder analysis

Generally speaking, a company has to hold itself accountable to all its different stakeholder groups including future generations that are not able to articulate their stakes today (Elkington/Zeitz 2014: 24). However, it will be impossible to please all stakeholders to the same degree as their interests often conflict. Following Chandler and Wehrter (2014: 57), the "businesses most likely to succeed in today's rapidly evolving global environment will be those best able to adapt to their dynamic environment by balancing the conflicting interests of multiple stakeholders. (...) Integrating a stakeholder perspective into a strategic framework allows firms to respond to stakeholder demands in ways that maximize both economic and social value. A central component of this strategic framework is the ability to prioritize among stakeholders—both in absolute terms, as well as on an issue-by-issue basis." (See also Strategic Snapshot 3.8 in chapter 3.5.1.5). Based on this understanding, analyzing a company's interaction with various stakeholder groups is considered a crucial foundational step for assessing a company's strategic situation. Analyzing and engaging stakeholders may reveal important insights about a potential gap between key expectations of relevant stakeholder groups and the type of value delivered to those groups by the company. In some firms, it may show that significant company resources and time are spent with "convenient" but less relevant stakeholders whereas key expectations of highly relevant but "inconvenient" stakeholders are neglected. For conducting a stakeholder analysis, the following four steps are suggested (Müller-Stewens/Lechner 2011: 160–168):

1. **Identifying relevant stakeholder groups:** First, the stakeholder groups that are expected to strongly affect a company's performance are identified and listed. Apart from economic stakeholders such as customers or competitors that are typically straightforward regarding a firm's performance, this may also include organizational or societal groups capable of initiating actions that directly or indirectly influence an effective strategy execution such as unions (e. g., strikes), regulatory institutions (e. g., product approvals), industry specific groups (e. g., setting standards for certain types of products that are often adopted by regulatory bodies), NGOs or action groups raising public interest and attention as well as other departments or areas within a company that need to be aligned to make a strategy come true (see chapter 4.3.3.2).
2. **Prioritizing stakeholders:** Not all stakeholders of a company are of equal importance for its strategy and performance. A company needs to identify the *relevant* stakeholders and understand to which stakeholder relationships it should dedicate particular management attention and resources. Figure 2-2 shows an approach in which stakeholders are classified according to their impact on and controllability by the company. *Playmakers (type A)* reflect stakeholder relationships with a high level of interdependence with the company. Examples are a firm's key customers or certain business units within the organization that are critical for a strategy.

Fig. 2-2

Stakeholder relevance matrix

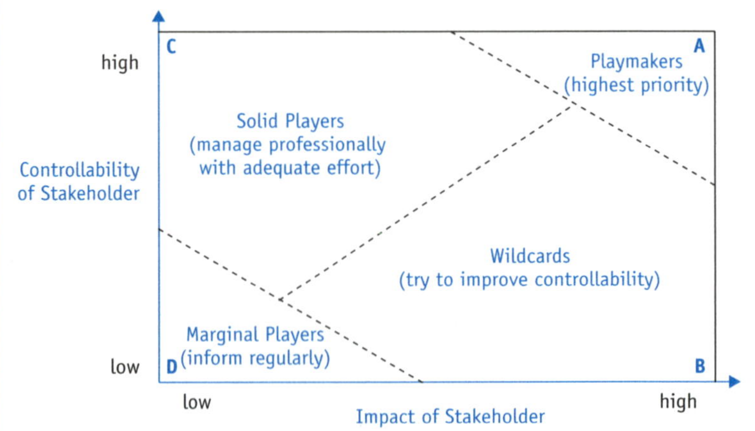

Source: Müller-Stewens/Lechner 2011: 162; adapted

A company may want to explicitly address these important stakeholder groups with specific policies, processes or strategies as well as regular dialogs. *Wildcards (type B)* such as certain regulators or specific strategic groups and competitors can have a strong impact on the company but are hard to control. They are labeled "wildcards" because a company typically tries to increase controllability and may thus gain unexpected influence by various means such as lobbying or cooperation. *Solid players (type C)* are stakeholders where the power clearly lies with the company. An example is a supplier relationship where the company is one of the supplier's key customers, and thus, may have a relatively high bargaining power. This type of stakeholder is managed professionally with an adequate level of effort depending on their corresponding impact. *Marginal players (type D)* reflect stakeholder relationships that are less important at the time of the evaluation. Due to their low impact and controllability they are typically managed with very limited attention and resources. The focus may be on providing regular information only. However, even marginal players need to be monitored carefully as they can quickly have a major impact on a company through corresponding campaigns or communication initiatives in public media or social networks.

3. **Capturing key stakeholder expectations and deriving value propositions:** In the third step, key expectations or claims of relevant—that is high-priority—stakeholder groups are systematically captured. This can be effectively facilitated with regular stakeholder engagement and stakeholder dialogs (O'Riordan/Fairbrass 2008; Unerman 2007). For stakeholder engagement, a company can leverage approaches such as surveys, focus groups, community panels, or various types of corporate advisory groups and boards. Systematic stakeholder involvement has been gaining significant importance for companies today due to increased public awareness and sensibility regarding a company's corporate social responsibility practices as well as evolving sustainability reporting guidelines (Global Reporting Initiative 2015a and b). Once the stakeholder perspective has been captured, the company's own ambitions regarding the different relevant stakeholder groups need to be clarified. By comparing stakeholder expectations with the company's ambitions, specific value propositions a firm wants to deliver for different relevant stakeholder groups are defined.
4. **Integrating stakeholder insights into strategy:** Finally, the need for integrating insights from stakeholder analysis in the company's strategic guideposts—such as its vision, mission, or values as well as its corporate and business strategies needs to be identified. At this point in a strategy process, this might be a more general identification of important strategic considerations, which need to be further elaborated on during strategy formulation (see chapter 3). Also, it might be necessary to deep dive into specific *issues* identified with stakeholder analysis or into specific *stakeholder groups* that are of particular importance, which will be explained in the course of this chapter.

Stakeholder issues, dynamics and deep dives

The relative importance of stakeholders is company-specific and needs to be determined on a firm-by-firm and issue-by-issue basis. In case particular stakeholder concerns or issues have been identified with stakeholder analysis, they need to be further evaluated to get a better understanding whether there is need for action. This *issue-related assessment* can be done based on the following three criteria (Chandler/Werther 2014: 58–61):

- **Strategic relevance:** importance of the particular issue to the company (i. e. proximity to a firm's core competency or competitive advantage), and thus, potential business impact
- **Stakeholder motivation:** importance of the particular issue to the stakeholder group, and thus, the likelihood of the stakeholder to act upon the concern
- **Operational impact:** the extent to which a particular stakeholder group and issue can impact a company's operations (e. g., damage reputation, influence profitability, etc.). Here the prioritization scheme shown in figure 2-2 may be leveraged.

Results of a stakeholder analysis need to be reviewed on a regular basis. For example, an evaluation that considers a certain stakeholder group to be less important for the company's strategy today may change over time as the strategic situation of the company changes. No company can afford to ignore the interests of relevant stakeholder groups over a prolonged period of time (Chandler/Werther 2014: 57 f.). Furthermore, stakeholder groups identified as highly relevant require analytical deep dives to get a comprehensive understanding of the current strategic situation. Key considerations of those deep dives regarding two of the most relevant stakeholder groups for a business—its customers as well as competitors—are provided in chapter 2.2.3.

2.1.2 Developing a Strategic Analysis Framework

Layers of strategic analysis

Stakeholder analysis is a foundation for determining a company's strategic situation and provides focal points for the subsequent areas of strategic analysis. Those areas of investigation need to be defined on a firm-by-firm basis and typically differ regarding their proximity to a company's business activities. Figure 2-3 provides a framework for strategic analysis with different layers of investigation. It covers both a firm's *external environment* and *internal configuration*. At the center of the framework is the *company* itself with its resources, capabilities and core competencies. This is the focus of internal strategic analysis in which the firm's strengths and weaknesses are identified. This internal perspective of the company is embedded in three different layers of its environment, which are analyzed to identify opportunities and threats. First, the *competitive arena* is the most immediate layer surrounding the firm. Key elements are the firm's markets (i. e. submarkets within an industry, market segments) and customers, its strategic groups and competitors as well as its suppliers and partners. The second layer captures the *industry structure* and its dynamics. It relates to factors of the relevant market the company operates in or considers entering (Hungenberg 2011: 99). The third layer and highest level of analysis is the *macroenvironment*. Here, broad environmental factors, trends or future events are identified that can influence almost all organizations and stakeholders and are typically not directly controllable by the company. Those general external developments can be used to construct alternative scenarios or pictures of the future (see Strategic

Fig. 2-3

Layers of strategic analysis

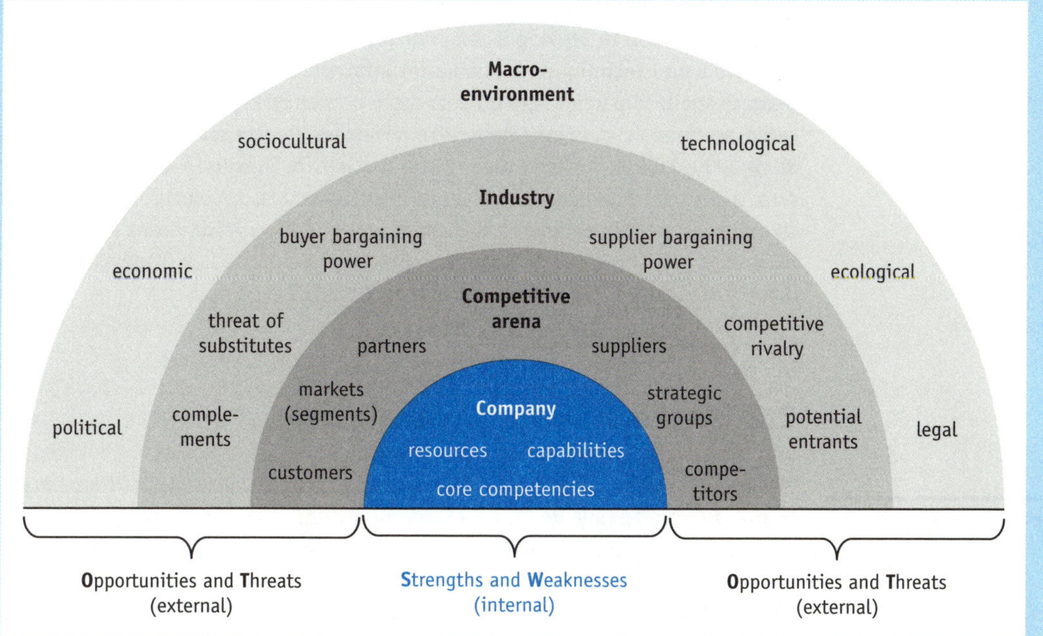

Snapshot 2.1). Various macroenvironmental factors are relevant for all elements in the firm's industry structure and competitive arena as well as its internal set up.

Customizing the general areas of analysis

The framework in figure 2-3 provides an exemplary overview of areas that a company may want to look at for the purpose of strategic analysis. The different areas in each of the layers are general and are supposed to provide a basic direction for gathering and exploring relevant data. Such a framework typically needs to be customized for company practice. In each layer, a firm may need to emphasize, refine, adapt, add or even skip areas of analysis based on the company's specific internal characteristics, the environment it operates in, and the time and resources it dedicates to the task of strategic analysis. For example, it may be important to deeply analyze a certain powerful stakeholder group which is currently not explicitly listed in the industry layer but strongly affects the industry attractiveness through its actions or regulations (e.g., unions, non-governmental organizations, and business or trade associations). In such a case, this stakeholder group becomes an element of the company-specific framework. Also, a company may analyze "technology" as a more general macroenvironmental factor that is not directly controllable whereas high-tech firms operating in technology-intense business environments are likely to deep-dive into this topic as an element of their immediate competitive arena. This is particularly the case for firms that are in a position to actively drive innovation and set new global technological standards themselves.

The goal of the following chapters is to provide a condensed overview of essential considerations for external and internal strategic analysis as well as its consolidation and integration. Some selected tools and methods will be briefly explained. It is not the intention to provide a comprehensive overview and discussion of all theories, tools and techniques available for strategic analysis. There are a number of texts specialized particularly on this topic or addressing selected tools with dedicated sections (e. g., Fink/Siebe 2011; Fleisher/Bensoussan 2015; Kerth/Asum/Stich 2009; Pillkahn 2008; Stöger 2013; Zeitschrift Führung + Organisation 2015). Also, many of the mainstream strategic management textbooks provide comprehensive discussions of strategic analysis methods and tools (e. g., Grant 2013; Hungenberg 2011; Müller-Stewens/Lechner 2011; Rothaermel 2013; Thompson/Peteraf/Gamble/Strickland 2014; Wheelen/Hunger 2010).

2.2 External Strategic Analysis

Systematically dealing with the future

Woody Allen once said: "I think a lot about the future because that's where I'm going to spend the rest of my life." At the beginning of this book it has been illustrated that even high performing companies who fail to adapt their *strategy* to the changing environment are likely to face a drop-down in performance and may ultimately vanish from the business landscape (see chapter 1.1). In general, the future is hard to anticipate or even shape. Although the long-term future cannot be predicted other than in highly generic terms, there are a variety of methods and tools to systematically gather, analyze, evaluate and leverage future intelligence (see Strategic Snapshot 2.1). Companies have to make strategic decisions today to be prepared and well positioned for future scenarios or—if possible—to actively shape the future. As past performance is no guarantee for future success, dealing with the future is a crucial task in any strategic management process. External strategic analysis is supposed to support this difficult task. Generally speaking, an effective external strategic analysis process allows the firm to systematically and continuously anticipate the future, discover opportunities and threats and ultimately make more informed strategic choices (see chapter 3.5). It helps to assess the attractiveness of the relevant market, better understand specific market segments and key customers, gain insights about the nature of competition, and identify opportunities for collaboration with certain stakeholder groups such as strategic partners or suppliers.

2.2.1 Macroenvironment

Identifying broad environmental factors that have a high impact on a firm's performance

A macroenvironmental analysis is focused on identifying broad environmental factors such as trends or future events that can influence almost all organizations and stakeholders and that are—for the most part—not directly controllable by the company. For example, there are a variety of important megatrends in areas such as demographic

Strategic Snapshot 2.1

Strategic Foresight and Pictures of the Future: Avoiding Future Myopia

Historically, methods companies applied for dealing with future developments in their environment evolved in different phases. The first phase tools can be characterized as *early warning systems* that were supposed to draw management attention based on deviations from thresholds of certain metrics. The second phase enhanced the scope of investigation from focusing primarily on threats ("warnings") to also capturing opportunities with *early detection systems* or *early intelligence systems* based on leading indicators. One of the most prominent examples is the strategic issue analysis which focused on so-called "weak signals" as early indicators for upcoming discontinuities in a firm's environment (Ansoff 1965, 1976). The third phase enhanced those approaches from early warning and detection only to creating actionable intelligence, sensitizing management and deriving strategies of how to deal with or even shape future developments. Nowadays, this type of management discipline is frequently labeled *strategic foresight* or corporate foresight.

Strategic foresight can be seen as an interdisciplinary management program, process or discipline to systematically identify new developments in a company's environment at an early stage and engage the entire organization with the most important trends and scenarios. It leverages principles and tools from trend research and future studies to systematically create and maintain a high-quality mid- to long-term forward view of the company's environment. Intelligence about the future is continuously gathered, analyzed and evaluated to support present-day strategic decision making. Hereby, strategic foresight may also include a company's initiatives to *actively create and shape* the future instead of just trying to foresee it. Companies practicing strategic foresight programs or processes are trying to avoid future myopia which could arise when dealing with future issues is seen as a secondary task conducted only if there is some time left after tackling the day-to-day issues. Companies practicing strategic foresight want to be continuously one step ahead of competition, and thus, ensure the long-term success and survival of the firm.

Strategic foresight needs to be a continuous process rather than a one-time program or project. It has to be based on unbiased information gathering and analysis (see chapter 1.2.2.3). Insights are derived after creating a holistic and comprehensive view about the future(s) instead of evaluating and judging each single piece of information. Rather than working with probabilities, strategic foresight is primarily based on working with scenarios about alternative "futures". Instead of just focusing on an extrapolation from the past, companies may want to leverage different elements of the future such as trends, uncertainties, contradictions as well as unforeseeable wildcards. By analyzing and combining different of those elements, different *pictures of the future* can be elabo-

Fig. 2-4: Smart Cities Scenario 2060: From Parking to Parkland

Continued on next page

Continued from previous page

rated to create a shared understanding about potential environments that a company might be facing in the near- or long-term future. Some of those pictures of the future are animated to illustrate the anticipated chronological future change. Figure 2-4 presents the end result of such an animation (Siemens AG/Pictures of the Future 2015). Another example of an illustrated future scenario is the "Rail Station 2020" by Deutsche Bahn (see fig. 4-75 in Strategy Practice Example 4.13; chapter 4.3.6.2).

(Source: Burmeister/Neef/Beyers 2004; Fink/Siebe 2011; Müller 2008; Müller/Müller-Stewens 2009; Müller-Stewens/Lechner 2011: 191 f.; Pillkahn 2008; Siemens AG 2015; Slaughter 2002; World Economic Forum 2015)

dynamics, globalization and future markets, scarcity of resources, climate change and ecosystem at risk, global knowledge society, etc. that are shaping our future (Roland Berger Strategy Consultants 2015). They fundamentally influence all areas of our society over a long period of time (e. g., several decades) and are thus reshaping the economic and commercial landscape companies are operating in. The goal of a macroenvironmental analysis is to identify factors and developments that have an impact on the company's performance, and thus, need to be addressed in its strategy (see Strategy Practice Example 2.1). Those general external developments—that may well have a high level of uncertainty—can be used to construct alternative scenarios or pictures of the future (see Strategic Snapshot 2.1). The key question to be answered is: "How will the world look like in, for example, 10, 20 or 30 years, and what do we need to do to be successful in this future environment?" Conducting such an analysis essentially requires two interrelated components that will be explained more in detail next (Johnson/Whittington/Scholes 2012: 21–25):

1. holistic *framework* that covers relevant areas of the macroenvironment to be analyzed
2. *scenarios* based on key drivers identified by applying the analytical framework.

PESTEL framework

A general structure for examining the various areas of a macroenvironmental analysis is provided by the PESTEL framework (also referred to as STEEP, SLEPT or other abbreviations). With this approach various political, economical, sociocultural, technological, ecological and legal factors in a firm's general environment are systematically investigated. This relates to the outer layer in the strategic analysis framework introduced in chapter 2.1.2 (see fig. 2-3). Present and potential developments in the six categories can create both opportunities and threats for a company. The areas are interrelated and may sometimes be overlapping when it comes to the allocation of certain development. In such a case the corresponding factor is placed in the areas where it fits best. Here are some examples of factors in the different categories that might have a significant impact on a company's strategy:

- **Political** factors are related to developments caused by political institutions, such as stability of governments, foreign trade regulations and free trade zones, special incentives and subsidies for companies, health care policies (e. g., regulations and reimbursement), dominance of political parties and philosophies, energy policies, and fuel prices.

Strategy Practice Example 2.1

Linking macroeconomic trends to business at BAYER AG

Figure 2-5 presents a practical example of how the BAYER Corporation translated selected global megatrends into business opportunities for its different divisions operating in the business fields of health care, crop science and material science. The presented trends apply to society in general as well as all organizations and stakeholders, though they provide different business opportunities for different companies.

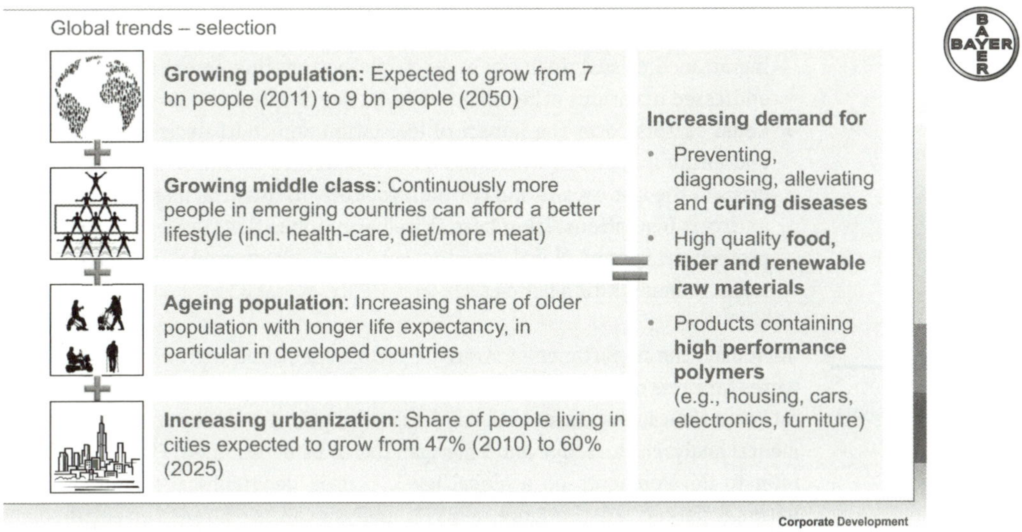

Source: Bayer 2013. Dr. Alexander Moscho, Head of Corporate Development, Bayer AG. Reprinted with permission.

Fig. 2-5: Global trends with implicit business opportunities for the BAYER AG

- **Economic** factors in the broad environment relate to economy-wide developments and have an obvious impact on business activities. They include gross domestic product trends, price stability (inflation and deflation rates), unemployment levels, work conditions, interest rates, currency exchange rate fluctuations, and disposable income including pension plans.
- **Sociocultural** factors are typically relevant for most companies as they affect the majority of current or potential customers. They include attitudes, values, norms, and lifestyle changes in society (e. g., increased ecological and social awareness) as well as demographic trends related to population size and composition (e. g., age, gender, family size, religion, ethnicity, socioeconomic classes, education level, life expectancies).
- **Technological** factors such as disruptive technologies may create new opportunities but also pose significant threats. They include utilization and advancements

of Internet technology and the World Wide Web, information technology (e. g., big data and predictive analytics), biotechnology, nanotechnology as well as robotics. New technologies impact customer behavior (e. g., wireless application technology for online services) and provide new ways for company operations (e. g., industry 4.0). For example, technology fundamentally changed the way people take pictures from using analog cameras to digital cameras and, nowadays, to smart phones, which significantly changed the business landscape in that area.
- **Ecological** factors are related to broad environmental issues such as global warming, scarcity of natural resources or sustainable development in certain regions. It also includes elements such as eco-competition and energy consumption controls. Increasing ecological as well as social problems of our society highlights the importance of sustainability and a triple-bottom-line approach that have been addressed in various other chapters (e. g., see chapters 1.1.3.4 and 3.4.4.5).
- **Legal** factors cover the impact of legislation and court decisions on companies and entire industries, which has gained increased attention after the economic crisis of the last decade and the high volatility of financial markets. They include antitrust regulations, labor laws as well as unionization, tax laws, environmental protection laws and global warming legislation, international law, liability regulations, or immigration policies.

Setting scope of time frame of a PESTEL analysis

Obviously, the importance of strategic factors identified with a PESTEL framework varies for firms operating in different industries and environments and also relate to different time spans. As an important first step prior to conducting a macroenvironmental analysis, the scope and time span has to be determined. First, the *scope* may refer to developments on a global level, certain geographical regions or specific national markets. Furthermore, a PESTEL analysis can be related to an entire company operating in various industries and strategic business fields or to only one industry and—as some industry boundaries tend to blur (see chapter 2.2.2)—also capture adjacent industries. It must be noted at this point that it might not be advisable to narrow down the scope for macroenvironmental analysis too much. This type of external analysis in particular may reveal completely new strategic business fields a company may want to look into in the future based on anticipated scenarios. If a PESTEL framework is applied only to relevant markets or industries currently addressed (see chapter 2.2.2.1), it may miss important trends and vital strategic adjustments. Second, the *time span* of the analysis needs to be determined, which, to a certain degree, depends on the industry dynamics. Future scenarios are typically built for a time span starting at five years as a lower limit to about 20–30 years as an upper limit (Müller-Stewens/Lechner 2011: 196).

Identifying key drivers for change and their time-relevance for company strategy

The number of possible factors in the six categories of the PESTEL framework tends to be very high. Furthermore, those factors usually do not affect a company or industry abruptly but in a creeping process over time. They typically become visible in the behavior of certain stakeholder groups. The sooner a company recognizes such key developments in its environment, the earlier it can evaluate their impact in terms of opportunities or threats. It is essential for the effectiveness of a PESTEL analysis to identify *key drivers* for change in the broad environment of a specific company based on all factors identified and determine the time when those key drivers need to be addressed in a company's strategy. A firm may decide to capture some key drivers in their current strategy whereas others might be monitored and addressed later on in the future. Figure 2-6 illustrates a macroenvironmental factor radar with an emphasis on sustainability trends. It shows key drivers and a corresponding timeline that highlights the potential relevance of those drivers for the company's strategy.

Building future scenarios

Drivers of change identified with a PESTEL framework can be interrelated and may have divergent or opposing outcomes. One way to deal with this situation is by building scenarios. They are supposed to provide different plausible views of how the company's business environment might develop in the future. Scenarios may be built based on trend extrapolation and probabilities in case the development of key drivers are relatively stable and certain. However, many long-term developments of environmental factors will have a high level of uncertainty and dynamics, which means that they cannot be forecasted with a probability. In such cases, the benefit of building scenarios is not to predict or forecast a single view of the future. Nor does it make sense to allocate relative probabilities or suggest that one scenario prevails over the other. The goal is to alert managers to think in and prepare for a range of alternative futures based on selected and highly uncertain key drivers that have been identified with the macroenvironmental analysis. Two different ways of building scenarios—one based on extrapolation and the other based on pure anticipation—are the scenario funnel and the 2x2 scenario matrix (see fig. 2-7):

Fig. 2-6

Macroenvironmental factor radar with strategy relevance timeline

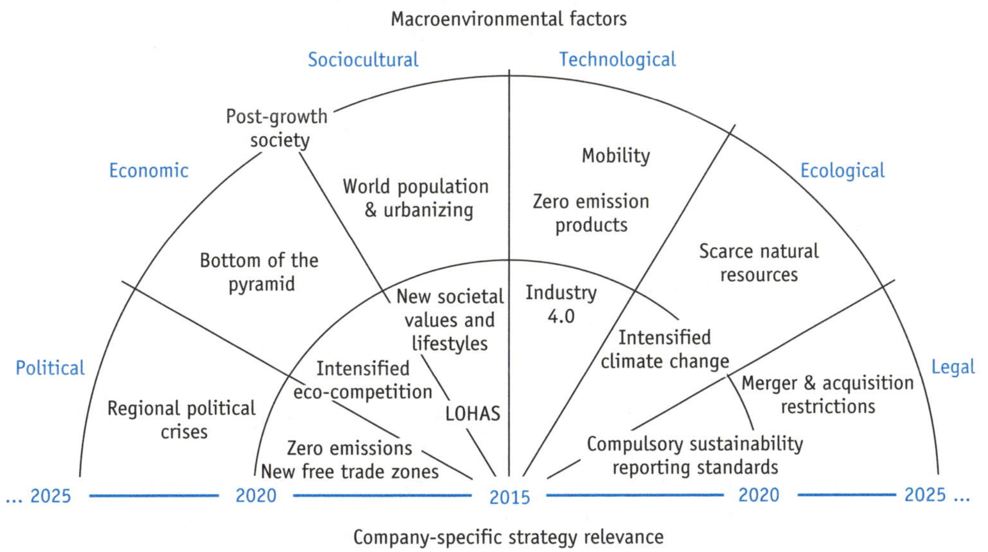

Source: Wunder 2014a: 70; adapted

- In a *scenario funnel*, two highly divergent scenarios are built based on trends of key drivers in the company's environment identified earlier. The two extreme scenarios provide the boundaries within which potential future scenarios are generated. Based on an extrapolation of various key drivers for environmental change, trend scenarios are developed that lie within the scope of the funnel. Once that funnel is established, disruptive events are continuously factored in, anticipated trend developments are adjusted and potential corrective actions are elaborated on by the company. Companies applying this approach tend to work with three scenarios. The longer the chosen time span, the more scenarios are possible. Although using more than three scenarios will provide a more comprehensive picture about possible "futures", it will also increase complexity, and thus, practicability of this method (Müller-Stewens/Lechner 2011: 195 f.).
- The *2x2 scenario matrix* is a practical way of building scenarios in a highly uncertain environment (Johnson/Whittington/Scholes 2012: 23 f.). Hereby, instead of incorporating a multitude of factors, only two key drivers are selected that are governed by a great deal of uncertainty and have a high potential impact on the company. For each of these two drivers two extreme opponent outcomes—the upsides and downsides—are determined, which leads to a 2x2 matrix with only four scenarios. None of these scenarios are considered more probable than others. They present four possible alternative futures that are supposed to sensitize decision makers and see the need for action and contingency plans. For example, a non-for-profit organization that uses this type of scenario approach is the World Economic Forum in Geneva. In its world scenarios series it developed scenarios for the global financial system based on two key drivers both with a high degree of uncertainty: the pace of geo-economic power shift (slow versus high) and the degree of international coordination of financial policy (harmonized versus discordant). The combination of these drivers led to four scenarios that were not linked to any kind of forecasted probability: re-engineered western centrism (slow shift and harmonized coordination), rebalanced multilateralism (rapid shift and harmonized coordination), fragmented protectionism (slow shift and discordant coordination), and financial regionalism (rapid shift with discordant coordination). (World Economic Forum 2009: 9; see also Johnson/Whittington/Scholes 2012: 24).

Consolidation and developing scenario stories

Basically, macroenvironmental scenarios are artificially created models of the future that are based on a variety of assumptions. Some developments—particularly in the near future—tend to be predetermined by the past and the status quo and will be considered more certain. They may be forecasted based on an extrapolation of trends including an allocation of probabilities of their occurrence. Other developments are highly uncertain, particularly when looking at time spans that lie further into the future. In such cases it is even risky to work with probabilities because they provide a translucent accuracy for something that cannot be predicted mechanistically. One way of consolidating all the different key drivers and environmental factors—including developments and trends that may be induced by the company itself in case it is in such a position—are *scenario stories* or *pictures of the future* (Johnson/Whittington/Scholes 2012: 23 and Pillkahn 2008). They make anticipated future scenarios tangible for internal and external stakeholders of the company and provide an illustrative way to support strategy formulation and strategy communication (see chapter 4.3.6.2). For example, the Siemens Corporation is continuously developing and publishing scenario stories in its magazine "Pictures of the Future. The Magazine for Research and Innovation" (Siemens AG 2015; see also Strategic Snapshot 2.1).

Impact analysis, robustness checks and contingency plans

The final step of scenario building within a macroenvironmental analysis is to evaluate the impact different alternative scenarios or key drivers will have on the company (i. e. *impact analysis*). What are the most important opportunities and threats resulting from the analysis that need to be considered in strategy formulation (see chapter 2.4.1)? How robust is a company's current strategy in the face of each plausible scenario and which strategic adjustments might be necessary? The *robustness check* shows whether a strategy is sufficiently flexible regarding other scenarios not addressed in case the lead scenario does not become true (see also chapter 3.5.1.1). A company may decide to develop or adjust a strategy based on the lead scenario only and craft *contingency plans* for each of the other scenarios identified. This way the company is prepared for a fast strategy adaption in case one of those other scenarios materializes.

Macroenvironmental analysis as ongoing management process

As mentioned earlier on, conducting a macroenvironmental analysis should not be considered a one-time event. Scanning and monitoring the environment has to be a continuous interdisciplinary process that might be embedded in a company's broader

Fig. 2-7

Methods for building scenarios: scenario funnel and 2x2 scenario matrix

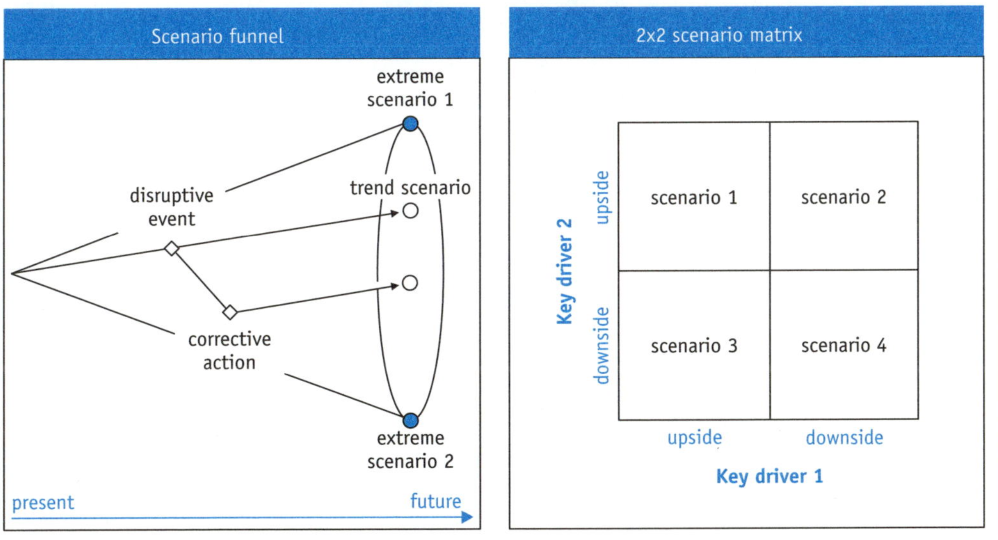

Source: Müller-Stewens/Lechner 2011: 195 and Johnson/Whittington/Scholes 2012: 24; adapted

strategic foresight management discipline (see Strategic Snapshot 2.1 and also chapter 4.3.7.2). Even though the importance of such a management process seems to be straightforward and a great variety of practical tools and methods are available, its application in company practice seems to be rather limited. The reason might lie in some significant challenges that need to be tackled when trying to implement such a process. Key considerations are (Müller-Stewens/Lechner 2011: 193):

- Avoid an "overkill" of methods and tools that divert management attention, discussions and resources from the future elements that matter most.
- Ensure a sense of urgency for dealing with the future to avoid "future myopia" (see Strategic Snapshot 2.1) caused by focusing only on current day-to-day issues.
- Emphasize the qualitative benefits of foresight for long-term survival of the firm even though they are hardly quantifiable. There are many examples of strategy failures due to missed trends (see also chapter 1.1).
- Ensure that strategic management is not only a calendar-driven "event" but also an ongoing process with elements such as environmental scanning and monitoring (see chapter 4.3.7.2).
- Sensitize the entire organization to environmental observation and the identification of weak signals instead of making it a private project of a few people or delegating it into a dedicated staff department.

2.2.2 Industry

Defining the relevant market and understanding its attractiveness

While macroenvironmental analysis is focusing on broad environmental factors that can influence almost all organizations and stakeholders and may not be strictly linked to a particular industry or market, industry analysis moves one layer closer to the firm (see fig. 2-3). The goal is to determine the attractiveness of an industry based on an understanding of competitive forces surrounding a company and the related profit potential. An important first step for conducting such an analysis is the determination of the industry boundaries or, in other words, the definition of the relevant market or "relevant industry" (Porter 2008: 14). It is a challenging task to decide how broad or narrow a market should be defined. For example, should a company producing chocolate bars focus its analysis on the industry or market for chocolate bars, chocolate snacks, sweets, snacks, or food? Should a firm planning to launch a new cola beverage look at the cola market, soft drink industry, beverage industry, or the industry for food and entertainment? The answer to those questions is of utmost importance for a company as it will determine the characteristics of competitive forces as well as deeper analysis of customers and competitors in particular (Hungenberg 2011: 99). "Defining the industry too broadly obscures differences among products, customers, or geographic regions that are important to competition, strategic positioning and profitability. Defining the industry too narrowly overlooks commonalities and linkages across related products or geographic markets that are crucial to competitive advantage. Also, strategists must be sensitive to the possibility that industry boundaries can shift." (Porter 2008: 14)

2.2.2.1 Defining the Relevant Market

Criteria for segmenting a company's environment and defining the relevant market

In the strategic analysis framework suggested in chapter 2.2, the industry analysis is separated from the market analysis conducted as an element in the competitive arena. The underlying rationale distinguishes between two areas of investigation: first, a broader *relevant market* (i. e. the industry perspective) that can be derived from a strategic business field segmentation of the firm's environment (see beginning of chapter 2.1); second, more refined *submarkets* (i. e. the competitive arena perspective) based on a market and customer segmentation conducted within the corresponding relevant market or industry. Depending on the scope and goal of analysis as well as how broad the industry boundaries are set, the relevant market and the competitive arena might be the same. There are a variety of approaches for building strategic business fields and defining a relevant market. Common criteria include (Bruhn 2014: 19 f.; Müller-Stewens/Lechner 2011: 195 f.):

- Products/services: On what products or services are companies competing?
- Customer needs/wants: For the fulfillment of which customer needs/wants are companies competing?
- Customer groups/market segments: Which customer groups are companies competing for?
- Technologies: Based on which technologies are companies competing?
- Geographies: In which regions are companies competing?

The relevant market can be defined based on an individual criterion or a combination of those. Typically, either two or three criteria are combined to define the relevant market. Hereby, listing all discrete variants (alternatives) for each criterion provides the entire market scope in which the relevant market is defined (see fig. 2-8 and Strategy Practice Examples 2.2 and 2.3). For example, to define the relevant market for an

Fig. 2-8

Illustrative principle of defining a relevant market

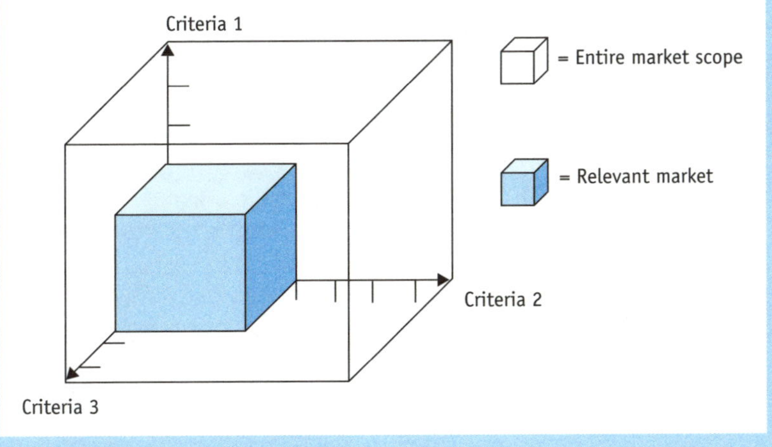

Strategy Practice Example 2.2

Defining and Expanding the Relevant Market at QIAGEN

Qiagen is a leading biotechnology company in the area of sample and assay technologies with more than $1,3 billion in sales and about 4,300 employees (end of year 2014). The company was founded in 1984 and is headquartered in the Netherlands. In its past strategy, Qiagen applied a three-dimensional cube to define and direct expansion of its relevant market. The starting point in 2004 was a strategic focus on sample preparation and the market segment of academic research. As illustrated in figure 2-9, the company expanded its market scope to four major *customer segments*:

- Molecular diagnostics (e. g., supporting health care providers in patient care)
- Applied testing (e. g., supporting government or industry customers in fields like forensics, veterinary diagnostics, or food safety)
- Pharma (e. g., supporting pharma and biotech companies in fields like drug discovery)
- Academic research (e. g., supporting researchers in exploring pathways of diseases).

Another dimension of the relevant market is the *workflow at customers* in which certain steps are supported with Qiagen products. The company expanded its scope from sample preparation only to offering also testing technologies. Furthermore, it is covering various areas of *system biology* such as DNA, RNA and Protein, which is used as a third criterion to define the relevant market (Source: Qiagen 2015a and 2015b).

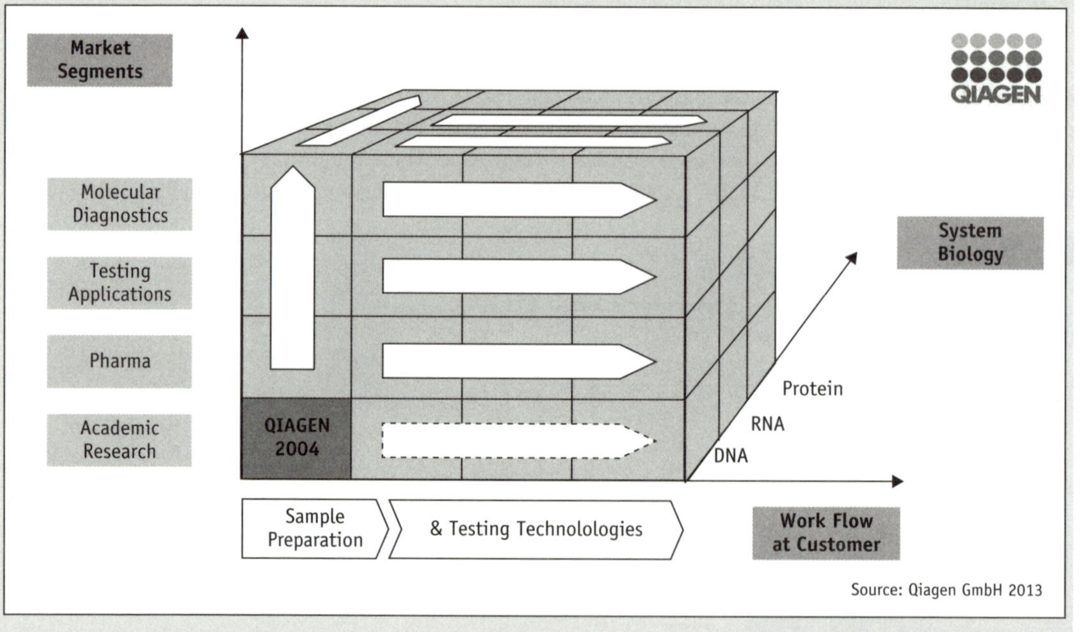

Fig. 2-9: Three dimensional market definition at QIAGEN

automobile manufacturer a criterion "product" might be used with discrete variants such as *small car, compact car, luxury car, van, SUV, convertible,* etc. Another criterion might be the "geography" with *America, Europe, EMEA, Asia Pacific/Oceania,* and *Rest of the World* (e.g., with corresponding countries) as alternatives. A third criterion could be the "customer group" with *private, commercial, governmental* customers.

Inside-out versus outside-in business field segmentation

Companies may select criteria to define their relevant market following an inside-out or an outside-in perspective. In an *inside-out approach,* a company's existing offering is used as a starting point. Hereby, existing products and services are typically combined with the different market segments or customer groups. This practical and simple approach results in a matrix of product-market combinations in which the currently addressed segments are highlighted. Those different product-market combinations are then combined to homogenous strategic business fields based on commonalities regarding customers, competitors, or suppliers. As explained earlier, those strategic business fields can be seen as relevant markets the company wants to focus on (see fig. 2-10). For example, a producer of printers may focus on the strategic business field of all-in-one printers for the market segments of large business customers and public sector customers and define this as its relevant market. This company is likely to face very different competitive forces including customer buying patterns and key competitors than a producer of small printers for the market segment of private households only.

Outside-in approach

A pitfall of the inside-out approach is the strong orientation only on business areas that are currently addressed by the firm. The concern of being too inside focused might be tackled by extending the relevant market definition in the matrix with product-market combinations that are not addressed yet by the firm. Furthermore, an inside-out market definition is focused on a company's existing solutions for certain customer groups, and thus, addresses customer needs and wants only indirectly. If the firm wants to capture those more directly, an *outside-in approach* is

Fig. 2-10

Defining a relevant market based on product-market combinations

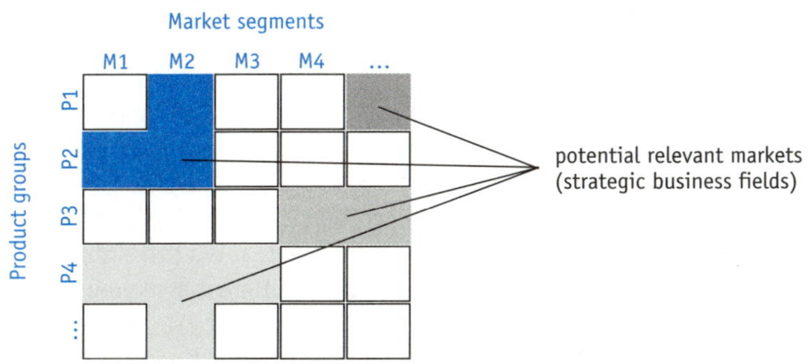

Source: Müller-Stewens/Lechner 2011: 146; adapted

Fig. 2-11

Outside-in definition of relevant markets based on two criteria (tourism sector example)

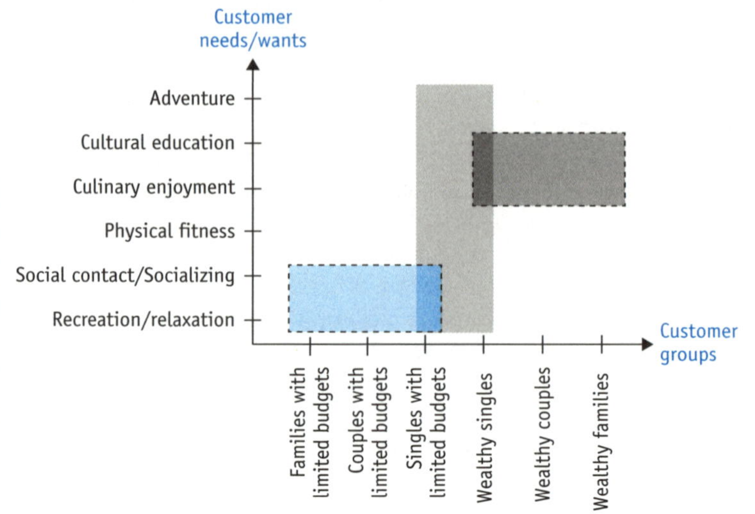

Source: Homburg/Kuester/Krohmer 2009: 5; adapted

more suitable for defining the relevant market. Here, customer needs or wants are used as one criterion for market definition, which is combined with one or two other criteria such as customer groups and alternative technologies to fulfill the need (Müller-Stewens/Lechner 2011: 145–148). Figure 2-11 demonstrates this type of market definition.

2.2.2.2 Industry Profitability Analysis and Competitive Forces

Structure-conduct-performance (SCP) model

Industries or markets can differ significantly regarding their average profitability level (Porter 2008: 5). If it is low, it will be much harder for a firm to succeed economically than if the average profitability were high. Once the relevant market or industry is defined, its structure needs to be analyzed to better understand the underlying mechanisms that influence industry competition and profitability. The underlying theoretical foundation for this rationale is the *structure-conduct-performance (SCP) model* developed in industrial-organization economics. According to this theory, the industry *structure* strongly determines firm *conduct*—that is a firm's ability to influence the prices it can charge through, for example, differentiated products and services—and thus, ultimately its *performance* (e. g., profitability).

Porter's five forces that shape competition in industry performance

Harvard professor Michael Porter applied his theories and findings and developed a highly influential framework that helps firms to understand the underlying forces that shape the attractiveness of an industry in terms of profitability: the Porter's five forces model. Its basic rationale is that the attractiveness of a relevant market—as

Strategy Practice Example 2.3

Differentiation of Strategic Business Fields in Diagnostic Radiology

Figure 2-12 illustrates how four companies defined their relevant market differently within the same entire market scope in the area of diagnostic radiology. Combining the discrete alternative variants in each of the three dimensions—application area, customer group, and technology generation (X-ray)—leads to 18 cubes (3 x 3 x 2), which represent the entire market scope. Each cube is a potential component of a strategic business field and is used to define the relevant market. For example, EMI defined its relevant market by offering second-generation X-ray technology for the application areas of the head and entire body to all three customer groups of large- and medium-sized clinics as well as X-ray physicians. With this definition, the company is facing Ohio Nuclear as a key competitor for medium-sized clinics whereas the other two companies (GE and Pfizer) are addressing slightly different segments based on the three dimensions used.

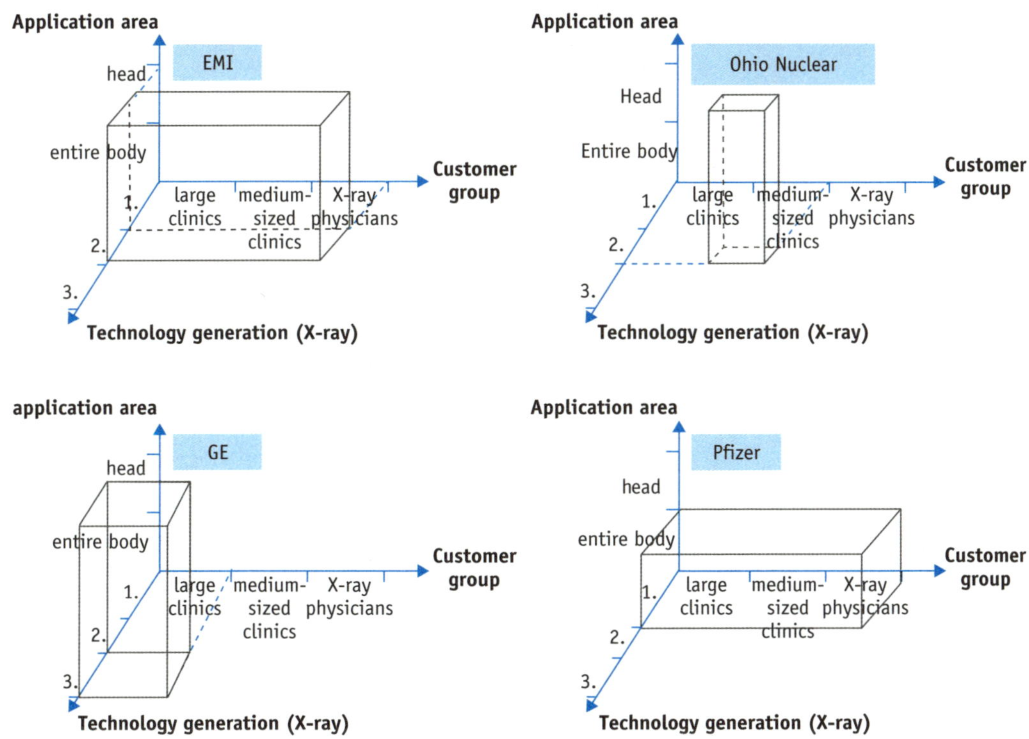

Source: Müller-Stewens/Lechner 2011: 148

Fig. 2-12: Framework for defining the relevant market (diagnostic radiology example)

measured by the long-term profitability of the average firm in terms of "return of invested capital (ROIC)"—is largely determined by five competitive forces that will be explained below (Porter 1980, 2008). Based on a comprehensive understanding of those forces, the goal is to effectively position a firm in its industry by influencing and controlling the forces to its advantage. "Managers need to position their company in an industry in a way that relaxes the constraints of strong forces and leverages weak forces." (Rotharmel 2013: 65) According to Porter, an industry is a group of companies that offer similar products or services within a certain geographic scope (Porter 2008:14). In general, the model can be applied to most organizations in any industry—including the public sector where profitability criteria may not apply (Johnson/Whittington/Scholes 2012: 25 f.)—as well as to a market or submarket within an industry (Aaker 2011: 66 f.).

Figure 2-13 illustrates Porter's model with its five competitive forces that determine profitability in an industry or relevant market. They present factors that a company needs to consider when analyzing the industry environment to identify opportunities and threats as a base for strategy formulation (Porter 1980, 2008).

Threat of entry:

If firms can easily enter an industry, competition will typically increase and it is harder for incumbents (existing competitors) to succeed. Above normal profits might be bid away quickly by the new market entrants. This phenomenon leads to the concept of *entry barriers* and its relationship with industry profitability. Entry barriers lower the threat of entry, and thus, make an industry more attractive. This is true

Fig. 2-13

The Porter five forces model of industry profitability

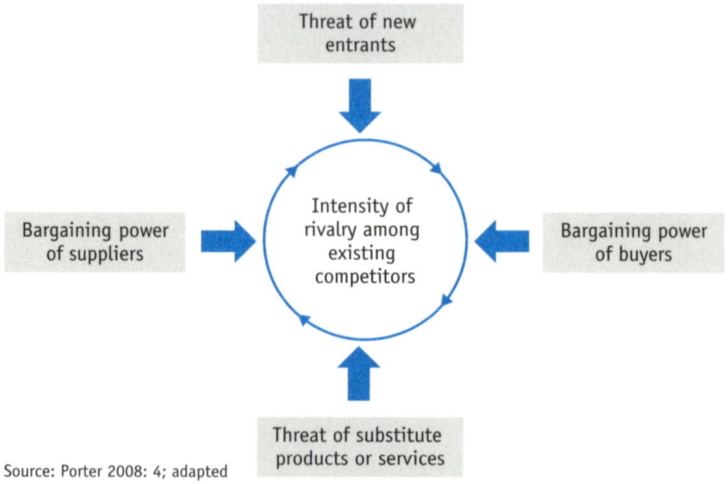

Source: Porter 2008: 4; adapted

whether the firm conducting the analysis is already in the industry or thinking about entering. What is needed, though, when a company is weighing the option of entering an industry is to have unique capabilities or core competencies (see chapter 2.3) that can make entry possible for the firm but unacceptably hard for other companies (Hax/Majluf 1996: 73). Examples of barriers to entry are

- *product differentiation* (e. g., new entrants have to overcome customer loyalties to existing brands or products in the industry)
- *high switching cost* of existing buyers (e. g., due to proprietary standards, exclusive customer intimacy or complementary products; see chapter 3.4.2.2)
- *economies of scale* (i. e. if a firm cannot produce the minimum efficient scale in a particular industry, it will be hard to succeed)
- *cost advantages* other than scale (e. g., learning advantages of existing competitors that have been operating in the industry for a while)
- *access to supply or distribution channels* due to, for example, direct ownership by existing producers (vertical integration) or their strong relationships with suppliers and distributors
- *government policies* (e. g., trade restrictions, patent protection, or regulations in favor of certain companies).

Bargaining power of buyers:

Buyers are the company's immediate customers, which are not necessarily the end consumers. For example, for a consumer goods company such as Nestlé or Procter & Gamble the buyers are large powerful trade chains and not the ultimate consumers of their products. Powerful buyers can 'squeeze' the focal firm's profitability by demanding cheaper prices or insist on higher quality and service levels, which might be costly, and thus, reduce profits of those firms. Industry conditions that facilitate buyer power include

- *small number of buyers* for the focal firm's output (e. g., there are only a few large customer accounts for the majority of the seller's sales)
- *lack of differentiation* of the focal firms product (e. g., weakly differentiated commodities)
- the purchased product *represents a high percentage of the buyer's costs* (e. g., the buyer is motivated and incentivized to shop around)
- the buyer *operates in a highly competitive market* leading to only low profits, and thus, making it very sensitive to costs and service differences
- the *buyer has the potential to vertically integrate backwards,* and thus, can decide to produce the required product itself and become its own supplier (see also chapter 3.2.2.3)
- the *purchased product is insignificant* to the final quality and price of a buyer's products or services and can thus be easily substituted.

Threat of substitutes

Substitutes fill the same need and offer a similar benefit to a focal firm's customer but in a different way. They are typically available from *outside* the defined industry or relevant market. For example, given the relevant market of cola beverages or—

somewhat broader—of soft drinks, Coke and Pepsi are competitors but not substitutes. Milk would be a substitute for both given that industry definition because it can also meet the need of current customers (e. g., quench thirst). Note that depending how broad or narrow the industry is defined, the same factor may be considered substitute or direct rival. In the previous example, a milk producer is a rival of Coca-Cola and PepsiCo. in case the relevant market is defined as "beverage industry". This again highlights the importance of thoroughly defining the relevant market (see chapter 2.2.2.1). Substitutes create a price ceiling because consumers can switch to the substitute if prices for the focal firm's products or services rise or the price/performance ratio of the substitute is significantly better than at the focal firm's offering (see also chapter 3.4.2.1). Consequently, the higher the threat of substitutes, the less attractive the industry will be. This factor requires managers to look outside the boundaries of their relevant market and also consider more distant threats and constraints.

Bargaining power of suppliers

As suppliers are those who supply the company with what it needs to produce its products and services, they can affect the firm's business by raising prices or reducing the quality of goods and services. Consequently, powerful suppliers or supplier groups can 'squeeze' the focal firm's profitability. Industry conditions that facilitate supplier power are somewhat converse to those of buyer power and include:

- *small number of suppliers* for the focal firm's purchases (e. g., supplies are dominated by only a few companies that sell to many firms)
- *highly differentiated supplier products* that are either unique or established high switching costs though lock-in mechanisms (i. e. buyers are willing to pay the premium to avoid the additional effort or cost; see chapter 3.4.2.2)
- *lack of close substitutes* for suppliers' products (i. e. buyers have no alternatives to solve their problems with other solutions)
- focal firm's purchases *represent only a small percentage of supplier's sales* (i. e. the focal firm is an unimportant customer of the supplier because it buys only a small portion of the supplier's products and services)
- *supplier has the potential to vertically integrate forward*, produce their present customer products themselves, and thus, compete directly with them (see also chapter 3.2.2.3).

Rivalry among existing competitors

Rivalry among existing competitors describes the intensity of direct competition between a company and its most immediate rivals. These are firms offering similar products and services (i. e. not substitutes) for the same market segments or customer groups in the same geography. Those competitors are interdependent in most industries, meaning that a competitive move of one firm may have a significant effect on its competitors that may consider retaliation—and vice versa. High rivalry means firms compete rigorously for market share and profit, which has a negative impact on the average profitability level in this industry. The more competitive the rivalry is, the more likely it is that above average profits are competed away and the less attractive it is for incumbents within the industry. As shown in figure 2-12, all four competitive forces discussed earlier influence the intensity of rivalry among existing competitors. In addition to those forces, there are a number of other factors that directly affect the intensity of rivalry. Industry conditions that facilitate rivalry include:

- *Few competitors that are roughly equal in size* (e. g., risk of intensely competitive behavior between rivals that are of roughly equal size. They watch each other carefully and match any competitive move to gain dominance with an equal countermove). Industries with more genteel competitor behavior tend to have one or two dominant companies and a number of smaller ones that do not challenge the big ones directly (e. g., by focusing on niches only).
- *Slow or declining growth* tends to initiate price wars and reduce average profitability because firms try to grow at the expense of others.
- *High fixed costs* (e. g., investments in capital equipment or R&D) require firms to compete rigorously as they will have to spread their costs by increasing their sales volume. This can lead to price cuts and corresponding competitor retaliations, which may ultimately result in price wars.
- *Capacity can only be added in large increments*, which may cause short-term over-capacity in the industry. This happens because the company making the addition will need to run their new plant at full capacity to keep unit cost as low as possible (i. e. economies of scale). Consequently, the selling price falls throughout the industry.
- *Low product differentiation* typically leads to intense rivalry as the only way to compete on commodity products and services—that is where characteristics are very similar regardless of who sells them (e. g., gasoline)—are prices (or location). Other than that there is little to stop buyers from switching between competitors.

Complements and relative power of other stakeholders

Additional factors to the five forces model

Some strategy scholars add additional "forces" or factors such as the availability of complements (Rothaermel 2013: 70 f.) or the relative power of other stakeholders (Wheelen/Hunger 2010: 113) to the five forces model explained before to shed additional light on the industry structure and leverage those insights for strategy formulation. A *complement* is a product or service that increases the value of the focal firm's primary product. Examples are certain software packages or processors that enhance the customer value of a computer. In this example, companies like Microsoft or Intel might be complementors to the hardware manufacturer as customers perceive more value in the focal firm's product when it is combined with the complementary product. Complements increase the demand for primary products, and thus, enhance the profit potential for firms in the industry of the primary product. Companies may want to actively seek complementors, which are typically found outside the focal firm's industry, and encourage collaboration to improve their competitive position within its relevant market. Complementary products may also increase switching cost of existing buyers, and thus, function as an entry barrier (see the force "threat of entry"). Another factor to be considered is the industry-specific *relative power of other stakeholders* such as governmental institutions, trade associations, NGOs, local communities or unions which can potentially 'squeeze' the focal firm's profitability.

Although key insights from a stakeholder analysis should have already been gained by a company (see chapter 2.1.1), it might be necessary to deep-dive into certain stakeholder groups other than buyers, competitors, complementors, and suppliers, too, in case they are of particular importance in the industry or relevant market.

Type of industry structure

The five forces explained above are influenced by the *type of industry structure,* which can be categorized along a continuum from fragmented (e. g., perfect competition) to consolidated (e. g., monopoly). On the one hand, there are highly *fragmented industry structures* that tend to generate a relatively low profitability level. They can be characterized with a large number of (small) competitors, commodity products, and low entry barriers. On the other hand, there are *consolidated industry structures* that are highly profitable. They are dominated by a few or just one company with considerable pricing power, highly differentiated products, and high entry barriers. This may also depend on the geographic region. For example, the frozen food industry in some countries is highly concentrated with a few dominating large multinationals whereas in other countries it is highly fragmented with hundreds of small family-owned firms. This illustrates again the importance of clearly defining the relevant market prior to applying the five forces model. Furthermore, it is important to note that an industry analysis must not be a onetime event as industry structures are not stable over time. For example, as consolidated industry structures tend to be more profitable than fragmented ones, firms are trying to change industry structures to their advantage through, for example, mergers and acquisitions (Rothaermel 2013: 61–64; Wheelen/Hunger 2010: 114).

Industry dynamics: industry life cycle and industry convergence

Furthermore, industry structures are typically underlying a life cycle similar to the concept of a product life cycle that has an impact on the five competitive forces. Based on the *industry life cycle* concept (Johnson/Whittington/Scholes 2012: 33 f.; Wheelen/Hunger 2010: 114), industries "start small in their development stage, then go through a period of rapid growth (...), culminating in a period of 'shake-out'. The final two stages are first a period of slow or even zero growth ('maturity'), and then the final stage of decline ('old age'). Each of these stages has implications for the five forces. For example, in the development stage where an industry is new, there are typically only few competitors and people tend to buy the product putting little importance on price because it is new and differentiated. In the following phase of high growth there tends to be low rivalry as there is plenty of market opportunity for the existing firms. Contrary to this, rivalry tends to be very high at the decline phase, particularly if there are high exit barriers, which leads to head-on competition to survive (Johnson/Whittington/Scholes 2012: 33). Finally, key drivers in the firm's macroenvironment may have a significant impact on shaping an industry over time. Traditional industry boundaries are blurring in a way that formerly unrelated industries start to address the same customer needs. An example is today's media industry where technological advancements in IT, telecommunications, and digital media unites formerly distinct media industries. Traditional stand-alone newspapers as well as old-line media companies are going out of business or are forced to adapt to these new industry structure driven by companies like Apple, Amazon, Google, or Yahoo. Also, *industry convergence* has been leading to the formation of entirely new industries such as neutraceuticals (pharmaceutical and consumer goods industries), catering (hotels, hospitality, and airline industries), biotechnology (pharmaceutical and agriculture industries), or automobile banks (automotive and financial industries). With a one-time snapshot, these types of industry dynamics, the speed of change as well as the level of innovation in an industry cannot be captured. Keeping a dynamic perspective on an industry requires a comparative industry structure analysis in which changes of the five competitive forces are analyzed and compared at different points-in-time (Rothaermel 2013: 61–74; Johnson/Whittington/Scholes 2012: 31 f.).

2.2.3 Competitive Arena

Deep-dives into specific segments and stakeholder groups of the relevant market

Industry analysis can be applied as a starting point for deep-dives into specific areas such as submarkets or particular stakeholder groups of the relevant market. It can be seen as a layer of analysis between the more general macroenvironment and the immediate competitive environment of a company. The latter will be referred to as competitive arena. Key areas of analysis within the competitive arena are submarkets and corresponding customers, strategic groups and competitors in the market, as well as suppliers and partners.

2.2.3.1 Markets and Customers

Market segmentation: dividing the relevant market into distinct and homogeneous submarkets

As a company can rarely address and satisfy everyone in the relevant market, it typically has to define promising target markets and target customers it wants to focus on strategically. After having defined the relevant market and having developed a basic understanding of its competitive forces, the buyers or customers as one of the firm's most important stakeholders have to be analyzed more in detail. Instead of just looking at their general bargaining power like in the industry structure analysis explained earlier, the company has to better understand specific customer needs and wants as well as some fundamental market characteristics. Key questions regarding the market are: Who are current and potential customers? What and when do they purchase? Who exactly makes the purchase? Why and how do they purchase? There might be many answers to those questions, which makes it hard for companies to approach an entire relevant market with only one common strategy. For example, in the frozen food industry the answers to those questions tend to vary strongly depending on the product category (e. g., frozen fruits and vegetables, frozen prepared meals, frozen seafood, ice-cream and frozen desserts, etc.) and on characteristics of the customer groups or segments (e. g., lifestyles, household size, disposable income, etc.) that are analyzed. For a more comprehensive understanding of the relevant market it needs to be further divided into segments which are then addressed with specific strategies. With *market segmentation*, a heterogeneous relevant market is divided into distinct submarkets, that is customer groups that are sufficiently homogeneous based on particular aspects—the segmentation criteria. (Homburg/Kuester/Krohmer 2009: 46 f.; Müller-Stewens/Lechner 2011: 170)

Prerequisites for an effective market segmentation

For a market segmentation to be effective and supportive for strategy formulation, the selection of the right segmentation criteria is of utmost importance. Market segmentation based on those criteria should meet the following requirements (Homburg/Kuester/Krohmer 2009: 48):

- **Clearly distinct customer needs and wants as well as behavior** such as buying patterns between the segments, which allows a clear segment differentiation
- **Customers accessibility and responsiveness** regarding a firm's market activities (e.g., customer groups have to notice and appreciate segment-specific product variants)
- **Sufficient market segment size and profit potential** that makes a segment-specific strategy formulation economically feasible
- **Measurability of criteria** and **stability of segments** ensures that the market segments can be objectively defined and are valid for a certain period of time.

Market segmentation criteria and approaches

Depending on the segmentation criteria, several approaches of market segmentation can be distinguished that vary for different industries or market types such as a consumer goods, industrial goods, service industry (Bruhn 2014: 60f.). For example, markets particularly in the consumer goods industry can be segmented based on the following five categories: *socio-demographic* (e.g., age, gender, marital status, income, education), *geographic* (e.g., place of residence, microgeographic areas within a city), *psychographic* (e.g., lifestyles, personality characteristics, attitudes, preferences), *behavioral* (e.g., price sensitivity, actual buying and consumption patterns), and *benefit* (e.g., individual benefit expectations derived, e.g., from a conjoint analysis). These approaches can either be applied alone or in combination (Homburg/Kuester/Krohmer 2009: 48f.). An illustrative example of a market segmentation is provided in table 2-1 which shows a segmentation of the Australian wine market using a wine-related lifestyle approach (i.e. psychographic and behavioral).

Dimensions of market analysis

Once a relevant market is divided into segments, there are a number of dimensions that are of interest for an analysis of these submarkets. They include (Aacker 2011: 60–72)

- emerging new customer segments such as a new psychographic customer group with a lifestyle of health and sustainability (LOHAS)
- market size such as market capacity, potential, volume, and shares (see fig. 2-14)
- market growth (e.g., growth rates and corresponding expected competitive dynamics)
- market profitability (e.g., based on competitive forces, see chapter 2.2.2.2)
- distribution systems (e.g., widths and depths of distribution channels, emerging channels)
- market and customer trends (e.g., technologies, unmet customer needs, product attributes)
- key success factors (e.g., resources, capabilities and competencies needed to succeed).

Concepts of market size: determining the magnitude of a business opportunity

All those dimensions must be considered now and also regarding their expected development in the future. Obviously, market size and growth are of particular importance for any market analysis. It needs to be determined for the relevant market as well as for the targeted submarkets (market segments) as there might be different

Tab. 2-1

Segmentation of the Australian Wine Market

Type	Characterization	Segment %
Enjoyment oriented social wine drinkers	▸ Focus on pleasure seeking and enjoyment ▸ Drink wine in pubs/wine bars or at home while just relaxing ▸ Purchase wine brands impulsively in specialist wine stores	14 %
Fashion/image oriented wine drinkers	▸ Drink wine because it fits the image they like to portray ▸ Low wine knowledge ▸ Pre-plan their wine purchases/engaging in safe brand buying	19 %
Ritual oriented conspicuous wine enthusiasts	▸ Drinking wine is an important lifestyle aspect ("connoisseurs") ▸ Drink wine often and in large quantities (home or social occasions) ▸ Like to browse around in specialist wine stores	18 %
Purposeful inconspicuous premium wine drinkers	▸ Inconspicuous wine drinkers ▸ Wine is regularly consumed in a social setting away from home ▸ Pre-plan their wine purchases	25 %
Basic wine drinkers	▸ Low level of involvement in wine and low wine knowledge ▸ Drink wine when engaged in a celebration occasion ▸ Buy mostly mainstream mass market brands and also cask wines	24 %

(Source: Bruwer/Li/Reid 2002 as quoted in Homburg/Kuester/Krohmer 2009: 243)

developments. Even when a relevant market is decreasing in size, selected submarkets might be increasing. An example is the beer industry in Germany. Whereas beer consumption has been declining for years, market segments such as beer-mix beverages and non-alcoholic beers have been increasing. When analyzing the market size, some key concepts need to be distinguished, which are illustrated in figure 2-14. All concepts presented are conceptually related to the same specific market and the same time period (Schürmann 2011: 66–68):

- The **market capacity** shows the total capacity of a market based on the number of people in a certain geographic area. Hereby, the purchasing power and the willingness to buy is not considered.
- The **market potential** is the maximum absorption level of a market. That is the maximum sales volume in monetary values or units (e.g., number of people or businesses, or consumption such as servings, kilowatt hours, hours of usage time, etc.) considering the buying power and exclusion criteria. Only potential buyers are considered (i.e. prospects). For example, for a certain car manufacturer only people with a driver's license and a corresponding buying power are considered. Children are excluded, although they obviously influence the purchasing deci-

Fig. 2-14

Hierarchy of market sizes

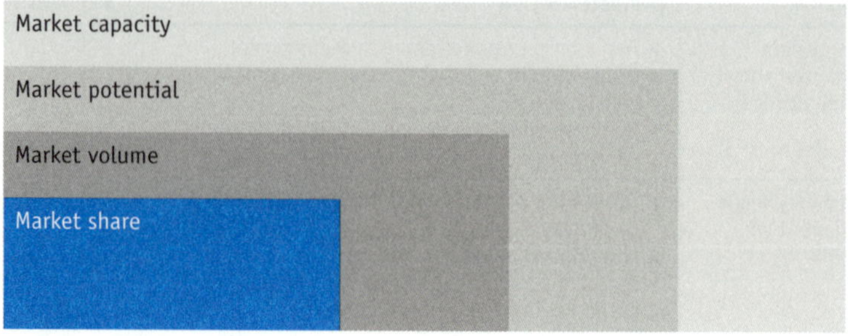

Source: Schürmann 2011

sion. In company practice the total potential of a market is often referred to as total addressable market (TAM)—also called total available market—which can be applied to the relevant market or a market segment (i. e. segment addressable market or SAM). It basically reflects the total possible demand for a firm's products or services. Companies may break out a TAM by global regions or countries, business units as well as certain customer groups such as global key accounts, which can be used for sales forecasts and target setting (see Strategy Practice Example 2.4). The TAM is an important number to get an idea of the magnitude of a business opportunity. Based on this number, the company has to estimate which portion (e.g., 1% or 5%) of the TAM it is realistically targeting (i. e. segment obtainable market or SOM) given the fact that competitors also share the total addressable market or other constraints such as distribution channels. The TAM may be calculated or estimated for today and, for example, in five years to capture also the dynamics of market growth. It can then be used as a baseline to determine how the revenue forecast looks like for the coming years and whether proposed strategies are likely to be profitable or not.

- The **market volume** reflects the actual sales number of all companies for a certain product in the market in a given time frame. With corresponding market activities, companies in this market could theoretically increase the market volume until it reaches the market potential. This would lead to a saturated market, which means that growth is only possible at the expense of other competitors. Consequently, the market saturation level shows the market volume as a percentage of the market potential. Markets with a low saturation level and the existence of market niches or market gaps are particularly compelling for companies.
- The **market share** represents the percentage revenue portion that a specific company has of the total sales in this market (i. e. market volume). The **relative market share** is typically defined as a company's sales in relation to sales of its biggest competitor.

Strategy Practice Example 2.4

Enterprise Business TAM Breakout at Hewlett-Packard (HP)

The total addressable market (TAM) of an information technology company like Hewlett-Packard (HP) can be seen as a portion of the total IT spend in a certain time period because not all IT-related needs of customers might be served by the firm's product and service offering only. Even large multinational corporations are not necessarily offering products and services to everyone. Figure 2-15 illustrates the general process how HP breaks down its total addressable market (TAM). In the first breakout, the total HP TAM is split for the printer, personal computer (PC), and enterprise business (EB). In this example, the total EB TAM is then broken out by global regions (Asia/Pacific/Japan, Europe, Middle-East and Africa, the Americas), and further by countries within Europe as well as to an account (customer) level. The illustration shows also a deep-dive into the revenue sales potential in Germany, which is broken out by business units.

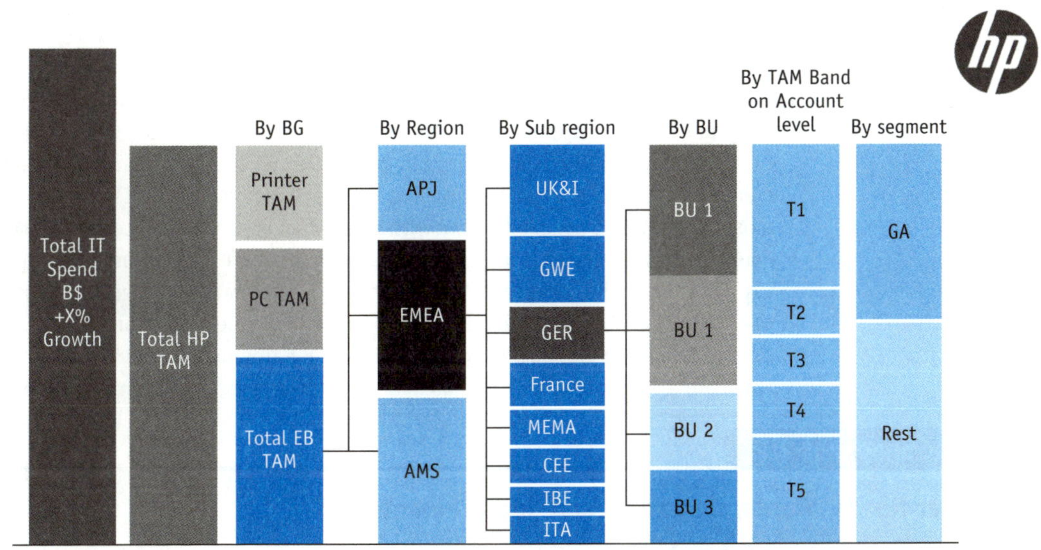

Source: Hewlett-Packard 2011. Reprinted with permission of Hewlett-Packard GmbH.

Fig. 2-15: Illustrative breakout of the total addressable market (TAM) at Hewlett-Packard

Customer needs, wants, demands, requirements

After promising market segments have been identified, the company needs to understand customer motivations, that is the underlying motives behind their purchase decision. The formulation of specific strategies for attractive market segments—as well as the definition of market segments based on psychographic, behavioral and benefit-related criteria—requires a comprehensive understanding of customer needs, wants, demands, and requirements. Some of these terms are used extensively and sometimes controversially in both colloquial and business language. Some people claim that companies and particularly marketing or advertisement practitioners "create needs" or "get people to buy things they don't want." From the perspective of these concepts, however, companies do not create needs as these are basic human requirements that preexist. Furthermore, customers typically do not buy things they don't want—at least if they are not forced to. Companies obviously try to influence wants with their product or service offerings and the corresponding value proposition. It is important to define and understand these core concepts that are explained in Strategic Snapshot 2.2.

Understanding customer requirements: the Kano model

Apart from needs and wants, understanding and effectively fulfilling customer requirements is foundational for many strategic approaches. Ultimately, the customer determines whether a company wins or loses in the marketplace. It is thus a key area of analysis within the competitive arena. However, not all customer requirements are of equal importance for a firm's strategy. A practical way to structure and analyze customer requirements is the *customer enthusiasm model* or *Kano model*, which is illustrated in figure 2-16 (Kano et al. 1984; Seidenschwarz 1997). The model is strongly related to the target costing approach where it is a key element for establishing market oriented cost management (Seidenschwarz 1997; Weber/Wunder 2002). It distinguishes between three types of customer requirements: basic, performance, and enthusiasm requirements:

- *Basic requirements* have to be fulfilled as otherwise the customer will not accept the product (e. g., cup holder or fuel gauge in a car). Companies have to completely capture these requirements with their product or services in the most cost efficient way. Missing a basic requirement can have serious consequences and will most likely cause a company to fail with its product or service in a certain market.
- *Performance requirements* are product or service attributes that a customer explicitly compares with a competitor's offering (e. g., acceleration or fuel consumption of a car). Based on these requirements a firm can differentiate from its competitors by designing products or services that offer a unique value proposition, and thus, are a source for achieving competitive advantage. The firm needs to put a

Strategic Snapshot 2.2

Customer Needs, Wants, Demands, and Requirements

"*Needs* are the basic human requirements. People need air, food, water, clothing, and shelter to survive. People also have strong needs for recreation, education, and entertainment. These needs become *wants* when they are directed to specific objects that might satisfy the need. A consumer in the United States needs food but may want a hamburger, french fries, and a soft drink. A person in Mauritius needs food but may want a mango, rice, lentils, and beans. Wants are shaped by our society.
Demands are wants for specific products backed by an ability to pay. Many people want a Mercedes; only a few are willing and able to buy one. Companies must measure not only how many people want their product, but also how many would actually be willing and able to buy it." (Source: Kotler/Keller 2009: 12)

Finally, *customer requirements* are specific product or service requirements verbalized in the "voice of the customer". Note that these are not product attributes. The "voice of the customer" usually does not cover technical solutions. It is the job of a R&D engineer or marketer to translate the voice of the customer into product functionalities and components. For example, the customer wants a Mercedes, is willing and able to buy but requires a particularly high level of safety for the family as well as warm hands when starting the car in winter. In this example, those customer requirements may translate into a certain number and configuration of air bags as well as a heated steering wheel.

Fig. 2-16

Customer enthusiasm model (Kano Model)

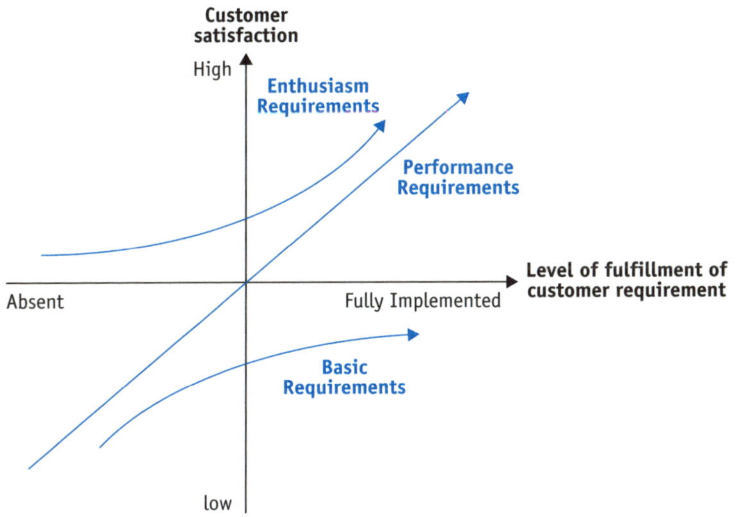

Source: following Kano et al. 1984; Seidenschwarz 1997: 61–63

special emphasis and allocate sufficient resources to elaborating on performance requirements. It needs to be noted, though, that over-fulfilling those requirements may be as problematic as under-fulfilling them as this can lead to a cost position that is not competitive—also referred to as "over-engineering".

- *Enthusiasm requirements* are often a key element for the customer's buying decision. They refer to the sometimes small but significant innovative difference a firm's product or service has compared to its competitors (e. g., a heated steering wheel in cold regions, a bird's eye parking system or a heads up display in a car). Contrary to basic and performance requirements, customers are typically not in a position to call out enthusiasm requirements in, for example, a survey as they may be beyond their imagination. They are typically initiated by R&D experts, based on comprehensive customer insights. It is an entrepreneurial decision whether a company wants to emphasize the development of enthusiasm requirements or focus its resources more on performance requirements.

Different perceptions and evolution of customer requirements

Depending on the relevant market and submarket, customer requirements may differ significantly. For example, for one of the customer groups an air condition in a car might be a performance requirement whereas a similar customer group in a different country may consider this element as basic requirement. Particularly when companies are addressing international customer groups with the same or similar products, it is important to be sensitive to these potential differences. Furthermore, customer requirements typically change over time. Product attributes that reflected an enthusiasm requirement at one point in the past may evolve to performance or even basic requirements later on. This can be seen with many innovations such as touch screens on smart phones or rear view parking cameras in cars. For many customers they have become performance or even basic requirements after a while.

Process of analyzing customer requirements

For strategic analysis the Kano model can be applied in four basic steps. First, customer requirements have to be identified. It goes without saying that the main source of information for identifying those requirements (except enthusiasm requirements) should be the targeted customer group. This process can be facilitated by market research initiatives such as conjoint analysis, customer surveys or focus groups. Second, the identified customer requirements need to be classified according to the three categories explained earlier. Third, performance requirements in particular need to be prioritized in terms of their importance to the customer. This can be facilitated by, for example, a pairwise comparison. Hereby, all customer requirements are compared with each other in pairs to find out which one is preferred. This is particularly important in case a long list of requirements has been generated that needs to be condensed to focus resources (e. g., R&D, marketing) on what matters most. Fourth, it needs to be evaluated to which degree relevant customer requirements are currently fulfilled by the firm's products or services. In case there are significant gaps or the offering has not been developed yet, insights about key customer requirements need to be leveraged for strategy formulation.

The explanation of customer requirements concludes the analysis of "markets and customers" in the competitive arena. Much more could be explained and discussed regarding those fields and types of analysis. This is one of the core subject matters in the marketing discipline. For further elaborations it is suggested to refer to mainstream texts of marketing or to specialized market research literature.

2.2.3.2 Competitors and Strategic Groups

Which competitors pose the biggest threats or opportunities?

Contrary to customers which reflect the demand side, competitors are related to the offering of products and services in a certain market segment. They are another key stakeholder in the competitive arena that needs to be analyzed more thoroughly based on the insights that have already been gained by applying the five forces model to the relevant market. Typically, competitors are continuously trying to outplay and outsmart each other with the objective to win in the marketplace, that is to gain a larger market share or a higher profitability than their rivals in the targeted market. The goal of competitor analysis is to gain insights about how the different competitor groups may impact a firm's performance, and thus, need to be considered when formulating successful strategies. The analysis should focus on the "identification of threats, opportunities, or strategic uncertainties created by emerging or potential competitor moves, weaknesses, or strengths" (Aaker 2011: 40). Conducting a competitor analysis can be facilitated by addressing three key areas:

1. Identifying competitors: Against who do or will we compete?
2. Evaluating competitors: What threat or opportunities are imposed by our competitors?
3. Monitoring competitors: What new threats or opportunities arise from competitors?

There are a number of factors to analyze and approaches a firm can apply in answering these three questions. Essential considerations will be explained in the following sections (Aaker 2011: 40–58).

Identifying competitors

Important insights based on an extended competitor perspective

Competitors can be identified in various ways. Typically, the most direct competitors come to mind first when using the term "competition". A firm may define its competitors as other companies in the same industry offering similar products and services to the same customers at similar prices. In reality, however, companies are facing a much broader range of competitors. Depending on how narrow or broad their relevant market and its specific market segments are defined, a firm may face less, more, or different direct competitors. Also, there are peripheral competitors adjacent or outside the particular market that can have a significant, yet more subtle and implied impact on the company. Finally, identifying competitors from the perspectives of products and services offered (supply-side) or based on customer needs (demand-side) may reveal different kinds of insights. Competitor analysis will typically benefit from an extension of perspective beyond the obvious competitors. Two essential classifications are direct and indirect competitors as well as product-based versus customer need-based competitors, which will be explained in the following (Aaker 2011: 21–46; Kotler/Armstrong 2012: 528–535).

Direct and indirect competitors

Companies can classify their competitors into direct or indirect. First, *direct competitors* are other firms within the relevant market or submarket that offer the same products and services targeting the same customer base. Note that firms that appear to be competitors when looking at the broad industry (e. g., BMW and Hyundai in the automotive industry) may not be direct competitors in the market segments they focus on because they are positioned differently (e. g., Mercedes or Audi are more direct competitors for BMW when looking at the premium car market). In certain product-market combinations or market segments (see chapter 2.2.3.1) a company may face very different direct competitors based on, for example, offered product or service categories, presence and market positions in different geographic regions, or targeted customer groups. Second, *indirect competitors* offer customers product and service alternatives that fulfill the same needs and provide similar benefits but in a different way (e. g., substitutes). Sometimes, they are also referred to as "replacement competitors". For example, depending on the market definition, providers of tea and milk are indirect competitors to Coca-Cola and Pepsi-Cola. Although they are not in the carbonated soft drink industry, they also provide liquids that quench thirst (see also chapter 2.2.2.2). Again, this situation may look very different when looking at market segments within an industry. Also, companies pursuing a different business model and offering very different products and services to the same market segment may be indirect competitors to a firm when their offering causes the firm's products to not be bought. For example, a food retailer pursuing revenue and profit growth based on selling fresh and frozen vegetables may be facing indirect competition from gardening centers promoting home growing of vegetables to the same customer group with the goal of selling gardening equipment, plants and landscaping services. Indirect competitors tend to come from *outside* the defined industry or relevant market. Strategically, this indirect competition can be highly important. For example, there are a variety of cases where disruptive technologies from outside an industry created new indirect competitors that eroded entire markets (e.g., based on internet or digital technology). Companies must avoid "competitor myopia", that is focusing only on their direct competitors and a product perspective (Kotler/Armstrong 2012: 529).

Product-based and customer need-based approach

Distinguishing direct and indirect competitors closely relates to two different approaches of identifying competitors. On the one hand, the firm can follow a *product-based* competitor identification that is closely connected to the industry the company is operating in. For example, it might look for competitors in the automotive industry, food industry, oil industry, or banking industry. To be successful, a firm has to understand the general competitive forces that shape the structure of an industry or relevant market (see chapter 2.2.2.2). This perspective strongly relates to the definition of direct competitors explained earlier. On the other hand, companies can also identify their competitors from a *customer need-based* point of view. Here, companies are identified as competitors of a firm when they are trying to satisfy the same customer needs and wants or build relationships with the same customer groups, which potentially lead to a decrease in demand for the firm's own products or services. This understanding closely relates to the identification of peripheral or indirect competitors. A firm that applies a customer-need based approach has to essentially analyze what choices customers are making, for example, by asking them what product or service providers they would have selected had the firm not offered the required item. Another approach for customer need-based competitor identification is associating products (or brands) that customers prefer in specific usage situations and applications (Aaker 2011: 41 f.). Hereby, a firm analyzes different use contexts and the products or brands that customer from a focus group consider appropriate for those use contexts. Thus, for example, if a certain nut mixture is perceived and used as a snack by the customer, the company is likely to face different competitors than if it was perceived as a party enhancer.

Potential competitors

The previous elaborations have been focusing primarily on current direct and indirect competitors. It is important, though, to include also potential new market entrants in a competitor analysis. These are companies that want or are forced to expand their market coverage to the market segments a company is operating in. Potential competitors may plan to enter new geographic areas that a firm is already serving (see chapter 3.3.1.1), expand their existing product breadth or depth to include products that a firm is already offering, or diversify into new businesses for various reasons where the firm is already in. Furthermore, customers are potential competitors if they are able to integrate backward and become their own supplier. The same rationale applies to suppliers that may consider a vertical forward integration for various reasons (see chapter 3.2.2.3). Also, mergers and acquisitions in an industry can create major competitors, particularly when the newly formed or integrated company is able to capture synergy potentials, overcome weaknesses, and thus, transform from a weak to a strong rival (Aaker 2011: 45 f.).

Strategic groups approach

Depending on the industry, identifying competitors may result in a large number of firms that would need to be analyzed individually. Often, it is simply not feasible to consider and analyze all those different competitors thoroughly. The concept of strategic groups can make a competitor analysis more manageable and will also provide additional insights that can be used for strategy formulation. The strategic group concept is a different approach toward competitor analysis in which competitors within the same industry or relevant market are classified and grouped according to similarities in their strategy. The main goal of strategic group analysis is to identify differences in firm performance (e. g., profitability) within the same industry based on the strategic group a firm belongs to. This implies that firm performance is not only determined by a firm's industry as discussed in chapter 2.2.2.2 but also by the strategic group membership of a firm (Porter 1980).

Strategic groups are typically mapped on two-dimensional charts. The dimensions can be based on many different characteristics. A common category for determining one dimension is the scope of an organization's business activities in terms of product range, geographic coverage, scope of market segments served, or breadth of sales channels. The second dimension is often related to the resource commitment such as branding intensity, marketing effort (e. g., advertising), extent of vertical integration, or product and service quality. For choosing appropriate dimensions, top performers in an industry are analyzed to identify common characteristics that are not shared by low performers. If those characteristics are not highly correlated they can be used to define the two dimensions (Johnson/Whittington/Scholes 2012: 36 f.).

Figure 2-17 illustrates the strategic group concept based on an example from the food industry. The four different strategic groups vary in profitability. For example, group 2 consists of large diversified branded consumer goods companies with top market positions in its various categories and above average profitability. Group 1 are

Fig. 2-17

Illustrative strategic groups in the food industry

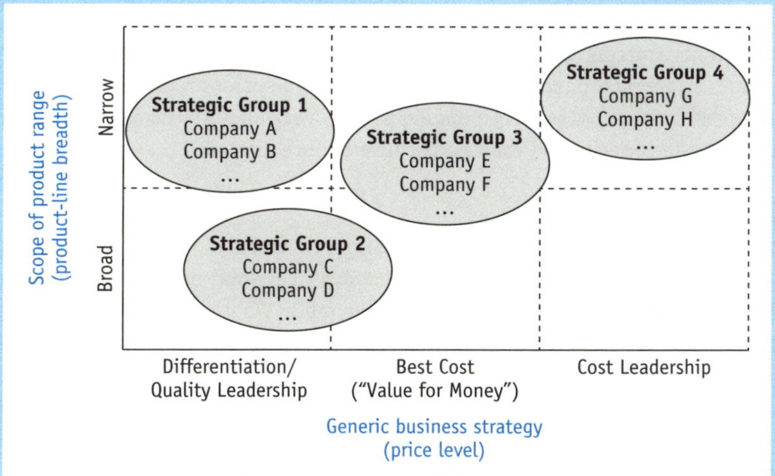

highly focused branded premium producers that offer only a narrow product range with above average to average profitability. Group 4 consists of private label producers and group 3 of firms pursuing a best-cost strategy with at relatively narrow product range, both of which are characterized by different profitability levels.

Key insights that can be derived from the strategic group model are (Aaker 2011: 44; Homburg/Kuester/Krohmer 2009: 55; Rothaermel 2013: 75 f.):

- First, each strategic group has mobility barriers that make it hard for companies to alternate between strategic groups. These are entry barriers (see chapter 2.2.2.2) as well as exit barriers such as extensive investments in assets and plants as well as specialized labor. For example, it would be very difficult for a generic pharmaceutical company with little R&D investment (dimension 1) and a low-cost/low-price strategy (dimension 2) to switch into the strategic group of branded pharmaceutical companies with high R&D investments pursuing a differentiation strategy in which they charge price premiums based on unique active ingredients under patent protection.
- Second, a firm may want to distinguish its competitive environment when conducting a competitor analysis: On the one hand, there is competition within the strategic group it belongs to (intragroup competition) where rivalry tends to be strong. On the other hand, there is competition with other strategic groups (intergroup competition). Hereby it needs be considered that developments in the macroenvironment (see chapter 2.2.1) as well as competitive forces in the industry structure (see chapter 2.2.2.2) can affect strategic groups differently.
- Third, some strategic groups are more profitable than others. Knowing the profitability levels of strategic groups can be considered for designing strategies to overcome the mobility barriers and switch from one strategic group to another in order to improve profitability. This shifts the focus of corporate strategy from the question which industries are most attractive to what strategic groups are most desirable. For example, a restaurant that belongs to the strategic group of companies with limited product-line breadth (dimension 1) and low prices (dimension 2) may consider to broaden its product range and lifting its price level to move into a different strategic group that enjoys a higher profitability. A firm competing across strategic groups usually has a disadvantage unless it positions its various product lines or brands correspondingly (e. g., a branded pharmaceutical company may have a generics division under a different brand name).

Evaluating competitors

Understanding competitor's strengths, weaknesses, or potential moves

Having identified the main actual and potential competitors, the company needs to analyze and evaluate them regarding questions like: What drives the competitor? What is it doing and can it do? Where is it vulnerable? What will provoke the greatest retaliation by the competitor? (Porter 1980) The goal is to understand their current strengths and weaknesses as well as potential competitor moves—possibly leading to new strengths or weaknesses—that impose a threat, opportunity, or a strategic uncertainty that needs to be closely monitored. Also, if a company considers certain strategy options it may want to anticipate corresponding competitor responses to prepare for or consider in its strategy option evaluation and choice process (see chapter 3.5). Figure 2-18 provides an overview of eight elements that need to be considered for competitor evaluation, which will be briefly explained next (Aaker 2011: 46–55).

- **Size, growth, and profitability:** High relative market share and the achievement of rapid growth indicate a strong competitor. Furthermore, profitable businesses typically find it easy to gain access to capital, both externally or internally, which they can invest to become even stronger. Note that a business belonging to a strong parent company may get financial support even if it is less profitable.
- **Image and strategic positioning:** To understand the strengths or weakness of a company, it is helpful to determine their image and brand perception in the market. Perceptions such as high quality, great value-for-money, high degree of innovation, or sensitivity to societal sustainability issues can be related to product or service attributes but may also reflect an intangible asset of the company in total that spans across its different businesses or offerings.
- **Objectives and commitment:** Analyzing objectives and commitments regarding sales, market share, profitability, or other performance dimensions are indicators for what drives competitor behavior in the future. It shows how satisfied a company is with its current performance level, how much pressure it is facing, and where it will focus resources on in the future. The same applies to a competitor's parent company. Here, the question is what role in the overall parent strategy a competing business is attached to given the parent's objectives and approaches for realizing parenting advantage.
- **Current and past strategies:** The competitor's current strategy including the articulated strategic guideposts (i. e. vision, mission, and values) has to be analyzed and compared to a firm's own strategic approach. This may include its corporate and international strategies as well as generic competitive strategies and business models. Also, understanding a competitor's past strategy patterns may help to anticipate future growth directions as well as areas where the competitor is likely to neglect or approach differently due to past strategies that failed (e. g., failed entry into a certain country market).

Fig. 2-18

Elements that influence competitor actions

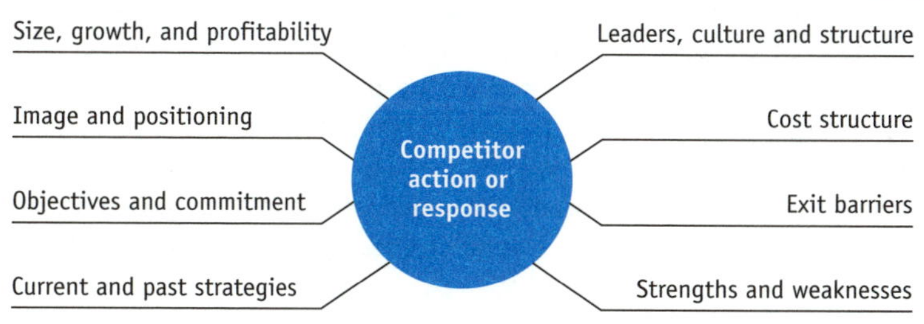

Source: Aaker 2011: 46; adapted

- **Leaders, culture, and structure:** The background and experiences in a competitor's present top leadership team may provide valuable insights about its future actions. Are the top managers drawn largely from competing firms in the same industry or from other industries? Do they have a technology, marketing, manufacturing, or supply chain background? Are they well-known for pursuing certain strategic approaches such as cost cutting and restructuring, aggressive growth with mergers and acquisitions, expanding the business scope through diversification or others in their former companies? Also, a competitor's organizational design—its culture and structure (i.e. organizational structure, business processes, and managerial systems)—can be used as indictors for its actions as it typically has a pervasive impact on the range of strategies a firm is likely to consider (see chapter 4.2). For example, a highly cost-conscious power culture with a highly structured organization and tight controls may find it difficult to foster innovation whereas a flat organization emphasizing entrepreneurship and risk-taking might struggle to pursue aggressive cost-cutting programs in the future.
- **Cost structure:** Analyzing a competitor's cost structure may reveal insights about its ability to react to economic downturns and competitive price pressures, as well as its own ability to impose those pressures on others. Considerations may include a competitor's investment in inventory, plants, and equipment, its number of plants, the sales level and the corresponding product capacities it needs to utilize for achieving economies of scale and competitive prices, the number of employees in relation to the revenues, its relative cost of supplies (e.g., raw material and purchased components), which can be derived from an analysis of a competitor's supplier structure, its automation level and the related number of direct labor, and its outsourcing strategies as well as shared services centers.
- **Exit barriers:** A competitor's barriers to exit its existing businesses provide an indicator for its commitment to these areas and the likelihood of certain action. Exit barriers may include specific assets which are costly to transform to another application (see also cost structure), its relationships with other business units in its parent company with regard to cross-selling, shared facilities, synergies in distribution channels and sales force, social barriers such as loyalty to workers, commitment to production sites in certain countries, or restriction imposed by unions, and government regulations (e.g., antitrust laws).
- **Strengths and weaknesses:** Understanding the strengths and weaknesses of key competitors does not only provide insights about their ability to pursue certain strategies but also about a company's own relative strengths and weaknesses. For example, a company may attempt to exploit a key competitor's vulnerability in one area with its own strengths. Evaluating a competitor's strengths and weaknesses requires an assessment of relevant competitor resources, capabilities, and core competencies in relation to other companies (see chapter 2.3). Hereby, insights gained from looking at the other seven elements explained before can be leveraged.

Competitor notebooks, profiles, and simulation

Companies may want to develop competitor notebooks and profiles for each competitor or strategic group they consider relevant. A competitor notebook contains the different elements that determine competitor action or response explained above and provides a framework to areas a firm may want to monitor continuously for relevant data. Thus, a competitor notebook not only provides the documentation of point-in-time analysis results but may also be continuously updated to capture changes in the local competitive environment. A relevant and up-to-date competitor notebook may become the first database consulted in a company for key competitor-related strategic considerations. A competitor profile summarizes the key insights from a competitor analysis on one page. Furthermore, to better understand competitors and their potential response patterns to own strategic moves from a dynamic and interactive perspective, a company can use simulation models such as "strategic business wargaming" (see Strategic Snapshot 2.3).

Monitoring competitors

Apart from identification and evaluation, competitor monitoring is the third essential area for analyzing competitors. This is often referred to as the field of "competitive intelligence", which will be explained next (Wheelen/Hunger 2010: 120–122).

Competitive intelligence: formal programs to continuously gather competitor information

In addition to just gathering and analyzing data once in a while, companies may want to monitor their key competitors continuously to detect actions that may affect a firm's own performance and strategy as early as possible. Corresponding formal programs of gathering information are called *competitive intelligence*—sometimes also referred to as business intelligence. Without continuously updated intelligence about its key competitors, a company runs the risk of missing crucial developments such as new technologies and incremental or radical innovation in the marketplace. There are a number of companies that have established corresponding teams or departments—competitive intelligence units—for conducting this task. Those teams are leveraging manifold publicly available information sources to learn about competitor moves as a basis for quick reactions by their own company. Corresponding information may be acquired from professional external service providers such as information brokers, market research institutes, and investigatory services, trade fairs and exhibits, benchmarking programs, or product reverse engineering (Wheelen/Hunger 2010: 120–122).

Evaluation and quality assurance of competitive intelligence

Strongly related to the information sources is the question of how to ensure that the information gathered is actually valuable for strategic management. Nowadays, the internet provides a wealth of information regarding competitors, which has

Strategic Snapshot 2.3

Strategic Business Wargaming

"Strategic business wargaming is a method of analyzing corporate strategies in highly competitive industries. It covers the development and use of market models based on game theory instruments, computer simulations and scenario analyses. The interaction between the teams (human-based simulation) and computer-based simulation with the model enables experiences to be collected for the development of central system parameters right up to key performance indicators for all competitors. The strategic business wargaming method is particularly well suited to highly competitive industries and a few dominant competitors." (Source: Center for Strategic Business Wargaming 2015) never been available and accessible so easily before. However, along with investigative opportunities comes the challenge of judging the available information in terms of its likelihood of being correct. A company needs to ensure that its competitive intelligence is valid, reliable, and accurate. First, *validity* refers to the relevance and appropriateness of information regarding a particular strategic question. Note that to focus strategic intelligence efforts, they are sometimes directed toward a specific problem or topic. For example, an automotive manufacturer may want to closely monitor how its competitors in a certain region are dealing with the question of sustainability or, in particular, how they are addressing the increasing environmental alertness of customers and governmental institutions. Second, *reliability* refers to how trustworthy, and thus, reliable the data source is. For this purpose, relevant data sources such as certain web pages may be ranked on a corresponding scale (e. g., extremely reliable, reliable, unknowns, somewhat unreliable, very questionable). However, it is not just the source that is relevant but also the piece of information obtained. Third, *accuracy* relates to the likelihood that the acquired information is correct. Just like the source, it may be ranked on a corresponding scale (e. g., correct, probably correct, unknown, doubtful, and extremely doubtful). One way to evaluate information is to compare it with separately obtained information from other sources.

At the end of the day it will be people that gather, evaluate, prioritize, and forward information to executives that need to decide on taking action. To make competitive intelligence programs effective, it is important that competitive intelligence teams are supported with clear processes, guidelines and checklists and are staffed with competent people who are able to efficiently gather and evaluate information. Furthermore, it must be noted that some practices for gathering competitive intelligence are illegal and others may be legal, but unethical. These tactics can include blackmail, trespassing, industrial espionage, headhunting competitor employees, or using certain types of investigatory services.

2.2.3.3 Other External Stakeholders

Various external stakeholder groups other than customers and competitors can also have a significant impact on a firm's performance. For example, *suppliers* are extremely important for companies such as Apple or Nike that have outsourced a great portion of their value chain such as manufacturing and are focused primarily on areas such as R&D or marketing. Furthermore, the social and environmental conduct and corresponding failures at a company's suppliers may have a direct impact on the firm's image or the quality of its final products. Other important stakeholder groups are the providers of capital. Companies seek funds to invest in and grow their businesses. They can get required capital from *banks* (borrowings) as well as from *private and institutional investors* (equity) such as private equity companies. Both, the capability of a firm to raise funds and its corresponding cost of capital, have a strong impact on a firm's performance. They depend on the risk, reward expectations, and a number of other factors. Based on a comprehensive stakeholder analysis (see chapter 2.1.1), a company has to get clear understanding what stakeholder groups it needs to put a special emphasis on in their strategic analysis. Those highly relevant stakeholder groups require deep dives to gain insights just like customers and competitors do (see also chapter 2.1.2).

2.3 Internal Strategic Analysis

Understanding a firm's internal factors and how they affect performance

For formulating strategy it is not only important to understand developments in a company's external environment but also key internal factors that affect performance. Ultimately, the firm's ability to capture opportunities and deal with threats in its quest for competitive advantage strongly depends on the kind of resources, capabilities, and core competencies it can leverage. This shifts the perspective from a firm's external environment and its impact on performance to its internal environment and its strengths and weaknesses. A firm must not only understand its current strengths and weaknesses but also create a "fit" between its resources, capabilities, and core competencies and its external environment from a dynamic point of view. This may require an adaption of those three elements to be better prepared for the future (see also chapter 1.1.2.2).

Resource-based view and competitive advantage

Emphasizing the role of internal strategic factors as an essential driver of competitive advantage is the rationale underlying the *resource-based view of strategy* (Barney 1991; Barney/Hesterly 2010; Peteraf 1993). This theory is based on two assumptions. First, resources and capabilities differ from one company to another even if those firms are competing in the same industry (i. e. resource heterogeneity). Second, resources and capabilities are hard to transfer from one company to another, and thus, difficult to replicate (i. e. resource immobility). Figure 2-19 shows the relation-

Fig. 2-19

Principle of the resource-based view of strategy

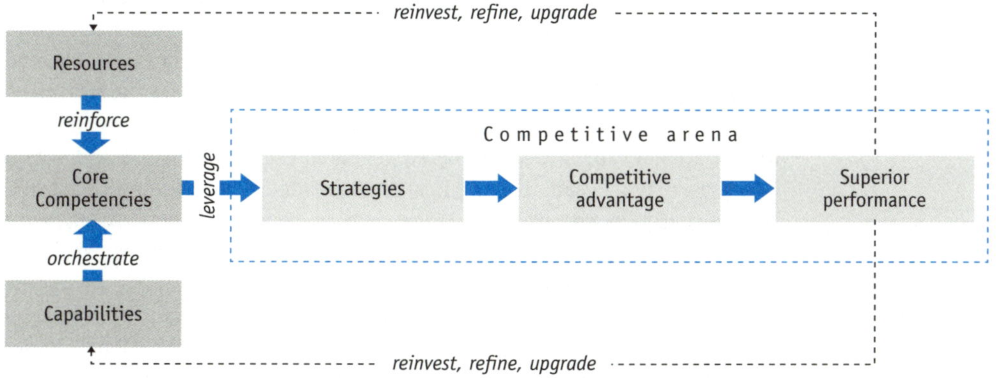

Source: following Hill/Jones 2010: 76 and Rothaermel 2013: 87; adapted

ship of resources, capabilities, core competencies, and strategy, which is the foundation of the resource-based approach. The right hand side in the figure illustrates that strategy is about gaining, sustaining, and renewing competitive advantage, which, in turn, leads to superior performance. This has already been explained earlier (see chapter 1.1.2.1). But what will be leveraged for crafting strategy? According to the market-based view, it is the competitive forces in an industry or relevant market that a firm is trying to understand and shape to position itself successfully in this environment (see chapters 1.1.2.2 and 2.2.2.2). Contrary to this understanding, the key "ingredients" for strategy in the resource-based view are a firm's resources and capabilities that are evaluated as to its strengths (or weaknesses) and impact. If these strengths fulfill certain requirements they can be expressed in terms of core competencies. The next sections will turn to those three elements more in detail.

2.3.1 Analyzing Resources, Capabilities, and Competencies

Tangible and intangible resources—"what we have"

Resources are assets the company can draw upon when crafting and executing strategy. They fall into two distinct categories. First, *tangible resources* are assets that have physical attributes and are typically visible. They can be related to production factors such as land, labor, and capital. Tangible resources in a company include human resources (e. g., number of employees and talent pool), plant facilities and equipment, buildings, supplies (including access to raw materials), finances as well as IT-related assets such as software, hardware, and networks. Second, *intangible resources* are assets that have no physical attributes and are thus not visible. A firm's intangible assets may include intellectual property such as patents and copyrights, reputation, brand equity, knowledge and managerial capacity, structure (organizational structure, processes, and managerial systems), organizational culture, trust, and capacity to innovate. Intangible resources in particular strongly affect a company's performance and are considered a key driver of market value of publicly traded companies.

Capabilities—"what we do"

Capabilities are intangible organizational and managerial skills built over time that enables a company to orchestrate and exploit its resources and to deploy them strategically (Rothaermel 2013: 86–90). They consist of processes and organizational routines. Capabilities reflect how well a company manages the interaction and coordination of resources to solve problems and ultimately turn inputs into outputs. Organizational capabilities are rooted deeply in the firm's history and accumulate over time. Often, capabilities in a firm can be related to specific groups in the organization that have specific knowledge such as certain functions (e. g., marketing capabilities, supply chain capabilities, research and development capabilities, or human resource management capabilities). For example, a company's marketing and sales capabilities can be based on the interaction among its market research specialists, branding experts, distribution channels, and sales force (Wheelen/Hunger 2010: 138). Organizational capabilities that are constantly being adapted to reflect changes in a company's external environment are referred to as *dynamic capabilities* (Schreyögg/Kliesch-Eberl 2007; Teece/Pisano/Shuen 1997).

Competencies—"how we integrate what we do"

Resources and capabilities are strongly related to specific areas within a company. For example, Walmart has strong logistics management capabilities, Procter & Gamble is renowned for its branding capabilities, the fashion company Zara has the capability to identify trends through point-of-purchase data collection methods, and chainsaw manufacturer Stihl is capable of designing and producing reliable cylinders that are a key component in their core products. However, what makes a company successful is typically not based on individual capabilities but on a cross-functional integration and coordination of capabilities—which will be referred to as organizational *competencies* (Wheelen/Hunger 2010: 138). Often it is the effective bundling of capabilities (and resources) across different organizational areas and divisional boundaries that makes a difference in the market. This perspective adds another important perspective for analyzing internal factors and provides a foundation for understanding *core competencies* (see chapter 2.3.2).

In the next section, three essential approaches for analyzing internal factors of a company (i. e. its resources, capabilities, and competencies) are presented that affect performance, and thus, need to be considered for strategy formulation. Each of these methods will reveal certain strengths and weaknesses of the company that will be prioritized and consolidated later on (see chapter 2.4). When applying all three approaches, some insights may be redundant whereas others may only be gained by applying the corresponding method and its underlying analytical logic. Although companies tend to apply only one approach for the purpose of internal strategic analysis, it should consider the potential value of combining different methods.

2.3.1.1 Strengths and Weaknesses Profile

Process for identifying strengths, neutrals, and weaknesses

One way of analyzing a firm's resources and capabilities in terms of strengths and weaknesses is the development of a strengths and weaknesses profile. This can be facilitated by the following four steps (Kerth/Asum/Stich 2009: 115–120):

1. **Identification and selection of relevant evaluation criteria:** First, strategic factors that are likely to have an impact on the firm's performance (i. e. success factors) are identified and selected. As those factors tend to be different from one industry to another, they need to be determined individually by the specific company conducting the internal analysis. This can be facilitated with a brainstorming exercise in which an interdisplinary team of managers propose areas that should be analyzed. Another approach is leveraging a generic checklist from which each factor is evaluated regarding its relevance for a company's success and selected for evaluation if highly relevant. An example of such a checklist is provided in table 2-2.
2. **Definition of evaluation process:** Second, the basic procedure for conducting the analysis needs to be determined. For example, should the selected evaluation criteria (step 1) be weighted? How should the weighting look like? Which evaluation scale should be applied? Should there be a dimension of "neutrals" included in the scale to position factors that are considered neither a strength nor weakness?
3. **Evaluation of success factors (S-N-W):** Third, each success factor is evaluated on the scale in relation to key competitors to determine whether it is strength, neu-

Tab. 2-2

Potential areas (success factors) for analyzing strengths and weaknesses in a firm

Area	Potential evaluation areas (selection)/success factors
Marketing and Sales	branding, pricing, advertising and promotion, distribution channels, innovation, market share, sales development, sales network/force, customer structure and retention, market research
Product (Services)	core product features, additional product features, basic services, value adding services, brand, packaging, product portfolio (breaths and depths), quality, image, product life cycle
Procurement	supplier structure and dependencies, purchasing costs, quality control, stock-holding and control, optimal order volume, ordering cycles, logistics, capital commitment, stock turnover, stock system
Production	plant and equipment, technology and automation, sites, productivity, cost structure, capacity utilization, flexibility, sales & operations planning, quality control, vertical integration, lead time
Research and Development	pipeline, speed to market, innovation processes, know-how, innovativeness, research intensity, cooperation, licenses, development potential, fulfillment of customer requirements, design to cost
Human Resources	qualification, talent pool, training and development, age structure, incentive structure, payment, motivation, willingness to learn, employee fluctuation, working atmosphere, employer image
Organization	organizational structure, operational structure, adaptability, strategy compliance, information management, IT usage, project management, knowledge management
Leadership	quality of decisions, leadership style, capacity for teamwork, goals, dynamics, inspiration, motivation, market orientation, transparency, delegation, age structure of managers, control
Finance	equity base, leverage (debt), financial power, liquidity, profit development, appropriation of earnings, profitability, cash flow, budget, capital budgeting, managerial control/business partner
Accounting	cost allocation, cost control, cost planning, profit and sales planning, efficiency of repetitive work, early-warning system

(Source: Kerth/Asum/Stich 2009: 117; adapted)

tral, or weakness. Note that identifying strengths and weaknesses is always comparative. For the purpose of documentation and traceability it is usually helpful to add explanatory comments to certain factors where understanding the evaluation requires additional information. The final assessment can be visualized with a profile line and comparative strengths-neutral-weaknesses profiles of key competitors (see fig. 2-20).

4. **Prioritization and consolidation of strengths and weaknesses:** The final step is the discussion of positive and negative outlier results. Based on key insights gained in these discussions, conspicuous strengths and weaknesses should be prioritized with regard to their relevance for strategy formulation. They are candidates for being included in the final SWOT consolidation (see chapter 2.4.1).

Benefits and pitfalls

The advantages of developing strengths and weaknesses profiles are the systematic and intuitively understandable procedure as well as the comprehensive, yet customized development of evaluation criteria. Instead of applying standardized sets of criteria, key success factors that are specific to the corresponding industry or relevant market are identified first. Although providing more specific and in-depths insights, this however may also create some sort of myopia as it narrows the perspective down

Fig. 2-20

Illustrative example of a strengths and weaknesses profile

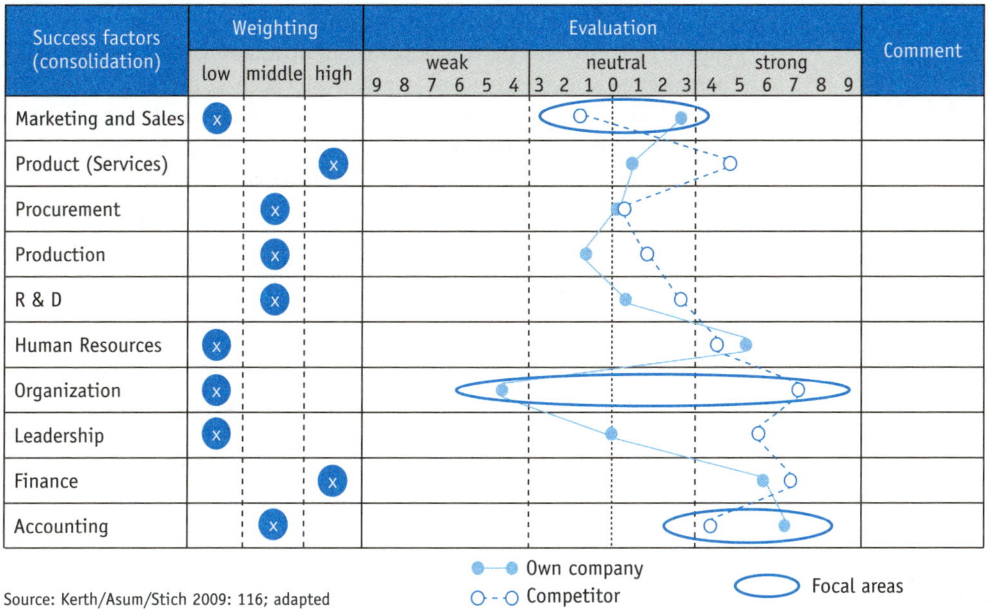

Source: Kerth/Asum/Stich 2009: 116; adapted

to the currently relevant market. There is a risk that unique strengths or core competencies that might be leveraged for expanding into new businesses outside the relevant market are neglected (see chapter 2.3.2). Another challenge of this approach is that it requires in-depth knowledge of the company and its key competitors to develop valid analysis results. Thus, it is important that the right people are involved and that additional analysis is conducted where required.

2.3.1.2 Value Chain Analysis

Upstream and downstream activities

Another framework for analyzing a company's strengths and weaknesses and identify sources for competitive advantage is the value chain model (Porter 1985). Generally speaking, a value chain is a linked set of value creating activities that, in case of a manufacturing business, begins with basic raw materials coming from suppliers and ends with distributors getting the final goods into the hands of the final customer. In company practice, the supply side of a value chain is often referred to as *upstream activities* whereas the distribution and sales side as *downstream activities*. For example, a high-level value chain of an oil company covers exploration and production (upstream), refineries, and marketing and sales through, for example, gas stations (downstream). The equivalent in the automotive industry is several tiers of suppliers (upstream), original equipment manufacturer (OEM), and retailers or dealers (downstream). The part of the value chain that is most important to the company is also referred to as center of gravity. Often, this is the point where the company started, where its core competencies lie, and where strategic moves forward or backward the value chain are initiated (Wheelen/Hunger 2010: 143–145).

Value chain model: primary activities, support activities, and profit margin

Figure 2-21 presents an illustration of a generic value chain framework. It is based on the renowned value chain model developed by Michael Porter (Porter 1985). The model provides a holistic framework covering all categories of internal activities a firm engages in when it creates and markets a product or service (i. e. transforming inputs into outputs). Elements of a value chain are value activities—which are combined to value chain categories—and profit margin. For example, primary value activities such as marketing management, advertising, promotion, sales force administration, technical support, and distribution control may be included in the value chain category "marketing, sales, and distribution". The model distinguishes between primary and secondary activities. The *primary activities* are directly concerned with the creating and delivery of a product or service and add value directly when transforming inputs into outputs. *Support activities* add value indirectly as they are necessary for ensuring effective and efficient primary activities. All value chain activities are supposed to directly or indirectly add value, and thus, are driving *profit margin*. They are a source for competitive advantage in terms of their impact for differentiation and cost (Porter 1985).

Fig. 2-21

Value chain model of a company

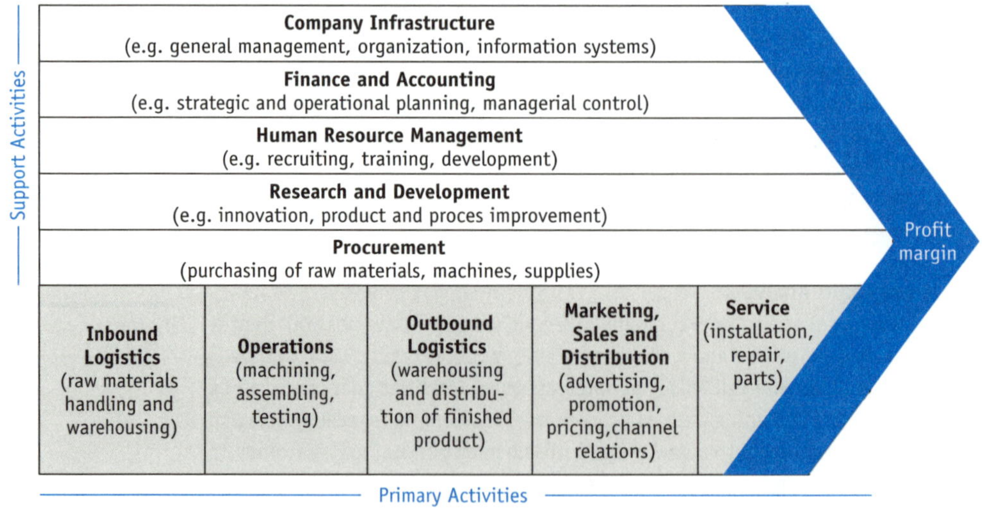

Source: Porter 1985: 167; adapted

Applying the value chain model to identify strengths and weaknesses

The value chain model can be used to identify strengths and weaknesses in each value activity or across various value activities. This involves the following steps (Kerth/Asum/Stich 2009: 76–79; Wheelen/Hunger 2010: 146 f.):

1. **Description (illustration) of the value chain:** First, the value chain of a company, strategic business unit, or product line needs to be described. This requires that the various primary and support activities involved in direct and indirect value adding processes are determined. Note that value chain categories and activities of companies in different industries or different strategic business units might look very differently.
2. **Examination of individual value chain activities:** Based on the specifically developed value chain model, a company examines which of its primary and support activities are considered strengths or weaknesses. This is typically done in comparison with key competitors, which requires that corresponding competitive intelligence has been gathered in advance.
3. **Examination of linkages within one value chain:** In addition to evaluating each value chain activity individually, the integration or combination of value chain activities is examined. This may reveal additional strengths or weaknesses and also core competencies (see chapter 2.3.2) as a base for sustainable competitive advantage. For example, a company striving for innovation and a high speed to market needs to integrate its R&D or new product development activities strongly with marketing management and operations.
4. **Examination of synergies among value chains of different strategic business units or product lines:** A cross-divisional analysis of value chains provides additional information of a company's strengths and weaknesses in terms of economies of scale and economies of scope (see chapter 3.2.1). In large corporations this may reveal additional areas for gaining competitive advantage from a more holistic perspective of the entire company and its different value chains.

Benefits and pitfalls

The value chain analysis has similarities to a process analysis and corresponding process categories and process models (see chapter 4.2.2.2). The advantage of the value chain model for internal strategic analysis as compared to the checklist-based strengths and weaknesses profile explained earlier is that it encourages managers to think of an organization in terms of a set of value adding activities or processes. This can help to better understand the pillars of a firm's competitive position and advantage. Furthermore, it allows the systematic and holistic identification of new sources of competitive advantage based on differentiation, low cost, or both (see chapter 3.4.2.1). One of its biggest challenges is the significant effort that is required to conduct such an analysis. First, the individual value chains of the company, its strategic business units or product lines need to be defined. Second, cost need to be allocated to activities, which is typically not supported by mainstream cost accounting systems but requires additional analysis (e.g., process costing or activity based costing). Third, one of the biggest challenges of conducting a value chain analysis is the lack of detailed value chain information of competitors, which is required to identify strengths and weaknesses and elaborate on sources of competitive advantage. Some corporations use the value chain framework only internally and compare their different subsidiaries or strategic business units with their internal peer group to identify strengths and weaknesses.

2.3.1.3 7-S-Model

Interrelated hard and soft factors that drive a company's performance

Another method for internal strategic analysis is the 7-S-model (Pascale/Athos 1981; Peters/Waterman 1982). It differs from the other two models explained before in so far as it explicitly addresses "soft factors" within an organization and emphasizes the interrelation of those different elements. The 7-S-model suggests that seven internal factors of an organization are core pillars for success. Furthermore, it states that these factors are interrelated and need to be aligned effectively in order to drive company performance. The model is shown in figure 2-22. Strategy, structure and systems are considered *hard factors* that drive performance whereas shared values, skills, staff and style are considered *soft factors* that are supposed to support the hard factors. The connecting lines between the elements are intended to illustrate interdependencies and highlight the need for alignment. The required "fit" of various hard and soft factors underlying this model strongly relates to the principle of matching strategy, structure, and culture, which is explained in chapter 4.2 (see also fig. 4-5).

Applying the 7-S-model for strategic analysis

For internal strategic analysis, the model is applied in two steps. First, each of the seven factors is assessed individually, for example, based on a comprehensive set of predefined questions that are discussed and answered in a management workshop. Some exemplary questions are illustrated in figure 2-22. Through discussing and

Fig. 2-22

Applying the 7-S-model for identifying strengths and weaknesses

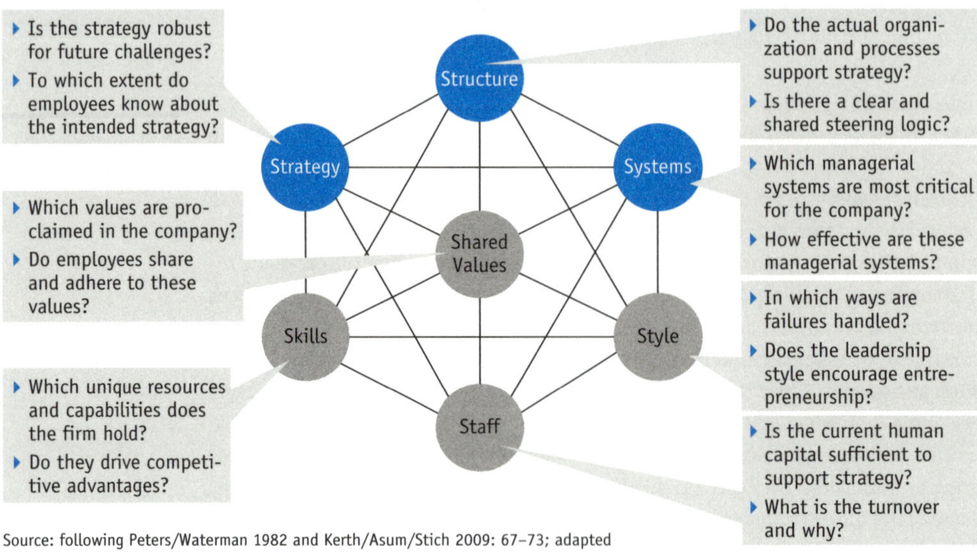

Source: following Peters/Waterman 1982 and Kerth/Asum/Stich 2009: 67–73; adapted

answering those questions, strengths and weaknesses are identified. Second, relationships between those factors are examined. This can be practically achieved by developing a matrix in which each single factor is systematically compared with all others. The goal is to understand whether there is a "fit" between those factors and to identify any serious conflicts. This provides insights about strengths and areas for improvement related to the linkages between factors with the ultimate goal to ensure a seamless integration of all seven elements. The insights derived from applying the 7-S-model are summarized in a SWOT framework (see chapter 2.4.1) and leveraged for strategy formulation and execution (Kerth/Asum/Stich 2009: 67–72; Müller-Stewens/Lechner 2011: 203–205).

Benefits and pitfalls

Compared to the other methods explained earlier, applying the 7-S-model provides two key advantages. First, it explicitly addresses not only hard but also soft factors which may be otherwise neglected though they are considered to have a strong impact on performance. Second, the model also emphasizes the interrelation of those factors, which can be analyzed based on the matrix explained. The major pitfall of this model lies in its very generic and high-level categories. They may not be sufficient for getting a comprehensive picture of all relevant internal strategic factors. Based on this evaluation, the 7-S-model may provide a valuable starting point or addition to an internal strategic analysis that captures a more comprehensive set of the different organizational areas that need to be considered.

2.3.1.4 Additional methods for internal strategic analysis

Examining specific areas with an organization

In addition to the three approaches introduced in this chapter, there are a variety of other methods for analyzing internal strategic factors of a company. Some of them relate to specific areas or certain functions within a company. Examples are *product life cycle analysis*, analysis of the firm's *customer structure and profitability* (e. g., based on the process costing or activity-based costing approach), *portfolio analysis* (see chapter 2.4.2), *experience curve analysis*, analysis of the firm's *cost structure*, *benchmarking*, *organizational culture assessments* (see chapter 4.2.1.1) as well as specific *customer, supplier, and employee surveys* (Kerth/Asum/Stich 2009). Those methods may support some of the more holistic approaches explained earlier.

Business models and business excellence models

Another approach that has gained tremendous interest in recent years is the *business model* framework that can also be used for identifying strengths and weaknesses in the context of an internal strategic analysis. It is discussed in chapter 3.4.4 as a primary approach for strategic ideation. Furthermore, *business excellence models* may be applied for identifying a company's strengths and areas for improvement. They typically come with a comprehensive self-assessment scheme that provides a set of questions covering all areas within an organization (see Strategic Snapshot 2.4). Additional methods for better understanding a firm's capabilities and how they relate to competitive advantage are corporate *skill mapping*, *opportunity matrix*, and the *analysis of skill clusters* (Müller-Stewens/Lechner 2011: 209). Finally, an approach for identifying core competencies based on an assessment of the competitive implications of a firm's resources and capabilities is the *VRIO model*, which will be explained next.

Strategic Snapshot 2.4

Applying Business Excellence Models for Internal Strategic Analysis

A practical way for facilitating the identification and analysis of strengths and weaknesses are business excellence models. Those models come with comprehensive frameworks, questionnaires for self-assessment, audits, and best practices that cover critical aspects of achieving excellence throughout an organization. Benefits or applying such models for internal strategic analysis are their

- holistic criteria that cover all relevant performance dimensions
- practical self-assessment worksheets that can be used for an individual analysis and as a reference for continuous improvement
- best practices such as reports about award winning companies and how they achieved performance excellence
- community access for best practice sharing and networking with peers.

Two renowned examples of business excellence models are the Baldrige Performance Excellence Framework (www.baldrigepe.org) in the United States and the EFQM Excellence Model (www.efqm.org) in Europe. The Baldrige framework is the foundation for the corresponding Malcolm Baldrige National Quality Award (MBNQA), which is awarded to U.S. companies that have demonstrated quality and performance excellence. Figure 2-23 presents the excellence model of the EFQM (European Foundation for Quality Management), which is applied as a blueprint for the EFQM Excellence Award (formerly known as the European Quality Award), Europe's most prestigious award for demonstrating total quality management and business excellence. The model is based on nine criteria. Five of these are "enablers" that cover what an organization does and how it does it. Four criteria reflect "results" and capture what an organization achieves. Based on these criteria, managers can assess their organization with methods such as a self-assessment to identify strengths and areas for improvement (source: Baldrige Performance Excellence Program 2015, EFQM 2015, Zink 2004).

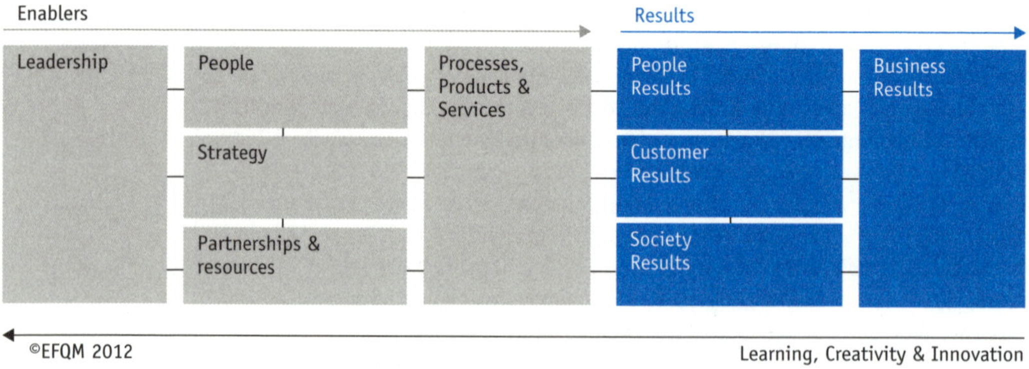

Source: EFQM 2015

Fig. 2-23: The EFQM Excellence Model

2.3.2 Identifying Core Competencies

What we do exceedingly well and what is superior to competition

Having a number of strengths is desirable for any company but might not be sufficient for gaining and sustaining competitive advantage in case other companies have similar strengths. The concept of core competencies adds another perspective to the analysis of strengths and weaknesses. A core competency can be seen as something the company can do exceedingly well and that is superior to comparable competencies of the competition. They are based on a unique combination of resources and capabilities (i. e. competencies) that, when exploited correspondingly, make a significant difference in the market. Core competencies are "unique strengths, embedded deep within a firm, that allow a firm to differentiate its products and services from those of its rivals, creating higher value for the customer or offering products and services or comparable value at lower cost." (Rothaermel 2013: 86) They are not only unique or rare but also difficult for competitors to imitate or substitute. For example, it was mentioned in chapter 2.3.1 that the European fashion provider Zara has the

Strategy Practice Example 2.5

Core Competencies at Canon

Founded in1937, Canon initially focused on cameras and x-ray devices. The company expanded its business scope into office products and developed its first copier for uncoated paper in 1970. Canon extended its business areas to inkjet and laser printers in the 1980s and successively developed other new products by leveraging its core competencies in the areas of *precision optics*, *precision mechanics*, *microelectronics*, and *laser technology*. Based on this, the company has been able to drive market share and profitability in various business fields of enhancing office efficiency. Figure 2-24 illustrates how Canon's broad product portfolio links to its core competencies. An example is Canon's product range of high quality photocopiers where it has achieved a very strong market position. This is based on excellent core products (e. g., cameras, copier, and laser) as well as core product components (e. g., optical lenses and drum units) at the next level. These building blocks of the final products are essentially based on combining core competencies in the various fields already mentioned. Leveraging those unique and outstanding competencies enabled the company to achieve strong market positions in the area of analog photo and video cameras as well as, later on, in the business field of digital photography and video (source: Keth/Asum/Stich 2009: 65).

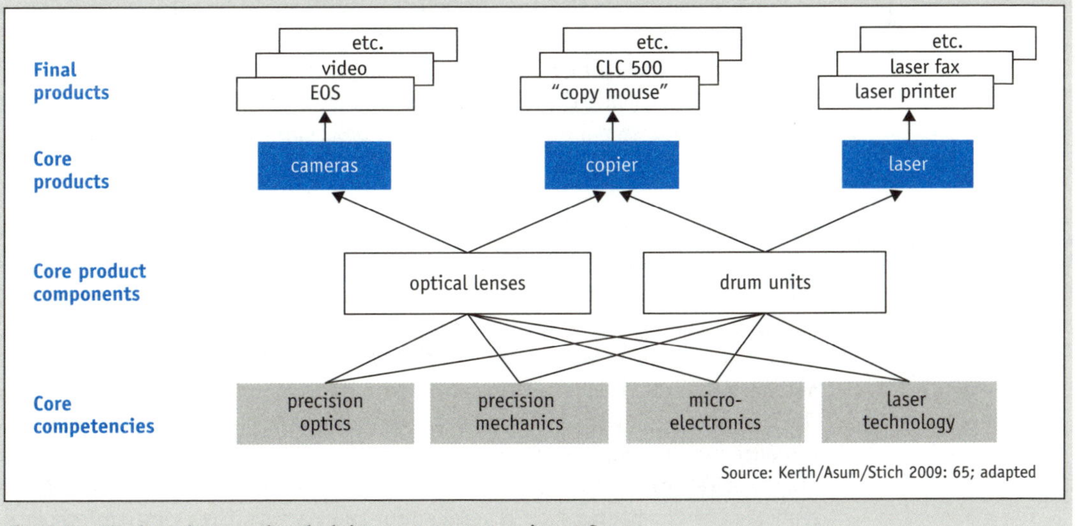

Fig. 2-24: Final products and underlying core competencies at Canon

capability to identify trends through point-of-purchase data collection methods in its stores. However, this alone is not sufficient for creating competitive advantage and driving performance. Its success lies in its complete control and effective integration and coordination of every part of its business. It starts at systematically identifying trends in the shop floors that are directly transmitted to design units that, in turn, are highly interconnected with production, and finally, distribution. This core competence enables Zara to react with an outstanding speed to ever-changing trends that appeals to fashion-conscious customers (Homburg/Kuester/Krohmer 2009: 62 f.). Another example of a company that is able to create competitive advantage based on its core competencies is Canon (see Strategy Practice Example 2.5).

Identifying core competencies: VRIO framework

How can a firm evaluate its competencies (i. e. resources and capabilities) to understand whether they represent a core competence and are thus a promising foundation for gaining and sustaining competitive advantage? The VRIO model (also referred to as VRIN model or VRINO model) provides a framework for analyzing a company's competencies with regard to their potential for providing an advantage in competition (see fig. 2-25). According to this model, a competency needs to be assessed with a set of essential criteria (Barney/Hesterly 2010; Müller-Stewens/Lechner 2011: 209; Rothaermel 2013: 91–94):

- **Valuable** means that the competence can help the company to increase the perceived value of its products and services in the eyes of customers.
- **Rare** means that the competence is unique, that is, only a few firms possess the competence in the same way as otherwise a differentiation in the market would not be possible.
- **Hard to imitate** means the competence cannot be developed or bought easily (i. e. only with significant effort and cost) by other companies as otherwise the advantage may be diluted quickly. This includes indirect imitation (i. e. substitution), which means that the competence is difficult to be substituted by rivals with other competencies that provide a similar value proposition to the customer.
- **Organized to capture value** means that the company's organization is designed in a way that enables the firm to capture and fully exploit the competence's value (e. g., organizational structure, business processes, managerial infrastructure, and organizational culture).

What makes a competency a core competency?

A competency is considered a core competency, and thus, a source for sustained competitive advantage if it is valuable, rare, and hard to imitate and substitute. Additionally, the firm has to be organized to capture the value. If the competence is not valuable, it is something the firm does well but which is not appreciated in the market. This leads to *competitive disadvantage* as it is a waste of resources. If it is valuable but not rare means that there is a diffusion of the underlying resources and capabilities in the market (e. g., knowledge about best practices), which leads to competitive convergence and *competitive parity* (see also chapter 1.1.2.3). If it is valuable and rare but easy to imitate or substitute, rivals are likely to duplicate the competency in order to replicate the company's success mechanism and improve their own performance. Depending on the nature of the competency, though, this may take a while allowing the firm to gain *temporary competitive advantage* until competitors catch up. Finally, if a competency is valuable, rare, hard to imitate and non-substitutable, a company may still struggle to leverage and exploit its full potential due to a lack of appropriate organization including its ability to execute (see chapter 4). In such cases, it may still gain *temporary competitive advantage* but lack sustainable competitive advantage given the firm's inability to execute and adapt to the dynamics in its competitive environment.

Compelling model logic versus what happens in practice

It must be noted that applying the model in company practice may sometimes lead to sobering results. Strictly speaking, a company should only label something a "core competency" if it fulfills all criteria in the VRIO model. In the business world, though, the term is often used to express that a certain competence or strength is considered particularly important. This, however, rarely stands up to an appraisal with the aspiration to fulfill all criteria. Companies may be left disillusioned and with the impression that they do not have any core competence. Of course, this might be true even though the company is performing well. There is no empirically grounded rule that says how many—if at all—core competencies a firm needs to have in order to be successful. Despite this pitfall, expanding the common analysis of strengths and weaknesses with core competencies provides an important perspective on how to create sustainable success and ensure the company's future, which can be considered a key benefit of the model. It enhances the horizon of internal strategic analysis and may provide insights that can hardly be obtained with any of the other models explained in chapter 2.3.1.

Fig. 2-25

Evaluating competencies for their competitive implications (VRIO analysis)

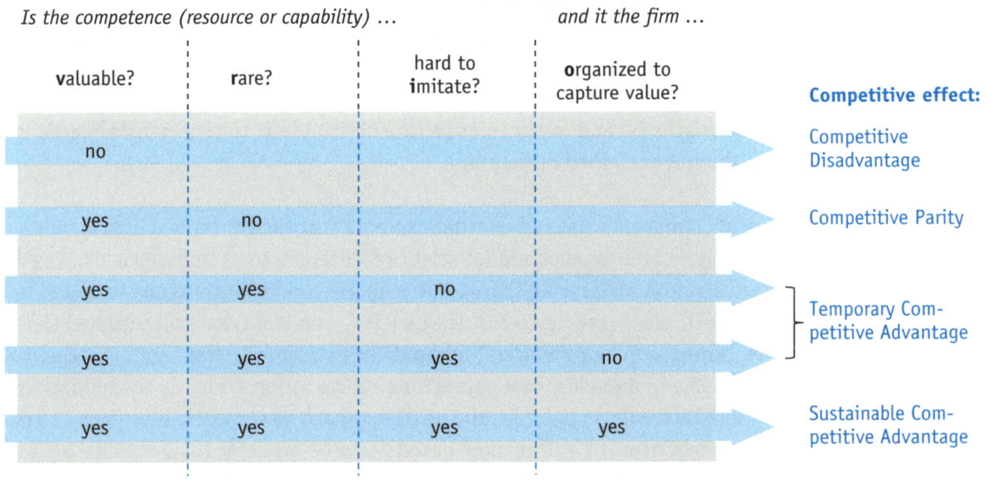

Source: Kerth/Asum/Stich 2009: 116; adapted

2.4 Consolidating Strategic Key Insights

Prioritizing and summarizing key factors from strategic analysis

A strategic analysis typically results in a large number of factors that determine opportunities and threats as well as strengths and weaknesses of a company. Once this wealth of information has been collected, it needs to be consolidated to get a comprehensive and actionable understanding of the company's current strategic situation. Managers involved in strategic ideation and decision making have to be able to "digest" and best exploit the insights from strategic analysis. The SWOT-based strategic analysis summary and the portfolio-based situation assessment are two ways of prioritizing and condensing information down to a few strategic key insights that can be used as practical reference for strategy formulation. They will be explained next followed by an approach for identifying initial focal points for strategy formulation at the end of this chapter.

2.4.1 SWOT-based Strategic Analysis Summary

Frame of reference for strategic analysis results: the SWOT approach

Each step and method of strategic analysis provides a wealth of information that needs to be structured and summarized effectively. The SWOT approach provides a useful framework for this purpose. It is one of the most popular and widely used methods in business management (Choi/Lovallo/Tarasova 2007). Using SWOT as an analytical framework for strategic analysis goes back to work conducted at the Harvard Business School and the Stanford Research Institute in the 1960s (Ansoff 1965; Andrews 1980; Christensen et al. 1982; Humphrey 2005; see also chapter 1.3.1). It can be seen as a foundation underlying virtually all concepts for rational and formal strategy development (Mintzberg 2000: 36). By applying a SWOT framework for strategic analysis, the strategic situation is elaborated on from two distinct perspectives:

- opportunities (O) and threats (T) are derived from the company's *external* environment, which is explained in chapter 2.2
- strengths (S) and weaknesses (W) are derived from the company's *internal* set-up, which is explained in chapter 2.3.

Insights from external and internal strategic analysis are derived from different areas of analysis and by applying a variety of different tools and methods. Key factors resulting from these analyses are now summarized, prioritized and gathered into the four SWOT areas (see fig. 2-26). The goal is to condense and synthesize all the available information in a way that highlights key insights on the firm's current strategic situation and provides focal points for strategy formulation. By consolidating external and internal perspectives, the SWOT approach can essentially be seen as a conceptual integration of the market-based view of strategy (i. e. outside-in) and the resource-based view of strategy (i. e. inside-out; see chapter 1.1.2.2).

Avoiding pitfalls underlying the supposedly "simple" SWOT approach

The SWOT approach appears intuitive and simple, but there are some serious pitfalls to be avoided for effectively applying the framework in business practice. Here are some essential points to bear in mind when trying to apply and utilize the SWOT

Fig. 2-26

Integrating strategic analysis results with the SWOT approach

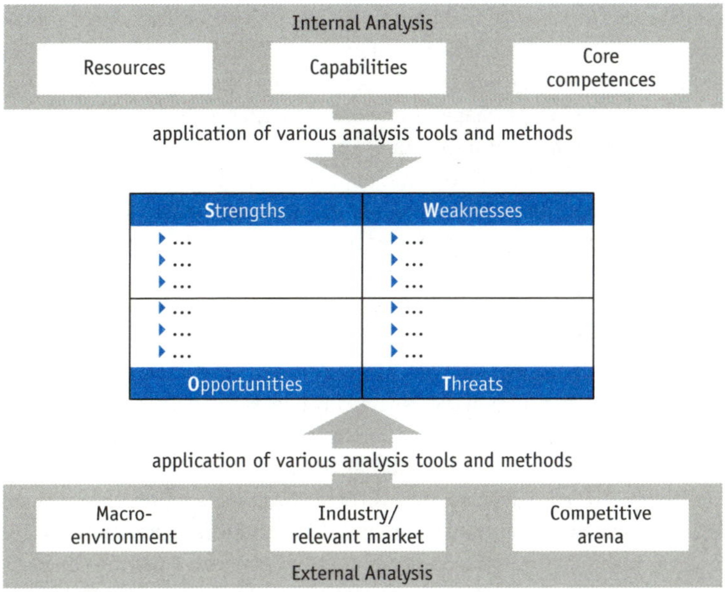

Source: Müller-Stewens/Lechner 2011: 210; adapted

approach to its fullest potential (Johnson/Whittington/Scholes 2012: 68; Kotler/Berger/Bickhoff 2010: 30–33; Wheelen/Hunger 2010: 176):

- First, the SWOT approach should not be seen as an analysis "tool" itself but as a *frame for structuring and consolidating* information gathered with various other analytical methods and tools applied in the strategy process. The quality and benefit of information and insights is not determined by applying a SWOT approach per se but by the sophistication and thoroughness of analysis used under the structure of the SWOT framework (e. g., effectiveness of methods and tools, reliability of data sources, validity and accuracy of information requiring focused and unbiased information gathering and evaluation, completeness of information, etc.).
- Second, information gathered in different fields of a situational analysis and consolidated in the SWOT framework should be *purely descriptive*. It might be tempting to directly jump to interpretation and recommendation at some individual bits and pieces of information, such as macroeconomic trends identified. However, this should not be done without having determined a more complete picture of the strategic situation. The value of SWOT lies in its integrated view combining an external with and an internal perspective. The synthesis of strategic analysis is ultimately done in strategy formulation leveraging key findings from the entire SWOT (see chapter 3).

- Third, the external perspective (opportunities and threats) and internal perspective (strengths and weaknesses) of a SWOT analysis are equally important and must be kept sharply distinct. Findings from external analysis are typically combined with internal findings for deriving strategic ideas (see chapter 2.4.3). Whereas strengths and weaknesses can be influenced by the company, opportunities and threats in a SWOT analysis arise from developments in the firm's external environment. Those typically cannot be influenced by the company, unless it is in a position to set own trends or shape its markets. Though simply worded, "opportunity" and "threat" is sometimes misapplied. For example, overcoming a weakness may seem to be an "opportunity", which is not in line with the SWOT principle. In SWOT logic, a company may need to overcome its weakness (internal analysis) to take advantage of an opportunity identified in its environment (external analysis).

Strategic factor analysis summary (SFAS)

An analytical and rigorous approach to synthesizing strategic analysis results is the development of a *strategic factor analysis summary (SFAS)* table. It reflects a systematic consolidation of the most important internal and external factors—key insights regarding its strategic situation that a company needs to consider for strategy formulation—to a manageable number. As illustrated in figure 2-27, the SFAS table is derived from two components: the internal factor analysis summary (IFAS) and the external factor analysis summary (EFAS). These two tables reflect the summary and evaluation of a company's current internal and external strategic situations. They are based on the most important factors (e. g., determined with a pairing comparison) a company has identified with its internal strategic analysis categorized into strengths and weaknesses as well as its external strategic analysis categorized into opportunities and threats (e. g., eight to ten internal and external factors). IFAS and EFAS show how well a company's management is currently responding to the corresponding factors in light of the perceived importance of those factors to the company (Wheelen/Hunger 2010: 126 f., 164 f., and 176–179).

- **Internal factor analysis summary (IFAS)**: To understand how well a company is currently dealing with its internal factors, they are weighted in terms of importance and rated in terms of response. First, a weight is assigned to each factor based on an evaluation of its impact on the company's current strategic situation, for example, from 1.0 (most important) to 0.0 (not important). Second, each factor is rated based on how the firm is currently responding to that particular factor, for example, from 5.0 (outstanding) to 1.0 (poor). Multiplying the weight with the rate leads to a weighted score for each factor. The total weighted score in the IFAS shows, how well a company is currently responding to the important factors in its internal environment with respect to the baseline of 3.0 (i. e. the total weighted score of an average firm in the industry).
- **External factor analysis summary (EFAS)**: To understand how well a company is currently dealing with its external factors, they are first weighted with regard to their probable impact on a firm's strategic position (importance). The higher the weight (e. g., 1.0 for most important and 0.0 for not important) the more important this factor is for a company's current and future performance, and thus, its strategy. Similar to IFAS, each factor is then rated from 5.0 (outstanding) to 1.0

(poor), which is supposed to demonstrate how well the company is currently responding to each factor. Calculating and adding up the weighted score provides an evaluation of the current external strategic situation of the company. A total weighted score higher than 3.0 shows that the firm is responding better than the average firm in an industry to the most important current and expected factors in its external environment. This should be reflected in the firm's current performance such as its profitability and market share.

The final SFAS table is derived from the most important internal and external factors identified in the strategic analysis (e. g., ten factors in total). This is illustrated in figure 2-27. Hereby, the highest weighted factors from the IFAS and EFAS tables are used. Note that not the weighted scores—which were required for understanding both the internal and external situation individually—but only the weights are used for this

Fig. 2-27

Synthesizing strategic analysis results

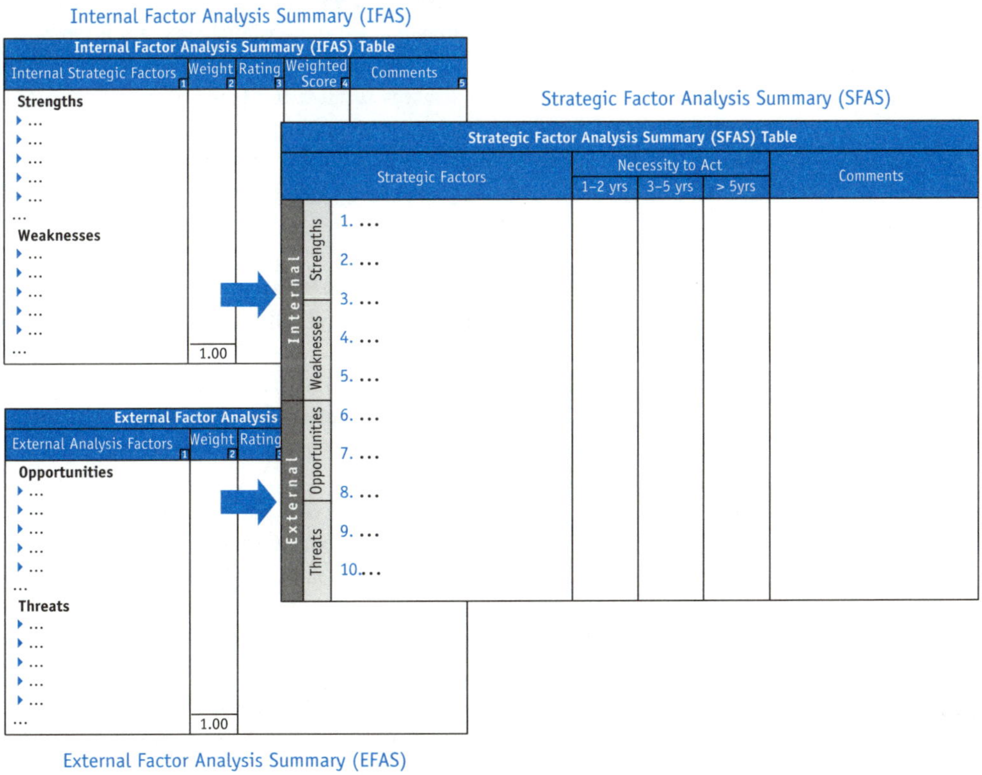

Source: following Wheelen/Hunger 2010: 126 f., 164 f., and 176–179; adapted

purpose. As those factors may relate to different time frames (e. g., a long-term opportunity versus a weakness that needs to be overcome soon), adding duration indicates the firm's necessity to act. Also, a company may want to apply the weighting and rating procedure similar to IFAS and EFAS again in the strategic factor analysis table to get an evaluation of the overall situation, which is not shown in figure 2-27. Although, the strategic factor analysis does not require to distinguish between internal (strengths and weaknesses) and external (opportunities and threats) factors anymore, companies may want to maintain the categories for reasons of traceability and further analysis such as developing a TOWS-matrix (see chapter 2.4.3).

2.4.2 Portfolio-based Situation Assessment

Summarizing the strategic situation in multi-business or multi-product firms

Another way to summarize the results of strategic analysis particularly in multi-business or multi-product companies is the portfolio method (see also chapter 3.2.2.1). Portfolio matrices used for the purpose of strategic management have several characteristics in common. Figure 2-28 shows the basic structure of a portfolio used for this purpose. First, a portfolio represents the graphical display in which the firm's businesses or product lines are allocated in a two-dimensional area that is typically separated into squares. The squares are usually tied to strategic suggestions or generic strategies that can be used for strategy formulation (see Strategic Snapshot 3.4 in chapter 3.2.2.1). Second, the businesses or product lines are illustrated with circles. The size of the circle can be used to indicate each business' revenue or profit contribution, thus showing their relative economic significance in the portfolio. Third, desired directional changes for the businesses or product lines are illustrated with arrows as well as corresponding circles at targeted future positions in the portfolio, which may also result in an anticipated change in their revenue or profit contribution (size of the circle; Hax/Majluf 1996: 300 and 1991: 192; Kotler/Berger/Bickhoff 2010: 38–43):

- The *external dimension* is based on criteria that are usually uncontrollable by the company. It captures the overall attractiveness of the market in which a business participates (e. g., market size, market growth, competitive intensity, market profitability, maturity, etc.).
- The *internal dimension* is based on criteria that are largely under the control of a company. It indicates a business' overall competitive strengths within its corresponding market. The internal dimension relates to the strengths and weaknesses of a business (e. g., relative market share or market position, degree of vertical integration, flexibility, capacity utilization, R&D/technology competency, etc.)

Single picture of the strategic situation based on few key criteria

For a portfolio-based situation assessment, external and internal strategic factors are significantly reduced to a few if not only two key criteria. Those criteria are not related to the company as a whole—this would be too generic, and thus, not valuable—but to its strategic business units or product lines/areas. Each criterion or group of criteria is supposed to represent an external as well as an internal perspective used to evaluate the strategic situation for each strategic business unit or product line in

Fig. 2-28

Basic structure applied in the portfolio method

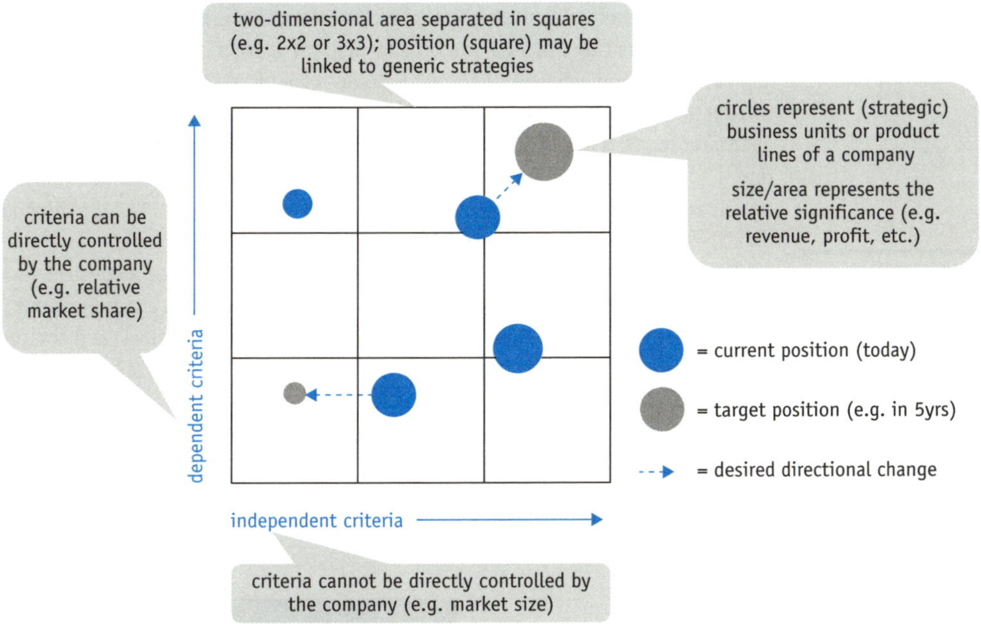

Source: Welge/Al-Laham 2012 following Bullinger 1994: 144; adapted

the portfolio. In addition to portfolios following such a market-oriented rationale, there are a number of portfolio approaches that are based on other dimensions that are used for other purposes (Welge/Al-Laham 2012: 475–508). Three popular market-oriented portfolio matrices that evolved in business or management consulting practice are listed next. The first two will be explained more in detail in the course of this chapter.

- *Market attractiveness–business strength* matrix (GE/McK-matrix) jointly developed by General Electric and McKinsey & Company
- *Market growth–market share* matrix (BCG-matrix) developed by the Boston Consulting Group
- Life cycle matrix with its two dimensions stages of *industry maturity* (embryonic, growth, mature, aging) and *competitive position* (dominant, strong, favorable, tenable, weak) developed by Arthur D. Little, Inc.

McKinsey/GE portfolio: market attractiveness–relative competitive position

General Electric with the assistance of McKinsey & Company developed a portfolio model that is based on multidimensional criteria in its external and internal dimensions. Contrary to the BCG matrix that uses only two criteria and will be explained later on, its basic principle is to determine the strategic situation of a business or product line with more comprehensive measures of strategic success. There are a number of different factors a company may use for designing a corresponding portfolio fitting to its specific industry and business context. Figure 2-29 illustrates this approach with an example. Hereby, the evaluation of *market attractiveness* (i.e. external dimension) is a result of the three factors market size, market growth, and competitive intensity. The internal dimension—*competitive strength*—is determined by a businesses' relative market share, its strengths to deliver customer value (e.g., its relative fulfillment of customer requirements) and its flexibility. Based on a simple scoring model, the different factors can be weighted and rated to determine the precise position of each business or product line in the matrix. (Perlitz/Schrank 2013: 221–225; Thompson et al. 2014: 234–240; Welge/Al-Laham 2012: 481–485)

Six steps for developing and applying the GE/McK methodology

The GE/McK methodology can be applied and customized to a specific company situation and used effectively to support strategic thinking in a management team. As a first step, the different criteria for rating market attractiveness and competitive strength need to be determined along with a scoring model for rating. For example, the scoring may be determined by using a scale from 1 to 5 for each criterion and corresponding quantitative values or qualitative descriptions for each number on the

Fig. 2-29

Illustration of McKinsey/GE portfolio with multi-dimensional criteria

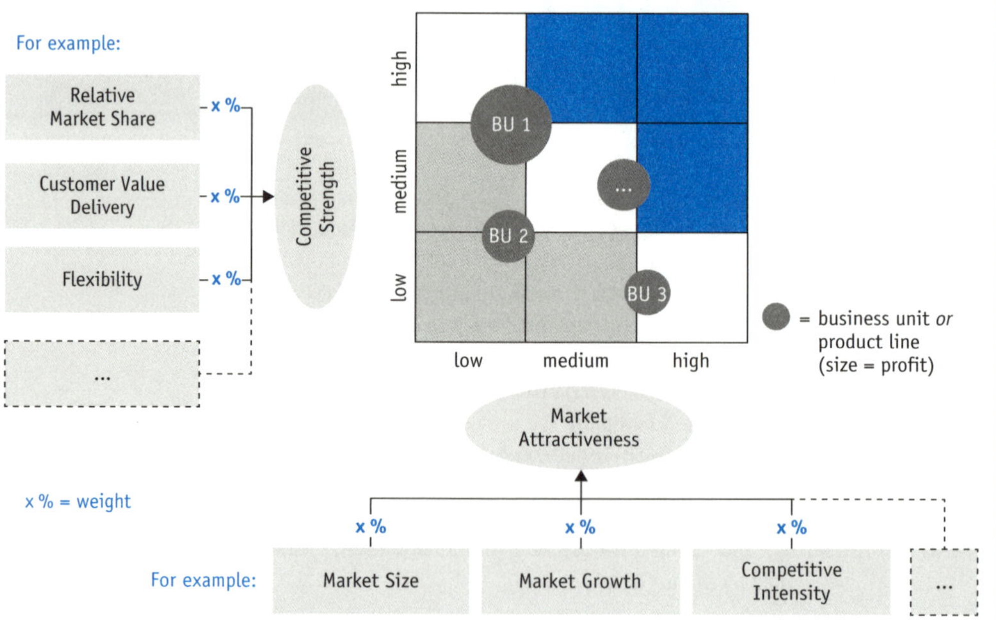

Source: following Perlitz/Schrank 2013: 225; adapted

scale. Once the scoring model is developed, the six steps illustrated in figure 2-30 are suggested for each business or product line to be able to plot them on the matrix and leverage the approach for future-oriented strategic thinking and ideation (following Hax/Majluf 1991: 182–198):

1. **Determine the current position** by assessing overall market attractiveness as well as competitive strength on the corresponding scales for each criterion in the external and internal dimensions (1 = very unattractive/very weak and 5 = very attractive/very strong).
2. **Anticipate the market dynamic** (i. e. the estimated future development of market attractiveness) by projecting each criterion to a future point in time (e. g., 5 years). As the external dimension (i. e. market attractiveness) typically cannot be controlled directly by the company, this should be done prior to anticipating the future competitive strength.
3. **Investigate the risk position.** This is the expected change in a business' or product line's competitive strength caused by the anticipated market dynamic (step 2) assuming that its current strategic approach is continued without any significant changes in resource allocation. The example in figure 2-30 shows that the business is expected to lose its competitive strengths in case it is continued "as usual" because the projected change in market attractiveness alters the situation in its competitive arena (e. g., new strong market entrants in an increasingly attractive market). In other words: If the company doesn't go forward, it goes backwards.
4. **Project the target position** by assessing what is possible and desired to achieve at each criterion for determining competitive strength. To avoid overly optimistic target positions it needs to be considered that sometimes significant efforts are necessary just to maintain the existing competitive strength given a certain anticipated development in the competitive arena (see step 3). Based on this, each internal criterion is projected to the future point in time resulting in a targeted position in the portfolio. A change in size of the circle represents the future revenue or profit expectation (e. g., rough order of magnitude) of the business or product line.
5. **Identify the gap** between the current position and target position in terms of what basic resources are required for funding, people, competencies, etc. to achieve the defined aspiration. For example, is the target position (e. g., the growth and profit aspirations) achievable organically or does it require external elements such as new capabilities, licenses, products, etc.? This can be assessed by looking at the gap for each controllable internal criterion that is supposed to be a key factor for determining the competitive strengths.
6. **Formulate strategy options** to close the identified gap and realize the desired target position based on insights from external and internal analysis. This topic will be addressed in detail in chapter 3. Some initial reflections on strategy may already be required earlier on in this process to get a more refined understanding for both achievable future revenues and profits (step 4) as well as required resources (step 5). This is one of the reasons why iterations between strategic analysis and strategy formulation in general and the steps 4, 5 and 6 in particular are likely and common in business practice.

Fig. 2-30

Guiding strategic thinking with the McKinsey/GE portfolio

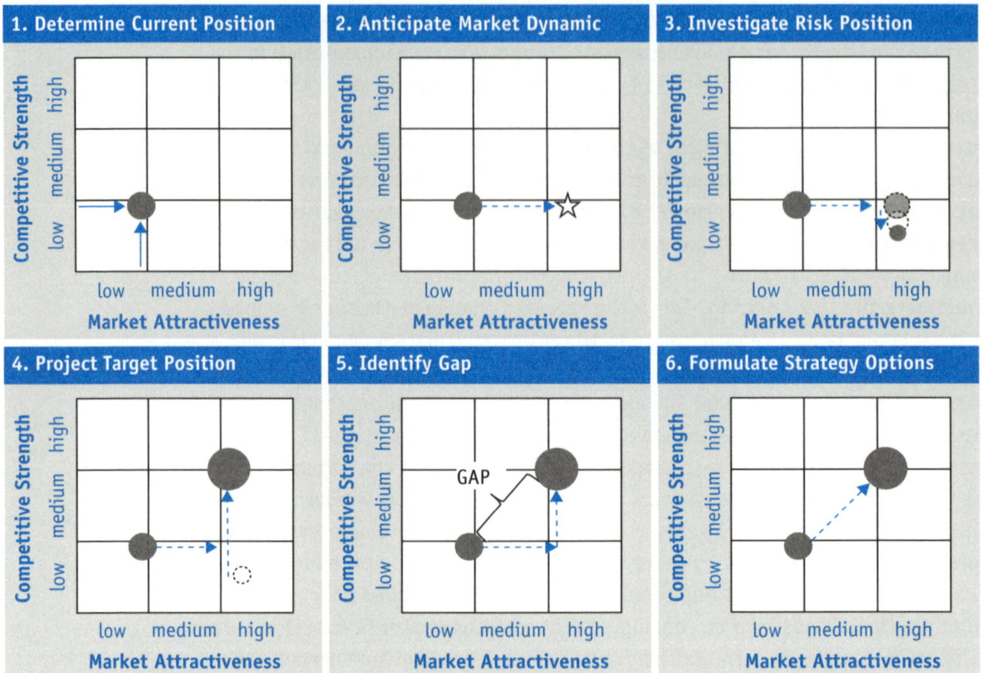

Source: following Hax/Majluf 1991: 182–198; adapted

BCG portfolio: market growth–relative market share matrix

The BCG portfolio or BCG matrix is probably the best known market-oriented portfolio method. Contrary to multiple-criteria approaches such as the McK/GE portfolio, the BCG matrix uses only one single criterion in each dimension to plot a business or product line in the matrix. One criterion is the *relative market share*, that is own sales in relation to the biggest competitor's sales, which relates to the internal analysis of strengths and weaknesses. This criterion is considered a strategic success factor and supposed to indicate the competitive strengths of a business based on insights from the experience curve model. The experience curve shows that—under certain assumptions—an increase of accumulated output leads to a reduction of unit cost due to learning effects. This in turn means that increasing the relative market share leads to competitive cost advantages, and thus, to increased profitability. The other criterion is the *market growth*, that is the percentage of overall sales increase in a particular overall market. Market growth is used as single indicator for the market attractiveness. It relates to the external analysis of opportunities and threats and is based on the product-/market life cycle model. Products in the introduction or growth phase as well as businesses in emerging markets are considered more attractive than mature or saturated markets as they have a greater growth potential that, however, requires greater investments to be captured. The BCG portfolio seeks to optimize free cash flows in a way that cash requirements and cash supplies are properly balanced. According to its stage in the life cycle and its competitive position in terms of market share, a business or product line is allocated to one of the following four categories, which are used to support funding decisions (Wheelen/Hunger 2010: 221–223):

- A **star** has a strong competitive position (leading or high relative market share) in a market that is still growing strongly (often at the peak of its life cycle). Its cash generated is consumed for keeping up with growth and maintaining its high share of the market. It is thus self-sufficient and usually contributes to the corporation's profit. Stars in markets where growth rates slow down typically become cash cows.
- **Cash cows** have high relative market shares but in mature markets. Because market growth is low, there is little need to invest in the business. Due to their strong market position they should be profitable and generate more cash than they need for maintaining their market share. By "milking", cash cows provide funds that can be invested in other businesses such as promising question marks. As they are in a declining stage of their life cycle, only maintenance investments are done.
- A **question mark** is a business that requires more cash than it can generate due to its lack of market share. It is in a growing market but has not yet achieved a strong market position. Question marks need to be investigated thoroughly regarding their future prospects and require management decision and action as they currently consume cash. If the management decides they are worth the investments, they are promoted to become future stars. Otherwise they are divested. If question marks fail to become stars—that is they are unable to obtain dominant market positions—they may decline into poor dogs as market growth rates will slow down over time. It is important that some question marks become stars as existing stars will eventually become cash cows and ultimately poor dogs.
- **Poor dogs** have low relative market shares in mature or declining, and thus, unattractive markets. They neither consume nor generate much free cash flows but tie up management attention and company resources, and are thus the worst of all combinations. They should be preferably sold off to other companies or liquidated and closed down, though a company may decide to manage them carefully for a small amount of cash they may be able to generate.

Benefits of portfolio analysis and watch-outs for practical application

Portfolios are supposed to help managers, particularly in diversified multi-business or multi-product companies, reach a better understanding of overall competitive standing. They provide a number of benefits, but there are also some pitfalls that need to be considered (following Hax/Majluf 1996: 312 f.; Müller-Stewens/Lechner 2011: 289 f.):

First, they are powerful practical tools for *communication and strategic ideation* by providing an intuitively understandable single picture that illustrates the overall strategic situation and its potential evolution of a firm's entire portfolio of businesses or product lines. They represent a pragmatic way of capturing the essence of strategic

Fig. 2-31

BCG portfolio and its conceptual foundations

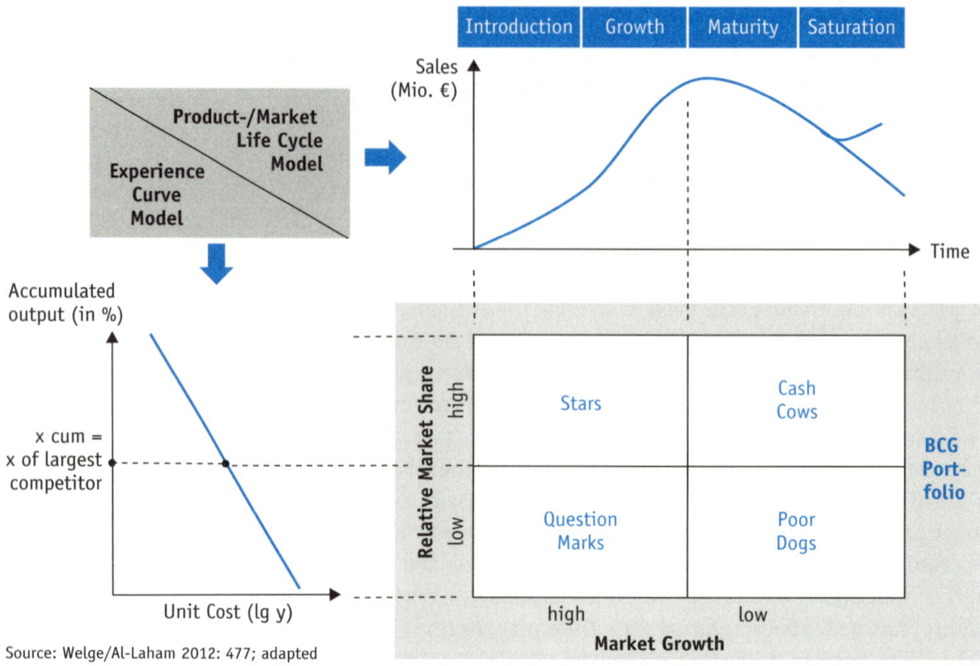

Source: Welge/Al-Laham 2012: 477; adapted

analysis without providing lengthy lists of SWOT-results. Hereby, illustrating the dynamic perspective in particular can effectively raise a manager's *strategic alertness* about the uncontrollable dynamics in the external environment (e.g., step 3 "risk position" in fig. 2-30), and thus, lead to more realistic aspirations regarding the determination of target positions. If portfolios are applied this way, they are not meant to provide supposedly perfect illustrations of the situation or even strategic solutions and "answers" to strategic challenges but to *facilitate and support strategic thinking* in management teams that have to come up with strategy options. Second, the consolidated view allows better *evaluating and comparing* different businesses or product lines with the *same* criteria. Third, based on the individual position in the portfolio, many approaches suggest normative or *generic strategies* which can be used by managers as a *guide for resource allocation* (see chapter 3.2.2.1) to ensure consistency between the businesses and product lines at different positions and to properly balance cash requirements and cash supplies. For example, a company needs to ensure that it has businesses in its portfolio that primarily generate net cash flows in order to invest in other businesses that want to enhance their market leadership positions. Fourth, portfolios may be used not only for the purpose of a one-time situational assessment but for showing *desired future projections*. They allow a structured and systematic elaboration on future scenarios. Furthermore, they are a practical way to *document a chronological evolution* of the portfolio ex post.

Even though the use of portfolio analysis for assessing the strategic situation appears compelling, there are some serious watch-outs to consider before applying this methodology. First, reducing the complexity of internal and external analysis down to only two (e. g., BCG matrix) or a few (e. g., GE/McKinsey matrix) criteria may result in an *inappropriate simplicity* for strategic decision making. For example, applying the BCG portfolio for the premium automotive segment has its limitations. First, the portfolio dimension "relative market share" provides limited value due to the oligopolistic competitive situation with very equal market shares. The other dimension, market growth, is not a purely uncontrollable external criterion but strongly driven by the automaker's product offering (see Strategy Practice Example 2.6). On the other hand, adding criteria and applying a more comprehensive portfolio version may provide a level of complexity that is not very practical for management. It can lead to people arguing more about the validity of criteria and weights instead of focusing on strategic opportunities and challenges of their company. Second, the supposedly analytical approach has the risk that managers may perceive the resulting portfolio as a true and precise representation of the strategic situation, and thus, may *rely blindly on the suggested norm strategies* as a sure formula for success. However, it must not be forgotten that in many portfolios the determination of a businesses' strategic position is a highly subjective assessment of mostly qualitative criteria into a quantitative scoring model. Third, *interdependencies of businesses or product lines* are not considered in the portfolio. Whereas a norm strategy may suggest divesting in a business, it may still be an important pillar in the company's strategy given synergies with other businesses (see also chapter 3.2.1). Fourth, the *theoretical-conceptual assumptions* underlying certain portfolio frameworks may not be fully valid and applicable for certain businesses or industries. For example, not all businesses or product lines follow a life cycle as assumed in the BCG portfolio (see fig. 2-31) but are successfully revitalized or renovated to avoid ending up as a "poor dog" (e. g., mechanical watches). Also, the experience curve assumes that increasing the accumulated output decreases unit cost, and thus, drives profitability. There are a number of businesses, however, that achieve high profitability with small market shares due to providing an outstanding customer value based on differentiation. Here, the assumption is not a positive-linear correlation of relative market share and profitability but a u-form (see chapter 3.4.2.1).

With reference to the various benefits and pitfalls a company has to decide whether it can apply standard portfolios like the ones explained to their particular situation. Some companies develop their own portfolio approach by customizing the principles explained earlier in a way that is more valuable for strategic ideation and decision making in their specific business context (see Strategy Practice Example 2.6).

Strategy Practice Example 2.6

Product Portfolio Methodology and Management at BMW

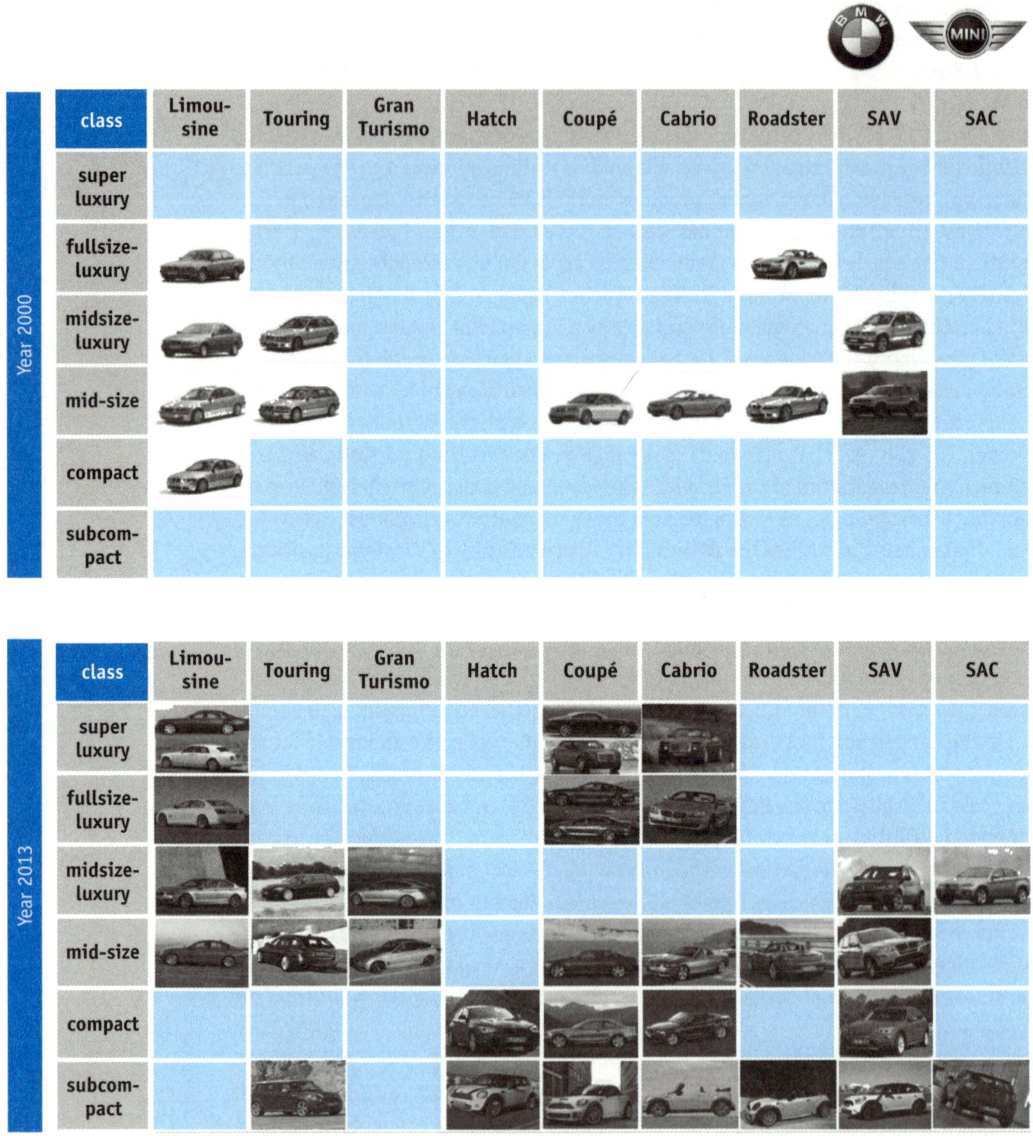

Source: BMW 2013. Reprinted with permission of BMW AG.

Fig. 2-32: BMW product portfolio in 2000 and 2013

Original equipment manufacturers (OEMs) in the automotive industry have been reacting to consumer preferences for more segmented vehicles by significantly expanding their product portfolios in recent years. Figure 2-32 illustrates this trend by comparing the BMW product portfolio in 2000 with 2013. It shows significant changes and expansions across all classes. Furthermore, the industry can be characterized by high initial investments in R&D and production for new models, relatively long product life cycles as well as new technological trends such as hybrid engines and e-mobility. Consequently, for premium carmakers such as BMW the question of how to best develop and optimize their product portfolio is crucial.

For supporting its strategic portfolio decisions, BMW applies a portfolio model, which is shown in figure 2-33. It is based on the two dimensions *average volume p. a.*, which is derived from an overall market growth projection and the aspired market share as well as *return in %*, which reflects the profitability. There are two important hurdle rates: a minimum

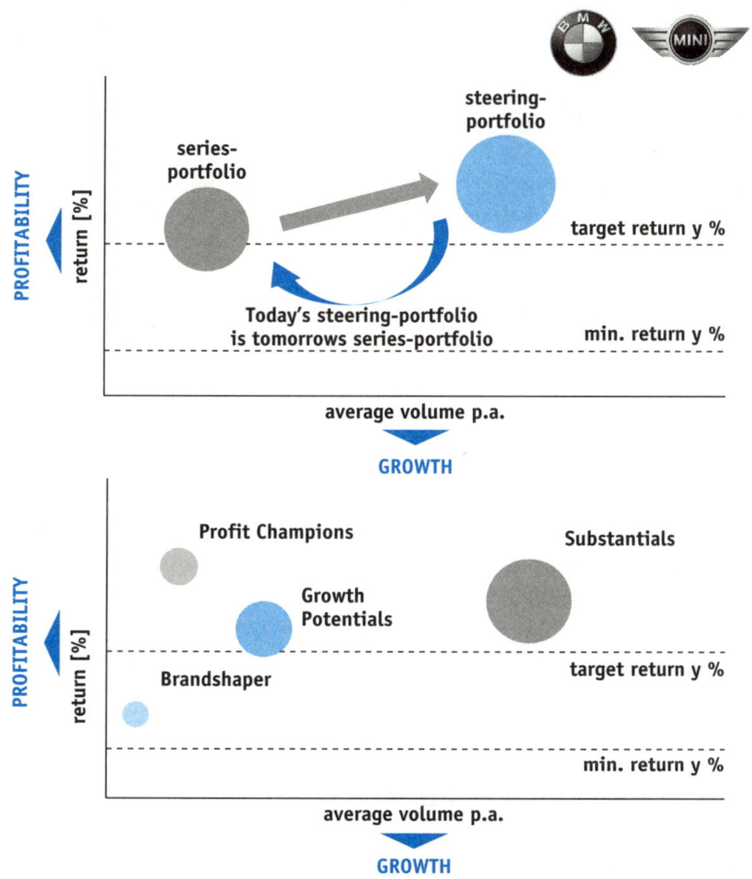

Source: Adelberger 2013: 31 f. Reprinted with permission of BMW AG.

Fig. 2-33: BMW product portfolio and clusters

profitability reflecting the cost of capital as well as the company-specific total target return. The size of the circles indicates the capital invested in the product or product group. Separating a *series portfolio* reflecting the products currently offered on the market from a *steering portfolio*, that is the expected situation after the launch of the new product generation, is the foundation for management discussions about how to optimize the overall situation. Hereby, today's steering portfolio is supposed to become tomorrow's series portfolio. The BMW product portfolio provides a single picture of the products or product groups, how much capital investments they require, how much profit they deliver and how much they contribute to achieving the desired overall sales volume (i. e. market position). To further deep dive into the various products, BMW developed the following categories:

- **Substantials** are the key pillars of the portfolio representing the core of the product offering that is well established in the market. They provide a large amount of cash and still deliver sufficient profit contributions. A product example is the limousine in the BMW 3-series.
- **Growth Potentials** are similar to the stars in the BCG portfolio model as they are in strongly growing market segments with attractive profitability and future volume growth potential. Usually these are new product concepts that require several product life cycles until they are fully established in the market. Recent examples are the SAV (Sports Activity Vehicle) concepts such as the BMW X1, X3, and X5.
- **Profit Champions** are vehicle derivatives strongly related to the core products (substantials). They are designed to fulfill specific customer requirements in certain market segments with relatively little additional effort (e. g., based on modular systems). Despite their limited volumes, those products can be highly profitable, and thus, provide additional cash flows to the portfolio. Examples are the BMW Touring models, convertibles ("Cabrio") or certain performance or sports versions such as the BMW M-class.
- **Brandshaper** are pioneer products or technological forerunners that are supposed to shape a company's image. Their profitability is usually very limited, but they are crucial for conveying the company's innovativeness and technological competence to the market as well as for gaining experience in new or future market segments (e. g., e-mobility). They are also foundations for the emotional core and the identity of a brand (e. g., certain sports cars).

Allocating all those different product categories in the portfolio based on the explained criteria provides a valuable picture of the current strategic situation. Following a similar rationale as in the BCG model, it shows how balanced and strong the overall portfolio is. More detailed assessments require a further drill-down to the level of individual vehicle derivatives to analyze the relative importance of individual models in the past, today and in the future. This type of portfolio approach is also used for analyzing different scenarios such as the impact of introducing a new vehicle derivative. (Source: Adelberger 2013; Adelberger/Haft-Zboril 2013)

2.4.3 Focal Points for Strategy Formulation

Exploiting strengths and overcoming weaknesses supplemented with an external perspective

It will be evident that strategies are generally based on exploiting strength and overcoming weaknesses. This basic principle applies in particular to current strategies. However, such an approach must not lead to strategic myopia where companies focus only on areas they are currently good at and markets that are attractive today. There is a significant number of examples of companies that have gone out of business due to such a short-sighted view (see chapter 1.1). To ensure long-term survival and success, companies need to follow another key principle that is capturing external opportunities (or avoiding threats) in the future (see also chapter 1.1.2.2). Consequently the simple rule of thumb to emphasize strengths and reduce weaknesses needs to be supplemented with a long-term and future-oriented external perspective (Pillkahn 2008; Welge/Al-Laham 2012: 448).

TOWS-matrix for matching external and internal strategic factors

A simple tool that combines these two strategic principles related to strengths and weaknesses as well as opportunities and threats is the TOWS-matrix (David 2013: 206 f.; Johnson/Whittington/Scholes 2012: 68–70; Weihrich 1982; Welge/Al-Laham 2012: 449; Wheelen/Hunger 2010: 182 f.). It is a framework that illustrates four ways of matching the results from external strategic analysis (opportunities and threats) with insights from internal strategic analysis (strengths, weaknesses and core competencies). This leads to four alternative strategic combinations:

- **SO strategies** prompt the consideration of strategy options that use a firm's strengths to take advantage of opportunities in its external environment. For most organizations the possibility of applying SO strategies will be an ideal situation that may be addressed with different types of growth strategies (see chapter 3.4.1).
- **WO strategies** are based on the necessity of overcoming internal weaknesses that prevent a firm from capturing opportunities in its external environment. For example, an automotive manufacturer may try to overcome its weakness in battery technology by forming a key partnership such as a joint venture with a firm having competency in this area to take advantage of the e-mobility opportunity in its external environment (see chapter 3.4.3.1).
- **ST strategies** suggest using the organization's strengths for reducing the impact of threats in the external environment. For example, if growth opportunities in a firm's core business are increasingly limited by imposed legal restrictions, the company may consider using its strengths and core competencies to diversify in other businesses facing a different legal environment (see chapter 3.2.2.2).
- **WT strategies** are defensive strategy options to protect the company against threats in its external environment by overcoming or minimizing internal weaknesses. Companies that see their strategy options primarily in this area are in a precarious situation. They have to carefully consider how to mitigate the risk of going out of business or actively choose a retrenchment approach (see chapter 3.2.2.1).

Initial focal points for corporate or business strategy formulation

Applying the TOWS-matrix during consolidation of the current strategic situation can help to identify initial focal points for strategy formulation. It is unlikely that the four generic strategic alternatives alone are sufficient for formulating a successful corporate or business strategy (see chapter 3). However, they are a pragmatic and effective way to focus strategy formulation on a corporate or business level, particularly on those future-oriented strategy options an organization is generally capable of supporting. Whereas systematically generating different strategic alternatives with a TOWS-matrix may widen the scope of thinking and ensure that all perspectives are considered, the resulting initial focus is supposed to increase the effectiveness of strategic ideation later on.

Fig. 2-34

Identifying focal points for strategy formulation with the TOWS-Matrix

		Factors from Internal Strategic Analysis	
		Strengths (S)	Weaknesses (W)
Factors from External Strategic Analysis	Opportunities (O)	**SO Strategies** use of strenghts to take advantage of opportunities	**WO Strategies** overcome weaknesses to take advantage of opportunities
	Threats (T)	**ST Strategies** use of strenghts to avoid threats	**WT Strategies** overcome weaknesses to avoid threats

Source: Johnson/Whittington/Scholes 2012: 68; adapted

2.5 Summary

There are a variety of approaches for determining a company's initial strategic situation as a starting point for strategy formulation. Foundational for analyzing a firm's external and internal environment is the identification and understanding of relevant stakeholders that affect performance. Based on this, a strategic analysis framework has to be developed and applied that consists of various layers with different proximities to the company. The most distant layer is the macroenvironment in which broad political, economic, sociocultural, technological, ecological, and legal factors are identified that have an impact on the firm's performance, and thus, need to be addressed in strategy formulation. Based on those factors, scenarios or "pictures of the future" are developed considering a certain scope (e. g., geography, industry) and time frame. Another area in a firm's external environment that needs to be analyzed is the industry, also referred to as relevant market. This needs to be defined carefully as a base for understanding competitive forces that shape the corresponding profit potential. The weaker those forces are, the more attractive is an industry or relevant market. Firms can shape the industry structure they favor with corresponding strategies. The third layer in a firm's external environment is the competitive arena in which key stakeholders such as customers, competitors and strategic groups, as well as other relevant external stakeholders have to be analyzed. To gain valuable insights about the current external strategic situation, this may require a segmentation of the relevant market into submarkets.

Analyzing the external environment is only one part of strategic analysis. Companies also need to understand their internal environment to identify strategic factors that represent strengths or weaknesses, which reflects the closest layer in the suggested strategic analysis framework. Hereby, a firm's resources, capabilities, and core competencies are analyzed. Following the resource-based view of strategy, those elements reflect key sources of competitive advantage, which can be evaluated by applying the VRIO model. There are various ways of analyzing strengths and weaknesses such as strength and weakness profiles, the value chain model, and the 7-S-model, which can be applied individually or in an integrated way. All the different areas of strategic analysis provide a wealth of insights in terms of opportunities and threats as well as strengths and weaknesses. Those need to be prioritized and consolidated to be valuable and manageable for strategy formulation, which can be done by applying a SWOT framework or a portfolio-based situational assessment. By combining key results from external and internal analysis in a TOWS matrix, initial focal points for strategy formulation can be identified.

3 Strategy Formulation

Learning Objectives

After studying this chapter, you should be able to:

- understand the relevance of strategic guideposts for strategy formulation
- explain vision, mission, values, and strategy as well as their interrelation
- describe corporate strategy in terms of corporate parenting and business scope
- explain drivers for going international and analyze international market potentials
- explain the four strategy postures in multinational corporations
- explain the concept of transnational strategies and how to execute them
- describe different intensive growth strategies
- describe different generic competitive strategies on a business level
- apply blue ocean strategies to identify uncontested market space
- explain, apply and evaluate business models for business-level strategy
- describe the relevance and principles of sustainable business models
- explain the six key criteria for evaluating and selecting strategy options

3.1 Strategic Guideposts

Framing and directing strategy formulation

The purpose of strategic analysis was to establish a clear understanding of where the company strategically stands today. In light of its strengths, weaknesses and core competencies as well as identified opportunities and threats in the external environment, the firm now has to decide if and how it wants to change its strategic position. As a first step, this requires clarity about the firm's fundamental strategic intent and identity. This will be referred to as *strategic guideposts* which are supposed to frame and direct strategy formulation. Clarifying strategic guideposts essentially comes down to three questions (Rothaermel 2013: 32):

- What is our purpose or raison d'être (**mission**)?
- What do we want to be or have accomplished in the future (**vision**)?
- What are our behavioral principles for interaction (**values**)?

Strategy is the approach of how to make the vision come true.

Based on those strategic guideposts, a **strategy** represents the approach or pattern of how the company wants to make its vision come true framed by the company's mission and values. In this understanding, strategy is only a means to an end (see chapter 1.1.2.4). The "end" is the vision as it creates a picture of the aspired future destination. A planned strategy shows how the vision is supposed to be achieved from today's perspective. Vision, mission, values and strategy are essential complements that need to be aligned with each other. It seems almost impossible to formulate a strategy if the organization's mission and vision as well as the behavioral principles and rules remain unclear. Hereby, strategic guideposts represent a kind of implicit

Strategic Snapshot 3.1

Illustrative Relationship between Vision, Mission, Values, and Strategy

The relationships between vision, mission, values, and strategy are illustrated in figure 3-1 by using the analogy of climbing a mountain. Imagine some extreme sportsmen have formed a team with the purpose of jointly seeking athletic limit experiences (*mission*). They define an aspiration of reaching the peak of one of the highest mountains in the world by a certain time in the future (*vision*). Their plan is to accomplish this by climbing a certain route with some base camps (*planned strategy*) based on thorough analysis of their resources and capabilities (the environment including anticipated weather conditions). Note that in case of unforeseen weather changes, they may need to alter their path (*emergent strategy*) driven by the same vision to reach the peak. In this case observers at the bottom of the mountain watching the team climbing with their binoculars see an unplanned strategy pattern. The team members jointly pursue their vision based on self-imposed behavioral rules, such as reaching the peak only together—"all or nobody"—or leaving no garbage behind on the mountain (*values*).

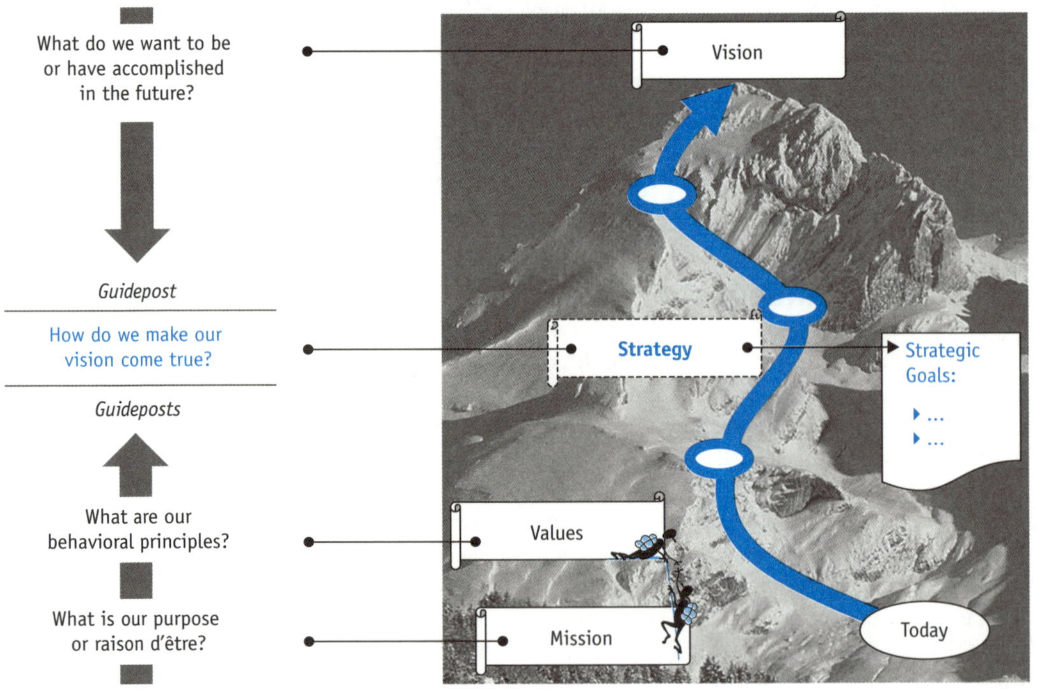

Source: following Horváth & Partners 2007

Fig. 3-1: Illustrative relationship between vision, mission, values, and strategy

"referee" for strategy formulation that defines the scope for strategic ideation (see Strategic Snapshot 3.1).

Precise strategic goals are defined after strategy formulation.

It is important to note that the described understanding of strategic guideposts does not explicitly capture precise **goals** as it is sometimes suggested (Müller-Stewens/Lechner 2011; Wheelen/Hunger 2010). If goals represent foundational key expectations that are supposed to direct strategy formulation they become part of the organization's vision or a broader articulation of the aspired strategic target position. Typically, few corporate or business objectives are set prior to strategy formulation. They often relate to key financial performance metrics such as revenue growth, market share, profitability, or shareholder value (see also chapter 1.1.3). Otherwise, precise strategic goals are derived later on in the strategy process to refine and align strategies once they have been formulated and selected (see chapter 4.3.2.1). Goals are more specific commitments over a specific time period and a result of strategy-oriented planning activities. This is also illustrated in figure 3-1.

Vision, mission, values, and strategy have to be clearly distinguished and articulated.

Unfortunately, vision, mission, values, and strategy have become some of the most overused and confused terms in corporate practice. Many managers will agree that those concepts are important but may have diverse understandings of what these terms actually mean and may communicate them with various meanings. For example, although visions and missions are different concepts and fulfill different purposes, many managers use the terms interchangeably. Sometimes it is surprising, how firms mix up and skip back and forth between messages related to vision, mission, values, strategy, goals, and other concepts in their written strategy documents. The result may cause more confusion than provide orientation. Many companies formally summarize and communicate their strategic guideposts with a corresponding statement that is supposed to provide orientation both internally as well as for external stakeholders. Those articulation statements usually capture the organization's vision, mission and proclaimed values but may also include key strategic goals or other strategic messages the company wants to communicate (Bleicher 2001: 264–266; Lombriser/Abplanalp 2004: 229). A word of caution has to be noted when using the term "mission statement" (see chapter 3.1.2) for articulating not only the mission but the different concepts in a single document (Wheelen/Hunger 2010: 17). Clarity about strategic guideposts requires distinctness regarding the applied terminology. The following chapters are supposed to add clarity and rigor to those trendy terms and provide some prescriptive practical guidance for their development.

3.1.1 Vision

A vision is a picture of what the company aspires to be in the future.

A vision is an imaginative future picture of what the company wants to accomplish in the long run. It is the description of a desired future state of the organization. A corporate vision is often strongly related to the explicit or implicit personal vision of its founder or leader and may survive those persons. Although there is no incontestable proof that companies with a vision are more successful than those without, there are many examples of successful companies applying this concept (Collins/Porras 2002). The key idea of a vision is not to narrow down the scope of action too much. It is about providing direction and sense in a way that unfreezes people's energy toward a shared future aspiration. On the one hand, it needs to be articulated broad enough to provide sufficient scope for formulating strategies and allow managers and employees to mentally integrate their personal ambitions and desires. Ultimately, it portrays an aspirational future state—a hope or dream—that is not yet realized. On the other hand, it needs to be specific enough for providing clarity about the pursued direction (see Strategic Snapshot 3.2) and convey a sense of inspirational concreteness and commitment. A vision creates awareness about a wishful dream that is typically limited in time ("dream with a due date"). If it is articulated too generic, a vision may become disconnected from reality and meaningless. If it is too precise, it might narrow down the scope of strategic ideas and action too much. Although employees are likely to embrace a more specific message about what the company they are working for wants to accomplish in the long run, leaders generally tend to stay more generic as they know they will be measured against their ambition in the future (Henzler 1988, Hinterhuber 2015: 85–99; Müller-Stewens/Lechner 2011: 227). Companies defining their visions are always facing this tension between specificity and generalness.

Six types of vision

There is a variety of ways of how a company may define its vision. Some vision statements are strongly related to a company's financial aspirations whereas others go beyond pure growth or profitability targets and may relate to other aspects such as competitors, role models, customer needs, etc. There are six fundamental types of vision that can be distinguished (following Müller-Stewens/Lechner 2011: 226 f.; Collins/Porras 2002: 232 f.):

- **Goal- or competition-oriented visions** articulate a desired competitive position that is often expressed in terms of financial targets or an aspired market leadership level (e.g., "become the globally leading ... company by the year ..." or "each business to become number one or two in every served market"). An example is the ambitious vision Walmart stated in 1990 when the company had $32 billion in revenues. At this time the founder wanted Walmart to "become a $125 billion company by the year 2000", a number that, in retrospective, was even exceeded by today's largest corporation worldwide (see also Strategy Practice Example 3.1).
- **Enemy- or rival-oriented visions** are directly addressing a particular competitor that the company wants to outpace in the future. This type may be particularly interesting for smaller companies that want to take on a larger and more estab-

Strategic Snapshot 3.2

Vision—Polar Star and Sign of the Zodiac Analogy

The function of a vision can be compared anecdotally to the polar star or signs of the zodiac providing direction to a caravan that is travelling through the desert. Given an ever changing environment caused by sand storms, the travelers may need to alter their route (strategy) from time to time but get reliable orientation by the starry sky. The stars are not the final destination of their journey itself, but they provide clarity about the direction, independent of where the caravan is currently located (current strategic position), their equipment (resources) and their immediate (operational) challenges. Once their destination such as an oasis is about to be reached, the view may need to be redirected toward other constellations of stars that guide a new journey. With this respect, a vision is also finite in time and needs to be advanced and adjusted. (Source: Hinterhuber 2004: 75) lished rival. Examples are "Crush Adidas" (Nike 1960s) or "Knock off R. J. Reynolds as the number one tobacco company in the world" (Phillip Morris, 1950s).

- **Role-oriented visions** emphasize and relate to a role model of an outstanding company that a firm aspires and wants to emulate. Examples are "Become as respected in 20 years as Hewlett-Packard is today" (Watkins-Johnson, 1996) or "Become the Harvard of the West" (Stanford University, 1940s).

Strategy Practice Example 3.1

Vision DB2020: On the way to becoming the world's leading passenger and logistics company

The "DB Group offers globally mobility and logistical services and operates in over 130 countries worldwide. Every day, more than 300,000 employees, of which about 196,000 are located in Germany, are committed to ensuring that customers are provided with mobility and logistical services and that the related rail, road, ocean and air traffic networks are operated and controlled efficiently.
During the 2014 financial year DB Group posted revenues (adjusted) of about € 39.7 billion, and an EBIT of a good € 2.1 billion after adjustments for special items." (DB Group 2015a)

The goal-oriented DB2020 vision is based on a triple bottom line perspective (see fig. 3-2 and chapter 1.1.3.4): "With its realigned strategy, DB is tackling tomorrow's challenges and capitalizing on their potential. The objective of the DB2020 strategy is to create a lasting balance between economic, social and environmental interests. In this way, the company aims to safeguard its long-term commercial success and secure social acceptance. The firm's vision is still to become the world's leading mobility and logistics company. To achieve this, DB has set ambitious targets for all three dimensions: it wants to be a profitable market leader, a top employer and an eco-pioneer by 2020." (DB Group 2015b)

Source: DB Group 2014. Reprinted with permission of Deutsche Bahn AG.

Fig. 3-2: DB2020: vision and strategic directions of the DB Group

- **Transformation-oriented visions** can be often found at long established companies that want to change their strategic identity significantly. This typically requires an internal transformational process. For example, Edzard Reuter, the CEO of Daimler-Benz AG from 1987–1995, was strongly driven by his vision to transform the company from an automotive focused corporation to an "Integrated Technology Group". Another example is the transformation from a public utility company to an integrated infrastructure service provider, which was articulated by some German utility companies after privatization in the 1990s.
- **Customer- or need-oriented visions are** expressing a picture of the future in which the best fulfillment of particular customer needs and wants is aspired. A famous example is the vision of Henry Ford that everybody should be able to afford an automobile ("Democratize the automobile", early 1900s).
- **Organization-oriented visions** are directed toward an aspired internal performance level, such as regarding Six Sigma, failure rates, quality management, or customer orientation. For example: "(...) have the strengths of a big company combined with the leanness and agility of a small company." (General Electric, 1980s)

Benefits of an effective vision

An effectively formulated and implemented vision is supposed to deliver at least four key benefits that are illustrated in figure 3-3. First of all, visions are supposed to provide **orientation** to people in the organization as well as to external stakeholders. Internally, a vision can create a shared understanding of where the company wants to go. The better it frames the organization's desired future development, the more everyone in the organization who is supposed to contribute knows what he or she can do to accomplish this aspiration. If this is in line with the personal aspirations and motivations of people, a vision may provide sense to employees that goes beyond pure job security, remuneration and other factors. Strongly related to this is the second benefit. An inspiring vision may have the effect to create **enthusiasm** for jointly making a desired future state come true. It fosters alignment and drives team spirit to join forces across departmental boundaries for achieving the shared vision. This can be illustrated with the legendary anecdote of John F. Kennedy visiting the NASA headquarters in the early 1960s. Apparently Kennedy interrupted his tour and walked over to a janitor who was holding his mop and asked him about what he is doing here. The janitor responded: "I am helping to put a man on the moon, Sir." Inspired by Kennedy's vision to land a man on the moon by the end of the decade, the janitor seemed to be strongly committed and understood the importance of his contribution as a valuable part of something bigger, the 1962 NASA space team. (Murray 2014: 100) Third, an effectively formulated vision provides **stimulation** for pro-active entrepreneurial behavior at all levels in the organization. This may result in valuable bottoms-up strategic suggestions and emergent strategies that can significantly help the company make the desired picture of its future come true. Fourth, visions create **security,** particularly in turbulent times and highly dynamic markets when strategic and operational plans have to be revised continuously. In such situations, important self-control mechanisms at different levels in the organization require a "secure" internal environment for people to take initiative. They need to feel at ease and have the freedom for learnings, knowing that decentralized strategic decisions and actions are generally in line with a commonly shared and meaningful vision (Hinterhuber 2015: 92–97).

Importance of vision in dynamic and volatile environments

Some managers question the benefit of a vision, particularly in highly dynamic markets and volatile environments. As the strategic circumstances are changing so fast, it is considered hardly possible to define a concrete long-term "destination" that provides reliable orientation over a prolonged period of time. The conclusion of this view is that the future is too hard to predict in such situations and therefore vision statements may become so generic and vague that they do not provide meaningful direction and are thus worthless. Contrary to this view, for a number of scholars the importance of a narrowly articulated vision and mission actually increases particularly in turbulent environments. In their view, a precise vision focuses the organization effectively on what the company wants to become in the future, whereas broad statements might be more appropriate in stable environments where identifying new growth opportunities requires a broader scope. In any case, a vision will set the focus and stretch the organization as it formulates, executes and continuously adapts its strategies (Henzler 1988: 20; Hinterhuber 2015: 95 f.; Lynch 2012: 221; Wheelen/Hunger 2010: 17).

Principles for developing a vision

For developing an effective vision it is suggested to follow four key principles (see fig. 3-3). A vision development should be (following Hinterhuber 2015: 88 f.)

- **open-minded** regarding changes in the external environment which includes the pro-active search for different strategic perspectives and opinions that may chal-

Fig. 3-3

Benefits of a vision and principles for its development

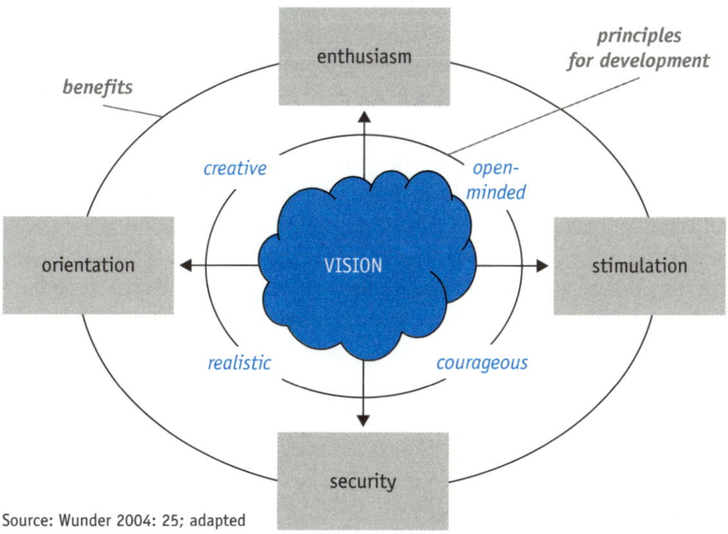

Source: Wunder 2004: 25; adapted

lenge the status quo as well as the ability for anticipative thinking and for empathizing with other views

- **courageous** in terms of the willingness to define a vision even if it breaks conservative thinking and attempts to "change the rules" in case the learning from being open-minded provides ideas for significant strategic adjustments
- **realistic** regarding its basic achievability in a certain time frame, which requires a good sense of reality. Utopian visions (wishful thinking) are unlikely to deliver the expected benefits such as enthusiasm and stimulation due to a lack of credibility, and may even harm a company.
- **creative** in terms of setting and articulating an aspirational "dream" that is open-minded, courageous and still realistic. It is about *creating* a desired picture of the firm's future, which requires creativity, and not about predicting one which might be done more analytically.

A vision may create great skepticism among outsiders and, at the same time, fuel enthusiasm among people inside the organization. Some companies formulate creative visions that are just on the boundary between realism and utopia, which means they are considered highly challenging but still not impossible (Drosdek 1998: 21 f.; Helm/Meiler 2003: 202).

3.1.2 Mission

Raison d'être of a company in terms of its value proposition for stakeholders

A mission articulates the purpose or reason of the company's existence, its raison d'être. Whereas a vision is the aspired picture of the future, a mission is more immediate, relates to the firm's present purpose and typically has no due date. It provides **identification** to employees as it defines the purpose of the business, and thus, also the greater purpose of the individual work of members in the organization. Before strategies are formulated and a vision is stated, it makes sense for a company to stand back and consider or even refine why it actually exists. This is a challenging undertaking as it usually involves not only an economic rationale based on customer value but may also include its general role and responsibility regarding employees, the society and community as well as other stakeholders (see Strategy Practice Example 3.2 and chapter 1.1.3.4). In this sense, the mission declares the purpose of a company in terms of the value it creates for selected stakeholder groups, and thus, **legitimizes** its existence in society (Müller-Stewens/Lechner 2011: 227–229).

The mission defines boundaries for focused strategic ideation.

A clearly articulated mission frames the scope for strategy formulation in a way that strategy options need to fit the company's purpose. With this respect, it emphasizes the products or markets the company focuses on strategically and its contribution to stakeholders. This creates also clarity about what is strategically *not* desired (see also chapter 1.1.2.3). Those boundaries may help to focus attention and energy as well as direct creativity on developing ideas and options that fit into the desired strategic scope. For example, if the mission of a global healthcare corporation is to provide "animal health" with corresponding pharmaceutical and biotechnological solutions, strategy options for expanding the market into new areas outside "health",

Strategy Practice Example 3.2

Redefining Business Purpose at PUMA

The early 1990s were a difficult time for the German sporting goods company PUMA, which was founded back in 1948 by Rudolf Dassler. Jochen Zeitz was appointed CEO of PUMA in 1993 when the company faced severe financial difficulties along with brand and price deterioration. At the age of 30, he was the youngest chairman in German history to lead a publicly traded company at that time. He spearheaded the restructuring of PUMA, including the redefinition of its business and brand purpose from selling primarily performance and functional sporting shoes to bringing lifestyle and fashion to sports. This was a major break with the traditional business and brand paradigm at that time. Later on, the mission was refined again into being the most desirable *and* sustainable sport lifestyle company. The idea was to be more than just a "responsible" company but capture sustainability innovation in an entrepreneurial way. This went along with a new PUMA vision of a better world that is more peaceful, safe, and more creative. (Source: Puma SE 2015; Zeitz 2014)

such as pet toys or pet food, are excluded from the strategy scope. If the leadership team wants to open the scope of strategies beyond this area, it would need to revise its mission. In such cases, it either broadens the statement to include new areas outside its mission prior to strategy formulation or signals openness and revises its mission later on if promising new strategic areas have been found.

Product-oriented versus customer-oriented missions

A business purpose is not just a description of the company's existing product or service lines and customer segments. Products and services as well as the targeted customer base might not exist anymore in a few years time due to changes in the external environment leading to corresponding strategy adaptations in the company. For identifying and articulating an enduring purpose it may be more appropriate to disconnect the firm's mission from the currently offered solutions and focus stronger on the fulfilled customer needs and wants (Gälweiler 2005: 254 f.). Whereas product-oriented missions may help a company to (re)focus on its core business, customer-oriented missions tend to provide greater flexibility and more strategic options when adapting to a changing environment. Companies that define the purpose of their business aimed at their customers are leaving the "how" a customer's needs and wants will be met open. Given an unknown future, they allow their organization to think about new and innovative ways for fulfilling customer needs and wants that may not have been thought of today (Rothaermel 2013: 34–36). Figure 3-4 provides some examples of product-oriented versus need-oriented missions.

Mission statements and claims

Many companies articulate their purpose in a brief, easy to understand and convincing mission statement, such as the one from SAP that states "to help every customer become a best-run business" (SAP 2015) or from Google saying "to organize the world's information and make it universally accessible and useful" (Google 2015). Some further examples are (Collins/Porras 2002: 225)

- "to make technical contributions for the advancement and welfare of humanity" (Hewlett-Packard)
- "to give unlimited opportunity to woman" (Mary Kay)

Fig. 3-4

Product-oriented versus customer-oriented missions

Company Type	Product-oriented Mission	Customer-oriented Mission
Railroad Company:	to run a railroad	to provide mobility (people & goods)
Toy Company:	to offer plastic toys	to raise children in a playful way
Motion Picture Studio:	to make movies	to entertain people
Metal Key Producer:	to offer metal keys	to provide security
Printer Manufacturer:	to help people print	to make office work more efficient
Petrochemical Company:	to offer gasoline	to deliver energy
Book-publishing House:	to sell books and magazines	to inform and educate
Soft Drink Producer:	to provide soft drinks	to refresh people
Life Sciences Company:	to provide cancer treatment	to save lives

- "to help leading corporations and governments be more successful" (McKinsey)
- "to preserve and improve human life" (Merck)
- "to make people happy" (Walt Disney).

Looking at those mission statements, a similarity to the concept of brand claims can be noted. In advertisement the term *claim* refers to a slogan—a full or partial sentence—that is often an integrated element of the company logo or brand logo and articulates the brand promise or value proposition (Adjouri 2014: 64). Some famous current or past brand claims that can also be seen as an indicator for the company's purpose are:

- Apple: "Think different."
- Audi: "Vorsprung durch Technik" (ahead by technology)
- BAYER: "Science For A Better Life" (see Strategy Practice Example 3.3)
- BMW: "Freude am Fahren" (pleasure in driving) or "The Ultimate Driving Machine"
- Coca-Cola: "Open Happiness."
- General Electric: "GE imagination at work"
- Henkel: "Excellence is our Passion" (see also Strategy Practice Example 3.4)
- Mercedes: "Das Beste oder nichts" (the best or nothing)
- Nestlé: "Good Food, Good Life"
- Nokia: "Connecting People"
- Volkswagen: "Das Auto."

Strategy Practice Example 3.3

Mission and Claim of BAYER

The Bayer Corporation articulates its purpose with the single statement or claim

"Bayer: Science For A Better Life"

"Bayer is a world-class innovation company with a 150-year history. Our scientific successes are intended to help improve people's lives. At the same time, our innovations form the basis for sustainable and profitable business activity and are the key to maintaining or achieving leadership positions in all of our markets. Our products are helping to address some of today's biggest challenges, including global population growth, an aging society and the need to make efficient—and, wherever possible, sustainable—use of natural resources." (Source: Bayer AG 2015)
The Bayer mission is based on the two pillars innovation ("Science") and customer value ("Better Life"), which is illustrated in figure 3-5.

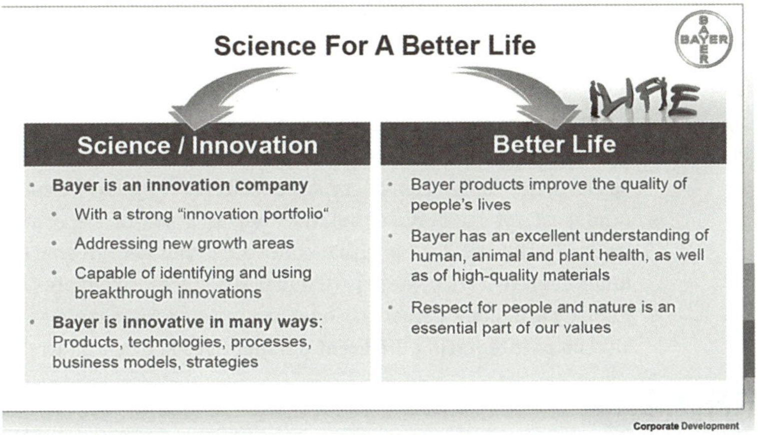

Source: Bayer 2013. Dr. Alexander Moscho, Head of Corporate Development, Bayer AG. Reprinted with permission.

Fig. 3-5: Innovation and customer value are major components of the BAYER mission statement

3.1.3 Values

Few essential and timeless behavioral principles for interaction

Organizational values are essential and enduring standards and norms of social behavior that are collectively held by members inside the company. Openly proclaimed values define the mode of conduct that is considered essential, valuable and preferred when interacting with each other and with external stakeholders. Although each person has his or her own set of values, proclaimed organization values are supposed to govern behavior and day-to-day decisions of individuals within their specific firm and are thus an important guideline when executing strategy in the quest for achieving the vision. Defining its core values is a highly company-specific process as each firm decides for itself which behavior is considered desirable and which is not. There is no such thing as a universally "right" set of values when it comes to defining the few—up to five or six—behavioral principles that are supposed to be applicable for all situations and enduring even in case circumstances change (see Strategic Snapshot 3.3). Although some values might look similar from one firm to another, there are usually also clear differences of what an organization emphasizes in terms of ethical standards, respect for the individual, teamwork, quality, agility and responsiveness, customer service, or ecological as well as social conduct (Collins/Porras 2002: 222 f.; Müller-Stewens/Lechner 2011: 23–235).

Benefits of proclaimed values

Defining and articulating core values is supposed to deliver a number of benefits (Müller-Stewens/Lechner 2011: 233–235):

1. Proclaimed organizational values are supposed to provide **clarity** about the core behavioral principles the organization stands for. This organizational "world view" is foundational for the firm's mission and supposed to guide the behavior when strategies are executed and visions are pursued.
2. Published organizational values provide employees reliability and predictability for what is considered good or bad behavior; everybody can use it as a **reference point** for one's own or other's practices.
3. A gap between proclaimed and lived values can be seen as an important starting point for **organizational design and development**. Values are supposed to have a positive impact on the long-term economic and social performance of a company as they are a foundation of corporate culture. They may be considered not just as guidepost for strategy formulation but as a source for competitive advantage itself. Systematically managing core values can help to shape corporate culture and integrate it with strategy (Glauner 2013; see also chapter 4.2.1).
4. Publicly articulated organizational value can be seen as a managerial control tool that helps **integrating different perspectives** and creating a shared identity in an

Strategic Snapshot 3.3

Identifying Core Values—an Anecdote from an Executive Leadership Team

"After you've drafted a preliminary list of the core values, ask about each one: 'If the circumstances changed and *penalized* us for holding this core value, would we still keep it?' If you can't honestly answer yes, then it's not *core* and *should be dropped*. For example, a high technology company we worked with wondered whether it should put 'quality' on its list of core values. The CEO asked: 'Suppose in ten years quality doesn't make a hoot of difference in our markets. Suppose the only thing that matters is sheer speed and horsepower but not quality. Would we still want to put quality on our list of core values?' The members of the management team looked around at each other and finally said: 'To be honest, no.' Quality stayed *off* the list as a core value. Quality stayed in the current strategy of the company—and quality improvement programs remained in place as a mechanism for stimulating progress—but it did not make the list of core values. Remember, strategies change as market conditions change, but core values remain intact in a visionary company. This same group of executives then wrestled with whether it should put 'leading-edge innovation' on its list of core values. The CEO asked the same question: 'Would we keep it on the list as a core value, no matter how the world around us changes?' This time, the management team gave a resounding, 'Yes! We always want to do leading-edge innovation. That's who we are. It's really important to us, and always will be. No matter what. And if our current markets don't value it, we will *find* markets that value it.' Leading-edge innovation went *on* the list of core values, and will stay there forever. A company should not change its core values in response to market changes; rather, it should change markets—if necessary—in order to remain true to its core values." (Source: Collins/Porras 2002: 222 f.)

organization, which may be particularly relevant for highly diverse and large multinational enterprises. Furthermore, clarity about core values is the foundation for aligning company priorities with perspectives of different stakeholder groups.

Articulation of proclaimed versus lived values

Companies proclaim their values in various ways. Some firms simply list and explain their core values in a corresponding statement that is often made not only internally but publicly available (see Strategy Practice Example 3.4). Other companies state their organizational values more in terms of behavioral principles. They define precisely expected behavior and attitudes for individuals in a refined way that allows an

Strategy Practice Example 3.4

Claim and Values of HENKEL

"Henkel was founded in 1876. People around the world trust Henkel's brands and technologies. In the fiscal year 2014, the company reported sales of 16.4 billion euros and an operating profit of 2.6 billion euros (adjusted for one-time gains/charges and restructuring charges). The DAX-30 company has its headquarters in Düsseldorf, Germany. Henkel employs almost 50,000 people worldwide, more than 80 percent of whom work outside of Germany. Henkel is thus one of the most international German companies." (Henkel 2015a)

Its claim **"Henkel—Excellence is our Passion"** and values (see fig. 3-6) are supposed to unite its employees around the world. "Our claim expresses our ambition: striving for the very best in everything that we do. This applies not only to the products and technologies that we develop for our customers but also to our internal processes. The goal to take leading positions in a highly competitive environment is shared by everyone at Henkel, irrespective of the business unit, function or country in which they work. And our employees will always put their passion, pride and enthusiasm into making this happen.

Source: Henkel 2013. Reprinted with permission of Henkel AG & Co. KGaA.

Fig. 3-6: Values and claim of Henkel

Conveying this attitude to our internal and external audiences is the task of our claim." (Henkel 2015b)
"Our values speak for themselves: They epitomize Henkel, because they reflect our corporate culture. These values are consistently lived, not least in the development of new products.

- We put our **customers** at the center of what we do. We anticipate, respond to and meet our customers' and consumers' expectations by providing the best value, quality, and most innovative brands and technologies. (...)
- We value, challenge and reward our **people.** We treat each other with respect and dignity and develop our capabilities. We expect everyone to take personal responsibility and perform to high standards. We rely on each other for our success as a company. (...)
- We drive excellent sustainable **financial performance.** We are a performance-driven company committed to growing the value of our business and providing a competitive return to our shareholders. (...)
- We are committed to leadership in **sustainability.** We provide products, technologies and processes that meet the highest standards. We are committed to the safety and health of our employees, the protection of the environment and the quality of life in the communities in which we operate. (...)
- We build our future on our **family business** foundation. We value the continuity of our purpose and vision based on our long history of success and a strong focus on our values. We are guided by our long-term vision which rests on a fair entrepreneurial spirit and a solid financial basis. (...)" (Source: Henkel 2015c)

examination whether actual behavior is compliant with those guidelines. The most binding format of articulating core values are so-called *codes of conduct,* which can be found in a wide variety of types in many companies today. They represent a voluntary self-imposed behavioral commitment that a company will comply with when dealing with certain or all stakeholder groups (Müller-Stewens/Lechner 2011: 235 f.). Independent of the type of value statement, there is always a risk that proclaimed values might not be expressed in the actual nonverbal behavior of individuals in a firm. Therefore, it is important that values are practiced and lived with integrity by top executives that tend to be seen as role models by employees in an organization. If there is no clear commitment of those leaders, value statements may easily become a meaningless public relations exercise that has no impact on behavioral practices when people try to accomplish their defined goals (Rothaermel 2013: 38).

3.2 Corporate Strategy

Corporate-level strategies versus business-level strategies

Once the strategic guideposts have been set, the next stage is to determine what strategic alternatives are available for selection both at a corporate level and at each individual business in case of a multi-business firm. Strategy formulation on both the corporate- and businesses level is a challenging task as it requires an analytical approach leveraging the results from strategic analysis as well as a highly creative approach for strategic ideation. On a business level, the strategic focus is on how to create competitive advantage for a single business in its competitive arena (i. e.

industry or product/market-combination) by providing greater customer value than competitors (see chapter 3.4). In case the company is operating or plans to operate in more than just one business, another strategy perspective needs to be added: corporate strategy. For multi-business companies the strategic focus is on creating parenting advantage by managing the different businesses in a way that has a positive impact on them. It is related to the overall orientation and approach of a company to strategically manage businesses operating in various environments. In addition, several business strategies are required on the business-level of those corporations. Figure 3-7 summarizes the key differences between corporate and business strategy.

Headquarters, divisions and strategic business units (SBU)

On a corporate level, strategy formulation is concerned with strategy options that are of interest to the corporation as a whole in case it operates with businesses in several industries or markets simultaneously. Usually these different businesses are addressed internally by specific organizational units such as divisions or strategic business units (SBU). They focus their activities on specific industries or product-/market combinations which they serve either with a single product or one or more product lines. A strategic business unit may cover only one single business area or segments or several ones that are related. In large corporations, several levels of those units can often be found ranging from divisions covering various strategic business units that cover various business areas and so forth down to a product line. From a corporate perspective, a SBU fulfills an individual market function (i. e. addressing a specific competitive arena such as an industry or segment) that may not be related to another SBU. Typically the SBU has its own SBU CEO and executive leadership team as well as employees. A SBU addresses its own customers, competitors and usually formulates and implements its own distinct strategic plan (see also chapter 2.1). It can be closely related to the idea of a "company within a company". In a corporation, all divisions and SBUs are under the overall leadership of the corporate

Fig. 3-7

Corporate strategy versus business strategy

	Corporate Strategy	Business Strategy
Scope	Multiple businesses	Single business
Goal	Parenting Advantage	Competitive Advantage
Advantage Creation	Corporate management has positive impact on businesses	Business management creates customer value proposition
Comparison	Other corporations as potential owners of businesses	Businesses of other companies in the same competitive arena
Primary Perspective	Owner/investor	Customer

Source: Hungenberg 2011: 410; adapted

headquarters (HQ) with corporate managers who typically facilitate and drive corporate strategy (Hungenberg 2011: 399–402).

Corporate strategy: parenting advantage and scope of businesses

The corporation with its corporate management needs to determine how it generally plans to create value among its businesses and what scope of businesses it wants to operate in. This can be seen as the main goals of strategic management at a corporate level. First, corporate level strategies should promote business level strategies in a way that they enable the different businesses to be more successful in terms of cost, differentiation or other factors than in case they were acting alone. Second, the corporate level needs to address the scope and structure of the company's business portfolio (Goold/Campbell/Alexander 1994: 5; Hill/Jones 2010: 285 f.; Johnson/Whittington/Scholes 2012: 134).

Strategy Practice Example 3.5

Corporate Strategy Elements Illustrated with the BAYER Group Strategy

Figure 3-8 provides a corporate strategy example of the Bayer group from 2013. The corporation articulates a variety of strategic approaches that it wants to leverage as a "parent" to create value with its different businesses: human, animal and plant health. It includes *corporate parenting* activities such as exploiting synergy potential between businesses, increased cooperation and licensing activities, driving economies of scope with the Bayer brand as well as productivity-related strategies. Furthermore, the group strategy also addresses the *scope of business* related to the three subgroups at that time, which includes targeted acquisitions as well as the realization of growth market opportunities. Being a multinational corporation, some of those articulated strategies will naturally relate to the company's *international strategy*.

Bayer Group strategy

Portfolio	Growth	Productivity
Strengthen Life Sciences: ▸ Exploit synergy potential of human, animal & plant health ▸ Maximize value of our product pipeline ▸ Targeted acquisitions ▸ Intensify cooperation and licensing activities Optimize MaterialScience: Further adjust business processes to continuous commoditization	▸ Invest in innovation capabilities ▸ Leverage Bayer brand more ▸ Realize growth market opportunities	▸ Reduce complexity – simplify structures and processes ▸ Continuous efficiency improvement

Source: Bayer 2013. Dr. Alexander Moscho, Head of Corporate Development, Bayer AG. Reprinted with permission.

Fig. 3-8: Bayer Group strategy 2013

These types of questions would naturally not be addressed by individual business units, which are concerned primarily with optimizing their own business strategy. It needs the corporation to take this view and formulate effective strategies (see Strategy Practice Example 3.5). Hereby, corporate strategy should provide answers to three main areas:

- the fundamental mechanisms the company plans to leverage for creating value in the interaction of its different business units (corporate parenting)
- the current and future portfolio of businesses and markets where the company operates or would like to operate (scope of business)
- the corporation's approach to internationalization both in terms of whether, where and how to go international and how to leverage its "internationality" in case of a multinational corporation (international strategy).

3.2.1 Corporate Parenting

Creating parenting advantage

In line with the famous view of Aristotle (384–322 BC), that the whole is greater than the sum of its parts, corporate headquarters (the "parent") have to ensure that the businesses under its stewardship are collectively more successful than if they were acting individually. Assuming that the company is managed with the goal to increase its value (see chapter 1.1.3.3), as it is typically the case with publicly traded multi-business corporations, the principle of corporate parenting can be illustrated as shown in figure 3-9. The corporation's entire value can be conceptually seen as the sum of individual strategic business unit values and as the value contribution of the corporation or corporate headquarters ("parent"). Hereby, it is assumed that a SBU is able to operate independently from other SBUs ("stand-alone value"). The corporate contribution is reflected in the positive and negative cash-relevant effects of having different SBUs under stewardship in one corporation. Whereas increasing the value of

Fig. 3-9

Illustration of parenting advantage concept

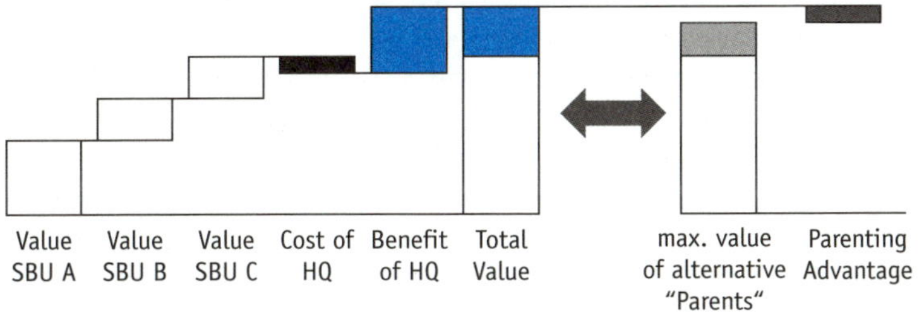

SBU = Strategic Business Unit
HQ = Headquarters (Corporate Management)

Source: Hungenberg 2011: 407; adapted

a SBU is a primary goal of business-level management and strategies, increasing the value contribution of the corporate headquarters ("parent") is the primary goal of corporate-level management and strategies. Ideally, the parent should not only create value more than offsets its costs but a greater value than the best alternative parent or competitor could do assuming it had the same strategic business units or portfolio of businesses under its stewardship. This so-called *parenting advantage* relates to corporate strategy in a similar way as competitive advantage relates to business strategy. The corporation's aspiration should be the best possible parent for its various businesses. Based on this principle of corporate-level strategy, corporate management would need to evaluate for each of its individual businesses whether it has a parenting advantage (Goold/Campbell/Alexander 1994: 7 f.; Hungenberg 2011: 399–408).

Value-adding parental strategies

The conceptual illustration of the principle underlying corporate parenting in figure 3-9 is not intended to provide an analytical model for calculating the different value creations. In company practice it is particularly difficult to determine the headquarters' value contribution because its benefit is hard to quantify. Though, in publicly traded companies, the difference between the corporation's market value and the stand-alone values of its strategic business units may provide an analytical orientation (Hungenberg 2011: 408). Achieving parenting advantage may appear particularly relevant in companies with related diversification (see chapter 3.2.2.2) where there are opportunities for realizing synergies within the portfolio of businesses. However, even in pure financial holdings (for different types of holding companies see Strategic Snapshot 4.5 in chapter 4.2.2.1) or highly diversified conglomerates, the headquarters has opportunities to add value through, for example, central financial management with more cost-effective or easier capital raising or transferring excellent managerial systems and best practices to all of its diverse businesses. Four key approaches of how the "parent" may add value are illustrated in figure 3-10. Although there is some overlap between those strategic approaches, they provide a useful taxonomy for understanding different roles of the "parent" for adding value in its business portfolio (Campbell/Goold 1997: 172–175; Goold/Campbell/Alexander 1994: 78–82; Hungenberg 2011: 405; Johnson/Whittington/Scholes 2012: 145–150).

- **Stand-alone influence** describes corporate activities that are supposed to enhance the performance of a strategic business unit viewed as a stand-alone profit center. The many ways of how the "parent" may influence it's singly businesses directly and in an isolated way vary regarding the degree of corporate involvement into strategic or operational activities of a business. For example, the headquarters may act as a portfolio manager from the perspective of an active investor and articulate only a desired aspiration with some financial targets to be accomplished by the SBU. Or it provides strategic orientation with strategic guideposts that are valid for all SBUs and are supposed to create commitment to a common purpose. Furthermore, it might exercise influence on a wider range by selecting executives for the SBU or imposing certain standards or best practices regarding specific business process in the SBU, for example, creating "a company way of doing things" for certain operations for all business units.
- **Linkage influence** describes an approach where the corporation creates synergistic value through facilitation of cooperation and sharing across the different busi-

Fig. 3-10

Four types of parental value creation

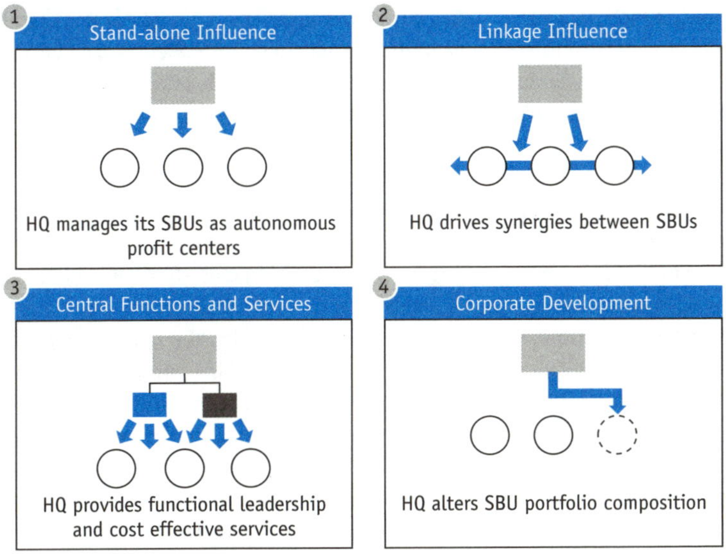

Source: Goold/Campbell/Alexander 1994: 79; adapted

ness units. The resulting synergies are a key for success for many multi-business corporations. Benefits are gained through leveraging complementary activities or assets of different businesses. Hereby, the value-adding parental strategies are influencing at least two SBUs. Compared to the more vertical and direct stand-alone influence, corporate value contribution with linkage influence is horizontally oriented and more indirect. For example, in its role as a synergy manager the headquarters may facilitate learning process and best practice sharing between different strategic business units for innovation, and thus, enhance the overall performance level of the business portfolio. The parent may also encourage or even demand from business unit leaders to invest in certain areas leveraging core skills developed in other strategic business units such as manufacturing systems, technologies or marketing capabilities. Hereby, the headquarters stimulates or imposes the utilization of know-how it had not directed to develop centrally, which is the case in the next value-adding strategy.

- **Central functions and services** are a way for corporate management to support the various strategic business units leveraging both *economies of scale* and *economies of scope*. Corresponding activities and capabilities may not be managed at all in the strategic business units but are provided more effectively and efficiently by central organizational units developed by the parent. For example, efficiency improvements and high-quality functional services may be provided by establishing *shared service centers* in areas such as Finance (e. g., centrally managed repetitive Finance processes like payroll, standard accounting, reporting, etc.) and HR (e. g., cross-SBU recruitment and personnel development programs). Furthermore, a corporate *operational excellence team* may support operational processes in various SBUs to gain efficiency improvements leveraging project management skills or expertise in Six Sigma or lean management, etc. Centralized purchasing of material or parts that are required in different SBUs may be used to increase negotiation power with suppliers and prices through consolidated purchase quantities. Furthermore, economies of scope can be realized in R&D with, for example, centralized developed technology platforms or materials that can be leveraged in various SBUs. Other examples for realizing scope economies are centralized Marketing (e. g., creating a powerful umbrella brand for products in all SBUs) or customer management (e. g., establishing cross-SBU global key account management).

- **Corporate development** is targeted toward altering the composition of businesses in the corporation's portfolio. Approaches for corporate development include acquisitions, divestitures, and alliances. It may also develop new strategic business units organically by, for example, combining core competencies from different other strategic business units. Furthermore, the corporation may redefine an existing SBU, for example, by integrating new resources and processes after a merger or acquisition of another company. Whereas the other three parental value-creating strategies are based on influencing businesses in an existing portfolio, corporate development may result in a very different set or redefinition of strategic business units.

Value-destroying activities

Although there are many ways in which corporate parents may add value to the portfolio of businesses in a corporation, it may also destroy value if managed in the wrong way. There are three main pitfalls of corporate parenting (Goold/Campbell/Alexander 1994; Johnson/Whittington/Scholes 2012: 146 f.): First, if the *corporate headquarters costs* are higher than the benefit it provides to its businesses, it destroys value. This is a most basic criticism frequently articulated in operating units that generate the revenues and profits with which highly paid top managers and staff functions as well as expensive facilities at the corporate center are financed. Second, in addition to those financial costs, the corporation's actual procedures, regulations and processes to influence its businesses in the quest for parenting advantage may cause a value-destroying *bureaucratic complexity*. It may slow down business manager's responses to customers or new competitive situations, and thus, negatively impact the businesses in developing or adapting their value proposition. This can result in missed entrepreneurial opportunities, compromises and tradeoffs, as well as frustration in strategic business units that are not able to fully exploit their business potential. Third, there is also a danger that strategic business units are not performing to their fullest potential because their competitive dynamics is backed with a kind of "safety-net" provided by the parent. Particularly in large and highly diversified multi-business corporations, weak performance of a business unit may be cross-subsidized by stronger units and are thus not transparent for shareholders. Corporate parents *obscuring financial performance* provide comfort zones for their businesses that undermine market mechanisms and can easily destroy value.

Creating parting advantage in multinational corporations

There are specific opportunities for leveraging value-adding parental strategies particularly in multinational corporations. They are related to the configuration, coordination and integration of cross-border value chain activities. Hereby, the corporations "internationality" itself can be a strong source for creating parenting advantage. This particular perspective on parenting advantage will be explained in chapter 3.3.2.

3.2.2 Scope of Business

Determining where to compete: businesses, value chain stages, regions

The question which parental approach a corporation wants to pursue for adding value to its business portfolio is strongly related to the question of the scope of business. Contrary to business-level strategy where the firm has to decide *how* to be successful in a single business, corporate-level strategy is concerned with *where* it wants to compete. It has to drive competitive advantage in several businesses simultaneously. The corporation has to determine its overall business scope or, in other words, has to decide which specific businesses or competitive arenas it wants to compete in. Hereby, three fundamental questions need to be answered (Rothaermal 2013: 202 f.):

- What range of businesses or industries should the corporation compete in (diversification)?
- What stages of corresponding industry value chains should it cover (vertical integration)?
- What international activities should it engage in (international strategy)?

Naturally, the answers to those questions strongly depend on the corporation's general strategic orientation toward growth. This strategic direction has to be clarified first, which will be explained in the following (Wheelen/Hunger 2010: 206 f.).

3.2.2.1 General Directional Strategies

As a first step for defining its scope of business, a corporation needs to clarify its general strategic orientation toward growth. There are three basic strategic directions to follow (Wheelen/Hunger 2010: 206–221):

- **Growth** is by far the most widely pursued strategic orientation. Corresponding strategies are often designed to achieve *top-line growth* (sales), *bottom-line growth* (profits), or a combination of both (profitable growth). They may also relate to a *growth of assets,* which is often required to accomplish the other two goals. Particularly in industries that are growing quickly or that are characterized by competitor consolidation (i. e. mergers and acquisitions that result in large and powerful rivals) firms may need to reach a "critical mass" to survive. This means that they have to grow to realize economies of scale vital for achieving competitive cost positions. Corporate growth may be driven on a business level by leveraging growth strategies within the existing business scope (intensive growth), which will be explained in chapter 3.4.1. Furthermore, the corporation may expand its business coverage along industry value chains both vertically and horizontally (integrative growth) or diversify into entirely new businesses (diversification growth). Those options are illustrated in figure 3-11 assuming a company has to close a growth gap. The gap results from the difference between a projected baseline growth (i. e. growth with no or only minor changes to current business operations) and a corporate five year growth aspiration of, for example, the shareholders or senior leaders (Kotler/Keller 2009: 44 f.). Another way of achieving growth is by expanding the firm's geographic market scope (e. g., going international), which can be applied to existing businesses (see chapter 3.3.1) or to enter into new businesses.

- **Stability** is an orientation where a company continues its current operations without any significant change. It is popular with many small business owners that are happy with their success and manageable size serving a niche. For large corporations choosing stability over growth is typically a deliberate temporary approach to overcome a current situation in the firm's environment such as an economic crisis or to digest a series of acquisitions in a non-crisis period. They may follow a pause ("wait and see") strategy to decide for growth or retrenchment later on or pursue a profit/cash-flow improvement strategy to get through a temporary difficult situation. For example, many large companies pursued stability strategies by strictly managing short-term discretionary expenditures or investments and

Fig. 3-11

Growth strategies for closing an identified growth gap

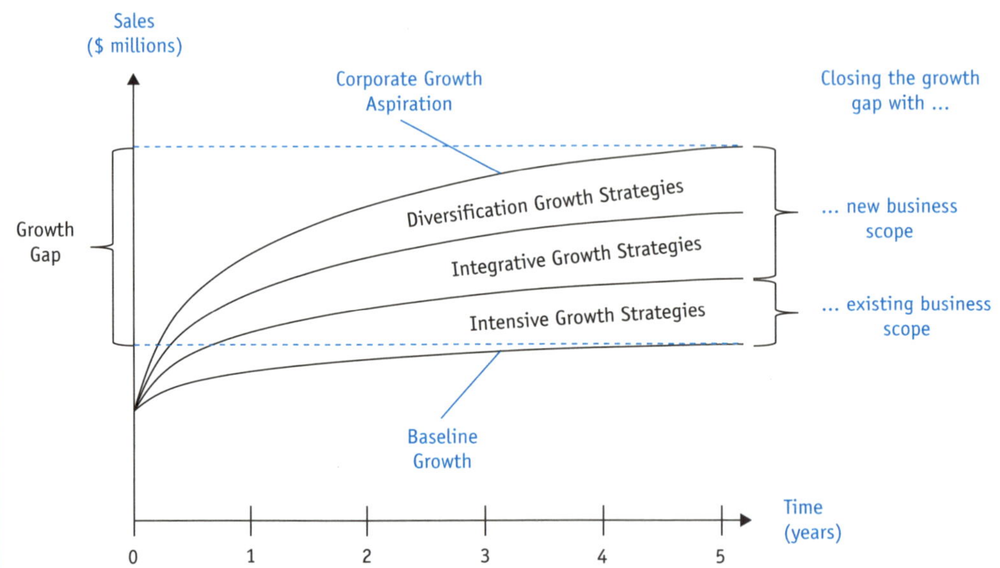

Source: Kotler/Keller 2009: 45; adapted

enhanced working capital management to get through the global financial and economic crisis within the 2008–2010 time frame (Wunder/Müller 2009).

- **Retrenchment** is a strategic approach where a corporation decides to cut down its business activities. Reasons may be poor performance or—even in case of significant sales and profits—an unreachable strong competitive position with some businesses that may be aspired by corporations that strive for market leadership. Furthermore, a business owner may decide for retrenchment in situations where the options are either growing to reach a critical mass or cutting down and focusing on a niche with a manageable size providing satisfactory profits. Retrenchment strategies range from *turnaround* (e. g., selling assets and cutting down cost by means of improving operational efficiency), becoming a *captive company* (e. g., selling production capacities to a single customer), *divestment* of certain divisions and businesses or *sell-out* of the entire company as well as *bankruptcy*—which still seeks to perpetuate a corporation—or *liquidation*, that is the termination of the firm. There is also a popular notion that due to dissynergies in a corporate parent some businesses may perform better and compete for capital and business investments on their own, which can results in a *spinoff strategy*.

Generic strategies applied to individual businesses in the portfolio

The three general directions toward growth of the entire corporation can also be applied to individual businesses or product lines in its portfolio. Hereby, various portfolio frameworks provide a way to visualize a corporation's entire business portfolio or portfolio of product lines in a single picture (see chapter 2.2.4). Portfolio methodologies typically suggest generic strategies for each business depending on its corresponding position in the matrix (see Strategic Snapshot 3.4), which can be related to the basic directions of growth, stability and retrenchment. First, a growth approach relates the generic strategy of investing into a business or product line for increasing its market share. Second, stability can be related to a generic business strategy with low investments for growth (maintaining overall position) but seeking to generate net cash flow, for example, for supporting other businesses in the portfolio in strengthening their market position. Third, businesses or product lines that the company wants to harvest, exit and divest reflect a retrenchment approach.

3.2.2.2 Diversification Strategies

Defining the scope of a corporation's business portfolio

Diversification is related to the question of how many different business areas a corporation gets involved in. It is reflected in the number and variety of distinct industries a company competes in with its products and services. This understanding of diversification relates to a multi-business model and does not include adjacent stages of an industry value chain covered by a firm, which will be referred to as vertical integration (see chapter 3.2.2.3). Whereas a highly diversified company generates its revenues from a number of different businesses that are either related or unrelated to the company's core business, a focused company gets its revenues from a single business or few dominant businesses. Furthermore, a company may also decide to diversify into different countries, which will be discussed in chapter 3.3.1. There are two fundamental types of diversification—related (concentric) and unrelated (conglomerate)—that can be realized in different ways, as explained in the following (Hill/Jones 2010: 313–328).

Strategic Snapshot 3.4

Generic Strategies Derived from Portfolio Analysis

Portfolio analysis provides a methodology that suggests normative strategies regarding the company's resource allocation. It turns the corporation in the perspective of an investor seeking to optimize its risk/return preferences based on a medium to long-term orientation. The positioning of each business or product line in the portfolio matrix (see chapter 2.4.2) is associated with a normative strategy suggestion that matches its competitive strengths (e. g., in terms of relative market share or based on multiple criteria) and the degree of attractiveness of its corresponding market (e. g., in terms of market growth or other criteria). These generic strategies can be used as initial reflection when formulating a specific strategy for each business unit or product line as well as an overall strategy for the portfolio. It will be evident for a corporation to invest in businesses that have a strong competitive position in an attractive market. Conversely, where a businesses' competitive position is weak in a market with low attractiveness, a corporation should consider divesting. Other positions require other solutions with sometimes more differentiated investigations. In addition to generic strategies for each business or product line, portfolio models also provide some guidelines regarding the entire portfolio to ensure an effective cash generation and allocation. Normative strategic suggestions regarding the BCG-portfolio are, for example, that a corporation's revenues should be composed with at least 10 % from question marks, 30 % from stars, 40–50 % from cash cows, and not more than 10–20 % from poor dogs (Perlitz/Schrank 2013: 218). The initial strategic suggestions for allocating resources need to be supplemented with a reflection of interdependencies between the businesses. For example, companies divesting in a certain business or product line that is part of a product package which provides significant customer value may face a negative sales impact on the remaining businesses or products that were part of the bundles. (Source: Hax/Majluf 1996: 304 f.; Hinterhuber 2015: 192–202; Müller-Stewens/Lechner 2011: 285–289)

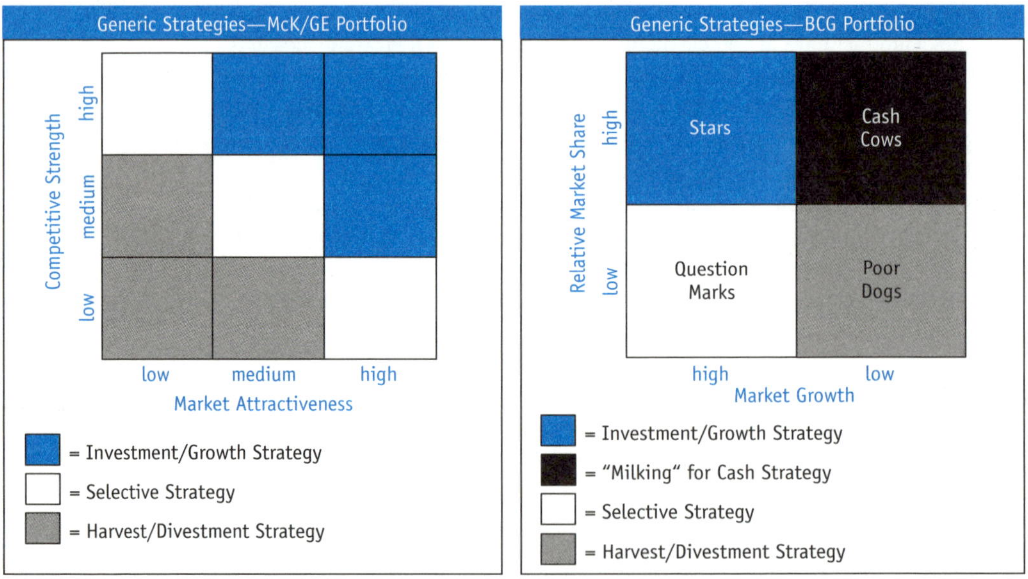

Source: following Hax/Majluf 1996: 305 and Müller-Stewens/Lechner 2011: 285; adapted

Fig. 3-12: Generic portfolio strategies

Strategy Practice Example 3.6

Related Diversification at Procter & Gamble (P&G)

Disposable diapers, toilet paper, and paper towels are products that need to have similar attributes for fulfilling customer requirements such as absorbing liquid, being tear-resistant as well as having a certain degree of softness. First, P&G can *share* R&D resources and capabilities associated with developing advanced material that fulfills these requirements across distinct businesses such as its paper towel and its disposable diaper business (see fig. 3-13). Second, as these products are sold via the same retail channels, the corporation can leverage the same sales force to support both businesses. Third, P&G can share, leverage or transfer its great marketing capabilities (e. g., in developing global power brands) from one business to another. Other companies focusing only on one of these products cannot realize those economies of scope, and thus, may have a competitive disadvantage. (Source: Hill/Jones 2010: 316 f.)

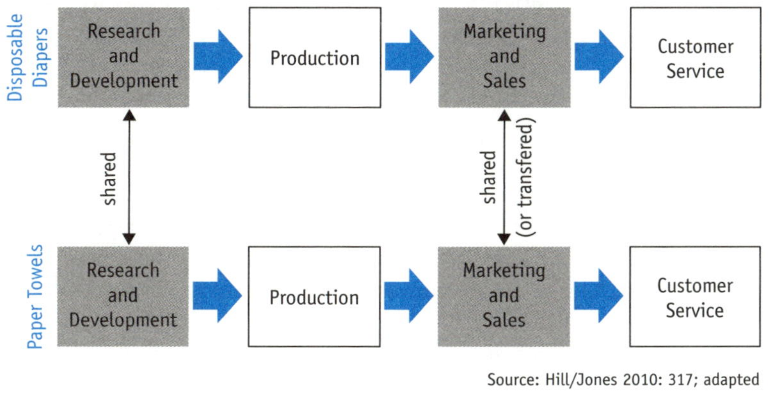

Fig. 3-13: Sharing resources for driving value in two related businesses

Related diversification

Related (concentric) diversification involves the entry into new businesses in different markets or industries that are related to the company's existing business by some form of significant commonalities or even linkages of the corresponding value chains. Corporations pursuing this multi-business model are trying to take advantage of their competencies or resources to establish business and drive competitive advantage in another related industry. They may *transfer competencies* from one established business to another or *leverage competencies* to create an entirely new business in a related industry. For example, a medical equipment manufacturer may decide to offer its products or variations to industrial customers based on the same technologies and competencies but addressing slightly different needs. Furthermore, a company may have the opportunity to *share resources and capabilities* across multiple businesses allowing it to diversify and realize economies of scope (see Strategy Practice Example 3.6). Another way for a company to diversify into related businesses is *product bundling*. Hereby, a firm may enhance their existing business portfolio with related businesses that allows them to offer product bundles at a special package price, and thus, to better fulfill their customer's needs and wants.

Unrelated diversification

Unrelated (conglomerate) diversification is a multi-business model where a corporation operates in many diverse industries or markets. Hereby, the distinction between related and unrelated diversification is a matter of the degree of relationships between businesses. A company that pursues unrelated diversification is also called conglomerate. It "derives less than 70 percent of its revenues from a single business activity and there are few, if any, linkages among its businesses" (Rothaermel 2013: 219). Unrelated diversification is typically based on general organizational and managerial competencies of the corporation such as outstanding entrepreneurial skills, experience in effective organizational design, superior strategic management capabilities, etc. They are used to successfully entering into new businesses with no connection to any of the company's current business activities and may be leveraged to increase the profitability of all businesses in the portfolio. Companies may want to drive conglomerate diversification in situations where they are facing an overall declining trend in their core business and have the corresponding organizational competencies as well as sufficient financial and capital resources to make an unrelated diversification happen effectively. This may result in a radical increase of an organization's business scope.

Diversify or focus?

In times of Schumpeterian creative destruction, continuously renewing and expanding, the scope of business is a key principle of modern business. A company might choose to diversify for a variety of reasons (Johnson/Whittington/Scholes 2012: 139 f.): First, facing growth limitations in their established core business leads companies to consider diversifying into new related or unrelated businesses to capture new *growth* opportunities. Second, expanding the business scope may also be driven by the desire to exploit economies of scope, and thus, realize *efficiency* gains as a source for competitive advantage. This may also include better utilization of available capacities. Third, diversification may reduce the overall business *risk* particularly in an era of highly volatile economic cycles (see Strategy Practice Example 3.7). Fourth, diversification may increase a corporation's overall *market power* and reduce vulnerability by, for example, the ability to cross-subsidize one business with cash flows generated by other businesses (see also Strategic Snapshot 3.4).

Although there seem to be good reasons for a firm to diversify, this approach also involves a number of pitfalls and disadvantages (Johnson/Whittington/Scholes 2012: 141; Lombriser/Abplanalp 2004: 301): First, diversification increases *complexity* and the *cost of coordination* for the parent. Diversification is generally supposed to create value by realizing synergies between businesses, that is the combined effect of having more than one business has to be greater than the sum of individual businesses (see also chapter 3.2.1). This synergistic value has to be greater than the addition cost, which can be quite challenging depending on the corporation's cost structure. Second, the more diversified a corporation is, the harder it will be to *align employees* around inspiring common strategic guidepost and create an overall strategy-focused organization (see chapter 3.1). Third, the *personal ambition* of senior leaders can sometimes cause inappropriate diversification. Some argue that the tendency to diversify does primarily serve egocentric interests of top managers: "(...) corporate growth enhances the social prominence, public prestige, and political power of senior executives. Rare is the CEO who wants to be remembered as presiding over an enterprise that makes fewer products in fewer plants in fewer countries than when he or she took office—even when such a course increases productivity and adds hundreds of millions of dollars of shareholder value." (Jensen 1989: 66) Fourth, diversification can easily lead to a loss of strategic focus and a *deterioration of a firm's strategic positioning* in the market (Porter 1996).

Strategy Practice Example 3.7

Diversification and Portfolio Management for Dealing with Volatility at ZEISS

Within the Carl Zeiss Group, diversification and portfolio management helps to balance different business cycles and volatilities. The six businesses in the portfolio react very differently to changes in the overall economic cycle, which requires targeted and diverse management decisions across the different businesses at the same time. Whereas the semiconductor business is facing negative changes in its specific competitive arena well before a downturn of the overall economy (leading), other businesses are impacted in line with the economic cycle (following) or later (lagging). This is illustrated in figure 3-14. For example, the medical technology and microscopy business is typically lagging an overall business cycle due to expenditure cuts in public institutions that are happening during the course of the downturn. In addition to their different phases regarding the overall economic cycle, the individual business cycles are also characterized by different frequencies and amplitudes (see fig. 3-15). For example, the businesses vision care and branded optics are stabilizing elements in the portfolio as they react less intense to overall economic cycle changes than other elements in the cycle such as businesses in the industrial goods industry. In addition to diversification of businesses, the company does also strive for regional diversification as it generates the greatest portion of its revenues in international markets. (Source: Spitzenpfeil/Schneider 2012: 106 f.)

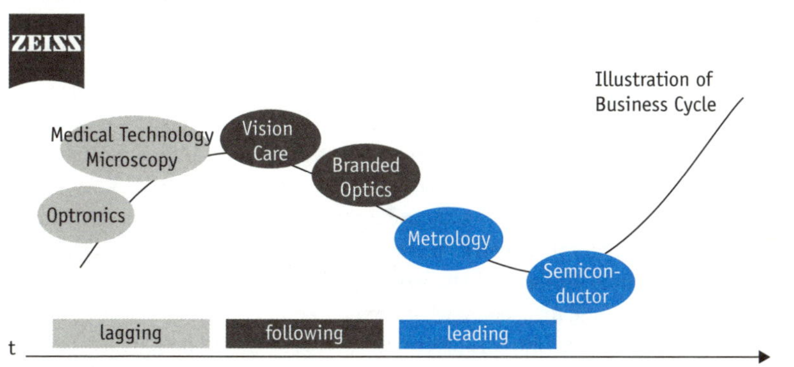

Source: Spitzenpfeil/Schneider 2012: 106 f. Reprinted with permission of Carl Zeiss AG.

Fig. 3-14: Portfolio of the Carl Zeiss Group in the business cycle

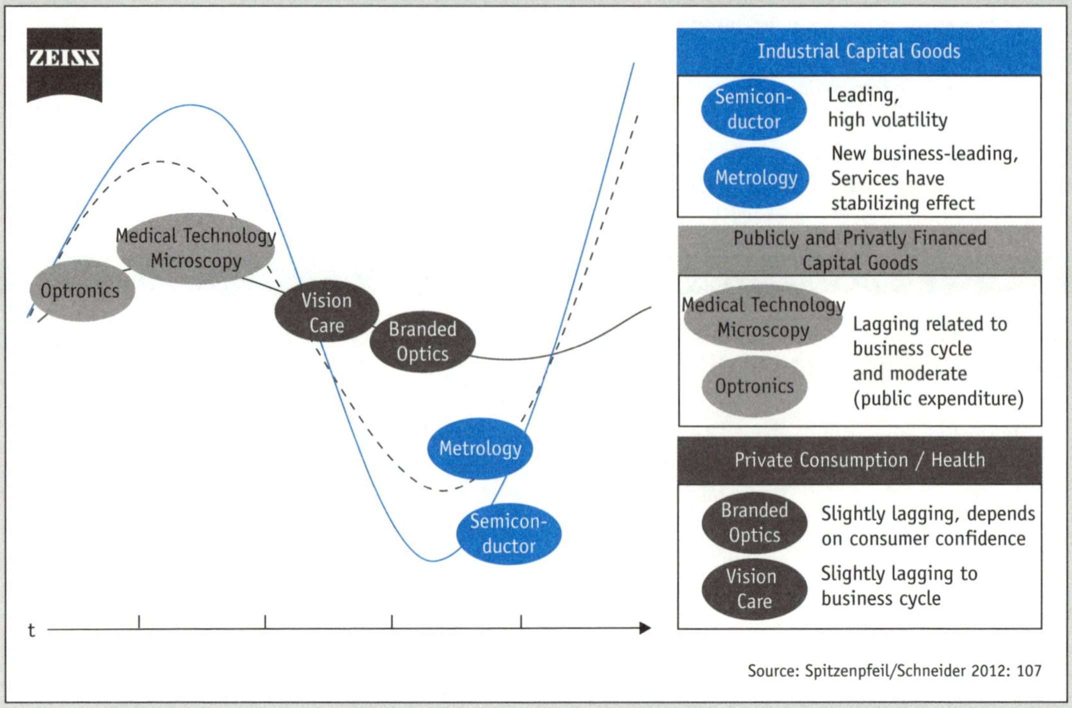

Fig. 3-15: Different volatilities of businesses in the Carl Zeiss Group

There are a number of companies that are successful by doing the opposite of diversification, that is focus (see Strategy Practice Example 3.8). Compared to diversified corporations, firms focusing on a single core business have some advantages (Hill/Jones 2010: 286 f.; Lombriser/Abplanalp 2004: 293):

- clear and unambiguous vision (and mission) and its corresponding benefits such as orientation, enthusiasm, stimulation, and security (see chapter 3.1.1)
- focus of all corporate resources and capabilities (e. g., managerial, financial, technological, functional) on a single business ("joint forces")
- building up deep and profound knowledge of the business and being able to perform corresponding activities more effective and efficient than rivals
- no distraction of the executive leadership team by other business activities and staying focused on what they know and do best ("sticks to its knitting")
- incentivizing corporate executives to think in longer terms and pursue a more sustainable strategic perspective as they do not want to jeopardize their only business with short-term profit maximization strategies.

There is no final answer to the question whether it is better to diversify or focus. "Diversification is a conundrum", states Robert Grant. "The quest to enter new fields of business has probably caused more value destruction than any other type of strate-

Strategy Practice Example 3.8

"We Are Skin Care"—Focus at BEIERSDORF

Beiersdorf AG, a global multinational consumer goods company with over 17,000 employees worldwide and headquarters in Hamburg (Germany), is an example of a highly focused corporation emphasizing skin care with three iconic global brands (see fig. 3-16): "For over 130 years, we have dedicated ourselves to meeting our consumers' individual needs and are considered to be the inventors of modern skin care. Our research and development expertise, innovative products, and strong brands are the reasons for our success. We want to be the number one skin care company." (Source: Beiersdorf 2015)

* *Euromonitor International Ltd.; NIVEA by global brand name in the categories Body Care, Face Care and Hand Care; in retail value terms, 2011*

Source: Beiersdorf 2013. Reprinted with permission of Beiersdorf AG.

Fig. 3-16: Focus on skin care with three brands at Beiersdorf

gic decision. Yet, diversification also offers growth options and the potential for a firm to free itself of the restrictions of a single industry." (Grant 2010: 402)

Horizontal integration: mergers and acquistions

Even when a company stays focused on a single business that it performs distinctively well, it may need to adjust its business model over time due to changing conditions in the business environment to drive profitability (see chapter 3.4.4) or create further growth opportunities if markets become mature. One way for corporate strategy to support and strengthen a single business is horizontal integration. This is "the process of acquiring or merging with industry competitors to achieve the competitive advantages that arise from a large size and scope of operations." (Hill/Jones 2010: 287) Whereas *acquisition* refers to the purchase of another company, a *merger* is the creation of a new entity by pooling the operations of two "equal" companies. Firms pursuing a horizontal integration strategy acquire or merge with rivals from the *same* industry or, more specifically, from the same stage in the industry value chain (see fig. 3-17). The benefits of horizontal integration include lower cost structures (e. g., increased economies of scale, pooling and reduced duplication of resources with larger scale tasks), increased differentiation (e. g., product bundling and cross-selling opportunities), increased bargaining power over suppliers and buyers due to larger scale output, stopping price erosion caused by intense competition as well as replication of successful business models. Some companies are very experienced in acquiring and integrating other companies from the same industry to replicate their successful way of doing business. Although horizontal integration can strengthen a business, there are also serious pitfalls. A great number of mergers and acquisitions have failed and actually reduced value due to various reasons such as failures to integrate different company cultures (e. g. acquiring a company if that mission has been to take you out of business) or high management turnover in case of hostile takeovers. Thus, implementing a horizontal integration strategy requires an effective integration process as well as a great deal of experience (Hill/Jones 2010: 286–292 and 335–338).

3.2.2.3 Vertical Integration Strategies

The scope of industry value chain stages the firm participates in

Another strategic approach for a corporation to strengthen an individual business, improve its competitive position, and grow is vertical integration. Hereby, a corporation expands its operations along stages of the *industry value chain* or value network. Regarding the value chain concept, three comments are important to note:

- At each stage of an industry value chain, *value is added* to the product by transforming inputs from previous stages in a way to increase its value to the "customer" who pays for the final outcome. Figure 3-17 shows the high-level value chains of the automotive industry and the oil industry.
- Each stage of an industry value chain can be considered a *separate industry* in which a company competes with different other firms of that specific industry. For example, a first tier supplier of a car manufacturer that supplies a certain compo-

Fig. 3-17

High-level value chains of the automotive industry and the oil industry

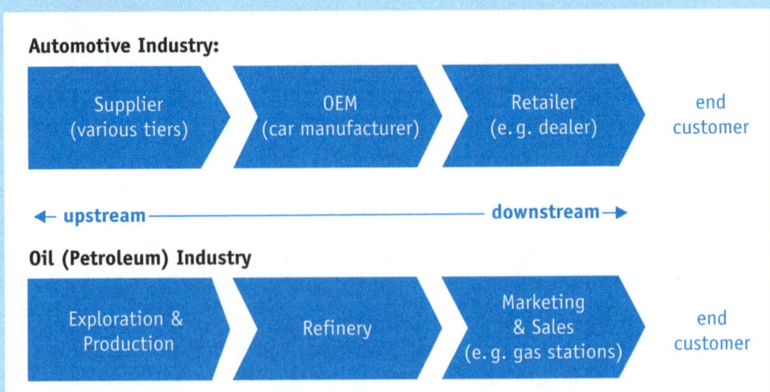

nent such as a seat competes in a very different industry than the supplier of an engine or the provider of car financing products. This means that vertical integration of a corporation can also be seen as some sort of diversification into new businesses and entering new competitive arenas.
- Every company within an industry has its own *company value chain* composed of primary value creating activities and support activities (see chapter 2.3.1.2) that should not be confused with an industry value chain that runs across different stages.

Vertical integration describes the firm's ownership of the production of supplies required for its core operations as well as ownership of channels for distributing its output. The degree of vertical integration indicates the extent to which a company is its own supplier or customer. This corresponds to the number of industry value chain stages that it directly participates in and adds value to. The two types of vertical integration as well as its benefits and pitfalls are explained next (Johnson/Whittington/Scholes 2012: 141–144; Rothaermel 2013: 209–214; Wheelen/Hunger 2010: 208 f.).

Backward vertical integration

With vertical integration, a corporation can expand its activities in either of two directions. First, *backward vertical integration* describes the firm's expansion into activities previously provided by a supplier, that is going backward on an industry's value chain into upstream stages. It is related to the inputs of the company's current business. An example is the acquisition of a component supplier by a car manufacturer (OEM) in the automotive industry (see fig. 3-18). This approach might be considered when the supplies (e. g., components or raw material) are crucial for the final product of the company and the suppliers are limited in number, appear to be unreliable or very expensive. In such situations, backward vertical integration helps the firm to reduce resource acquisition costs and inefficient operations and gain better control over supplies. Furthermore, the company may also generate some revenues with backward vertical integration, in case the products or services of this stage in the value chain are also offered to customers outside the company in the external market. To successfully master such integration, and thus, improve its competitive position, the company has to have the general organizational and technological competencies as well as financial resources. As a pitfall, backward vertical integration may significantly reduce a firm's strategic flexibility. There is a risk of expensive assets creating an exit barrier that prevents a firm from leaving a particular industry when demand drops. Also, it may create some potential conflicts of interest and loss of business if a supplier that is integrated has some of the firm's competitors as its customers.

Forward vertical integration

Second, *forward vertical integration* means that the firm is gaining control or ownership of activities previously provided by a distributor, retailer or a firm's direct customer (e. g., business-to-business), that is going forward on an industry's value chain into downstream stages. It is related to the outputs of the company's current business. For a car manufacturer (OEM) forward integration would be to own car retailers or repair and servicing garages (see fig. 3-18). A company might consider forward vertical integration when the distributors or retailers are charging high prices and the company wants to capture those margins itself. Also, the company may

Fig. 3-18

Integration and diversification options: car manufacturer (OEM) example

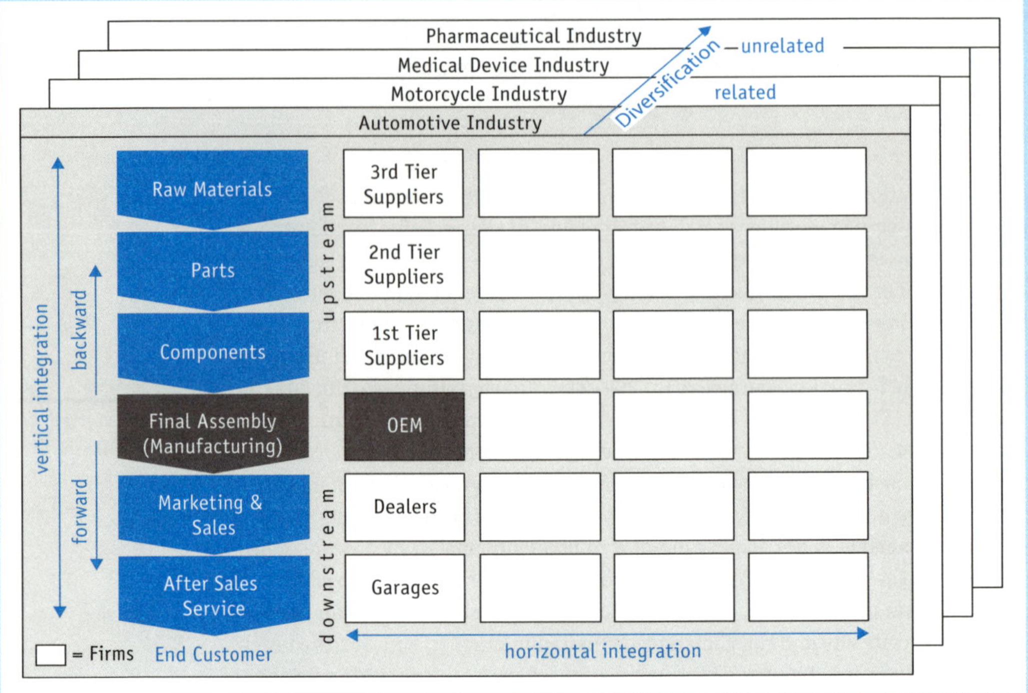

want to gain more control of their downstream value chain stages moving ownership of activities closer to the end customer because distributors or retailers may not be reliable or may not dedicate enough attention and resources to selling the company's products. Forward vertical integration is also a way for companies to improve customer intimacy and get direct feedback for improving their core products or services.

Figure 3-18 illustrates the different types of vertical integration using the example of the automotive industry. It also shows horizontal integration, which could be a merger of two OEMs as it was the case with Daimler and Chrysler in 1998 (followed by the sale of Chrysler in 2007). Also, an own first tier supplier buying a competitor in the same industry (i. e. stage in the automotive value chain) would be considered a horizontal integration. If the OEM decided to expand its business into the motorcycle business it would be considered a related diversification or—in case it expands the firm's business scope in other industries such as the medical device or pharmaceutical industries—an unrelated diversification. As related diversification requires bringing together different value chains to realize synergies, it is sometimes also described as horizontal integration (Johnson/Whittington/Scholes 2012: 142).

Benefits and pitfalls of vertical integration

In general, corporations often favor vertical integration as strategic approach because it seems to allow the firm capturing more of the profits created in the value chain. Hereby, not all industry value chain stages are equally profitable. Depending on the industry, backward integration may be more profitable than forward integration because of low margins in retailing, or vice versa. To decide which stages of an industry value chain a company should cover and which ones should be better outsourced it needs to understand the possible benefits and pitfalls of vertical integration. Vertical integration—either backward or forward—can have a number of benefits including (Hill/Jones 2010: 286–292 and 294–300; Rothaermel 2013: 212–214)

- enhancing or ensuring stable *product quality* to strengthen differentiation by, for example, better control of the quality of supplies or after-sales service standards
- lower *cost* (or better differentiation) of products through investment in specialized assets (e. g., manufacturing equipment, human resources with specific know-how and skills, etc.)
- improved *scheduling and planning* enabling a company to make transfers at different value chain stages quicker, easier and more cost-effective as well as responding better to sudden changes in supply or demand.

Those benefits may help a company to improve its profitability. However, the opposite may occur when firms become excessively large and bureaucratic. First, vertical integration may result in a *competitive cost disadvantage* as in-house suppliers are not exposed to market competition and companies may continue to buy from them even though low-cost independent supplies exist. Second, *administrative costs* arise from the need to coordinate various stages in the industry value chain. Third, vertical integration reduces a corporation's *strategic flexibility* particularly in environments with strong technological change where the company may be locked into an old inefficient technology it has invested in. Also, industries with unpredictable and highly fluctuating demand may provide a disadvantage for companies with a high degree of vertical integration. Thus, an efficient management of volume and working capital along the different value-chain stages are made difficult. It may also lead to an increased cost structure if the company cannot produce at full capacity at upstream businesses when downstream market demand falls.

Outsourcing and strategic alliances

Recently the trend has reversed from vertical integration toward decreasing own operations along the industry value chain stages particularly by means such as outsourcing (vertical disintegration). The question whether and to which degree a corporation should vertically integrate cannot be answered in general because it strongly depends on the company's specific situation (Wheelen/Hunger 2010: 209). Among popular alternatives to vertical integration are outsourcing as well as taper integration where a firm vertically integrates up- and downstream activities but also orchestrates specific outside firms (Rothaermel 2013: 215). Another alternative is to pursue more cooperative relationships such as *strategic alliances* with relatively stable long-term agreements allowing companies to obtain the same kind of benefits provided by vertical integration but without some of its disadvantages such as bureaucratic cost or technological lock-in (Hill/Jones 2010: 300–303). Furthermore, there is a variety of other specific strategies regarding the *value chain architecture* that will not be further elaborated on here (Müller-Stewens/Lechner 2011: 369–375).

3.3 International Strategy

Creating and exploiting international opportunities

Another area a corporation has to consider when defining the scope of its business is related to the international scope that it wants to address and utilize. The accelerating pace of globalization driven by factors like internet and information technology, decreasing trade barriers and increasing international exports and foreign investments as well as the rise of global standards offer companies new attractive business opportunities but may also create new competitive pressures and other threats. International strategy is dealing with strategic choices that managers have to make regarding their corporation's international business activities covering more than one country. Hereby, *international business expansion* and *transnational management* are two key areas that will be addressed next. They relate to strategic decisions both on a corporate and a business level.

3.3.1 International Business Expansion

The global trend of cross-border economic integration allows or even compels domestic as well as multinational companies to effectively expand business activities beyond their existing regional boundaries. For formulating strategies that deal with an international expansion of a company's business activities, the following three major questions need to be answered (Kotler/Keller 2009: 600):

- *Whether* to go international (motives)?
- *Where* to go international (country market selection)?
- *How* to go international (entry and go-to-market strategy)?

3.3.1.1 Whether to Go: Internationalization Motives

Multiple reasons for the international expansion of business activities

A fundamental decision of international strategy relates to the degree of internationalization of a company or business. In general, a firm's decision to expand its scope and "go international" will be targeted toward gaining or maintaining competitive advantage. There are different sources for achieving competitive advantage through internationalization. Reasons why corporations pursue a geographic expansion are often related to sales and profit growth though they are not limited to these two financial goals. Figure 3-19 shows key drivers or motives for going international in three main categories. (Bartlett/Beamish 2011; Deresky 2014: 196–200; Homburg/Kuester/Krohmer 2013: 427 f.; Morschett/Schramm-Klein/Zentes 2010: 71–82)

- **Opportunity-oriented motives** are related to *pro-actively* improving either sales or profitability (margin). A key driver for many companies to go international is the opportunity to leverage new sales sources. Most firms are likely to find the greatest portion of potential customers as well as purchasing power outside their domestic market. Furthermore, there is a great opportunity to improve profitability (margin) by, for example, capturing a higher willingness to pay for existing products. This is currently the case with German automakers selling their premium cars in China with premium prices that exceed by far their domestic price level. Another opportunity-oriented motive is to achieve cost advantages through

Fig. 3-19

Motives for corporations to "go international"

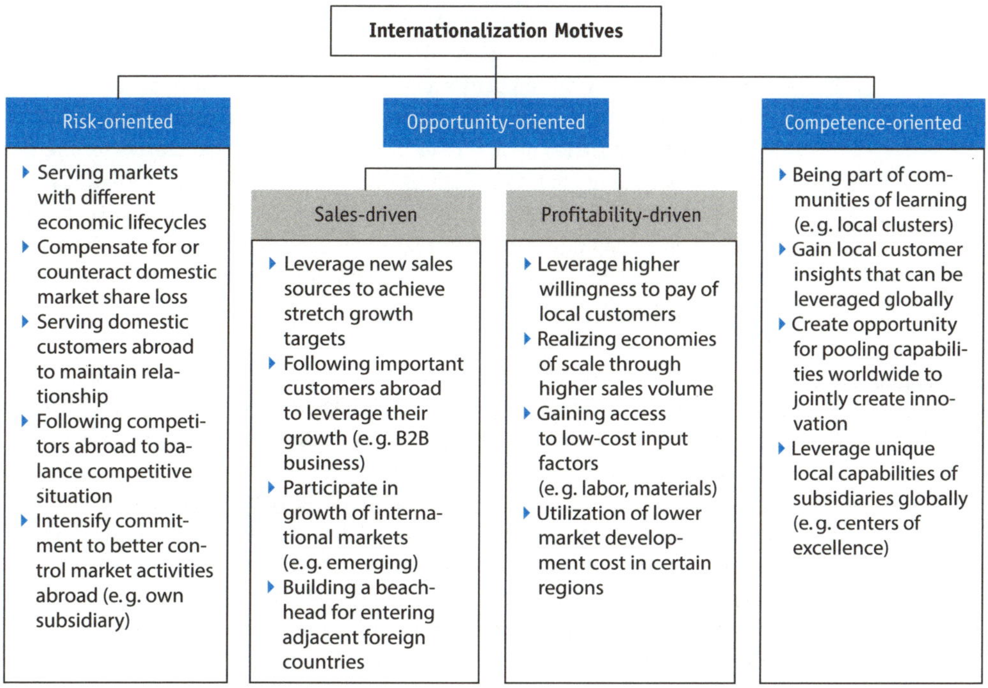

Source: following Homburg/Kuester/Krohmer: 428; adapted

either scale economies by increasing output for a greater international market or by getting access to and exploiting low-cost input factors.

- **Risk-oriented motives** refer to a more *reactive* or *defensive* mode of international expansion. It includes the creation of a portfolio of regional markets with different economic lifecycles to balance the risk of operating internationally. This follows a similar rationale as explained before related to the diversification of businesses or industries (see chapter 3.2.2.2). With a high degree of volatility in many markets these days, considerations of currency stability and reduced economic dependence on few regions have become very important drivers for companies planning to go international. Furthermore, companies in mature markets and developed countries are sometimes forced to look for new sales opportunities outside their home market to counteract slowing domestic sales growth and business risk. Also, key stakeholders such as a supplier's customers or competitors going global may create the need for a domestic company to "follow" or otherwise face competitive disadvantage.

- **Competence-oriented motives** for going international are centered on the idea of accessing and leveraging dispersed local knowledge, capabilities, ideas, and best practices to better drive innovation in the entire corporation. Hereby, the diversity of environments that a company is operating in is considered a key asset of multinational corporations that can be leveraged to drive competitive advantage through, for example, cross-border knowledge management and innovation processes. For example, company's may want to be present and participate in regional knowledge clusters such as Silicon Valley for information technology-related high-tech industries or the Basel (Switzerland) regional cluster for life sciences (Morschett/Schramm-Klein/Zentes 2010: 124–132). Going international may also be motivated by improving the R&D and innovation capabilities of the company based on managing locally leveraged or globally linked innovation processes. This typically requires corresponding transnational management processes of the "parent" company (see chapter 3.3.2).

Pitfalls of going international

Although there are many reasons for a corporation to pursue an international expansion of its business scope, there are also pitfalls for going international. First, in addition to the marginal cost of adding operations there may be some supplemental costs of doing business in distant geographic markets. Operating in unfamiliar cultural and economic environments may require additional efforts not just regarding go-to-market approaches (e. g., product adaptations) but also for effectively conducting business in general. There are many successful domestic or multinational companies who learned their lessons when trying to just replicate their national "success logic" such as a domestic value proposition or business model in a foreign country. Walmart's failure to establish a successful business in Germany is a famous example of corresponding pitfalls. The world's biggest and highly successful retailer entered the German market in 1997 but had to withdraw in 2006 after struggling for years due to mainly cultural and legal issues as well as not understanding the local customer needs and wants leading to an insufficient value proposition (Knorr/Arndt 2003). Furthermore, there are additional costs of coordination across different geographic markets, particularly when a corporation wants to drive cross-border knowledge management and innovation. Also, some of the advantages expected from various internationalization motives explained in figure 3-19 have to be assessed very carefully. For example, employing low-cost labor may have side effects such as a lack of quality or productivity. Furthermore, expanding overseas with advanced technologies may require firms to expose intellectual property to a degree that causes issues such as copyright infringements (Rothaermel 2013: 276 f.). A company has to be aware of those pitfalls and should develop a solid understanding of the following items before making the decision to go international (Kotler/Keller 2009: 600):

- Foreign preferences and ability of offering a competitively attractive product
- Foreign country's business culture
- Foreign regulations to avoid unexpected costs
- Availability of managers with international experience
- Risk that the foreign country might change its commercial laws, devalue its currency, undergo significant political changes, or expropriate foreign property.

3.3.1.2 Where to Go: Country Market Selection

Developing an effective country market evaluation and entry process

If a corporation decides to go international for expanding its market scope, the first major step is to identify, prioritize and select potential geographic target markets—typically countries—based on the company's objectives or motives. To support the entry decision process for a given product, companies can adopt a variety of different approaches (Meffert/Bolz 1998: 116–122; Perlitz/Schrank 2013: 332; Holtbrügge/Welge 2010: 96–103). They range from a relatively pragmatic evaluation and selection processes based on simple checklists to far more comprehensive methods (Kumar/Stam/Joachimsthaler 1994; Russow/Okoroafo 1996). All methods are supposed to help identify countries in which market opportunities exist. Figure 3-20 shows a high-level sequential evaluation approach as part of an international market selection and entry process. Effectively designing and conducting such a process is very important because international market entry decisions have a strong impact on the company's performance and are difficult to revert (Kotabe/Helsen 2008: 284).

Identifying specific evaluation criteria

Key elements in this process are the evaluation criteria based on which potential country markets are prioritized and selected. Unfortunately, there is no common set of criteria that can be applied as a general framework for all types of country market screening. Evaluation criteria need to be defined based on the company's product and market characteristics as well as its strategic motives for international market expansion. Elements in the different areas of external strategic analysis (see chapter 2.2) such as macroenvironmental factors, important stakeholders (e. g., institutions affecting local competition in the foreign market), or elements from industry, market and customer as well as competitor analysis are a source for identifying criteria to evaluate and compare countries for entry. For determining relevant criteria for the different steps in the evaluation process, the following guidelines are suggested (Kotabe/Helsen 2008: 284; Meffert/Bolz 1998: 120 f.):

- **Do not define too many criteria.** The number of criteria usually does not impact the quality of decisions significantly but makes the process complex. It becomes hard to trace back results of the evaluation and understand how they were actually computed.
- **Align evaluation criteria with product and motives.** The definition of evaluation criteria has to be strongly linked to the product considered for international market entry. Also, the company's risk-, opportunity- or competence-oriented internationalization motives will have a strong impact on which criteria are more or less relevant.
- **Select criteria where secondary data is available.** To drive an effective and reliable evaluation process, there should be secondary data (e. g., population growth, gross domestic product, per capita purchasing power, etc.) available for the different criteria. For example, most information on socioeconomic or political country indicators can be collected from publicly available data sources. Primary market research is usually too expensive and time consuming, particularly for early steps in the sequential evaluation process.

Country notebooks as frame for analysis and documentation

A variety of companies develop country notebooks for each country where they are doing business or which they consider to enter. It contains a variety of generic cultural, economic, market- and competition-related topics that can be used as a frame of reference for further investigation as well as the basic structure of a preliminary marketing plan. A country notebook framework provides a general direction to areas a firm may want to further explore for relevant data. Not all areas will be equally important. Depending on the company's internationalization motives, product characteristics or current business model and the targeted country market, some topics may not be relevant, and thus, should be ignored. Other topics might be highly significant for supporting an entry decision into a foreign country market and need to be covered in depth with relevant data. Some topics of the general framework may need to be adapted and customized to better reflect the specific company situation including its competitive arena and products. A country notebook does not only provide a checklist of potential areas for analysis. It may also be used for documenting the results of a country market analysis that should be continuously updated to capture changes in the local competitive environment. A relevant and up-to-date country notebook may become the first database consulted in a company for country-related strategic considerations. (Cateora/Gilly/Graham 2009: 584–591; Ghauri/Cateora 2010: 512–517)

Country market selection with sequential three-step process

Figure 3-20 shows a systematic global environmental scanning and assessment process that is part of international market entry planning. It is a sequential heuristic evaluation approach that zeroes down from a global or multinational to a regional and finally national level in three steps (Deresky 2014: 202–209; Holtbrügge/Welge 2010: 100).

1. Definition of initial country pool

First, an *initial country pool* needs to be defined as a starting point for the assessment. This might be the entire globe or a specific predefined region that is emphasized by the company (e. g., European countries, Northern Africa, neighboring countries to the domestic market, etc.). Some companies define cutoff or "must" criteria to already eliminate certain countries at this stage (Kotabe/Helsen 2008: 286). Those very few criteria (sometimes only one single criterion) are used for a broad assessment. Examples of potential "must" criteria are a minimum population size, the absence of certain legal restrictions imposed by local governments, a pre-defined threshold of import duties not to be exceeded, or the availability of specific technological standards and infrastructure conditions (Meffert/Bolz 1998: 110–113). As a result of this first broad assessment, countries that are somehow closed for market entry, insignificant or do not have favorable entry conditions regarding the firm's motives and products are eliminated from the list. The remaining list covers regions or countries that are considered appropriate for market expansion at first glance and shall be evaluated further.

2. Preliminary screening

The second evaluation step is a *preliminary screening* of the remaining regions or countries in greater detail. It is usually based on more criteria and a more refined evaluation process that, for example, is supported by a scoring model with weighting and rating of the different criteria. The goal of preliminary screening is to identify

Fig. 3-20

Three-step country market selection as part of international market entry planning

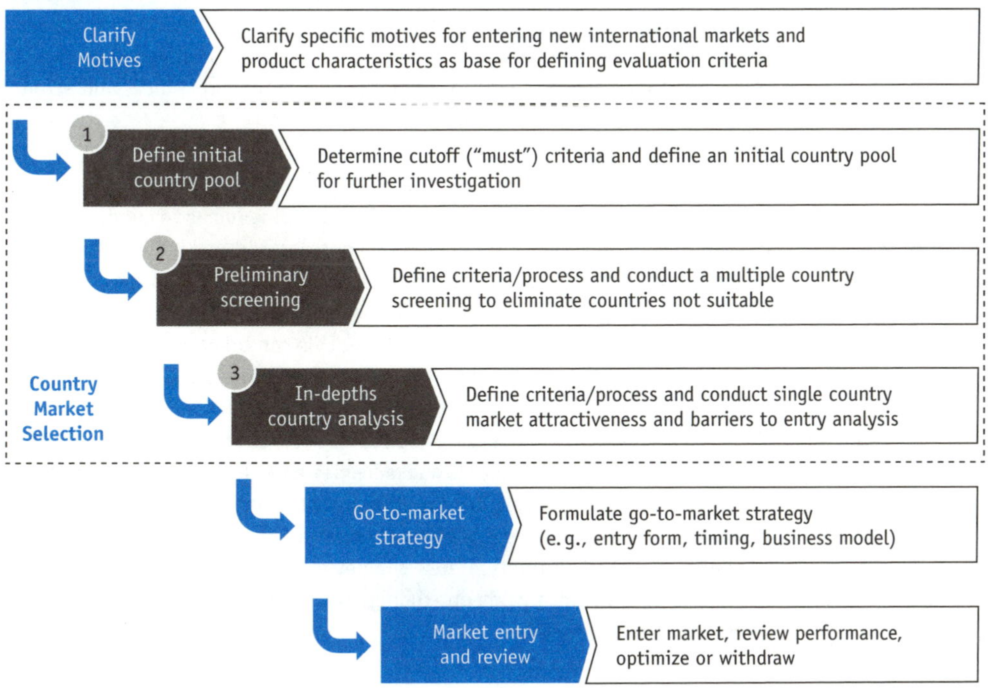

Source: following Deresky 2014: 206 and Holtbrügge/Welge 2010: 100; adapted

countries and markets with promising conditions for doing international business and that provide particular opportunities for the corporation given its motives and products. Criteria may be related to economic, socio-economic, natural and technical, political/legal or sociocultural factors of regions and nations. For example, a criterion to eliminate countries in preliminary screenings might be the political instability or business risk. This can be assessed by indicators such as the widely used Business Environmental Risk Index (BERI S. A. 2015). This index reflects a general cross-industry as-is perspective of a country's risk profile for conducting business based on several subindices and multiple criteria subjectively assessed by international experts. Although using this index might have some pitfalls (Holtbrügge/Welge 2010: 99), it may well function as a first pass indictor for preliminary screening. Other criteria may include climate or geographic location (e.g., seaside countries), gross domestic product, language and religion, cultural distance to the company and its products, or the political system (Holtbrügge/Welge 2010: 97; Meffert/Bolz 1998: 110–113). A company may select some relevant subcriteria that they want to emphasize for their specific situation from broader indices such as BERI and only focus on those for preliminary screening (Perlitz/Schrank 2013: 335). In addition to purely external factors, the company may also want to look at a general *match* between a potential country and the firm. Some countries may be more mismatched or distant than others and are thus candidates for elimination in preliminary screening although they may rank higher on other criteria. A way to analyze this generic country "fit" with a company is the CAGE framework which covers the dimensions cultural distance, administrative and political distance, geographic distance, and economic or wealth distance (Ghemawat 2001).

3. In-depth country analysis

Multi-criteria evaluation of country market attractiveness and entry barriers

After zeroing in on one or more regions, the firm finally has to conduct an in-depth analysis of each specific country market it considers for international expansion. The goal of this third step is to get a comprehensive understanding of the national level of a targeted country to make a final entry decision and to have a solid foundation for crafting an effective go-to-market strategy. This can be done by looking at the two dimensions *market attractiveness* and *market entry barriers*, in which specific multiple evaluation criteria have to be defined (Homburg/Kuester/Krohmer 2013: 429). Although market entry barriers are used as an own dimension of investigation in this book, it has to be noted that it can also be viewed as a subcriteria of market attractiveness. High entry barriers increase the attractiveness of a certain country market because they lower the threat of entry of other companies, and thus, help to sustain a certain level of profitability. If new competitors can easily enter a country market, profits will be bid away quickly due to increased competition (Hax/Majluf 1996: 72 f.; Porter 1980: 7–22; see also chapter 2.2.2).

Preferring attractive markets with high entry barriers

Figure 3-21 shows a matrix that can be used for weighing options which country markets should be entered, based on the two dimensions explained. It leads to portfolio view of country markets with four categories (Backhaus/Büschken/Voeth 2003: 124). Markets that are not very attractive and have high entry barriers are usually not considered for entry *(abstinence markets)*. Country markets with both low attractiveness and low entry barriers are considered *opportunity markets*. They may be targeted in case they offer compelling onetime opportunities or longer-term business prospects such as emerging markets that are likely to become very attractive in the future. Countries with highly attractive national markets and low entry barriers are labeled *gateway markets*. Some companies may appreciate and prioritize those countries that are attractive and easy to access (Homburg/Kuester/Krohmer 2013: 428). They may decide to enter those markets because there international expansion may be faster and may require less effort. In such cases, the firm may want to establish entry barriers later on through, for example, exploiting pioneer advantages as part of its go-to-market strategy (see chapter 3.3.1.3). Finally, countries with a high market attractiveness *and* high market entry barriers are considered *high prospect markets*. High entry barriers are preferable considering the sustained profit potential in this market. What the company needs, however, is to have unique resources and capabilities that can make an entry possible for the firm but unacceptably difficult for other companies (Hax/Majluf 1996: 73).

Fig. 3-21

Country market evaluation with market attractiveness/market entry barriers portfolio

Source: following Backhaus/Büschken/Voeth 2003: 124; adapted

In the following, market attractiveness and market entry barriers will be looked at as different dimensions for reasons of practicability. Explicitly addressing each of the two dimensions provides an important perspective and information for formulating the go-to-market strategy later on. Evaluation criteria in each of the two dimensions are structured in terms of *demand-based* and *competition-based* as well as *institutional* factors (Homburg/Kuester/Krohmer 2013: 427–429).

Assessing country market attractiveness

Figure 3-22 shows selected criteria for evaluating the market attractiveness of a country market. From an economic perspective, market size and growth as well as the buying power per capita are important *demand-based* factors. The total market capacity is the maximal receptivity of the market without considering the people's buying power (e. g., population size of a country). It may have already been considered in an earlier stage of the country market selection process. Now at the in-depth analysis, market potential may be of greater interest as it refines the number of potential customers considering their ability to buy and to use a product (e. g., people without driver's license and children as well as low-income classes are not considered for analyzing the market potential for premium cars). Furthermore, the market volume and market segment volume show how much of a certain product or product category is actually sold in a country or segment considering sales of all providers in a certain time frame. Although segment sales might be low, a high overall market potential might offer huge sales opportunities with an effective business model

Fig. 3-22

Assessing market attractiveness for entering a new country market

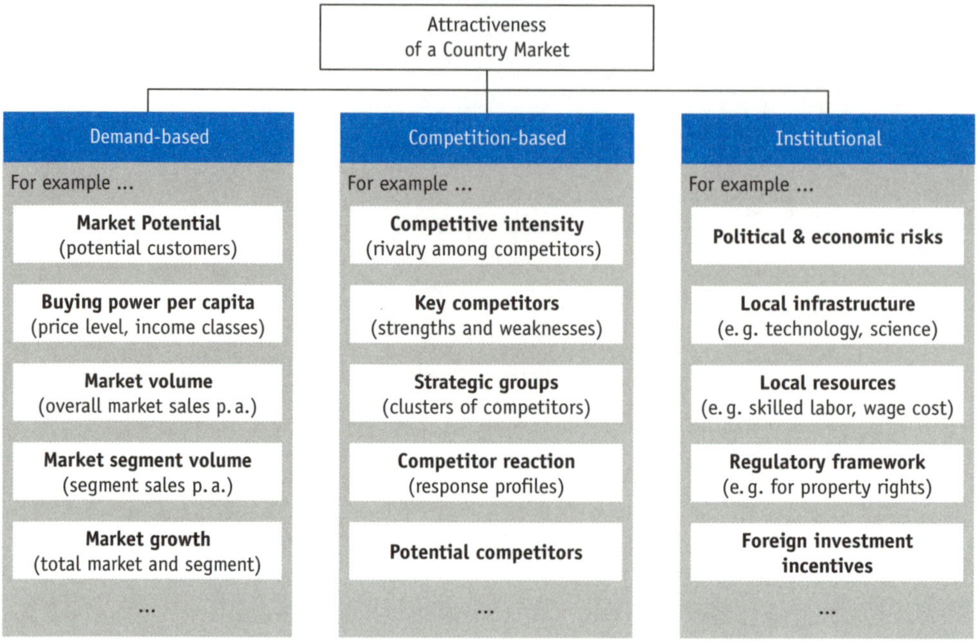

Source: Homburg/Kuester/Krohmer 2013: 428; adapted

including required product line adaption. Whereas some companies may look primarily at the pure market potential or volume they may be able to address in a new country market, other firms pursuing less but highly profitable sales may be more interested in the buying power per capita. A company entering a new country market may face a competitive situation that differs significantly from its domestic market. *Competition-based* criteria include the degree of competitive intensity (e. g., number and strengths of competitors), and the presence of global or local key competitors or strategic groups. The fact whether those competitors or potential competitors are local or not can be highly relevant as governments may favor domestic firms and customers may prefer domestic products. Furthermore, the company may want to investigate potential responses or retaliations of established firms to the company's market entry as well as potential new competitors that are not present yet but may also want to expand their international business activities to the same foreign country market in the future. From an *institutional* perspective, political and economic risks are of particular relevance for assessing market attractiveness (e. g., inflation, exchange rate, general political stability, etc.), as well as local infrastructure and available local resources, the regulatory framework (e. g., protection of intellectual capital, steps and requirements for product approval) and incentives for investing and doing business in the new country market provided by the local government or other institutions (see also chapter 2.2.3).

Assessing country market entry barriers

Figure 3-23 provides an overview of some selected criteria for evaluating barriers to an entry into a foreign country market. Customer characteristics such as consumption patterns or certain product, brand or provider preferences of customers as well as specific cultural factors are potential *demand-based* market entry barriers that have to be considered when evaluating a foreign country market. For example, customers may prefer products with different tastes or favor offerings from local domestic companies (ethnocentrism phenomenon). Also, targeted customers may be loyal to products and brands from local or multinational competitors that are already present in the market and provide a compelling value proposition. Furthermore, customers may face high switching cost due to certain lock-in mechanisms established by established competitors (see chapter 3.4.2.2). From a *competition-based* perspective, a company planning to enter a foreign country market may have relative cost/price-disadvantages it cannot compensate due to scale economies or other factors of established firms. This may include additional logistics cost that are required for the com-

Fig. 3-23

Assessing market entry barriers for entering a new country market

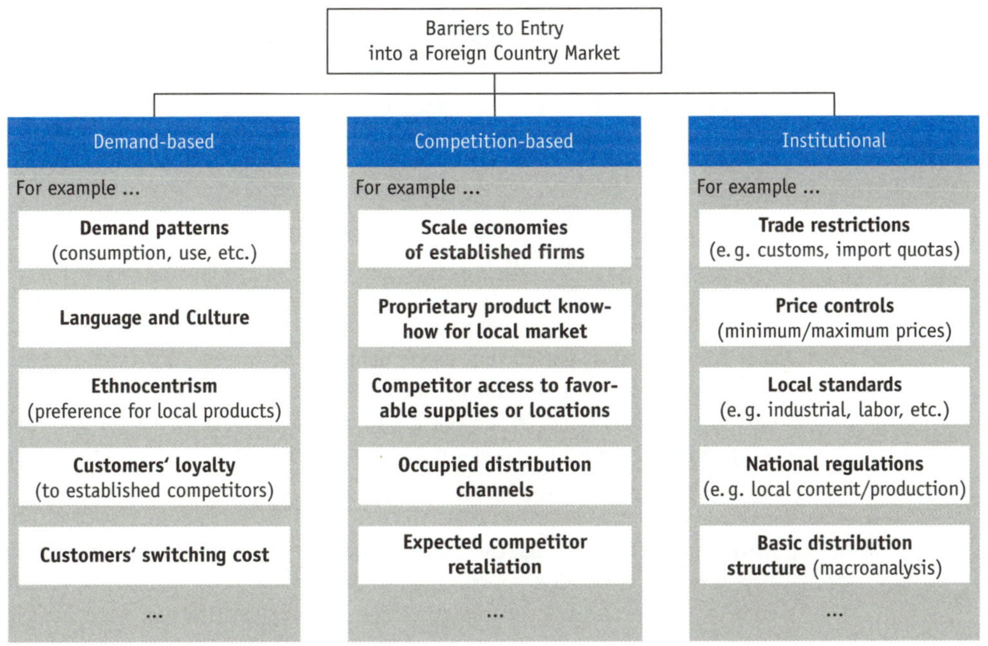

Source: Homburg/Kuester/Krohmer 2013: 428; adapted

pany to enter the foreign market, proprietary technology of competitors providing them a cost advantage, government subsidies given to established firms in the market, certain physical locations (e.g., for manufacturing) or distribution channels already occupied by competitors. Finally, formal institutions such as governments or various associations have a strong impact on a firm's constraints to enter a specific country market (Deresky 2014: 203 f.). *Institutional* factors include any kind of trade restrictions such as high customs, price controls, local industry or labor standards, national regulations such as local content requirements in production, the need to form joint ventures with local companies for getting market access, as well as legal regulations requiring significant product adjustments. Furthermore, the basic distribution structure may reflect an entry barrier. For example, sales partners or customers may be only reached through certain middlemen or require specific logistics solutions for the company.

As mentioned earlier on, it can hardly be overemphasized that the criteria for evaluating market attractiveness and market entry barriers have to be customized according to the company's motives for going international as well as its specific situation such as the type of product to be introduced and the degree of volatility in its markets. Particularly risk considerations have come into focus in certain industries recently that were perhaps underappreciated before. The final country assessment leads to serious entry planning resulting in a go-to-market strategy that will be explained next.

3.3.1.3 How to Go: Entry and Go-to-Market Strategy

Existing, latent, and incipient markets

Entering one or more foreign country markets with an existing product concept reflects a market development strategy (see chapter 3.4.1.3). Hereby, there are basically three types of market opportunities (Doole/Lowe 2008: 106–108): First, *existing markets* are already serviced by competitors with similar products. They are difficult to enter unless the company has a highly differentiated product or business model. Second, in *latent markets* potential customers may have been identified, but there has never been a similar offering to the company's product. Entry may be easier compared to existing markets. Entry timing decisions should get a special emphasis as competitors may have also identified this market opportunity. Third, *incipient markets* do not exist at present but may emerge in the future. At present there are no direct competitors, but other firms may be positioning themselves to take advantage once the time has come. In such cases, long-term competitive advantage is possible but requires significant market development efforts once those hard to find market opportunities are identified.

Assuming the company has decided why and where it wants to expand internationally, it has to determine *how* to go and identify the most appropriate method for entering and addressing the foreign market. Hereby, the firm can select from several strategy options in the following three key areas:

- Entry *timing* and *sequence* of international market entries
- *Mode* of international market entry
- *Business model* for entering the specific country markets (including *marketing mix*).

Deciding on the entry timing: pioneer or follower strategy

Companies planning to introduce an existing product in one or more foreign country markets have to decide when to enter the market. Particularly in case where the company's product has a high degree of newness to the foreign market, being the first or **pioneer** can be highly rewarding. Potential pioneer advantages include price premiums due to temporary monopoly, dynamic scale economies in terms of experience curve effects, establishing customer loyalty and establishing a certain brand and company image, and creating other barriers to entry such as secured key distribution channels. Furthermore, companies which are able to set industry standards with their products create lock-in mechanisms for customers (see chapter 3.4.2.2).

A pioneer strategy, however, does not necessarily lead to advantages, and thus, cannot be seen as generally superior compared to a **follower** strategy. A number of companies have been burned badly by entering a foreign market too early (Kotabe/Helsen 2008: 308). The pioneer has to spend extensive resources to develop the market, from which successive followers can benefit ("free riding"). Also, pioneers carry a great deal of business risk as there is no or only limited experience regarding the particular business model which might be leveraged in the new country market (e. g., demand patterns, price sensitivities, etc.) Furthermore, the provided customer value proposition has not been proven to be valid in the new country market and might for some reason not be embraced by the local target group. Ultimately, the decision whether to opt for being a pioneer or follower strongly depends on the company and competitive situation and needs to be decided on a case-by-case basis (Holtbrügge/Welge 2010: 129 f.; Homburg/Kuester/Krohmer 2013: 127–129; Kotler/Keller 2009: 280–282).

Deciding on the entry sequence: waterfall or sprinkler strategy

In case the company targets more than one country in its international expansion strategy, which typically happens with many global product launches like smart phones or automobiles, the following key alternatives regarding the entry sequence exist (Homburg/Kuester/Krohmer 2013: 430–433):

- With a **waterfall strategy**, the company enters new international markets gradually over time, which allows a successive development of resources and know-how. Experiences and pitfalls of early market entries can be leveraged for a more effective entry into other country markets later on. However, there is a risk of being outpaced by competitors that may enter markets the company has not yet addressed, and thus, prevent the company from gaining pioneer advantage in all targeted countries.
- With a **sprinkler strategy**, a company pursues a more rapid international expansion by entering a number of foreign markets simultaneously. This has the advantage of gaining pioneer advantages on broad regional scope including the creation of entry barriers for followers. It is also a way to diversify risks across different regions, which, however, requires a higher short-term resource commitment as compared to a waterfall strategy. If the go-to-market approach fails, the company is facing major losses as there is no opportunity to leverage the learnings.
- Many companies chose a **combined waterfall-sprinkler strategy** to get the best out of both strategy options. Hereby, similar country markets are grouped and entered simultaneously following a sprinkler strategy. Clusters or groups of (simi-

lar) countries as well as single countries are entered sequentially following a waterfall strategy.

Determining the best mode of entry

A company has different options for the mode of entry into a foreign country market. Market entry modes differ in their degree of *commitment* (e. g., internal resources allocated to dealing with a particular country market and degree of organizational involvement in a particular location), *risk* that may impact the company's business prospects, degree of *control* over its foreign operations (e. g., decisions on product positioning, pricing, distribution, promotion, etc.) as well as the *profit potential*. Five classic entry modes are shown in figure 3-24. Here is a snapshot overview of the different approaches (Kotabe/Helsen 2008: 291–305; Kotler/Keller 2009: 606–609; Wheelen/Hunger 2010: 211 f.):

- **Export** is one of the oldest forms of internationalization. Hereby, products produced in one country are sold in another country. It is often used as initial option for starting international expansion as it allows a company to "test the waters" for conducting business in a specific country without allocating extensive resources. For small companies that want to sell their products in foreign countries, export is often the sole alternative due to the relatively low commitment required. A firm can decide to *indirectly export* their goods by using a middleman located in its home market. These intermediaries typically have the required know-how and international contacts as well as a corresponding export organization and an overseas sales force. The company may also decide to *directly export* in several ways. It sets up its own export organization within the company, which may be supported by an own overseas sales branch, traveling export sales representatives or foreign-based distributors or agents. Nowadays, export can be effectively supported by a *global web strategy* that allows companies of any size to extend their geographic reach to new customers outside their home countries easily. Depending on the industry they can use online retailers to drive down transaction costs significantly through leveraging virtual shelf space and attract even distant customers in market niches (Kotler/Keller 2009: 607; Rothaermel 2013: 187 f.).
- **Licensing** is a contractual agreement in which the company (licensor) grants rights regarding its proprietary assets to a foreign company (licensee) for a fee or royalty. The rights can be related to certain technology, manufacturing processes, trademarks, patents or other items of value. Licensing can be a very profitable mode for entering a foreign market, particularly for smaller companies that are not willing or capable of investing much resource in foreign facilities or who want to navigate around import barriers they would have to face in case of exporting their goods. Major pitfalls of licensing is the absorption of potential profit by the licensee, a possible lack of commitment of the foreign company, and the risk to create one's own competitor abroad, which is particularly tricky if proprietary product ingredients or components are involved. Some variations of licensing include *contract manufacturing* (i. e. outsourcing production of the firm's product parts or the entire products to a local manufacturer) and *franchising* (i. e. providing a complete business model including brand concept, operating systems, training concepts, etc. to the foreign company).

- **Joint ventures** are a mode of entry where a company (foreign investor) often joins with a local company to create a new legal entity in the target country. The joint venture partners can be a mixture of one or more local and foreign companies as well as local government authorities. There are different ways to share ownership as well as control of the new company. Sometimes, foreign governments may forbid wholly owned ventures in certain industries and require joint ownerships with a local partner as a compulsory condition for market entry. In such cases, the joint venture may have to be majority owned by the local company to get market access. Also, a joint venture may be desirable due to its high return potential driven by a share of profit instead of receiving just royalty payments in case of licensing. Furthermore, joint ventures are a compelling entry option in case of synergies derived from complementary know-how, assets, expertise, etc. Compared to many other entry modes such as exporting and different forms of licensing, joint ventures entail relatively strong control over the operations. Major disadvantages of international joint ventures include the high risk of serious conflicts and friction-driven inefficiencies due to shared control and ownership as well as difficult or unclear exit strategies. Also, like with licensing, there is a risk for a company to develop its own local competitors by providing joint venture partners access to their know-how, technology, etc.
- **Direct investment** is the entry form with the strongest form of commitment and control. Some multinational companies prefer to enter new country markets through wholly owned subsidiaries. This approach is realized either through *acquisition* and *mergers* (i. e. buying up or merging with an existing local company) or *green-field development* (i. e. building an own manufacturing plant and distribution system). Purchasing another company already operating in the local market is a relatively quick way for international expansion. Hereby, the company gets all the profits and has full control over its operations, and thus, avoids all the risks and complexity associated in working with partnerships. It can be particularly beneficial in case there are synergies through complementary product lines and an effective local distribution system of the acquired company. The major disadvantage of direct investments is the burden of carrying all the mar-

Fig. 3-24

Five modes of entry into foreign markets

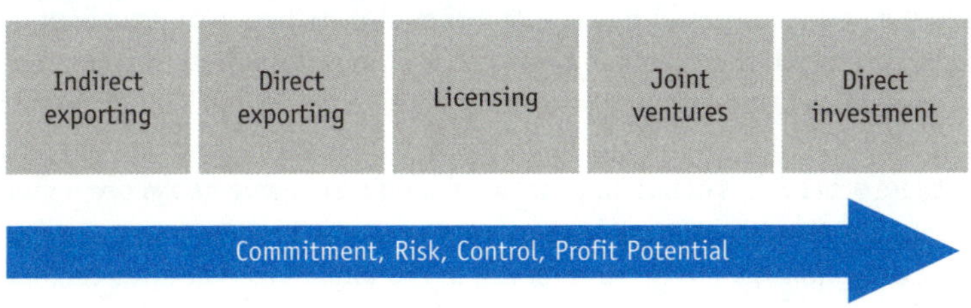

Source: Kotler/Keller 2009: 607; adapted

ket-related, political and economic risks alone. There is also a huge commitment of administrative resources required. Companies have to bring in their own finance, IT, HR, and legal infrastructure of which there is a minimum amount to invest even if the operation is small.

Several *external* decision criteria such as market size and growth, risk, government regulations, competitive environment, cultural distance, or local infrastructure do not only influence the country market selection but also the choice of entry mode once a target country for market expansion has been identified. Furthermore, *internal* criteria such as the company's motives or objectives, the need for control over its foreign operations, its available internal resources, assets, and capabilities, as well as the desired flexibility to adapt the go-to-market approach to local environmental changes need to be considered (Kotabe/Helsen 2008: 286–290). Determining the best entry mode may also be supported by comprehensive quantitative assessments leveraging more comprehensive calculation methods such as the net present value approach from capital budgeting. (Perlitz/Schrank 2013: 342–360; see also chapter 3.5.1.3)

Go-to-market strategy: marketing mix and business model

The final step of planning an international market entry is related to the local business strategy. Although it might be desirable, it is generally very unlikely that a company can replicate its domestic go-to-market strategy 1:1 in a new foreign market. Usually there is at least some adaption required to match business strategy with local conditions of the foreign market. To systematically elaborate on a go-to-market strategy for a new country, a classic **marketing mix** can be used as a frame of reference. Hereby, the question of how to adapt specific elements in the company's product concept, pricing approach, communication program, and sales strategy has to be answered (Homburg/Kuester/Krohmer 2013: 436–447; Kotler/Keller 2009: 609–621). Another way of elaborating on a go-to-market strategy in a foreign country is the **business model** approach (see chapter 3.4.4). Hereby, the description of an existing domestic business model is used as a baseline. Required adjustments of value proposition, value creation and delivery as well as value capture to the local market conditions have to be analyzed along the different building blocks, which may result in different types of business model innovation (see chapter 3.4.4.3). Questions that need to be clarified include:

- Does the customer *value proposition* need to be refined to better address the targeted local *customer segment* (e.g., different tastes, packaging, services, brands, etc.)?
- Does the new country market require different distribution and communication *channels* to reach the customer segment (e.g., new distribution structures with different sales channels, emphasis on other forms of media with different messaging)?
- Are there other forms of building and managing *customer relationships* required as compared to the domestic market?
- How can *sustainable profits* be achieved in the foreign market given factors such as different price level and cost structures?
- Are there any new or changed *key processes* required for the desired go-to-market approach?

- What *key resources* or *partnerships* are necessary to make the business model for the foreign market effective?

To support this process, business model patterns of foreign companies that are already present and successful in the local market may be leveraged (see chapter 3.4.4.4).

3.3.2 Transnational Management

Best exploiting "internationality" of multinational corporations for competitive advantage

International expansion driven by motives such as additional sales and profits as well as the need to manage existing revenues is one of the reasons why multinational corporations evolve. However, it does not explain why some of them continue to grow and flourish whereas others struggle in their global competitive arenas. In addition to the question of international market expansion, multinational corporations can take another perspective in their quest for competitive advantage. Those companies are in a position to create competitive advantage by exploiting their "internationality" better than rivals are able to do. Multinational enterprises (MNEs) or multinational corporations (MNCs) are referred to as companies that "have substantial direct investment in foreign countries and actively manage and regard those operations as integral part of the company, both strategically and organizationally." (Bartlett/Beamish 2011: 2) In other words, these corporations are already operating on an international scope and managing cross-border value chain activities with a number of subsidiaries (see Strategy Practice Example 3.9). Transnational management relates to leveraging international strategic management approaches to identify and exploit specific sources of competitive advantage that those multinational corporations have. The focus shifts from optimizing a single country market entry to an integrated view of pro-actively managing and coordinating business operations located in different countries with respect to achieving the company's overall strategic aspiration. It reflects a type of corporate parenting (see chapter 3.2.1) that is characteristic for multinational corporations (Bartlett/Beamish 2011; Bartlett/Goshal 2002; Cullen/Parboteeah 2008: 205–215; Holtbrügge/Welge 2010: 132–154).

3.3.2.1 Strategy Postures in Multinational Corporations

Forces of global integration or standardization

Managers of multinational corporations have to respond to two sets of external demands that vary depending on the specific industry (Bartlett/Beamish 2011; Bartlett/Goshal 2002; Morschett/Schramm-Klein/Zentes 2010: 29–32): On the one hand, they face forces of *global integration*, which requires them to configure their value chain so each activity is located in the country with the highest utility, that is where the company can do them best or cheapest. Consequently, the multinational corporation needs to coordinate and link different countries in which they operate. It also needs to conduct business similarly throughout the world to achieve global efficiency. Companies who are able to better realize global integration of their worldwide operations can realize competitive advantage. Corresponding sources include

- achieving **economies of scale**, for example, through high volume production of globally standardized products at global production plants from which they are

Strategy Practice Example 3.9

Competitive Advantage in Multinational Corporations Illustrated with Nestlé

Figure 3-25 illustrates four sources of competitive advantage at Nestlé that are strongly related to its international scope reflecting both elements of global integration as well as local responsiveness: first, a combination of strong trusted global and regional brands; second, the global Nestlé research center in Lausanne (Switzerland) as well as a number of specialized de-centralized research units across the globe allowing both global efficiencies as well as location advantages; third, a comprehensive geographic presence that evolved over many decades and can be leveraged for gaining specific competitive advantages; fourth, a highly developed global culture with principles and values that are congruent with all cultures.

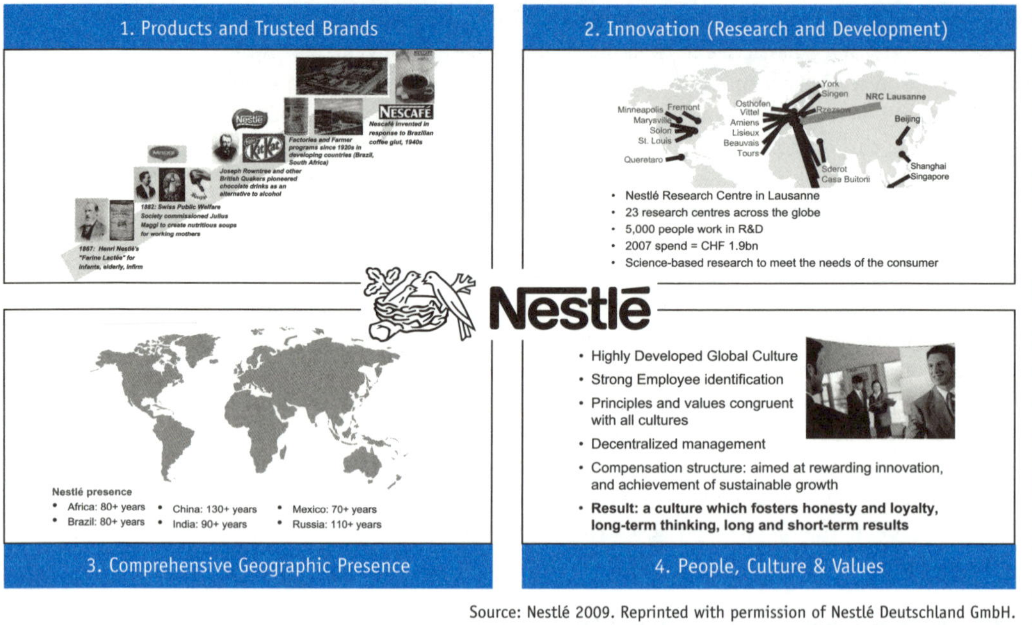

Source: Nestlé 2009. Reprinted with permission of Nestlé Deutschland GmbH.

Fig. 3-25: Four competitive advantages at Nestlé

exported to the different country markets. This leads to static and dynamic scale economies and other advantages that ultimately result in highly competitive prices.

- realizing **economies of scope**, for example, through global brand names, servicing global customers with global key account managers, global sales channels (e. g., Amazon, eBay), the use of existing global sales and distribution networks for multiple products, or shared R&D activities (e. g., global technology platforms, standardized products that appeal to customers in many countries)

- exploiting **national differences** in terms of factor cost and availability, for example, through leveraging comparative labor cost advantages by locating global production facilities in certain countries, particularly in less capital-intense industries (e. g., textiles, shoes, etc.) or by ensuring access and control of raw materials in case of limited or no home-country sources (e. g., oil and gas, rubber, etc.)
- leveraging **global learning and innovation,** for example, through companywide exploitation of knowledge created in one country or pooling resources from various countries to drive speed to market and enhance the business contribution of innovation in the multinational corporation.

Forces of local responsiveness or differentiation

On the other hand, there are forces of *local responsiveness* that require the multinational corporation to adapt its operations to the differences of various national markets. There are many examples of companies that tried to transfer their successful products or business models to foreign markets and failed. In case there are forces for local responsiveness, multinational corporations have to sense and respond to those demands. Sources for gaining competitive advantages through local responsiveness are strongly related to the idea of overcoming country market entry barriers (see fig. 3-23) and include

- **sociocultural differences** such as various consumer tastes and preferences, country-specific ways of doing business successfully, differences in income levels and income distribution in various countries, etc.
- **national infrastructure** such as different technical standards (e. g., voltage, TV broadcasting system, etc.), distribution channels (e. g., supermarkets versus bazaars or small local kiosks)
- **government demands** such as national product laws and trade regulations, host country demands regarding a degree of local content or the need to join with a local company
- **local competitive dynamics** such as number and profiles of specific local competitors as well as differences in local distribution systems that require an adaption of go-to-market strategies in different countries

Four strategic orientations in multinational corporations

Global integration and local responsiveness provide a frame for deriving the four generic strategy postures in multinational corporations shown in figure 3-26. Companies pursuing a *global strategy* view the world as a single unit of analysis. They centrally configure, link and coordinate their value chain activities in different countries with the goal to achieve worldwide efficiency through a high degree of global integration. For example, many multinational corporations in the consumer electronics industry offer globally highly standardized products and treat the world as a single integrated strategic unit. A *multinational strategy* is based on an entirely different approach as it views foreign markets as a portfolio of local opportunities that need to be captured with a high degree of local responsiveness. Rather than linking and integrating value chain activities in different countries, multinational corporations following this approach tend to manage their international operations as a decentralized federation. For example, some firms offering branded packed products in the consumer goods industry pursue multinational strategies by providing locally adjusted

Fig. 3-26

Strategy postures in multinational corporations

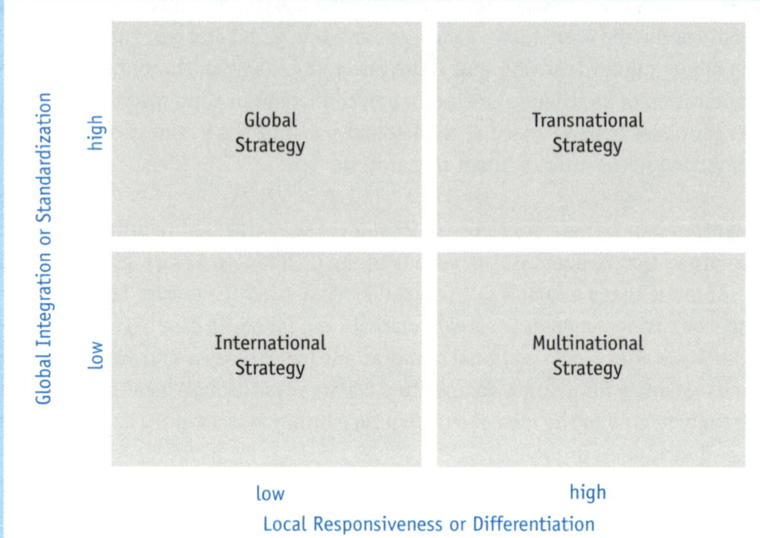

Source: Holtbrügge/Welge 2010: 132; adapted

offerings in almost every country they serve. Companies pursuing an *international strategy* try to replicate their domestic strategy in its different international markets with no or only a minor degree of local responsiveness. They "export" their domestic capabilities such as products, processes, or business models worldwide to less advanced foreign markets and are managed as a coordinated federation with an ethnocentric mindset. Finally, companies pursuing a *transnational strategy* are simultaneously responding to local national needs, global demands, and cross-border learning opportunities and are managed as an integrated network. It is a hybrid strategy that tries to leverage different and partially conflicting sources of competitive advantage at the same time (Bartlett/Beamish 2011; Bartlett/Goshal 2002; Cullen/Parboteeah 2008: 205–215; Holtbrügge/Welge 2010: 132–134).

3.3.2.2 Transnational Strategies

Leveraging different sources of competitive advantage simultaneously

A transnational strategy can be considered an evolving mentality that is becoming increasingly important for cross-border business management in multinational corporations (Bartlett/Beamish 2011: 124–127). Whereas each of the three classic strategic orientations—global, international, and multinational—have their own specific merits, a transnational strategy seeks to leverage the different sources for gaining competitive advantage simultaneously. A firm adopting a transnational strategy considers the diversity of environments in which it operates as a key strategic asset that can be exploited to create the three layers of worldwide competitive advantage shown in figure 3-27. They recognize the value of each of the more traditional strategy postures in multinational corporations but try to integrate them to one holistic transnational approach for achieving opportunity-related (sales and profitability), risk-oriented and competence-oriented strategic goals (see also chapter 3.3.1.1). Each of these three layers is considered more in detail next (Bartlett/Beamish 2011: 210–216).

Global efficiency: reducing costs and enhancing revenues

Global efficiency relates to both lowering costs and increasing revenues in a multinational corporation, and thus, drives profitability as well as growth. This is based on the simplified view of the firm as an input-output system that can be optimized through (1) increasing the value of outputs (i. e. revenues), (2) lowering the value of inputs (i. e. costs), or doing both (see also fig. 1-11 in chapter 1.1.3.1). This simple logic can be applied to the global integration-local responsiveness framework introduced earlier (Prahalad/Doz 1987). On the one hand, a firm that globally integrates its cross-border business activities can significantly lower its *cost* through scale and scope economies as well as leveraging comparative cost advantages in certain countries such as labor, infrastructure or transportation cost (location economies). On the other hand, local responsiveness translates primarily in higher *revenues* through effective differentiation and adaption of offerings and business models to local market requirements such as demand patterns and tastes, competitive structures, or government regulations.

Multinational flexibility: managing risk and exploiting opportunities

Multinational flexibility is based on exploiting the diversity of operating environments within a multinational corporation. Macroeconomic, political, competitive as well as resource risks vary strongly across countries and can significantly change over time given the high volatility underlying the global business environment today. To exploit the diversity among its internationally dispersed operations, a firm needs to

Fig. 3-27

Three layers of worldwide competitive advantage

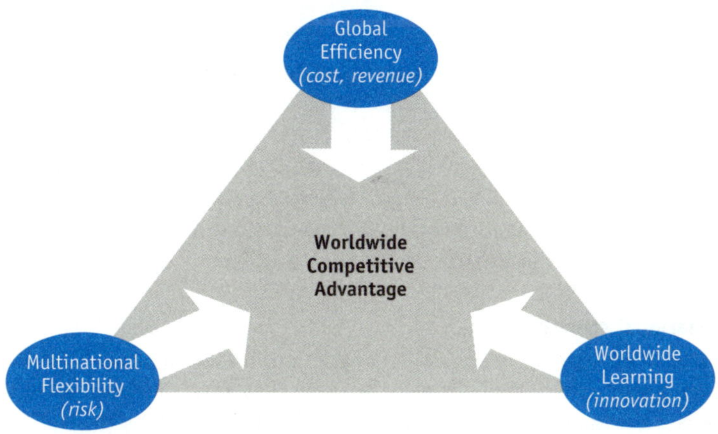

Source: following Wunder 2004a: 106 and Bartlett/Beamish 2011: 210–214

design its organization in more flexible terms as compared to a stable environmental situation. A multinational corporation operating with a network of globally integrated international subsidiaries that built in this flexibility is in a position to better respond to changes and discontinuities in its broad environment. For example, it may shift production from countries facing national political changes (e. g., exchange or interests rate adjustments driven by local policy) or changes in economic conditions (e. g., increasing labor cost) to other or more stable sites. Furthermore, being present in country markets with different risk profiles may create competitive advantage compared to rivals with a more local or regional focus. It allows multinational corporations to counterbalance revenue losses in countries being strongly burdened by global or regional economic crises with revenues from other country markets less impacted.

Worldwide learning

Developments such as shorting life cycles of products and technologies, increasing importance of global standards as well as new consumer trends emerging anywhere in the world increase the need for multinational corporations to fast and effectively innovate. "Today, the newest consumer trend or market need might emerge in Australia or Italy, and the latest technologies to respond to these new needs may be located in Japan or Sweden. Innovations are springing up worldwide, and companies are recognizing that they can gain competitive advantage by sensing needs in one country, responding with capabilities located in a second, and diffusing the resulting innovation to markets around the globe." (Bartlett/Beamish 2011: 123) Hereby, the headquarters is not the central competence center but acts more as an orchestrator creating conditions and processes for globally coordinated innovation. A key source of competitive advantage is the skill of a multinational corporation to learn by leveraging the diversity of the environments it operates. The goal is to utilize its access to worldwide information as well as expertise and knowledge for creating and diffusing innovative product, processes or business models in a better way than multinational or domestic competitors are able to do.

Executing transnational strategies

As with many other strategies, one of the biggest challenges for a transnational strategy to come true is not its formulation but its execution. Up to now it is still an ideal-typical strategy type that is hard to find fully implemented in company practice (Holtbrügge/Welge 2010: 153 f.). Effectively executing transnational strategies requires both a corresponding organization (see chapter 4.2) as well as an effective strategy execution system (see chapter 4.3) that is designed to support and drive the three layers of competitive advantage: global efficiency, multinational flexibility and worldwide learning. Key considerations of organizational design to make a transnational strategy come true are (Wunder 2014a: 106–112):

Organizational structure

- Optimizing the global configuration of interlinked value chain activities. Instead of conducting each activity at every location, they are dispersed anywhere in the world where they provide the optimal strategy contribution for the multinational corporation in total (i. e. they are performed best or at lowest cost).
- Flexible and mainly process-oriented allocation of responsibilities and decision making authority across the boundaries of business units, subsidiaries, or func-

Strategy Practice Example 3.10

Global Integration vs. Local Responsiveness at UNILEVER

Figure 3-28 illustrates a global integration-local responsiveness framework (Prahalad/Doz 1987), which is applied to various businesses, core functions as well as tasks at Unilever. It shows that the two dimensions vary according to different organizational dimensions within the same company. In this example, the chemicals business is facing strong forces for global integration. Unilever's packed food business, on the other hand, is characterized by a high level of local responsiveness. The detergent business is in an intermediate position regarding the two dimensions. Within the detergent business, research is strongly controlled by a global perspective to gain efficiency through global integration. Marketing decisions, on the other hand, are strongly influenced by the national subsidiaries to ensure a sufficient level of responsiveness to local markets. Among the marketing tasks, product policy is highly integrated, which means that it is coordinated from an overall global perspective. Though, promotion decisions are entirely decentralized and left to the discretion of national subsidiary managers.
(Source: Bartlett/Ghoshal 2002: 111)

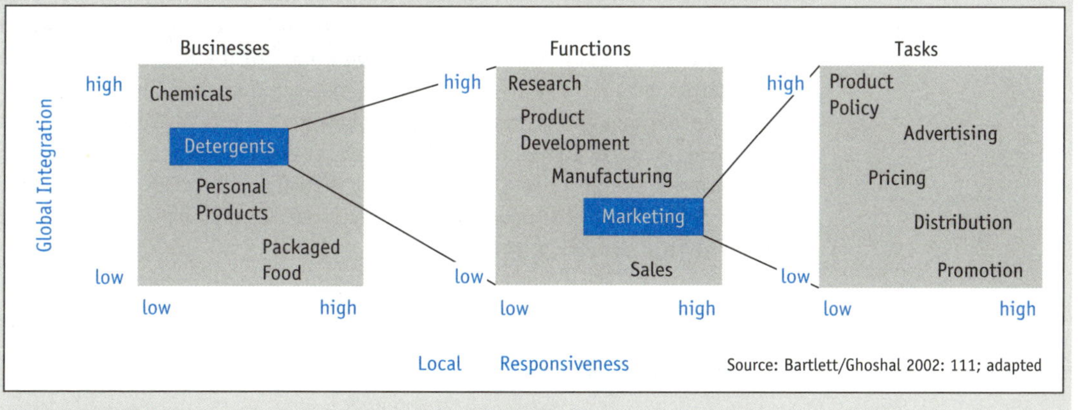

Fig. 3-28: Integration and differentiation needs at Unilever

tions. Furthermore, different global roles or worldwide strategic mandates are allocated to international subsidiaries (e. g., leading, contributing to, or only implementing global strategic themes). Country subsidiaries are controlled differently depending on their strategic role (see also fig. 4.57 in chapter 4.3.3.5).

- Individual assessment of each business unit, subsidiary, function, and value chain activity in terms of global integration and local responsiveness to identify the best lever of creating competitive advantage for the multinational corporation in total (see Strategy Practice Example 3.10).

Organizational culture

- Establishing a transnational management mentality, particularly among ethnocentric leaders and international subsidiary managers that are dominated by their siloed departmental or national perspective. Realizing the different layers of worldwide competitive advantage does not only require structural configurations but also a corresponding mindset, that is a geocentric-holistic leadership perspective of managers at different levels in a multinational corporation.

- Anchoring of commonly shared global strategic guideposts (see chapter 3.1) as orientation. International managers need to be aligned around a global vision, mission, values as well as key corporate strategic themes to better understand and buy in their specific strategy contributions. Nestlé provides an example of a highly developed global culture (see Strategy Practice Example 3-9 with the Nestlé strategic performance framework (4x4) providing strategic orientation worldwide, see fig. 4-73 in chapter 4.3.6.2).
- Building and practicing a culture of open communication and learning. Effectively executing transnational strategies requires the willingness to share and leverage information such as best practices, customer insights, and competitive intelligence within the company on a worldwide scale.

One way to effectively support the realization of competitive advantages derived from transnational strategies is a global management process for strategy execution based on the balanced scorecard approach (see chapter 4.3). With an appropriate design and application it can help to drive global efficiency, multinational flexibility as well as worldwide learning and match those strategic elements with both organizational structure and—in particular—organizational culture. It allows articulating, aligning and managing strategic priorities related to the different sources of competitive advantage across the global network of international subsidiaries in a flexible and practical way (Wunder 2004a: 257–260).

3.4 Business Strategy

Making a single business successful

Whereas corporate strategy is about how to deal with many businesses or markets that a company is operating in, business strategies relate to an individual business level. A business strategy is targeted toward improving the competitive situation of a single business that operates with its products and services in a specific competitive arena. It may serve a particular industry or market segment either with a single product or one or more product lines. From an organizational perspective, a *strategic business unit* fulfills an individual market function that may or may not be related to other business units in the corporation. It has its own customers as well as competitors and formulates and implements its own distinct strategic plan. Business strategy may also be applied for smaller companies with essentially only one business. There are specific strategic questions that have to be addressed and a variety of strategic approaches and insights that can be used as orientation for strategy formulation. The main business-related themes addressed in this chapter are derived from the following four questions:

- What are main approaches to grow a business (*intensive growth strategies*)?
- Which *generic competitive strategies* can a business apply to win in the market?

- How can a business break out of traditional competition (*cooperative strategies and blue ocean strategies*)?
- How can a business refine and innovate its strategy and success logic *(business model)*?

3.4.1 Intensive Growth Strategies

Improving the competitive position within an existing business

In chapter 3.2 different approaches have been introduced in the context of corporate strategy, which are essentially based on *expanding* a corporation's scope of business. The corporation may add promising businesses that are either related or unrelated to its current business (diversification strategies). It can also increase sales or profits by expanding its business scope across its industry's value chain through vertical backward or forward integration (vertical integration strategies). Additionally, the corporation may look at new business opportunities from a geographic perspective and expand its business activities to new regions (international strategies). Another way for managers to improve their firm's competitive position is by looking at opportunities for improving its *existing* businesses, also referred to as *intensive growth* or *intensive strategies* because there are intensive efforts required if a firm wants to improve its competitive position within an existing business (David 2013: 171; Kotler/Keller 2009: 44–46).

Product-market matrix

A useful framework for elaborating on intensive growth opportunities is the product-market matrix illustrated in figure 3-29, which is also called Ansoff matrix with reference to its original developer (Ansoff 1957). The framework can be leveraged as orientation to elaborate on corporate-level strategies (Johnson/Whittington/Scholes 2012: 135) as well as on business-level strategies. With the exception of diversification, the framework essentially relates to the question of which market

Fig. 3-29

Product-market matrix with three intensive strategies for business growth

	Current Products	New Products
Current Markets	1. Market-penetration strategy	3. Product-development strategy
New Markets	2. Market-development strategy	(Diversification strategy)

Source: Kotler/Keller 2009: 45; adapted from Ansoff 1957

segments are addressed with which products or product lines in an *existing business* (Hill/Jones 2010: 197; Lombriser/Abplanalp 2004: 274 f.; Müller-Stewens/Lechner 2011: 255 f.): One dimension of the matrix relates to markets or market segments addressed by a business. It is centered on the question whether the business wants to grow only in markets or market segments already served or add new market segments. This is combined with a product dimension addressing the question whether the two market approaches are pursued either with existing product lines or with new product lines. The three resulting intensive growth strategies are explained next. The fourth option—diversification—is described in chapter 3.2.2.2.

3.4.1.1 Market Penetration

Intensifying sales of existing products in markets currently served

With a market penetration strategy, the company "penetrates" a market already served with products currently offered through greater marketing and sales efforts. Compared to the other options, this strategy has a relatively low degree of innovation and low level of risk as it does not require the business to move into unchartered territory. Market penetration can be driven by measures such as intensifying sales by an increased sales force, spending more funds on effective advertising (e. g., print, TV, radio, internet), or driving sales promotion initiatives such as bonuses, giveaways, and retail displays. In general, companies seeking to grow a business may find a greater market penetration particularly effective in the following situations (David 2013: 171):

- low saturation level with the particular product in current markets
- significant room for increasing usage or consumption rate of current customers
- declining competitor market shares in a growing market
- the company can demonstrate a historically satisfactory correlation of their market expenditures (advertisement, sales promotion) and corresponding sales increase
- major competitive advantage can be expected from increased economies of scale.

Limits of a market penetration strategy

A company pursuing a market penetration strategy needs to be prepared for potential competitor retaliation and an increase in industry rivalry. The risk of price wars and expensive marketing battles need to be considered before deciding on aggressive sales initiatives to penetrate the market (Johnson/Whitington/Scholes 2012: 136). Also, market penetration alone is typically not sufficient for a business to survive, and thus, often combined with other growth strategies as the product life cycle indicates that even a highly promoted product will ultimately reach maturity (Homburg/Kuester/Krohmer 2013: 76).

3.4.1.2 Product Development

Most product development is incremental innovation.

A business pursuing a product development strategy for driving growth focuses on developing new or modified products for the same markets or market segments that the company is already serving. This strategy option is particularly important for high-tech firms such as Apple or Samsung that are continuously launching new and improved products to (re-)gain competitive advantage. Also in other industries such as food, driving new product development is a key success factor (Wunder/Bausch

2014a). Hereby, two types of innovation need to be distinguished. *Radical innovation* or breakthrough innovation refers to entirely novel products that may shape existing and create new markets. *Incremental innovation* or proximal innovation—also called product renovation—targets slight improvements of an existing product concept for existing or new markets leveraging existing technology (e. g., additional features, improved ease of use of packaging, etc.). In some industries, such as fast moving consumer goods, continuously providing incremental innovation can be regarded as a standard "way of business life" driven particularly by the demands of large trade chains and consumers. Also, other industries that have been traditionally driven by radical innovation and blockbuster thinking, such as branded pharmaceutical companies, are starting to emphasize incremental innovation for driving profitability and growth due to a lack of productive R&D pipelines. In general, the vast majority of innovations are proximal (incremental) though breakthrough (radical) innovations typically receive the greatest public attention (Rothaermel 2013: 182 f.; Wunder/Bausch 2014a: 47 f.).

A systematic innovation process drives product development.

Figure 3-30 provides an example of an innovation process that can be leveraged for driving a product development strategy. Its general logic can be applied to both radical and incremental innovation. Key success factors to ensure an effective and efficient innovation process are (Wunder/Bausch 2014a: 49):

- a clear *strategic focus* (e. g., provided by new strategic areas to be investigated or clear problem statements) to drive effective idea generation
- a *broad input of ideas* in line with the strategic focus leaving enough room for target-oriented creativity coming from within or outside the company (i. e. open innovation)
- an *effective and efficient prioritization* based on clearly defined decision gates (or *stage gates* related to the different phases or stages in the innovation process) with a predefined and consistent management committee and clear decision criteria to drive speed to market and ensure a strong business impact.

At the end of the process, post launch results are evaluated and product concepts are optimized based on new customer and market insights. Also, it is important to note that promising but dismissed ideas that did not make it to the development or launch phase must not be entirely forgotten. They may be reviewed and resubmitted in the future. Sometimes, a market is not ready yet for a new innovation and the company needs to be alert to a better window of opportunity.

Conditions in favor of a product development strategy

In general, a product development strategy that seeks to increase (profitable) sales may be especially effective when a company has *successful products in the maturity stage* that can be improved to attract already satisfied customers, and thus, extend their product life cycles. Furthermore, a firm operating in a competitive arena characterized by *rapid technological developments* or companies facing major competitors that have competitive advantage based on *better-quality products at comparable or lower prices* may need to pursue a product development strategy. Another condition that encourages an intensive growth strategy based on product development is when the company has particularly strong capabilities in research and development as well as marketing (David 2013: 173).

Fig. 3-30

Innovation process for driving successful product development strategies

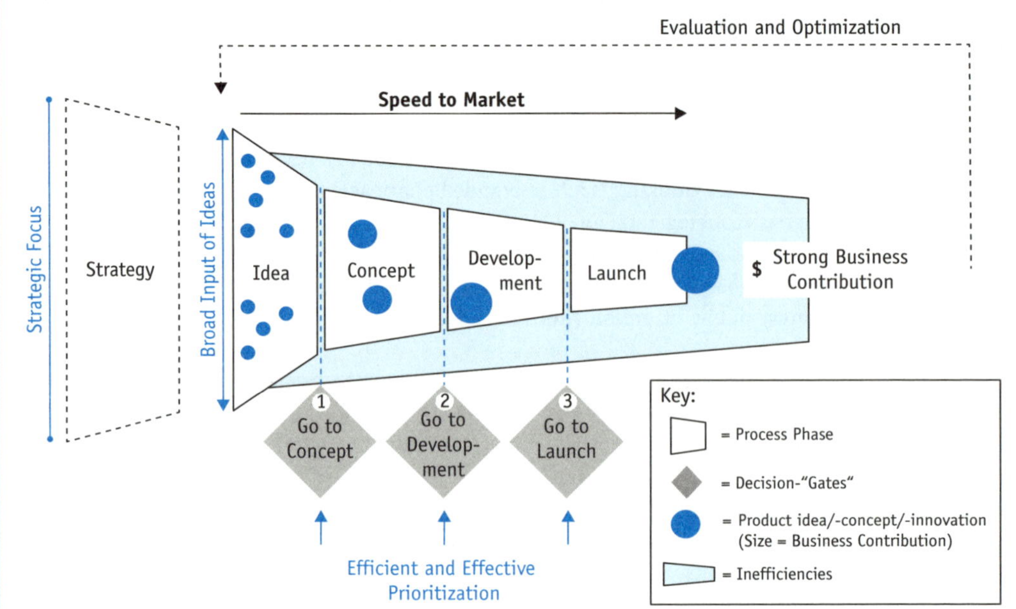

Source: Wunder/Bausch 2014a: 49; adapted

3.4.1.3 Market Development

Leveraging existing products in new markets

With a market development strategy, a company tries to sell existing products in new market segments in existing geographic markets (e.g., new customer groups that may be reached with new sales channels) or to entirely new geographic markets such as new countries (see chapter 3.3.1). Companies considering a market development strategy may want to leverage one of the best product strategies introduced in chapter 3.4.2.1. First, some firms pursue a differentiation strategy and try to exploit product differentiation advantages created in their current competitive arena with superior quality, compelling brands, outstanding features, etc. in new markets. Premium German carmakers offering their existing models in a variety of countries provide numerous examples of this approach. On the other hand, Japanese automakers historically provided examples of leveraging a low-cost strategy for successful market development. Once they had entered new international markets with selected existing products, these firms upgraded certain models over time, and thus, became also differentiators (Hill/Jones 2010: 198; Hungenberg 2011: 209). This approach is also referred to as outpacing strategy (Gilbert/Strebel 1987) and is further elaborated on in chapter 3.4.2.1.

Market development may require product adjustments.

Typically, a market development strategy involves also some degree of product development or adjustments as product concepts can rarely be just "off-loaded" in a new market. Particularly introducing existing domestic products in a new international country market often requires adaptations of the current product derived from a solid analysis of local demand-, competition- and institutional factors. Those local adjustments can range from changing only packaging or service to adjusting core product features or the entire business model to local preferences and customer requirements of the new target group (see chapters 3.3.1.2 and 3.3.1.3).

Conditions in favor of a market development strategy

Market development can be a particularly effective intensive growth strategy in the following situations (David 2013: 172): First, the *availability of new sales and distribution channels* that are reliable, inexpensive and of good quality are in favor of market development (e. g., internet technology combined with effective logistics processes have provided companies a way of driving sales by reaching new target groups also for niche products). Second, if a company is very successful and good at what it does (e. g., superior product) it may be able to successfully leverage these *core competencies* in other markets. Third, new and unsaturated or *untapped markets* may exist for the companies' products. Hereby, the key challenge is to identify those latent or even incipient markets and address them with an effective business model (see also chapters 3.3.1.3 and 3.4.3.2). Fourth, a business has to have sufficient *capital, human resources and experience* to manage and succeed with expanded operations. Fifth, a firm with excess production capacity may drive market development to fully utilize it, and thus, realize *scale economies*. It may also want to increase its capacity for further scale economies required to achieve competitive cost and prices domestically, which is one of the drivers why some companies seek to expand in new geographic markets. Finally, if the existing competitive arena where the company competes is rapidly becoming *global*, the firm may have no other choice than to expand its markets internationally if it wants to stay competitive (see chapter 3.3.1.1).

3.4.2 Generic Competitive Strategies

Creating competitive advantage in an industry to outperform competitors

Companies can draw from a number of generic competitive strategies to guide strategy formulation on a business level for achieving competitive advantage and ultimately superior performance. Those strategies are called "generic" because they reflect basic approaches that can be applied by any type or size of organization across different industries. Two of the most popular generic competitive strategies to create competitive advantage are *cost leadership* and *differentiation* (Porter 1985). They are widely accepted both in business practice and the academic world alike. Those classifications have been adopted, refined, altered and supplemented in a variety of ways. A number of additional strategy types have been empirically identified and conceptually developed that go beyond those fundamental approaches of how to position a product or service offering. They all have in common that customers ultimately have to see sufficient value for what they are paying for and the firm's business has to be able to create greater customer value than competitors (see chapter 1.1.3).

Triangle framework of generic business strategy

Figure 3-31 provides a frame of reference for distinguishing generic business strategies that is derived from the so-called *delta project* (Hax 2009; Hax/Wilde 2001) as well as the *value leadership disciplines* for market leadership (Treacy/Wiersema 1995). It has been partially adapted in this book both regarding terminology and strategy types covered. The framework is intended to provide orientation for different key types of business strategies including the two generic competitive strategies of cost leadership and differentiation mentioned before. The dark triangles provide different ways of how a business can strategically position itself based on three distinct strategy options. Additionally, three fundamental ways of how each of those strategy options can be created are shown in the light grey triangles (Gaiser/Wunder 2004: 461; Hax 2009; Hax/Wilde 2001; Kaplan/Norton 2004: 319–346; Treacy/Wiersema 1995):

- **Best Product** strategies are based on customer value derived primarily from products, either by low total cost (i. e. product price or total cost of ownership with industrial goods), differentiation (e. g., through product quality, functionality, design, brand, etc.), or a combination of both.
- **Lock-In** strategies are based on achieving dominant market positions by creating high switching cost for customers through various mechanisms such as creating proprietary standards, creating an exclusive customer intimacy, as well as offering complementary products.

Fig. 3-31

Triangle of generic business strategies

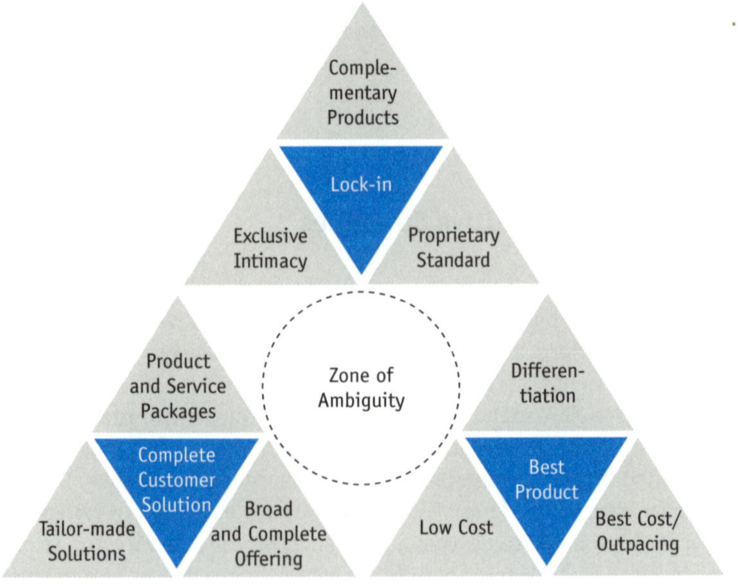

Source: following Hax 2009: 6 f.; Hax/Wilde 2001; adapted

- **Complete Customer Solution** strategies are targeted at building long-lasting relationships with customers based on broad and complete offerings, creating compelling product and service packages as well as developing tailor-made solutions.

Zone of ambiguity

For achieving outstanding results, each of the different strategy options requires the development of a creative value proposition based on a solid understanding of customer needs and wants in the targeted segments. A business is advised to select a preferred strategic positioning and align its core processes and activities to drive this strategy effectively. A clear and relevant strategic positioning is considered to be a key factor for achieving competitive advantage and ultimately superior market success (Porter 1996; Treacy/Wiersema 1995: 44 f.; Hax/Wilde 2001: 13 f.; see also chapter 1.1.2.3). Businesses with an unclear strategic positioning are referred to as being in a *zone of ambiguity*. "Not choosing means ending up in a muddle. It means hybrid operating models that are neither here nor there, and that consequently cause confusion, tension, and loss of energy. (...) Not choosing means setting yourself up to be overtaken by another player that is committed to unmatched value and focused on how to achieve it." (Treacy/Wiersema 1995: 45) There are companies that effectively combine several of those strategy options in their business models (see chapter 3.4.4). For example, Apple has been successfully pursuing a best product strategy based on differentiation with its iPod and, at the same time, applied a lock-in strategy with iTunes (see Strategy Practice Example 3.12 in chapter 3.4.4.2).

3.4.2.1 Best Product

Product-/service-related competitive advantage

Best Product strategies, which are centered on the product (or service) itself, relate to the most classic form of competition. They raise three fundamental questions for a business:

- Should we compete on the basis of *low cost*, and thus, *low price* in a fairly undifferentiated product category, or should we *differentiate* our product and services based on some unique features which are valued by our customers paying a price premium?
- Should we strive for providing our customers *"value for money"* and deliver some essential desirable features at a relatively low price? Or should we alter between a low-cost and differentiation approach?
- Should we compete with major competitors in a *broad market* for achieving a big market share but may face intense rivalry, or should we focus on a *narrow target market or niche* that provides profitable segments and is less sought-after by major competitors?

These three questions give rise to different competitive strategy options shown in figure 3-32. The classical four strategy options are broad/focused low-cost and broad/focused differentiation. They result from a combination of the type *of competitive advantage* pursued and *the competitive arena* or scope of a company (Porter 1985). The four generic strategies have been supplemented by a hybrid strategy approach with two different strategy types in the center (Thompson et al. 2014: 121 f.).

Fig. 3-32

Generic Best Product strategies

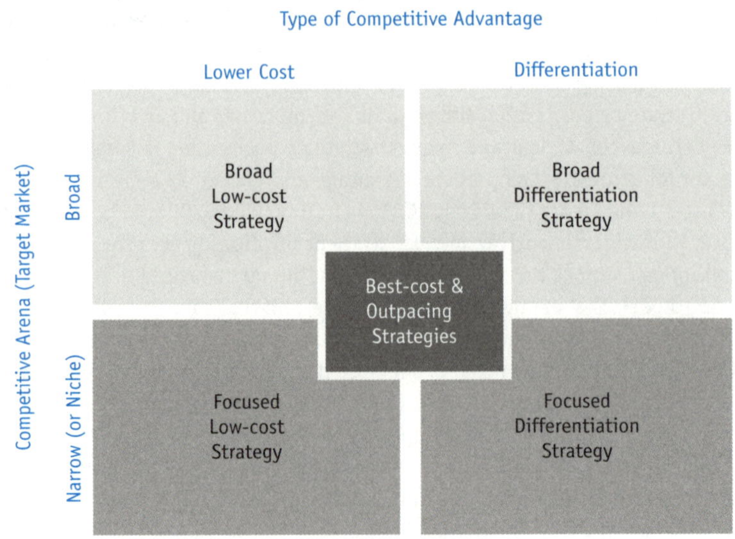

Source: Thompson et al. 2014: 122 and Porter 1985: 12; adapted

Differentiation

A differentiation strategy seeks to differentiate the products or services from competitor products or services in a competitive arena by superior attributes that are valued by customers and allow a price premium. The attributes that are relevant for a successful differentiation such as quality, features, value adding services, brand, etc. vary by industry and market. Firms typically select the attributes they want to focus on and leverage for a differentiation strategy very carefully based on comprehensive customer and market analysis (e. g., conjoint analysis). Firms practicing successful differentiation can generally expect to realize premium prices and corresponding profit margins, increase sales due to additional buyers that are attracted by the differentiating features, and ultimately gain customer loyalty by, for example, differentiating brands that create emotional bonds. A focused differentiation strategy is appropriate when there is a buyer group in a niche market that is willing to pay a price premium for getting products or services with special attributes. Most markets are likely to have a segment of upscale buyers that can be appealed with differentiated products (Thompson et al. 2014: 129–136). Differentiation may be achieved by a variety of individual or combined elements in a product's value proposition (Johnson/Whittington/Scholes 2012: 116). Some examples are engineering design and performance (BMW, Mercedes), the brand (Jägermeister, Fiji Water), luxury and prestige (Breitling, Rolex), product reliability (Bosch, Miele), multiple features (Microsoft, Apple), unique taste and effect (Coca-Cola, Red Bull), or other factors such as superior service (Singapore Airlines, Ritz-Carlton), lifestyle and fashion (Zara, PUMA) or health and sustainability (Interface Inc., Rapunzel, Frosta, Vaude).

Important company drivers for realizing a differentiation strategy can be found in almost any area of a firm's value chain (see chapter 2.3.1.2). Corresponding processes include (Porter 1985; Thompson et al. 2014: 130 f.)

- new product development driving superior product features, design and performance
- excellent basic and additional customer services
- flexible and reliable manufacturing and supply chain systems
- radical and incremental innovation processes based on deep customer insights
- excellent quality management and continuous quality improvement
- strong marketing and brand-building activities as well as market research
- high-quality inputs with reliable and excellent suppliers
- effective human resource management to get talent and improve employee skills, expertise and knowledge.

Market conditions that favor a differentiation strategy include (1) a high level of *diversity among customer needs*, wants and product uses; (2) *few or no strong rivals* that are likely to follow a similar differentiation strategy; and (3) *fast-paced technological change* influences customer requirements and heightens their interest for differentiated products. Serious pitfalls of differentiation lurk when new attributes are easily and quickly copied or imitated by rivals, which leads to just very temporary competitive advantages, or when the effort to differentiate is so high that it erodes profitability. Companies need to make sure that their differentiating features are driven by relevant customer requirements and are appreciated by the market and not "over-engineered" or too expensive by customer standards. Companies may also switch from one form of differentiation to another. For example, a number of world-class engineering companies that used to differentiate solely on product quality have started to increasingly emphasize their brand as an attribute of differentiation. One reason is low cost competition from certain regions entering the market with significantly improved product quality that now provides an alternative for value-/price-conscious buyers (Thompson et al. 2014: 134–136).

Low cost

Firms pursuing a low-cost strategy try to attract customers by offering lower prices than competitors. This can be realized by achieving lower overall cost than competitors with comparable products. Because of the relative total cost difference, a company can charge lower prices for its products than rivals are able to do and still realize a satisfactory level of profitability. Often, this generic competitive strategy is also referred to as "cost leadership" strategy. This, however, requires that a company becomes the lowest-cost (price) provider in the entire industry rather than just being one of several firms of a strategic group with comparatively low cost. Achieving a market leadership position in terms of total cost requires a discipline of operational excellence (Treacy/Wiersema 1995: 49–63; Wunder/Bausch 2015). It may be supplemented with product reliability and durability for achieving a product's lowest total cost of ownership for the customer. Companies that are successfully following a low-cost strategy include Primark (clothing), Ryanair (airline), Tata (automotive), Walmart (retail), Alamo (rental cars), and Oettinger (beer brewery).

Like with differentiation, important company drivers for realizing a low-cost strategy can be found in almost any area of the firm's value chain (see chapter 2.3.1.2). All different areas and activities in an organization have to be aligned around a low-cost approach to achieve cost leadership in an industry. Corresponding processes include (Porter 1985; Thompson et al. 2014: 123–125):

- increasing scale of operations to drive economies of scales (e. g., product standardization)
- fully leveraging experience and learning curve (including continues improvement)
- operating facilities at full capacity (fixed cost absorption)
- improving efficiency in supply chain (e. g., ordering, purchasing, materials handling, etc.)
- using lower-cost supplies
- gain all available concessions by using the firm's bargaining power
- direct R&D activities toward lowering cost (e. g., design-to-cost, target costing, total cost of ownership)
- drive operating efficiency with technology and information and communication systems (e. g., global ERP systems)
- strive for operational excellence in production with advanced methods (e. g., Lean Management, Six Sigma, activity-based costing, etc.) and technologies
- realize cost advantages by means of outsourcing, vertical integration, shared services centers, etc.
- encourage employees to adopt cost conscious behavior through incentives and culture.

Pursuing a low-cost strategy can be particularly important when (1) there is an intense *price competition* among rivals and there are only few ways to achieve product differentiation to escape the intense competition on prices; (2) *products of rivals are fairly identical* and customers use those products in the same ways; (3) there are *low switching cost* for customers to purchase from another provider (Thompson et al. 2014: 128 f.). However, companies need to watch out for situations of overly aggressive price-reductions that may lead to a spiral of continuously decreasing profitability. Particularly in industries where firms are facing price pressures from both sides, upstream (e. g., supply) and downstream (e. g., retail), finding ways to differentiate and innovate the business model might be more appropriate (Wunder/Bausch 2015 and 2014b). Also, efficiency programs and cost improvement methods and technologies are essentially available to all companies. Relative efficiency improvements that have been gained may be easily copied by rivals (see chapter 1.1.2.3).

In addition to the competitive advantage and profitability effects, a cost leadership strategy provides a company that is addressing a *broad* market a way for defending rivals: "Its lower costs allow it to continue to earn profits during times of heavy competition. Its high market share means that it will have high bargaining power relative to its suppliers (because it buys large quantities). Its low price will also serve as a barrier to entry because few new entrants will be able to match the leader's cost advantage. As a result, cost leaders are likely to earn above-average returns on investment." (Hunger/Wheelen 2007: 84) Furthermore, by keeping their costs for advertising, promotion or general management low, companies driving a low-cost strategy in a *narrow* market might be able to provide better prices to its niche customers than large competitors are able to.

Stuck in the middle?

As a general claim regarding the two generic competitive strategies of low cost and differentiation, companies in many industries can expect higher profitability if they chose *one* of the two approaches and develop a clear strategic positioning. According to this recommendation, both companies with a low market share derived from differentiated or focused offerings as well as companies with a high market share derived from being cost/price leader can be highly profitable. Firms that blur their choice by doing neither differentiation nor low cost strategies very well are positioned in between those two approaches and are labeled "stuck in the middle". They are expected to be less successful in terms of profitability. Therefore, business managers should decide on one of the generic competitive strategies and then rigorously stick to it (Porter 1980: 41–44). The underlying rationale is reasonable: "For a company seeking advantage through low cost (...) it makes no sense to add extra costs by half-hearted efforts of differentiation. For a differentiator, it is self-defeating to make economies that jeopardize the basis for differentiation. For a focuser, it is dangerous to move outside the original specialized segment, because products or services tailored to one set of customers are likely to have inappropriate costs and features for the new target customers." (Johnson/Whittington/Scholes 2012: 120)

Hybrid competitive strategies: outpacing and best cost

There are a number of companies though which are quite successful with hybrid strategy approaches of effectively combining the two generic competitive strategies of low cost and differentiation. Figure 3-33 shows the two classic as well as hybrid strategies where products and services achieve a satisfactory and above-average performance and are offered at highly competitive prices. Whereas at a best-cost strategy (Thompson et al. 2014: 140–142; Williamson 2010; Williamson/Zeng 2009) this is accomplished simultaneously, an outpacing strategy is based on alternating between those two dimensions over time (Gilbert/Strebel 1987; Müller-Stewens/Lechner 2011: 262):

Best-cost strategy

Companies pursuing a best-cost strategy are focusing on providing their customers more value for money than their competitors are able to do. It is a hybrid approach that balances a low-cost advantage with a differentiation advantage and may address both a broad market and a narrow market niche. "Companies pursuing best-cost strategies aim squarely at the sometimes great mass of value-conscious buyers looking for a good-to-very-good product or service at an economical price. Value-conscious buyers frequently shy away from both cheap low-end products and expensive high-end products, but they are quite willing to pay a 'fair' price." (Thompson et al. 2014: 140) Hereby, the price expectation and the perceived price fairness of the transaction have a significant impact on the customer's willingness to buy (Homburg/Kuester/Krohmer 2013: 174). A best-cost strategy can be particularly effective

Fig. 3-33

Classification of best-product strategies

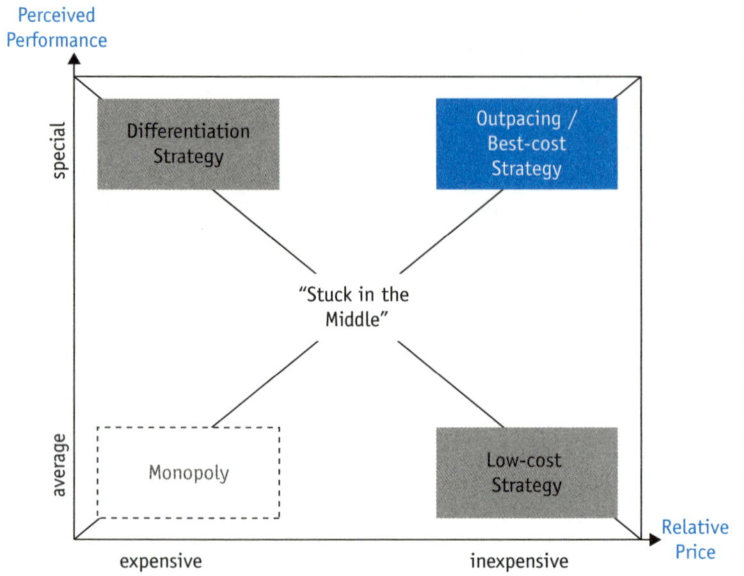

Source: following Hungenberg 2011: 211; adapted

in a competitive arena with a high degree of product differentiation and a large target group or market of value-conscious buyers. It can also be an effective way to successfully enter new international country markets (Johnson/Whittington/Scholes 2012: 121; see also chapter 3.3.1.3). However, a company is only likely to succeed with this type of business strategy when its resources and capabilities are sufficient to create products and services with a "special" and differentiated value proposition at lower cost than rivals are able to do. To avoid being "stuck in the middle", and thus, being highly vulnerable from both low-cost as well as differentiated competitors, a "best-cost provider has to achieve significantly lower costs in providing upscale features so it can outcompete high-end differentiators on the basis of a *significantly* lower price. Likewise, it has to offer buyers *significantly* better product attributes in order to justify a price above what low-cost leaders are charging" (Thompson et al. 2014: 141) It is evident that this approach requires a particularly well-designed business model where all building blocks are strongly aligned around a value for money proposition for a specific target group (see chapter 3.4.4). Firms that are effectively pursuing a best-cost strategy include Swatch (watches), Ikea (furniture), Dell (computers), Dacia (automotive) and Motel One (hotel). Another example is Aldi or Trader Joe's (food retail discounters), which offer competitively priced standard goods normally found in supermarkets as well as a number of differentiated low-cost products which are only available at those retailers. Examples are out-of-the-ordinary international culinary treats at special weekly campaigns or regular but unique standard goods such as special potato chip flavorings.

Outpacing strategy

An outpacing strategy is similar to the best-cost approach in terms of combining low-cost and differentiation. However, with an outpacing strategy, a company follows a pattern where it switches between those two generic approaches *over time*. First, the firm creates a competitive advantage with either a cost leadership strategy or a differentiation strategy. Once competitors catch up and are about to erase that competitive advantage, the firm switches to the other of the two strategy options. This, however, requires that enough profit and cash from the first approach has been generated that can be invested in pursuing the other strategy. If the new strategic thrust succeeds, it is continued until changes in the competitive arena call for another switch. Such a generic strategy approach requires a company to be strategically highly adaptable. It has to explicitly develop resources and capabilities that allow the company to alternate between providing a value proposition that is strongly based on unique product or service features and lowering total cost to achieve cost leadership. If the firm has this specific flexibility, it is able to outpace competitors depending on the competitive situation. An example of an outpacing strategy is provided by Japanese automakers in the 1980s that originally expanded their global market share by offering inexpensive small cars utilizing their local factor cost advantage. Once competitors caught up, they started to differentiate by adding special features or entirely new high-quality models. For example, with its luxury brand Lexus Toyota was able to establish a high-quality brand image and compete head-on with Mercedes and BMW. It had clearly differentiated the product but to a much lower price representing a best-cost approach. The challenge of an outpacing strategy, though, is to maintain this strategic flexibility. In the automotive industry, German and US luxury brand manufacturers reacted with a strong emphasis on achieving more competitive cost structures and improved product features, which—along with other factors—led to Japanese automakers losing their outpacing status in the 1990s. (Hungenberg 2011: 209 f.; Müller-Stewens/Lechner 2011: 262)

3.4.2.2 Lock-in

Strong customer bonding with high switching barriers

With a lock-in strategy, a strong form of bonding with the customer is created that is not focused on products or services only (best product). The main mechanism of the lock-in approach is driving long-term customer loyalty through increased switching cost, that is substantial economic costs or efforts (e. g., time to transfer data or technology), opportunity costs (e. g., abandoned benefits) or even penalties (e. g., clauses for costly termination fees) for a customer when changing to a competitor's offering. Customers are "locked-in" because the company makes it hard for them to switch providers through, for example,

- contractual arrangements such as long-term supplier and service contracts or price guarantees
- technological standards (e. g., incompatibility of the customer's own product with competitor products)

- strong economic incentives such as financial rewards or service warranties
- compelling exchange opportunities for transactions between buyers and sellers (e. g., eBay).

Lock-in can be an effective way for the company to capture value from a long-term customer base that yields enormous profit potential. However, it may also be frustrating to customers and can damage the reputation of a company if customers are locked-in against their will without providing them a convincing value proposition. In such as case, the strategy is somewhat short-sighted and unlikely to provide sustainable success for the company. Also, a word of caution has to be noted as lock-in strategies may cause legal issues such as antitrust allegations if they are designed for a firm to achieve or maintain monopoly power. Crafting a lock-in strategy requires a dominant market position, which can be established in various ways such as proprietary standards, exclusive customer intimacy and complementary products (Gassmann/Frankenberger/Csik 2014: 207–211; Kaplan/Norton 2004: 332–346; Hax/Wilde 2001: 81–104; Scheuss 2012: 340 f.):

Proprietary standards

One way of crafting a lock-in strategy is by establishing de-facto industry standards that may be based on a company's proprietary technology. For example, customers who want to switch from a technological standard such as a Microsoft Windows-compatible computer to another standard may lose access to certain application programs that run only or best on a Windows system. Companies with extensive databases that run on software packages from SAP may face significant cost as well as complexity and uncertainty regarding the technological compatibility when they want to purchase certain software packages from another provider such as Oracle. Also, customers who want to switch from globally dominant credit cards such as Visa- or Master-Card brands may risk not being able to pay with their other card at stores that only accept the industry standards. Another way to craft a lock-in strategy based on standards is by providing a platform of exchange between buyers and sellers (e. g., eBay, credit cards, etc.) that becomes a de-facto standard for the specific type of transaction. Firms participating in that platform are locked-in because there is no alternative to meet the same number of potential buyers (or sellers).

Exclusive intimacy

Another form of lock-in is the creation of exclusive customer intimacy by means of incentives. For example, large global hotel chains and airline alliances are providing their key customers with special customer- or bonus-cards that provide a number of benefits but are only valuable within their company network. The same principle applies to so-called payback cards that are tied to a certain group of merchants. The more a customer uses the company's services or products, the higher are the economic benefits derived from such customer cards. Frequent airline travelers, for example, appreciate an environment where they can efficiently conduct their business and also get some rest in between business meetings. Based on their frequent traveler status derived from the use of one particular airline or airline alliance they may get access to airport lounges as well as in-flight seat/class upgrades that provide significantly higher customer value. This type of lock-in can be very effective. Some competitors are trying to break the strong bond their target customers have established with their existing providers such as hotels or airlines by matching their customer status and providing the same level of comfort for a certain time period. This provides customers the opportunity for trying another service provider without the switching cost.

Complementary products

A classic approach of a lock-in strategy is based on complementary products, also known as the *razor and blade* model. Hereby, the core product is sold at a relatively low price whereas the compatible complementary product is priced higher and may be patented. Customers are attracted and won by a relatively low-cost approach for the core product, which makes an initial purchase decision likely. Profit is generated with the complementary products such as accessories, particularly when they are frequently used. Making this mechanism work requires that competitors cannot imitate the core product due to its proprietary design or technology or its complex and continuously upgraded construction that makes it hard to copy. A famous example of this model is Gillette who, for decades, has sold its razors for low prices and offered patented razor blades at much higher prices. Also, there are a number of industries such as elevator, telecommunication, medical equipment, and automotive where companies tend to make much more profits with after-sales services such as repairs and maintenance, accessories, or procedures than with the initial core product sale. For example, some companies in the medical device industry sell their equipment at a somewhat low price but make customers pay for every procedure required. Companies may combine different lock-in strategies as well as supplement this approach with generic competitive strategies. For example, a strategic pillar of Nespresso was the sale of coffee machines that only worked with the company's patented coffee capsules (complementary products). It also offers its customers membership in an exclusive Nespresso club (exclusive intimacy). Both lock-in elements are supplemented with a number of other strategic components in the firm's business model (see Strategy Practice Example 3.13 in chapter 3.4.4.4).

3.4.2.3 Complete Customer Solution

A complete customer solution strategy is targeted at building long-lasting relationships with customers based on tailored offerings that provide a total solution to the customers' needs and wants. The customers' value proposition neither comes primarily from the products with the lowest price nor from the most differentiated ones driven by quality, performance or uniqueness as it is the case with best-product strategies. Also, customers may not be locked-in with certain providers due to different mechanisms that prevent them from switching to competitor's products. With the generic strategy of complete customer solutions, the core of the value proposition(s) is derived from the "completeness" of the solution based on a deep customer understanding and relationship. Here are three key approaches of how a complete customer solution may be provided:

Broad and complete offering

Offering many different goods in one shop provides customers a "complete solution" to their needs and wants. A huge and competitively priced product range makes shopping more convenient and less time-consuming. For the company, it can significantly increase the share of wallet it gets from one customer, which is considered more important than the company's market share with one specific product. A classic example is Walmart with its stores offering a broad product range from food to electronics to clothes at comparatively low prices. There are other examples that are based on internet technologies. A number of companies provide a broad and complete offering in their online stores. As this technology enables even a domestic firm to reach very specific customer segments worldwide, such complete offerings typically cover not only best-selling or mass products but also highly specific niche products in various categories (long tail strategy). One of the most popular examples is the company Amazon.com that provides a kind of one-stop shopping experience online that includes also very specific niche products.

Product and service packages

Many customers look for somebody who provides a complete solution to their often very specific individual problems. Instead of buying bits and pieces of standardized products or services that contribute but do not address the customers' needs entirely, they want to hand over their issue and get it totally solved. One way to drive a complete customer solution strategy is by providing *packages of products and services* that are designed and integrated to entirely tackle a customer's problem. For example, instead of selling hardware, software, installation and maintenance separately, IT providers may bundle those elements and provide complete customer solutions for the information technology needs of their customers. Those solutions are based on taking a customer perspective to fully understand their needs and then providing multiple products and services integrated in one solution package to fulfill those needs. The solution provider as a single point of contact reduces the overall customer effort and gains cross-selling opportunities. Service business can also be a huge lock-in particularly with installed base products where equipment purchase is infrequent. Customers buy a machine and are locked into a service contract. This makes it very difficult for new competitors to come in and dislodge a current provider (see chapter 3.4.2.2).

Tailor-made solutions

In a complete customer solution approach, the customer value proposition requires a company to fully understand the customer's business or personal issues. Firms want to become a trusted partner for tailoring a complete solution together with their customer. Elements to those solutions may not come from one company but require an orchestration and integration of various suppliers and complementors. Developing tailor-made solutions requires more than just putting together a bunch of products and services. It includes integrating those elements in a way that assists the customer better than it would be done if the customer purchased and used each element separately. This may also require that a company develops customized technical components to enhance the final customer solution as it is often the case in highly technological environments. Developing solutions may also involve the customers themselves by transferring knowledge to enable them for driving their own performance enhancements. The profound understanding of the customer and the desire to make a customer more successful instead of just selling a product or service can create a strong and long-lasting bond between a company and its customers driven by a mutually beneficial relationship.

3.4.3 Breaking Out of Traditional Competition

The emphasis of business strategies so far has been on developing competitive approaches for winning against rivals in a specific competitive arena. However, there are also approaches for driving success of a business that try to restrain or break out of "traditional" head-on competition, and thus, avoid the negative and sometimes widely destructive consequences of intense rivalry. Two of such approaches are cooperative strategies as well as blue ocean strategies, which will be explained next.

3.4.3.1 Cooperative Strategies

Coopetition (cooperative competition) and collusion

A company may gain competitive advantage within a competitive arena not only by directly taking on rivals but also by collaboration with other companies such as competitors, suppliers or customers based on common market interests. Collaboration with competitors is often referred to as *coopetition*, a term coined to describe the idea of cooperative competition (Brandenburger/Nalebuff 1996). It can be generally applied to both a corporate- as well as a business level. Collaboration of companies within a certain market may allow them to create and capture higher value than when they were acting alone. Cooperative strategies may provide them with additional sources of competitive advantage over other companies in the same competitive arena. Although collaborating firms restrain competition among each other, they usually join forces to enhance their competitive strengths relative to other companies they are competing with. A word of caution has to be noted regarding illegal forms of cooperation such as tacit or explicit collusion. Typically, competitors are not allowed to directly or indirectly cooperate to offset normal market mechanisms, such as pro-actively limiting supply and raising prices to increase their profits (Wheelen/Hunger 2010: 195 f.).

Types of strategic alliances

A common and widely applied cooperative strategy is the concept of strategic alliance where two or more firms operating in the same competitive arena join forces by sharing resources and capabilities to craft strategy for mutual economic gain (Johnson/Whittington/Scholes 2012: 218–225; Wheelen/Hunger 2010: 196–199): A strategic alliance may be formed between a company and its key suppliers, key distributors or even rivals that usually compete for market share in the same competitive arena. Cooperative arrangements underlying a strategic alliance range from pure collaboration agreements of otherwise independent companies (non-equity alliances) to the formation of new legal entities with some sort of ownership arrangements between the partners involved (equity alliances). There are various types of alliances which include (Morschett/Schramm-Klein/Zentes 2010: 281 f.; Wheelen/Hunger 2010: 197 f.):

- **Mutual service consortia** are fairly loose partnerships of companies pooling resources to gain mutual benefit which would otherwise be too expensive to achieve alone (e. g., research alliances working on an advanced technology).
- A **joint venture** is the creation of a new business entity for strategic purposes. The two or more independent joint venture partners have certain ownership arrangements, managerial and operational responsibilities as well as pre-defined risk and reward allocation.
- **Licensing arrangements** are contractual agreements between firms in which the licensing company grants rights to the licensee regarding its processes, products,

distribution, or brand. For example, the licensee may produce (e.g., contract manufacturing) or sell a firm's product in another geographic market. It has to pay a royalty fee to compensate for the technical expertise or brand it is allowed to produce or market.

- **Value chain partnerships** are strong alliances in which a company works closely together with key suppliers or distributors based on long-term arrangements for mutual benefits.

Benefits of strategic alliances

Expected benefits of engaging in a strategic alliance include mutual *learning* of new competencies and getting *access* to specific markets, which would not be possible or too expensive alone. This may include the cooperation with a key distributor in the local market or working with an international alliance of otherwise competing companies to serve customers globally without the need to be present in all countries such as global airline or management consulting alliances (see Strategy Practice Example 4.4 in chapter 4.2.2.1). Other benefits are realizing *economies of scale* in the production of outputs or in terms of inputs (e.g., stronger bargaining power), reducing the *financial risk* for highly expensive activities such as research and development in investment intense industries like aerospace, and reducing *political risk* by joining forces with local partners for an international market entry (see also chapter 3.3.1.3). (Johnson/Whittington/Scholes 2012: 218–225; Wheelen/Hunger 2010: 196–199)

Pitfalls and success factors of strategic alliances

All types of strategic alliances involve uncertainty, complexity and risk for the participating companies. Although the underlying strategic concept might be logical and compelling, the fact that neither partner is in total control causes many issues that need to be dealt with on an operational level. This is one reason why mixed experiences regarding the success of strategic alliances have been reported. Although there are examples of companies that were able to improve their profitability and company value, about half of all alliances are reported to perform unsatisfactorily (Wheelen/Hunger 2010: 196). For avoiding pitfalls, the following success factors or guidelines may be considered by companies planning to form a strategic alliance (Wheelen/Hunger 2010: 199):

- clarify the strategic purpose of the alliance that is integrated in each partner's strategy
- ensure the compatibility of goals and complementary capabilities of cooperating companies
- identify and deal with potential risks when the alliance is formed
- clarify tasks and responsibilities of partnering companies
- make sure you have the right incentives and rewards to compensate parties for their strengths and results
- manage differences in corporate cultures and organizations by means of alliance-specific incentives, human resource exchange, communication programs, reporting processes, etc.
- avoid direct competition in the market
- establish mutual trust through effective leadership and avoid individual egos that dominate
- manage international alliances with people who have cross-cultural knowledge

- establish shared long-term time horizons to overcome short-term conflicts
- counterbalance failures with successes through developing multiple joint projects
- establish a monitoring process and clear rules for information sharing to build trust
- ensure flexibility for adjusting to environmental changes and new opportunities
- establish and execute an exit strategy in case the alliance is not effective.

3.4.3.2 Blue Ocean Strategies

Creating uncontested market space

A different approach to business strategy as compared to the more classic generic strategies explained before is the concept of blue ocean strategy (Kim/Mauborgne 2005): Whereas generic business strategies are typically applied in industries that are already well established and frequently driven by intense rivalry ("red oceans"), blue ocean thinking looks for strategic gaps that have not been identified, and thus, are not being exploited by competitors. Creating this uncontested market space is at the core of blue ocean strategies. Those are compelling opportunities because there are no competitors yet allowing the company—at least temporarily—to drive business without constantly elaborating on how to beat rivals. Instead of exploiting existing demand in a spiral of ever increasing differentiation or cost competitiveness, blue ocean strategies aim at creating value innovation by providing a highly differentiated customer value beyond existing industry success factors. This requires business strategy formulation to be reoriented away from competitor offerings to innovative alternatives as well as from existing customers to the question of how to create value for non-customers of the business. Therefore, blue ocean thinking sees only limited value in conducting extensive customer research as (potential) customers are probably not able to see uncontested market space and will rather emphasize their require-

Fig. 3-34

Red ocean versus blue ocean strategy

Red Ocean Strategy	Blue Ocean Strategy
Competition in existing market space	Creation of uncontested market space
Beating the competition	Making competition (temporarily) irrelevant
Exploiting existing demand	Creating and capturing new demand
Making the value-cost tradeoff	Breaking the value-cost tradeoff
Focus on creating temporary competitive advantage	Focus on creating (customer) value innovation
Business model for achieving differentiation *or* low cost	Business model for achieving differentiation *and* low cost

Source: following Kim/Mauborgne 2005: 18; adapted

ments within the context of what the industry currently offers. The key characteristics of the blue ocean strategy as compared to a more classic generic strategy approach in established industries (red ocean strategy) are shown in figure 3-34.

Strategy canvas helps to identify new market opportunities.

A practical framework that can effectively assist blue ocean thinking is the strategy canvas. First, it is diagnostic in a way that captures critical success factors in the current competitive arena and analyzes the performance of existing market players in relation to those factors. This results in *value curves* that are typically drawn for relevant strategic groups in a competitive arena and for the focal company. Second, the strategy canvas can also be used for strategy formulation as it provides a framework to systematically elaborate on which success factors can be raised, reduced, and eliminated as well as which novel factors can be created to reconstruct buyer value, and thus, establish new rules of competition in this particular business. A practical application of this methodology is provided in the Strategy Practice Example 3.11.

Benefits and pitfalls of the blue ocean strategy

Blue ocean thinking provides a different view on competition than the more classic approaches to business strategy. Looking for uncontested market space is a compelling idea to escape a competitive situation centered on intense rivalry. There are a number of companies that seem to be very successful following this strategic rationale (Kim/Mauborgne 2005). Although "uncontested market space" is strategically highly attractive but might be rare. Few companies will be in a position that allows a greenfield approach to formulate business strategies. What happens to established businesses and market segments remains fairly unclear in this approach. Even for competing in "red oceans", smart business strategies and innovative business models are often required (Scheuss 2012: 324). Also, there is not necessarily a trade-off between differentiation and cost. Some companies are pursuing hybrid strategies that combine differentiation and low cost approaches also in established industries (see chapter 3.4.2.1). Nevertheless, blue ocean thinking can be facilitated by a variety of practical and easy to apply tools and methods provided with this approach. Conceptually, it is strongly linked to the area of business model innovation such as finding new-to-world business models (see chapter 3.4.4.3). Thus, it essentially reflects the strategy principle of "being different" (see chapter 1.1.2.3) that is applied to creating a unique business model for developing uncontested market space. The business model approach will be explained in the next chapter. It can be applied to refine business strategies derived from both classic generic strategy approaches as well as more unconventional strategies.

3.4.4 Business Models

Many of the various business strategy concepts proposed so far in this chapter provide ideas for strategy formulation but do not explain how a concrete business strategy can actually be crafted. This can be supported effectively by using a business model approach that facilitates the development or refinement of business strategies. A number of business model patterns introduced in the coming chapters are actually based on generic strategy types explained before. There are, however, additional patterns providing strategic approaches on a business level that go far

Strategy Practice Example 3.11

Blue Ocean Strategy of Casella Wines and "yellow tail"

Within only two years in the early 2000s, the wine brand "yellow tail" emerged as the fastest growing brand in the histories of both the U. S. and the Australian wine industry. Despite strong wine labels from California, by 2003 it became the number one red wine sold in a 770-ml bottle in the United States. The company Casella Wines that created "yellow tail" developed a product with a new strategic profile centered on the idea of offering a social drink that is desirable to everyone from beer to cocktail drinkers and consumers of non-wine beverages. The profile had a clear, focused and compelling tagline: A "fun and simple wine to be enjoyed every day." The value curves of two traditional strategic groups of the U. S. wine industry in the late 1990s is shown in figure 3-35. On the horizontal axis in the red ocean area are the critical success factors in the industry, that is factors the industry competed on and invested in at that time. The vertical axis shows the performance level or offering level of the competing companies from a buyer's perspective. The resulting value curves illustrate the company's relative performance across the critical success factors. "Companies A" represent premium wine making companies that provided a high level of perceived offering across all the key factors. Although their individual products are different, from a broader market perspective they are different in the same "premium" way. Therefore, they are considered a strategic group that represents a classical differentiation strategy primarily based on product quality. "Companies B" are budget wine producers that—as a strategic group—essentially repre-

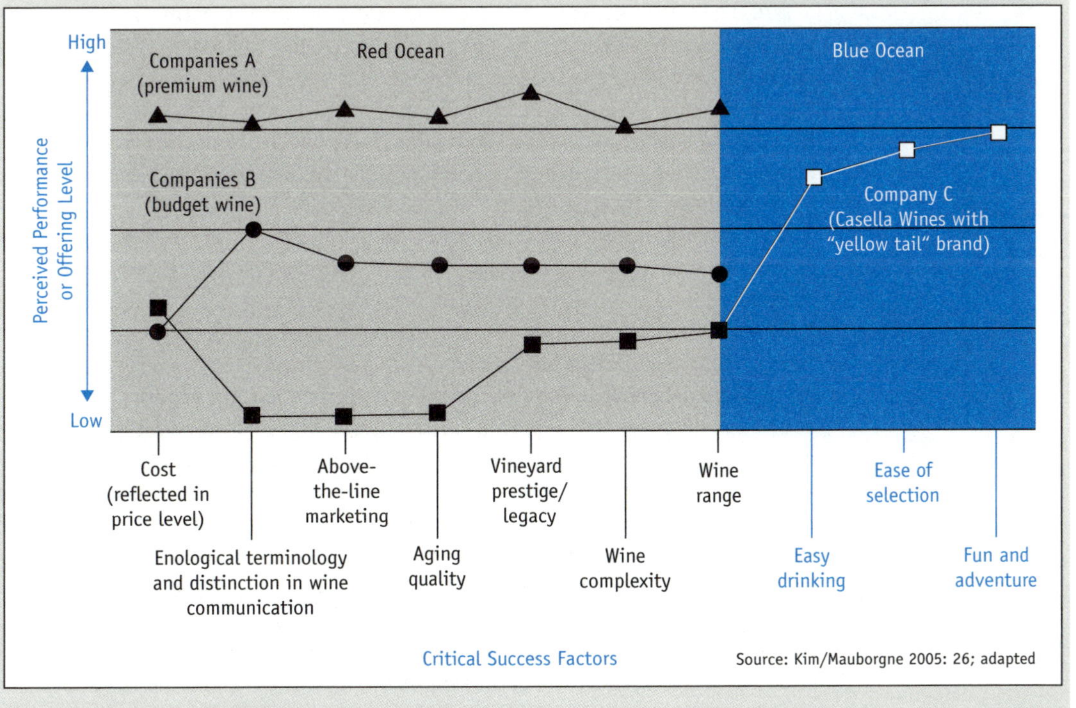

Fig. 3-35: Strategy canvas for the U.S. wine industry

sent a low-cost strategy. Casella Wines reconstructed buyer value elements and crafted a new value curve based on the following four actions:

- They *created* new factors like "easy drinking", "ease of selection" and "fun and adventure" outside the traditional industry boundaries (blue ocean area).
- They *eliminated* factors such as "enological terminology and distinctions", "aging qualities" and "above-the-line marketing" the industry took for granted at that time.
- They *raised* factors like "price versus budget wines" and "retail store involvement" well above the industry's standard.
- They *reduced* factors such as "wine complexity", "wine range" and "vineyard prestige" well below the industry's standard.

(Source: Kim/Mauborgne 2005: 25–37)

beyond the scope of strategies explained so far. Business modeling can also foster ideation about completely new types of business strategies such as sustainable business models.

Refinement of strategy on a business level

A business model is a simplified expression of the anticipated mechanisms how the business is supposed to be successful in its competitive arena. It can be considered as a refinement of strategy on a business level (Wunder 2013: 354). In companies with multiple businesses or strategic business units there are typically also multiple business models (Trapp 2014; see also fig. 4-41 in chapter 4.3.2). If the company has only one business, as it is frequently the case in small companies or start-up businesses, there is in general only one business model for the entire firm. For a new venture, a business model may represent a plan of how to become profitable in the marketplace. Contrary to the perspective of a strategy being more refined as a business model (e.g., Teece 2010: 179), the upcoming chapters follow a different logic. Hereby, the business strategy is considered more generic than the business model. Managers having a certain business strategy in mind can actually sharpen it by developing and describing a corresponding business model.

Creating and delivering customer value

This strategy-oriented business model understanding is centered on the concept of creating and delivering customer value (Casadesus-Masanell/Ricart 2010; Chesbrough 2010; Kagermann/Österle 2010; Osterwalder/Pigneur 2010; Teece 2010: 179; Zott/Amit 2010). In a business model the firm describes the fundamental principle of how a product and service offering for certain target customers (value proposition) is created and delivered (value creation and delivery), and thus, benefits for the company or business—typically profits—are realized (value capture). This way the business integrates upstream activities (supplier-related) in its value chain with downstream activities (customer-related) and the financial model. The three basic pillars can be found in most business model frameworks and are illustrated in figure 3-36 (Bocken 2014: 43; Gassmann/Frankenberger/Csik 2014: 6; Osterwalder/Pigneur/Tucci 2005: 17; Richardson 2008: 138).

Challenging the sustainability of competitive advantage

Questions of how to design a business model and how it functions have attracted tremendous interest by companies across different industries, popular business press as well as academic literature in the last few decades. The business model concept was first popularized during the e-business boom (the so-called "New Economy") in the 1990s when new business models for commercial activities appeared in the internet

Fig. 3-36

Three basic pillars of a business model

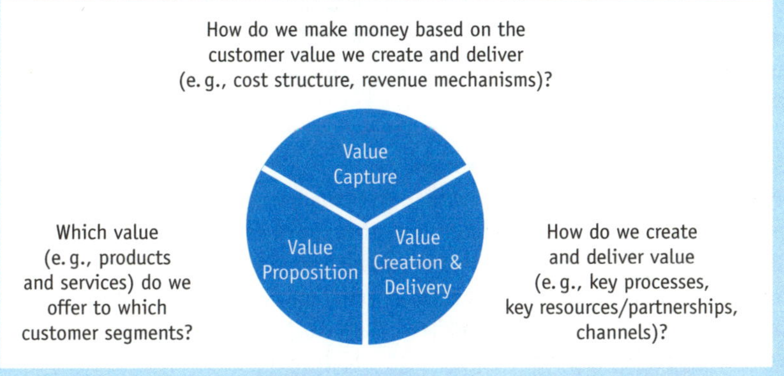

(Wirtz 2010). At that time, both start-ups and established firms looked for new internet-based ways of gaining competitive advantage and superior profits. Since then, dynamic and complex markets as well as shorter product life cycles have forced more and more companies even outside e-commerce to review the current success logic of their businesses critically and challenge the sustainability of their competitive advantage given an environment of ever changing strategic parameters (D'Aveni 1994; Wiggins/Ruefli 2005; Wirtz 2011: 3). Here are some typical questions companies should ask themselves related to their business model (Wunder/Bausch 2014b: 58 f.):

- Will our customer value proposition still be attractive in the future (e. g., are there any anticipated new solutions for our customers' needs and wants in or outside our industry)?
- Can we enhance the currently delivered value for our customers in any way?
- Are there untargeted customer segments that we should serve today or might become highly relevant in the future?
- How can we strengthen our customer intimacy, and thus, increase customer retention?
- Is there business development potential in areas adjacent to our core business?
- Can we join forces with other companies to deliver more value to our customers?

Differentiation with business models

Well-thought-out and effective business modeling is considered essential for strategic ideation as it helps to translate high-level strategic intent into competitive advantage. Furthermore, it is seen as an important and practical linking pin from strategy formulation into strategy execution (see fig. 4-41 and fig. 4-44 in chapter 4.3.2). Business modeling can be particularly relevant in competitive environments where companies pursue similar strategies and want to escape a competition that is solely based on operational excellence and competitive prices (Wunder/Bausch 2014b: 58 f.; Chesbrough 2010). Rethinking the comprehensive logic of how the business is successful can help to find new ways for differentiating the company in its competitive arena. Hereby, the business model does not only support the go-to-market approach for innovative products or services but stands for an innovation and source of competitive advantage itself. The firm differentiates not only based on its products or services but on the entire business model with well-designed and mutually enforcing elements that are hard to imitate (Porter 1996: 70–75; Wunder 2013: 354). Same products or services may yield different financial outcomes based on the business model they are embedded in.

3.4.4.1 Business Model Frameworks

Experiential versus structured process

Developing or advancing a business model is usually a creative and experiential process. Some successful entrepreneurs state that they never "planned" their business model but evolved it over time through trial and error. In those cases, business models might have never been formally written down but only communicated in an anecdotal way or just observed once they had been realized. A simple method of illustrating a business model is provided in figure 3-37, which shows a so-called "activity-system

Fig. 3-37

Activity-system map of IKEA

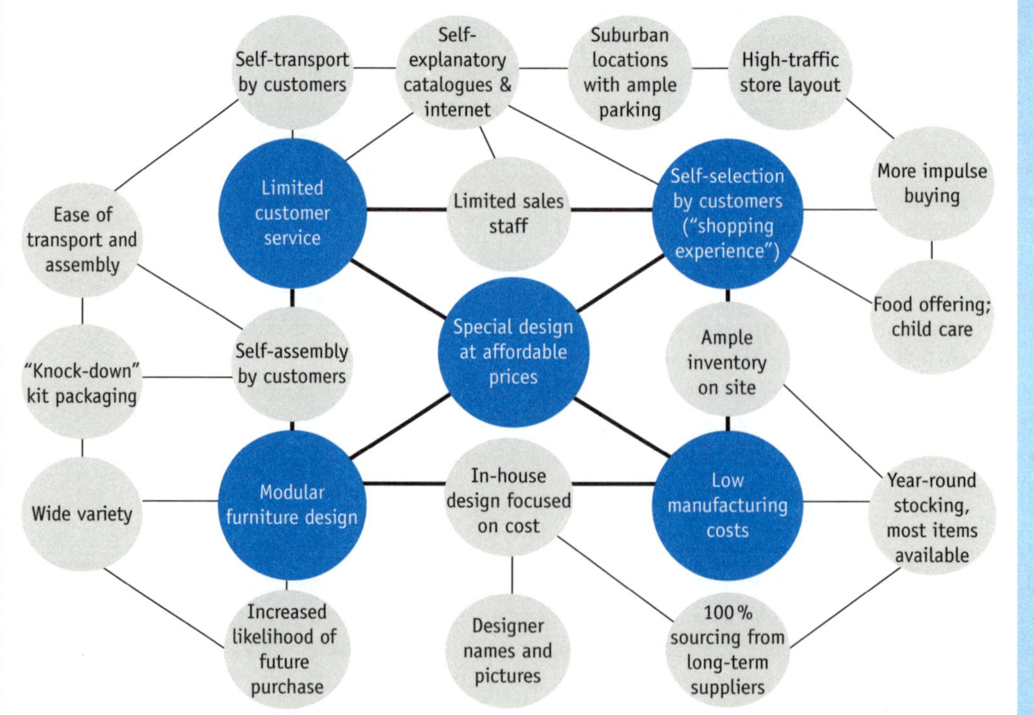

Source: Porter 1996: 71; adapted

map" (Porter 1996). In this example, observable key elements of the realized Ikea business model are collected and logically linked without using a formal structural framework. The five blue hubs in the center of the grid represent key elements of the Ikea business model whereas the other components are explanatory or adjacent subelements.

Structure for systematic business model elaboration

A variety of frameworks have been developed in recent years to provide a conceptual structure for systematically articulating and elaborating on key elements of a business model and their logical interrelations. Those frameworks are not supposed to replace entrepreneurial spirit and business acumen. Instead, they provide a conceptual scheme for describing a current business model in a more formal way and for guiding strategic ideation to further develop it. The various frameworks differ regarding their characteristics such as the type and number of building blocks, level of detail, logical flow, application area, and their theoretical foundations (Bieger/Reinhold 2011: 18f.; Schallmo 2013: 108 and 118). Figure 3-38 provides an overview of some selected business model frameworks sorted by authors in alphabetical order. Although those frameworks vary in layout and structure, they all explicitly or implicitly capture the basic business model elements *value proposition*, *value creation and delivery* as well as *value capture,* which are illustrated in figure 3-36.

Static and dynamic business model components

In addition to describing the as-is situation, some business model frameworks include a dynamic component that captures how the company plans to further develop its value proposition. This additional basic business model element of *value development* is considered particularly important in research-intense dynamic industries where filling the innovation pipeline with future technologies or solutions is as important for the business model as delivering the current value proposition (Bieger/Reinhold: 2011: 33; Greiner/Wolf 2010; Horváth & Partners 2007: 127 f.). Companies using more static frameworks typically capture the dynamic perspective not as a separate element but relate it to the entire framework as part of their business model innovation process (see chapter 3.4.4.3).

Selecting a business model framework that is practical and acceptable

When selecting a business model framework for practical application it is important to consider that the method is comprehensive as well as intuitively understandable. To be effective, the applied business model framework has to be accepted by the corresponding managers who have to work with it. On the one hand, the framework should cover all relevant areas in which strategic business decisions have to be made, and, on the other hand, it needs to be simple and compliant with other strategy tools applied in the strategic planning process (Wunder 2013: 357). Experience has shown that strategy processes in which people have to switch mentally from one strategy method, tool, or terminology to another can easily fail as those processes are confusing, and thus, tend to be neither practical nor acceptable. The following requirements can be applied as a practical guidance for selecting a business model framework:

- **comprehensive**, that is the business model building blocks should cover all relevant areas in which strategic business decisions have to be made
- **integrative**, that is the framework should provide some kind of guidance or logical flow of how the different building blocks are interrelated

- **practical**, that is the business model framework should be simple and intuitively understandable, and thus, easy to apply by management teams in company practice
- **compliant,** that is the framework should be compliant with other methods and tools in the strategy process particularly when it comes to effectively executing the business model.

Fig. 3-38

Overview of selected business model frameworks

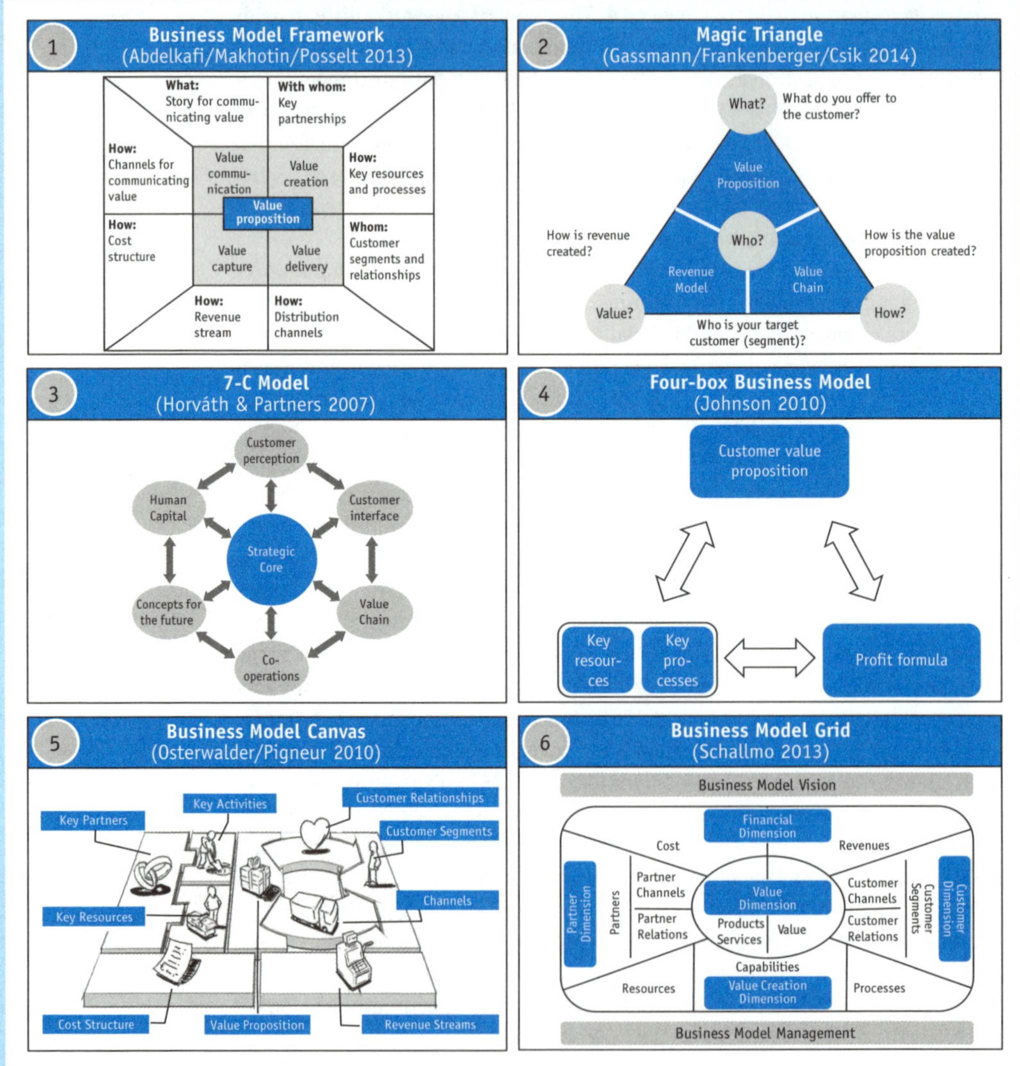

3.4.4.2 Key Building Blocks of a Business Model

Top-down logic

Following the business model canvas approach (Osterwalder/Pigneur 2010), seven key building blocks of a business model have been defined and arranged in the generic framework shown in figure 3-39. Those building blocks are ingrained in many business model frameworks today and have proven to be practically applicable. They are arranged in a way that follows a top-down logic based on the four perspectives of the classic balanced scorecard approach (see chapter 4.3). At the top is the building block "sustainable profit", which is supposed to articulate how value is captured in the business model and typically relates to *financials*. It is followed by *customer/market*-related building blocks such as customer segments, customer value as well as customer relationships and channels, which articulate the value proposition, how it is delivered and to whom. At the next level, which relates to the *internal process* perspective of the balanced scorecard, there are the key processes. At the bottom-level of the business model framework, there are key resources and partnerships that—along with the key processes—are required to create and deliver the customer value. Key resources and partnerships can be related to the *potentials* perspective of a balanced scorecard. Through the top-down arrangement, business model changes that require special management attention and resources can be easily transferred into strategy maps and related components for effective strategy execution (see fig. 4-44 in chapter 4.3.2.1).

Strategic choices in each building block

Business modeling as part of the strategy formulation is about strategic choices. On a corporate level, decisions have to be made about the businesses a company

Fig. 3-39

Generic business model framework with seven building blocks

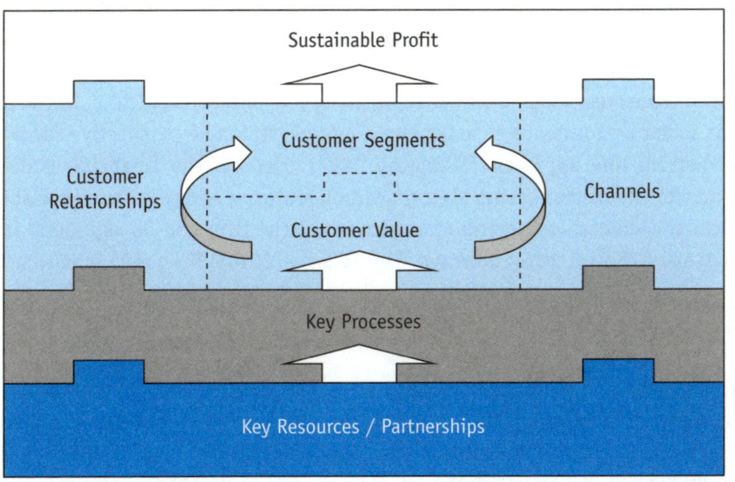

Source: Wunder 2013: 355

wants to compete in and which basic value creation activities it wants to perform in those businesses. On a business level, decisions need to be made on how each building block of the business model should look like and is integrated. This is not about setting goals and targets as it is the case in strategy maps or balanced scorecards (see chapter 4.3) but about foundational strategic "yes/no"-decisions (Casadesus-Masanell/Ricart 2010: 197–201; Horváth & Partners 2007: 117). In a business model, for example, it needs to be decided which customer segments or (international) markets are targeted and what kind of potential customer groups or markets are purposely excluded, which channels are used and which are not, what attributes of the product or service offering will be enhanced because they are considered most valuable to the customer, which alternative solutions the company pursues to fulfill the customer needs and wants in the future and so forth. "Yes/No"-decisions means that it is not necessary to define precise goals, key measurements and targets at this point. The focus is on strategic choice following the principle that strategy is about making trade-offs in competition and choosing what to do and what *not* to do (Porter 1996: 68–70; see also chapter 1.1.2.3). Having made those decisions is a prerequisite for effective strategy execution. Corresponding aspiration levels and expectations are derived later on as part of the strategy execution system (see chapter 4.3). The seven building blocks of the business model framework in figure 3-39 are explained in the following and are illustrated in Strategy Practice Example 3.12.

Sustainable Profit

Commercial model for generating sustainable profit

This building block describes the commercial mechanisms that a company wants to utilize for realizing profits for itself and its stakeholders. They are derived from the customer value delivered to the defined customer segments. It shows the success or failure of the entire business model in terms of sustainable profit. From an economic point of view it considers sources for revenue streams (e. g., sales, membership fees, rent/leasing, license fees, etc.), pricing mechanisms as well as cost structures that arise from the six other building blocks. The mechanisms for the financial performance of a business may be enhanced with aspects of its ecological performance as well as social performance to articulate a more sustainable perspective following a triple bottom line approach (Elkington 1997). Hereby, the financial model also accounts for the firm's ecological (e. g., carbon footprint and waste) and social (e. g., physical and mental well-being of employees of the firm and its suppliers) impact (Hengstmann/Seidel 2014; Sawczyn/Fink/Werner 2014). Economic, environmental and social achievements are simultaneously considered and integrated to one holistic perspective on performance (see chapter 3.4.4.5).

Customer Value

Satisfying needs/ wants and solving customer problems

The component customer value is about the added value or benefits the company creates for its customers with its proposition. This customer value proposition (CVP) represents the core element of the business model, as it is the primary reason why customers turn to one company and not to its competitors. It is necessary to clarify which customer needs and wants can be satisfied with which product or service concepts. Whereas needs are basic human requirements like food, mobility, etc., they become wants when customers direct them to specific objects for satisfying them such as a car for being mobile (Kotler/Keller 2009: 12; see also chapter 2.2.3.1). When elaborating on the value a business may provide to its targeted customers, it is helpful to disconnect it from the existing solution and think about the foundational need or "original customer problem" (Gälweiler 2005: 254 f.). For example, if the company is producing cars, the underlying customer problem is being individually mobile. Instead of selling a car to the customer (current solution), the original customer problem can also be solved in many other ways such as car-to-go, car-on-demand or car-sharing concepts, limousine services, rental services, other means of transport. Furthermore, the customer value may not only come from the product's or service's core function but also from additional features, its packaging or tangible service delivery environment, basic and value-added services as well as from the brand (Homburg/Kuester/Krohmer 2013: 108 f.). In this context, customer benefits may include both material benefits (e. g., price, quality, speed) and immaterial ones (e. g., prestige, status, image, customer experience). For a business model to be successful, it is important that (1) the added value or benefit a company tries to create is also perceived and experienced this way by its customers, and (2) that the added value stands out in relation to competitors.

Customer Segments

Targeted customers, segments and markets

In the building block customer segments, the company has to answer the question for which customers it will primarily generate the value. The definition of target groups can be based on the demographic, socio-economic, geographic, behavioral and psychographic criteria or on the combination of them. Different customer groups or segments have different wants or requirements and are typically related to different profitability levels (e. g., private consumers versus company customers, mass market versus niche market, premium versus low-price segment, domestic versus foreign markets, etc.). Different targeted segments need to be satisfied with different means in the sense of customer specific solutions, channels or relationships. It is particularly important to clarify in the business model which customer groups will get special attention and resources of the company and which segments the company is not planning to serve and are explicitly excluded (see also chapter 2.2.3.1).

Channels

Reaching and addressing customers

In the channels component of the business model, the company has to define ways through which it will reach and address its defined customer segments to transmit the value proposition. This covers

- communicating the value proposition in way that creates awareness and allows the customer to evaluate what is offered (e. g., through different forms of marketing communication measures)
- the selection of sales channels for allowing the customer to purchase specific products or services (e. g., own shops, trade, catalogue, telephone, online, etc.)
- the distribution and delivery of the value proposition
- the post-purchase support (e. g., after sales service).

Own and partner channels as well as direct and indirect sales can be distinguished. Important elements of indirect sales are, for example, the depth of a sales channel (the number of sales stages between company/supplier and customer) and the width of a sales channel (i. e. the number of sales partners used in parallel). In addition, the width of the entire sales system (i. e. the number of the sales channels used in parallel) needs to be considered for the question of how the value proposition is best transmitted to the customer segments.

Customer Relationships

Types of desired relations with specific customer segments

In the customer relationships building block the company needs to clarify which kinds of relations it is going to create with which customer segments. Defining and realizing those desired relationships is an important lever for customer loyalty and is thus supposed to drive sustainable business activities and ultimately profits with focused customer segments. The categories of customer relations range from personal contact to complete automation. Starting points for customer relationship management can be, for example, interaction with the customers (e. g., involvement into the company's R&D activities, 24/7 service call numbers, etc.) or the creation of switching barriers for the customers such as price-quality ratio guarantees, joint advertising/co-branding, incompatibility of own value proposition with competitor products, etc. (see also chapter 3.4.2.2).

Key Processes

Critical activities along the value chain

The component key processes describes how the value proposition is to be created for the defined customer segments and which activities play a critical role in the successful realization of the business model and all its building blocks (e. g., channels, customer relationships, etc.). Here, it is not the idea to list all the processes and activities in the company, but to identify and carefully design the few processes that have an outstanding importance for the realization of the business model. This may include, for example, key activities to identify customer wants and requirements, processes like innovation management, life cycle management, operations and supply chain management, or key activities for customer relationship management and environmental/sustainability management. Depending on the industry and business logic, key processes required to create the customer value differ significantly.

Key Resources/Partnerships

Utilizing tangible and intangible resources as well as partners

This building block describes which tangible or intangible resources are essential to establish the value for customers and whether they are provided internally or acquired from partners. Employees play a central role with their professional knowledge, their management skills, innovation power and experience. Additionally, physical facilities (e. g., production equipment, IT, logistics and sales structures, technologies, etc.), intellectual resources (e. g., patents, company knowledge, copyright, etc.), aspects of corporate culture or financial resources may be critical for the success of the business model. In addition to the internal resources, establishing and coordinating partnerships such as R&D alliances, sales partnerships as well as a collaboration with or an orchestration of key suppliers, etc. may play a central role for the creation of the value proposition. Key resources and partnerships have to be considered in conjunction with key processes. Not their pure existence but the firm-specific capability to coordinate and utilize its resources and partnerships are the foundation for establishing the customer value proposition.

Strategy Practice Example 3.12

Business Model Framework Illustrated with the Apple iPod

Until the 1990s Apple was a niche provider in the IT-industry with a relatively small market share. The launch of the iPod in 2001—a fiscal year in which the company had to report a loss—brought a breakthrough and initiated an outstanding success story. In January 2015, Apple reported a net profit of $18 billion for the first quarter of its fiscal year, the biggest quarterly profit ever made by a public company. Did a new to the world product innovation kick-off this success story back in 2001? The answer is probably "no". The idea of a portable music device had been already implemented earlier, for example, in the 1980s by Sony with their famous Walkman®. At the launch time of the iPod, there was a broad range of portable digital music devices (so-called MP3-players) in different designs and price categories on the market. Nevertheless, what was missing at that time was a unique business model built around the product idea and technology that provided such an outstanding value proposition based on several building blocks complementing each other in an intelligent way. The business model of Apple around the iPod included the following elements (extract):

- high margin price model that led to a great revenue and profit growth and improved the market capitalization significantly (**Sustainable Profit**)
- simple and user friendly design (ease of use), strong brand, high storage capacity as well as iTunes (since 2003), a new sort of customer interface which provided the customers with a very special added value through the simple downloading of digital music and other content (**Customer Value**)
- focus on young people of a generation who quite naturally grew up with iPods and subsequently had become prospective customers for other Apple products; this was enhanced by an interim claim from Apple to be able to satisfy each customer segment (**Customer Segments**)
- specially designed sales facilities like the Apple Stores at exclusive locations, Apple shops in selected consumer electronics retailers as well as the Apple Online-Store (**Channels**)
- protected proprietary technology, that is software (iTunes) that was compatible with Apple products only; it represented a switching barrier (lock-in) for customers and thereby fostered strong customer loyalty (**Customer Relations**)
- orchestration of specialized suppliers in the value chain to achieve cost and quality advantages compared to in-house production as well as high "speed to market" with new products (**Key Processes**)
- innovative R&D team, strong patents as well as attractive partners such as the sporting goods producer Nike (**Key Resources/Partnerships**).

Even though, at first glance, the pure product idea seemed relatively easy to imitate, the business model with its different integrated elements presented a unique differentiation feature which—all in all—has led to a substantial competitive advantage for Apple. (Source: Johnson/Christensen/Kagermann 2008: 50 f.; Osterwalder/Pigneur 2010: 47, Wunder 2013: 356)

3.4.4.3 Business Model Innovation

Types of innovation

An innovation is an invention (i. e. something new) that is made with the goal to be *economically exploited* in the marketplace or utilized within the company. In this understanding, the successful market introduction or utilization is used to distinguish an innovation from an invention (Hauschildt/Salomo 2007: 7). The ability to innovate generally tends to be considered essential for a companies' long-term survival. Hereby, various types of innovation can be distinguished. Traditionally, people tend to connect the term innovation with launching new products and services. Additionally, innovation may be related to processes or organizational structures (Burr 2004: 23; Hauschildt/Salomo 2007: 9; Möller/Menninger/Robers 2011: 3). In recent years, the term innovation has been increasingly used in relation to business models (Amit/Zott 2010; Chesbrough 2010; Johnson 2010; Johnson/Christensen/Kagermann 2008; Mitchell/Coles 2003 and 2004; Teece 2010; Zott/Amit/Massa 2011). Many senior executives see business model innovation as top priority following a rationale that it might be harder for competitors to imitate a unique business model than pure product or process innovations. Thus, business models are more likely to promise longer-term competitive advantage for the firm (Amit/Zott 2010).

Radical versus incremental (proximal) innovation

In addition to distinguishing innovation with regard to its application, it can be generally categorized according to the level of innovativeness into radical and incremental (Burr 2004: 24). Here are some basic differences (Nagji/Tuff 2012; Wunder/Bausch 2014a: 47 f.): Radical innovations are transformational in a sense that they do not exist in the market yet (i. e. disruptive or breakthrough innovation). From the perspective of the customers, those innovations provide significant benefits. Radical innovations are systematically pursued primarily by large companies as they are complex and typically require significant long-term investments (e. g., R&D or technology centers, large innovation teams, cooperation with research institutes, etc.), have a relatively high risk, require a long development time but promise high returns if successful. Frequently, they are a combination of different types of innovation such as product and process. For example, Nestlé transferred its competency in the pulverization and high solubility of milk powder to coffee and launched its instant "Nescafe" in 1938 in Switzerland (Maucher 2007: 129). Incremental innovation (or "renovations"), on the other hand, are evolutionary, represent an optimization of what is already existing (e. g., improved taste, easier to use packaging, minor quality improvements, etc.), provide proximal benefits to customers, have relatively low complexity, are associated with little investments and can be managed by small teams in a relatively short period of time. Successful incremental innovation may have a high return on investment, though their overall economic contribution tends to be lower than with radical innovation. Whereas radical innovations are often R&D-driven, incremental innovations are also marketing-driven.

Although it is sometimes difficult to draw a distinct line between these two types of innovation, they are highly relevant for company practice. Depending on the industry, incremental innovation tends to be the dominant form (Stockmeyer/Weindlmaier 1999: 3; Wunder/Bausch 2014a: 47 f.). The differentiation is considered a useful taxonomy when elaborating on business model innovation, which will be explained next.

Degree of newness

There are a variety of definitions of what business model innovation means (Schallmo 2013: 26). For some authors business model innovation occurs when two or more elements of their corresponding business model framework are altered to create, deliver and capture value in a new way (Gassmann/Frankenberger/Csik 2014: 9; Lindgardt et al. 2009: 2). A different view of business model innovation can be derived by referring to the degree of newness. Any type of innovation features a certain degree of newness regarding the object it is related to. Following a typology used for product innovation, the degree of newness can be determined from the perspective of the market and from the perspective of the company (Homburg/Kuester/Krohmer 2013: 112 f.). By combining these two perspectives and following the previously explained concept of radical versus incremental innovation (Wunder/Bausch 2014a: 47 f.), the four types of business model innovation illustrated in figure 3-40 are proposed (the examples provided for each type relate to Strategy Practice Example 3.13 in chapter 3.4.4.4):

- A **new-to-world business model** is a radical business model innovation with a high degree of newness regarding both the market and the company. A source for this type of breakthrough innovation might be business model patterns from other industries that are transferred across industry boundaries and adjusted accordingly (see chapter 3.4.4.4). As a fictive example, a coffee capsules-based business model may be transferred from the consumer goods market to the pharmaceutical industry. A pharmaceutical company might develop a unique capsules machine system for personalized health providing medicine, vitamins and other

Fig. 3-40

Types of business model innovation

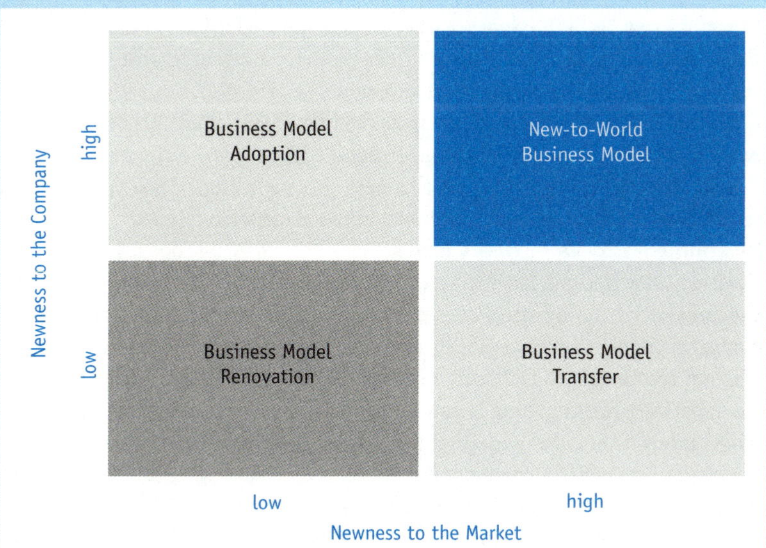

nutrition supplements, for example, for elderly care based on a completely new business model.

- A **business model adoption** occurs when another firms' business model that is already existing in the marketplace, or parts of it are adopted by the company and adjusted to be utilized effectively. The degree of newness to the market is low but to the company is high. An example is a company that used to sell only ground coffee in traditional packages and decides to additionally introduce an own coffee capsules system for its customers.
- A **business model transfer** occurs when a business model that is already existing in the company (or selected elements) is transferred to new markets. For example, Nestlé transferred the highly successful Nespresso business model to the baby food market and developed the BabyNes-system that has a variety of similarities to the coffee capsules business model.
- A **business model renovation** is an incremental or proximal business model innovation with only minor adjustments to an existing business model. The result is neither highly new to the company nor to the market but creates and delivers additional value that can be captured by the company. An example might be an additional new coffee flavor supplementing the existing coffee capsules portfolio or the introduction of capsules that can be recycled.

Developing a business model innovation

Business model innovation is either a greenfield approach (e.g., for an entrepreneurial venture based on a new business idea) or it is based on an existing business model. If the firm wants to further develop its already established business, the underlying implicit or explicit success mechanisms can be analyzed and advanced with reference to the seven business model building blocks (see chapter 3.4.4.2) and their mechanisms. Business model development is done in three steps:

1. **Describing an as-is business model:** This might be facilitated in a management team by, for instance, having people write down their key ideas related to the seven building blocks of the current business model (or a virgin business idea) on sticky notes and arranging them on a canvas (see fig. 3-39) showing the applied business model framework (Osterwalder/Pigneur 2010: 44–46). Afterwards, important interdependencies between the elements are identified. Finally, the business model draft is discussed and sharpened in the management team.
2. **Ideation on a to-be business model:** Once the "DNA" of the current business is clarified, new options for business model elements need to be developed within the context of the existing corporate or business strategy. For example, if the company wants to be strategically positioned as a premium provider in the market, it is not reasonable to elaborate on a value proposition for a low-price segment. An important question here is how narrow or wide the strategic focus for business model innovation is defined. On the one hand, clearly defined strategic directions or "challenges" of the company can be seen as a success factor for an efficient innovation process (Wunder/Bausch 2014a; see also chapter 3.4.1.2) and help to focus ideation regarding new business model options. On the other hand, if the strategic corridor for ideation is too narrow, it might hinder creativity for coming up with breakthrough ideas and various (innovative) business model options.

3. **Evaluating business model options:** Ideas to advance the business model need to be evaluated regarding different criteria, such as their expected economic benefit (rough order of magnitude), competitive situation, strategic fit for the company, feasibility, risk, etc. (see also chapter 3.5). Hereby, ideas for an advanced business model are put forward to a more detailed concept phase based on an initial rough evaluation. Once the business model concept has been refined, it is subject to a more detailed evaluation based on a detailed analysis of the needs and wants of the targeted customer segments, the value proposition and the other building blocks. Hereby, a key success factor is the degree to which key customer and market insights are built into the business model (Wunder/Bausch 2014a). It also needs to be checked how far the new business model could serve as a source for sustainable competitive advantage in the future. What kind of *value* for the customers and for the company does the business model deliver? Is it unique or at least unusual or *rare*? Is it difficult to *imitate*? In which way can it be replaced by other business models or is it non-substitutable? Only if all questions are answered with "yes" a sustainable competitive advantage might be achievable (see VRIO-model in chapter 2.3.2).

Business model patterns support business model innovation.

Creative ideas and the ability to imagine doing things in a completely new way are crucial for successful business model innovation. Great ideation is often seen as a gift given to only a few creative geniuses and as an unstructured process that cannot be influenced. However, the fact is that there are multiple methods that stimulate creativity and divergent thinking of individuals or groups, and thus, can effectively support business model innovation. By consulting relevant literature more than 1,000 creativity techniques can be identified. Essentially, most of them can be traced back to four principles of ideation (Birkenmeier/Brodbeck 2010: 130–141; Homburg/Kuester/Krohmer 2013: 115 f.):

- **Association or connotation:** To solve a certain problem as many ideas as possible are generated without constraints (e. g., brainstorming, brainwriting, 6-3-5 technique). This method is typically applied in a group setting based on the assumption that the quality of ideas increases with the number of ideas expressed.
- **Abstraction:** The problem to be solved (e. g., develop a new type of car) is abstracted to a next higher level of what "really matters" (e. g., provide individual mobility). This is supposed to unleash new ways of thinking disconnected from current mental solution models (e. g., progressive abstraction).
- **Decomposition:** This creativity technique is based on the insight that a complex problem can be better solved step-by-step (e. g., morphological box). Hereby, the problem (e. g., develop a new coffee maker) is decomposed to its individual parts (e. g., principle of operation, raw coffee material, water intake, output quantity, type of power supply, etc.). For each "subproblem" all possible solution variants are listed (e. g., raw coffee material: beans, ground coffee, capsules, tablets, bags).
- **Analogy:** Ideas are generated based on analogies derived from, for example, random inputs. Hereby, an input such as a word or picture derived randomly from an encyclopedia or newspaper is used to open new lines of thinking. Instead of using the random input itself (e. g., tree) it might be more effective to use its attributes

for generating new ideas (e.g., grounded, firm, old, changing, etc.). Another approach of using an analogy for encouraging creativity is the application of similar or comparable solution *patterns* (see Strategic Snapshot 3.5).

Although all four principles with its different techniques for promoting creativity can be generally applied to business model innovation, the pattern method (analogy) has become particularly popular among various business modeling approaches. The next chapter will explain how business model patterns can be used to support business model innovation.

3.4.4.4 Business Model Patterns

Patterns of consistent strategic behavior

A "pattern" can be defined as "a reliable sample of traits, acts, tendencies, or other observable characteristics of a person, group, or institution <a behavior pattern>" (Merriam-Webster 2014). The analysis and understanding of patterns helps to solve certain problems in ways others used successfully in comparable situations in the past (see Strategic Snapshot 3.5). Patterns do play a significant role in the area of strategic management. In chapter 1.1.1.2, a strategic perspective was described that considers strategy as a pattern in a consistent stream of decisions and actions (Mintzberg 1978). In this view, strategy constitutes itself when it happens, independent of whether it was previously planned or not. What matters according to this definition of

Strategic Snapshot 3.5

The Role of "Patterns" for Problem Solving

The application of patterns for problem solving has a long tradition and can be found in many disciplines today. An example is the TRIZ method that was developed for solving technical problems in **mechanical engineering** and that is used nowadays as a creativity technique for problem solving in many different areas. "TRIZ is an international science of creativity that relies on the study of the patterns of problems and solutions, not on the spontaneous and intuitive creativity of individuals or groups. More than three million patents have been analyzed to discover the patterns that predict breakthrough solutions to problems." (Barry/Domb/Slocum 2014) In **software engineering**, behavioral design patterns are used for designing comprehensive software systems (Buschmann et al. 1996). Patterns are also used in **architecture**. Hereby, existing architectural design ideas are captured and used as archetypal and reusable descriptions for planning, designing, and constructing buildings and other physical structures (Alexander 1979; Alexander/Ishikawa/Silverstein 1978). In **chess**, the ability of learning and applying patterns makes the difference between winning and losing. "Chess is a game of patterns; patterns about how the game has unfolded, about where the game stands at the moment, and, most importantly, about where the game is heading. In chess, the player with the best skill at pattern recognition has a critical advantage." (Slywotzky et al. 1999: 3) In **professional team sports** such as soccer, patterns are used to describe complex structures that underlie sporting contests between opposing soccer teams. The corresponding patterns support coaches in anticipating possible movements and adapt their own game plan such as attacking or defensive strategies accordingly (Camerino et al. 2012). In **strategic management** patterns have been exploited for strategic ideation and development in various ways. An example is the concept of *profit patterns,* which describe frequently observable changes in business and provide recommendations how to deal with these changes. They shall support companies in detecting threats in their environment as well as potential market opportunities, and in transferring existing risks into opportunities and related profit (Slywotzky et al. 1999). Strategic management approaches following a similar logic are *case-based strategy development* (Greiner 2006) as well as the work with *business model patterns* (Gassmann/Frankenberger/Csik 2014, Johnson 2010, Osterwalder/Pigneur 2010).

strategy is the fact that actual strategic decisions and activities of an organization lead to consistency in behavior, which can be observed and documented (see also chapter 1.1.2.5).

Business model pattern, prototype, archetype, analogy

On a business-level, these patterns of consistent strategic behavior can be examined, described and generalized with business model frameworks and their building blocks. This results in business model patterns, which are generalized shortcuts of an organization's business logic, that is the essential characteristics and principles of how an organization is successful in its competitive arena. In other words, business model patterns are descriptions of business models with similar observable characteristics and arrangements that can be found over and over again in our environment. The terms business model prototype, business model archetype, business model analogy, or generic business model are sometimes used synonymously.

Challenging the dominant logic of a business

The business model pattern approach is based upon the finding that most business model innovations are nothing but an imitation, variation or recombination of existing business model elements successfully executed by other companies, in other industries or in a different context. Consequently, for coming up with a successful business model innovation, managers do not necessarily have to "reinvent the wheel". Moreover, it needs to be investigated whether the core logic of already existing business models can be transferred to one's own business without doing it exactly the same way. As explained previously, this may result in business model renovations, adoptions, transfers, or even to the development of a new-to-world business model (see fig. 3-40). To effectively support this kind of strategic ideation process, descriptions of successful business model patterns—ideally with the same framework that is applied within the company—can be leveraged (Gassmann/Frankenberg/Csik 2014; Osterwalder/Pigneur 2010: 56–119). Those business model patterns are used for opening new lines of thinking and challenging or even breaking the dominant logic of one's business, which may result in anecdotal questions like

- What can a financial institution learn from the way Toyota is doing its business (e. g., strong process orientation and continuous improvement)?
- Can the Gillette's razor and blade strategy (lock-in) be utilized for our business?
- What can we learn from the business models of Amazon, Google or Apple?
- How would we conduct our business if we were Starbucks?
- What can we learn from the business purpose and business model transformation at PUMA?

Collection of business model patterns

The starting point for applying a pattern approach to business model innovation is a collection of business model patterns that describe specific successful compositions of business model elements in a standardized and universal way. Those success mechanisms are usually built around one or only a few dominant business model elements. Table 3-1 provides an overview of 28 common business model patterns as well as exemplary companies applying them, which is derived from relevant literature on business modeling. For each of the seven building blocks explained in chapter 3.4.3.2, four business model patterns are presented. Hereby, the corresponding business model building block is considered a dominant element, that is considered the key success driver for each of the four business model patterns described.

Tab. 3-1

Selected common business model patterns

TITLE	DESCRIPTION	COMPANY / INDUSTRY EXAMPLES
Process Patterns		
Integration	Responsibility for great portions of the entire value chain is assumed by one company, the integrator.	**Zara** has control over all value creation activities from production to distribution. This allows the company to identify trends and establish new fashion collections with a high speed to market.
Deintegration	Focus on own competencies (center of gravity) and outsourcing of the remaining value creation activities to specialists that are orchestrated by the company.	**Apple** outsourced manufacturing and focuses on its competencies in R&D, design and marketing.
Self Service	Outsourcing of parts of the value creation process to the customer (self-service), which reduces cost for the company and increases flexibility to the customer.	**IKEA's** customers pick their goods from the warehouse and assemble their furniture themselves. In **Banking**, traditional bank transactions (e.g., wire transfers) are now performed by the customer.
Focus on Blockbusters	Business activities (e.g., R&D, marketing, etc.) create a system that consistently delivers few high profit product hits (blockbusters) instead of a balanced product portfolio with moderate profits.	Following the model of branded pharmaceutical companies, some **film studios** and **computer game providers** devote much effort to creating a consistent series of blockbuster.
Resource / Partnership Patterns		
Franchising	Franchisees are granted the use of the franchiser's business concept and assets, e.g., trademark, products, corporate identity, business systems, and processes.	Most of **McDonald's** restaurants are run by franchisees at their own risk and cost; McDonald's is responsible for providing conditions that support a successful business of their franchisees.
Multiple Use of Assets	The potential of the company's assets (e.g., brand name, film characters, knowledge, persons, photos, information, product platforms, etc.) is turned into profit in numerous different ways.	**Disney** leverages the value of its film characters and makes additional profit by offering related consumer goods, musicals, ice shows, and theme parks.
Product Pyramid	Shift from a single product to a product pyramid with different levels for different customer and price segments. Usually each level has its own brand with profit maximizers at the top.	**Accor** runs economy class hotels (represented by the ibis brand), as well as midscale (Novotel, Mercure) and luxury (Sofitel) hotels.
Support of Weak Partners	Weak partners in the value chain that limit the company's ability to create the desired customer value are exclusively supported to avoid quality variations and improve the entire system performance.	To make its quality-driven business model work in Russia, **McDonald's** invested in the training of local suppliers, e.g., to grow the right kind of potatoes required for making long french fries.
Customer Value Patterns		
Cross Selling	Bundle the original product together with complementary products or services in order to create positive synergies.	In addition to their traditional business, **Gas Stations** offer a range of food and other everyday products. Also, **Tchibo** supplemented its traditional coffee business with special non-food items.

Tab. 3-1

Selected common business model patterns (Continued from previous page)

Branding	Establishment of a product brand which offers customers intrinsic benefits such as a feeling of security, luxury, hipness, etc. The brand becomes much more important than the core product.	**Fiji Water** is a brand that reflects clear and healthy water preserved from external elements and pollution. This story transcends other purchase decision criteria such as price or ingredients.
No Frills	Reduce the offering to its core; extras can be selected but for an additional charge. Customer pay only for what they want and get.	**Ryanair** offers relatively inexpensive flights with no extras, i. e. additional services such as seat assignments, meals, blankets or pillows are fee-based.
Solution Provider	Instead of selling a product, an overall solution for the customer's problem is offered with related goods and services in one package. The supplier as single point of contact reduces overall customer effort.	Instead of just selling its power tools, **Hilti** offers a leasing model for fleets of tools along with a comprehensive tool fleet management at a regular administrative fee.
Customer Relations Patterns		
Experience Selling	The product is linked with unforgettable impressions and special events that have a positive impact on customers' loyalty toward the seller.	**Harley Davidson** regularly organizes events such as rallies, festivals and tours for the "Harley Davidson community", thus creating a "Harley Davidson lifestyle" and high customer loyalty.
Lifetime Warranty	A 100 % commitment and conviction to the quality of a company's products are articulated by offering a lifetime warranty to its customers.	**Customers of Land's End** can return products they are not satisfied with at any time for repair, replacement or refund.
Lock-In	Creation of barriers to switching providers, e. g., through an incompatibility of the own product with competitors' products, the creation of a de-facto standard, joint advertising, or long-term contracts.	**Nespresso** machines are supposed to be operated only with the corresponding capsules exclusively available from Nespresso. **Gillette's** razor is only compatibel with its own blades.
Loyalty Club	Improvement of customer retention through bonus programs that reward customers for their loyalty through attractive offers or benefits exclusively available for club/program members.	With **Lufthansa's** frequent flyer program called "Miles and More", customers can collect miles on each flight, which can be set off against future flight tickets, upgrades, lounge access, etc.
Channel Patterns		
Direct Selling (Compression)	Avoidance or elimination of redundant distribution tiers between seller and customers. This reduces time and cost while increasing flexibility and gaining customer insights through direct relationships.	Through channel compression, **Dell** sells customized products directly (online or by phone) to its customers leading to both lower cost for Dell and lower prices for its customers.
Intermediation	Use of new intermediaries between sellers and customers that offer a novel service or improved business processing, and thus, additional benefits for one or both parties involved.	**PayPal's** online payment system simplifies payment processing for both sellers and buyers. Through its specialization, PayPal has become a value-added intermediary in the distribution system.
Multiplication	Use of multiple sales and distribution channels (e. g., online and offline channels) in order to meet the various shopping preferences of different customer segments.	**WMF** products can be bought at company owned WMF specialty stores in premium downtown locations or shopping centers, various department stores and retailers, and the WMF online shop.

Tab. 3-1

Selected common business model patterns (Continued from previous page)

One-Stop Shop	Offering many different goods in one shop to provide customers a "complete solution" with a huge competitively priced product range; it makes shopping more convenient and less time-consuming.	**Walmart** stores offer a broad product range from food to electronics to clothes at comparatively low prices. **Amazon** evolved from selling books to an online market that sells almost everything.
Customer Segment Patterns		
Customer Data Leverage	Systematic collection of big customer data for predictive analytics (e.g., personal purchase proposals), customer insights for product development and promotion, or customer data selling to third parties.	**Amazon** offers individual product suggestions often resulting in impulse buying and anticipatory shipping (speed) based upon an analysis of the specific buying behavior of customers.
Micro Segmentation	Breaking down a customer base in small, homogenous segments with distinct needs and wants that are fulfilled by customized product variants. This might be an internet-based niche provider ("Long Tail").	**Axel Springer SE** publishes magazines such as different types of Auto Bild and Computer Bild, Bilanz, Metal Hammer, Rolling Stone, me.style, or Travelbook for many smaller customer groups.
Redefinition of Customer Base	Redefining who the customer should be to create new opportunities for sustained profit growth. The focus might be on new profitable customer segments or on stakeholders influencing customers.	Traditionally consumed by older people as after-dinner drink at home, **Jägermeister** overcame its old-fashioned image and gained cult status as trendy party drink for a young customer base.
Open Source	The borders between supplier and customers become blurred as anyone from a public community contributes to product development and improvements, design, marketing, etc.	**Mozilla i**s a free software community developing and continuously improving open source software. With its popular Firefox web browser it generates revenues from its search providers.
Sustainable Profit Patterns		
Advertising	The traditional service can be used free of charge while income is generated by selling advertising space to advertisers.	**Google** offers its traditional service, the search engine, for free and generates income through selling advertising space to advertisers. Advertisements or links to the advertisers' homepage are placed on the users' search result page.
Freemium	The basic offering can be used free of charge and is financed by the fee that premium customers pay for the premium offer.	**Skype** offers free internet and video calling for registered users, premium services such as voicemail or calls to phones can only be used against payment of a fee.
Leasing	Renting a product instead of buying it allows more customers to use formerly unaffordable products or replace one-time purchase expenses with a pay-per-use model.	**Car2Go** offers registered users access to a fleet of cars that can be rented for a fee per minute.
Subscription	Customers pay a constant (e.g., monthly) subscription fee and can regularly take advantage of a specific product or service.	**Blacksocks** provides socks on a subscription fee basis, i.e. customers regularly receive a certain quantity of socks at a predetermined delivery interval, e.g., two pairs every four months.

(Source: Gassmann/Frankenberger/Csik 2014; Johnson 2010; Osterwalder/Pigneuer 2010; Slywotzky et al. 1999; Slywotzky/Morrison/Andelman 1998; Teece 2010)

Strategy Practice Example 3.13

Nespresso—Leveraging Multiple Patterns in one Business Model

Nespresso's success story started with a simple but revolutionary idea: a system for portioned coffee consisting of a coffee machine and corresponding capsules with high-quality coffee. After its foundation in 1986 and a challenging starting phase, Nespresso succeeded in breaking the previous industry logic by creating a new desire for classy coffee consumption among coffee consumers. The company turned the simple and highly commoditized coffee market at that time into something new: coffee enjoyment as status symbol. Today, Nespresso is one of the leading coffee brands in the world. It is positioned in the premium coffee segment and distinguishes between two customer groups: business customers and private households. The business market includes, for example, offices, upscale restaurants and hotels, and various airlines. In the private domestic segment, the company mainly concentrates on pleasure-loving and lifestyle-oriented coffee lovers of every age who have a comparatively high income. Based on own research, figure 3-41 provides an illustrative overview of the Nespresso business model and some of its actual and potential advancements. Various types of business model elements and their logical connections can be distinguished in each of the seven building blocks: original core (node) and detailed elements, current advancements and possible new options. The business model combines business model patterns of almost each category from table 3.1, which will be explained in the following (patterns are listed in brackets).

The core of Nespresso's business model is the capsule system, which consists of a coffee machine and corresponding portioned capsules. The coffee machines can only be operated with the corresponding patented capsules. The machines are offered at a comparatively low price while the coffee is priced relatively high with around € 0.30 per capsule, that is corresponding to € 40 per kilo (**Lock-in**). The company and its partners continuously elaborate on improvements to extend the lifespan of the coffee machines, and thus, increase both customer satisfaction and retention. For example, gaskets, which often led to a breakdown of the machines, are no longer built into the machines but were integrated in the coffee capsules. High customer loyalty does not only generate sales and sustainable profits but allows the company to build strong relationships and trust toward the brand.

Depending on their lifestyle and preferences, Nespresso's customers can choose from a range of more than 40 different coffee machines that have stylish compact designs and are easy to operate. In order to meet various taste preferences, Nespresso offers a range of more than 20 "Grand Cru" coffees for the home segment and around 10 "Grands Crus" for the business customer market, each of them with distinct aromas and flavors. Furthermore, Nespresso regularly supplements its product portfolio with limited editions of rare and exclusive flavors. Overall, the value proposition is based on variety, convenience, lifestyle, and best quality. The latter is promoted as Nespresso's top priority, which is achieved by sourcing only premium coffees from the best coffee-growing areas of the world and reflected in very high quality and sustainability standards that are met by only an estimated 1 % to 2 % of the world's coffee production. By labelling the different flavors "Grands Crus", Nespresso deliberately links its business to the context of premium winemaking, and thus, reflects excellent quality, a sense of connoisseurship, and a coffee culture covering coffee making, serving and enjoyment. All this allows a pricing policy based on which coffee is sold at the multiple price of coffee beans. Later, the premium segment was supplemented with another brand (Dolce Gusto) for the middle-range segment (**Product Pyramid**). Thanks to the exceptional success that Nespresso had with its business model Nestlé leveraged the capsules system and the corresponding competencies for business development in their tea (Special.T)

and baby food (BabyNes) categories (**Multiple Use of Assets**).
Nespresso coffee machines are broadly available and—depending on the country—can be bought in selected retail stores and online-shops (**Multiplication**). In contrast, the capsules are exclusively available directly from Nespresso (**Direct Selling**), either in one of the approximately 320 Nespresso boutiques (2013), in the online-shop or via mail-order. The Nespresso boutiques serve as sales channel and are an important factor for acquiring new customers as well as maintaining contact with existing customers. Usually located in major urban centers with high people traffic, they are a point of contact for club members who require advice or guidance. Furthermore, they serve as a place where club members and prospective customers can experience Nespresso with all their senses, taste different flavors, meet other club members and relax in a pleasant and elegant atmosphere (**Experience Selling**). Last but not least, the boutiques serve as showcase for the Nespresso brand and shall communicate the brand image of understated elegance, simplicity, luxury, exclusiveness, innovative spirit, and style. The brand icon, the product design, packaging and the boutiques, as well as all promotional activities consistently emphasize this image (**Branding**). The fact that Nespresso's products are only available through a small number of sales channels emphasizes this image. In order to drive brand image and recognition, Nespresso sponsors exclusive events such as the Cannes Film Festival, the America's Cup, Le Bocuse d'Or, or the Ryder Cup.

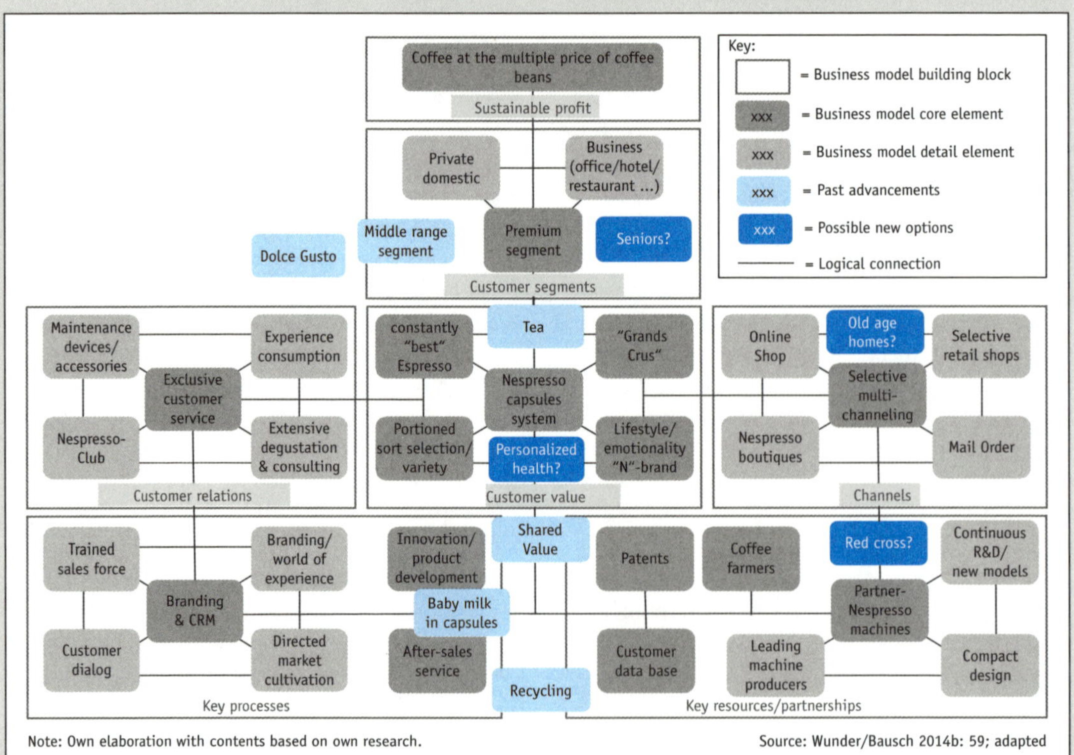

Note: Own elaboration with contents based on own research.
Source: Wunder/Bausch 2014b: 59; adapted

Fig. 3-41: Illustration of business modeling using the example of Nespresso

The Nespresso Club is the central element for establishing customer loyalty (**Loyalty Club**). Everybody buying a Nespresso coffee machine automatically joins this club and is welcome to an exclusive relationship with benefits from special offerings and unique experiences. The resulting customer database is an important asset for the company as it contains all relevant data of its premium clients. It supports continuous relationship management as well as personalized direct marketing measures such as coffee mailings or the offer of limited editions (**Customer Data Leverage**). Furthermore, analyzing the database provides customer insights used for launching appropriate new products. Customers can enjoy Nespresso's world of experience through various ways (**Experience Selling**). First, Nespresso has a homepage full of information for coffee lovers and regularly informs its customers with newsletters. When a new flavor is launched, high quality brochures present emotional stimuli for the new Grand Cru. Every flavor is promoted by an entertaining and educating story informing the customers about the origin of the coffee and establishing a small world of experience. Furthermore, Nespresso organizes coffee tasting events for club members or gives them the chance to apply for a role in Nespresso commercials with George Clooney. Such events, on the one hand, improve customers' loyalty. On the other hand, they serve as marketing activities encouraging participants to promote the brand later by telling friends about their unique experiences with Nespresso. More than half of Nespresso's new club members got introduced to the brand through existing members who, in turn, profit from coupons, rewards or a bonus. An optimal customer service tops off the customer relationship management. A 24/7 service allows club members to connect with coffee specialists in Nespresso's Customer Relationship Centers. The specialists do not only take orders, they also provide advice and guidance regarding the use of the products. Furthermore, Nespresso offers a selection of coffee-related accessories and maintenance devices (**Cross Selling**). Manufacturing of the coffee machines is outsourced to leading coffee machine producers (**Deintegration**), which are key partners in Nespresso's business model. And so are the coffee farmers. In order to ensure the supply of highest quality coffee while improving the lives of coffee farmers, their families and communities, Nespresso follows the shared value approach and puts high effort on supporting its suppliers and improving the social conditions in the areas where they live, e.g., by training them how to improve their productivity per square meter (and thus, to increase their income), building schools, and child labor monitoring.

(Source: Gassmann/Frankenberger/Csik 2014; Homburg/Kuester/Krohmer 2013: 581–587; Nespresso 2014; Nestlé Nespresso 2014; Nespresso/CIMS Sustainable Markets Intelligence Center/INCAE Business School 2014; Hanser 2009; Reinecke 2010: 63–78; Sidler 2004)

Creative adaption

It is very unlikely that an existing business model pattern can simply be copied and successfully exploited in the long term. To be successful, it typically requires an approach of creative adaption and calibration to one's own business situation, which has to be based on an in-depth understanding of the patterns' elements and their mutually enforcing interrelations. Differences between the business model pattern and one's own situation need to be identified and selected elements need to be adapted and customized accordingly (Gassmann/Frankenberger/Csik 2014: 21). Additionally, available lessons learned of companies who used to or presently apply the corresponding business model pattern may be leveraged to avoid pitfalls and get it right the first time. In this sense, business model patterns do not only support ideation but also provide guidance for realization.

Combination of several business model patterns

Companies have the choice either to adapt a single business model pattern to their own situation or to combine several business model patterns in a unique way. Simultaneously applying, integrating and advancing several of those patterns in one business model can be a powerful formula for sustainable success, which has been proven by various companies across different industries (Slywotzky et al. 1999: 383). For example, when analyzing the highly successful Nespresso business model, at least 12 of the business model patterns described in table 3.1 have been identified (see Strategy Practice Example 3.13).

3.4.4.5 Sustainable Business Models

Combining economic value creation with value creation for society

The business world is undergoing dramatic change due to pressing ecological and social problems that our society has been facing. Companies play a significant role in driving necessary progress toward sustainable development, that is a "development that meets the needs of the present without compromising the ability of future generations to meet their own needs." (WCED 1987: 43) Following the so-called *triple bottom line* approach (Elkington 1997), more and more companies simultaneously pursue economic, ecological and social goals and integrate those to one holistic perspective on performance (see also chapter 1.1.3.4). Hereby, corporate sustainability is not only a question of social responsibility (license to operate), moralities or ethical integrity, all being important elements of the solution. It has evolved to a highly strategic question and breakthrough management paradigm of how to combine economic value creation (*business case*) with value creation for society (*societal case*) for being successful in the long run, resulting in a *corporate sustainability case* that pro-actively integrates economic, environmental and social elements (Elkington/Zeitz 2014). This understanding is illustrated in figure 3-42. Hereby, the corporate sustainability case reflects a contemporary understanding of *Corporate Social Responsibility (CSR)* (Schmidpeter 2015a and 2015b). Key drivers for this management perspective are relevant or material stakeholders who have been imposing new requirements and expectations on companies (Schulz 2014: 8–22; Wunder 2014a; see also Stakeholder Analysis in chapter 2.1.1). Depending on the specific company these expectation include *socially responsible*

- *end consumers* supplementing their evaluation of the value they gain from a company's product or service offering with ecological and social considerations in their purchase decisions (upstream)
- *suppliers* that provide an excellent environmental and social performance of their supplies, which is driven by business customers managing sustainability along the entire supply chain (downstream)
- *investors* that not only evaluate the economical performance but also environmental, social as well as corporate governance concerns of a company or business when assessing the investment risk (ESG rating)
- *employees* that evaluate the attractiveness of employers with due consideration of how they authentically position themselves in terms of sustainability
- *senior executives* that perceive sustainability principles not as an add-on but as a key pillar of excellent management and as a key element in their core values

Fig. 3-42

Corporate sustainability case as combination of a business and a societal case

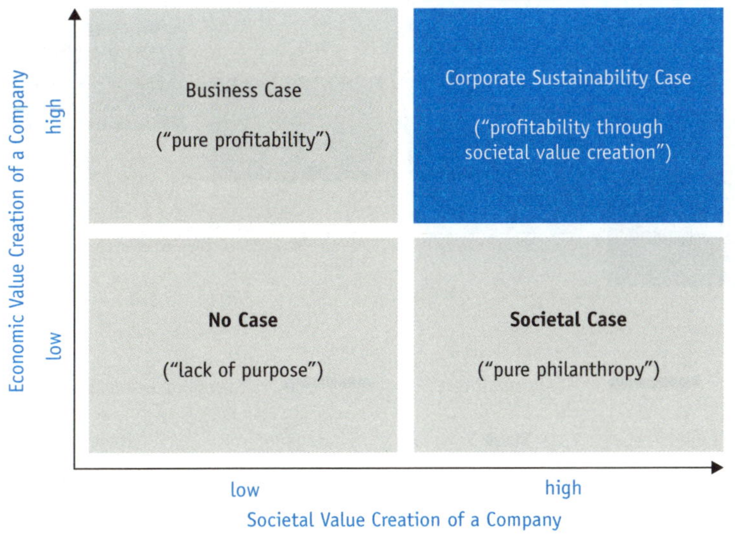

Source: following Schmidpeter 2015a: 39; adapted

- *supervisory boards* that consider sustainability considerations when deciding for or against certain strategy options proposed by the management to reduce their liability
- *governments* imposing new environmental and social laws and regulations that are binding for companies and may significantly impact their business models
- *non-governmental organizations* (NGOs) and *non-profit organizations* (NPOs) that are not only pushing for sustainable development on a governmental level but—aided by advances in information and communications technology—increasingly challenge companies to provide transparency about and manage the environmental and social impact of their business activities.

Stages of how a firm deals with sustainability strategically

In the new age of sustainability awareness, companies have to address environmental and social issues when assessing their business opportunities and make business-critical decisions of how they position themselves in the long run. Figure 3-43 provides five stages of how companies may deal with sustainability (following Esty/Winston 2006 and Nidumolu/Prahalad/Rangaswami 2009). The higher the stage, the more a company is trying to leverage sustainability *strategically*, that is the more it impacts their core business in the quest for competitive advantage. Whereas the focus in stage 1 and 2 is primarily on compliance with legal requirements and standards, sustainability is considered more as an entrepreneurial opportunity in stage 4 and 5

Fig. 3-43

Five stages of sustainability

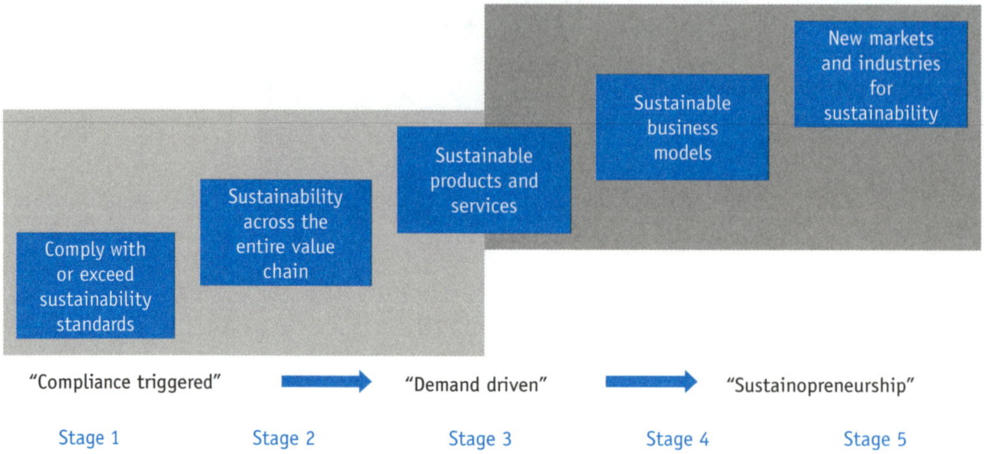

Source: following Nidumolu/Prahalad/Rangaswami 2009; ICV 2011: 4; Wunder 2014a: 73; adapted

(sustainopreneurship) as well as partially in stage 3. In this basic model, companies in a higher stage do not necessarily need to fulfill lower stages. Firms may apply the model as high-level framework for managing sustainability and pursue sustainable initiatives simultaneously in various stages (ICV 2011; Gänßlen/Kraus/Ette 2011: 461). Also, firms may pursue highly effective sustainability initiatives on lower stages that have a strong environmental or social impact but are managed outside the core business. In such cases companies may drive well-intentioned sustainability projects but without a strong link to corporate or business strategy.

Sustainable business models create extended customer value.

A sustainable business model (stage 4) can be interpreted as a "business model that creates competitive advantage through superior customer value and contributes to a sustainable development of the company and society (...). Since customer value is the strategic nexus of any business model, sustainable business models are crucial for creating extended customer value for individual customers and society, that is private and public benefits." (Lüdeke-Freund 2010: 23) The core of a sustainable business model is the idea of deliberately co-creating societal and business value by voluntarily integrating environmental, social and economic entrepreneurial business activities. Hereby, providing a substantial long-term contribution for solving environmental and social problems of our society is considered a strategic success factor that contributes to the economic performance and long-term survival of a company. One of the underlying goals of such a sustainable business model is to create a so-called "business case for sustainability" (Lüdeke-Freund 2013; Schaltegger/Lüdeke-Freund/Hansen 2012; Schaltegger/Wagner 2006; see also Strategic Snapshot 3.6). This is referred to as a corporate sustainability case in figure 3-42.

Strategic Snapshot 3.6

Business Cases for Sustainability and Business Model Effects

Figure 3-44 shows an illustrative model that attempts to describe the potential impact of voluntary social and/or ecological engagement of a company on its economic performance. It is based on a certain level of economic success (ES_0) as starting point after complying with social and environmental regulations. Moving to the right on the dotted curve ES_0-D-E-F represents a traditionalistic mindset where voluntarily driving social and/or ecological performance of a company is considered conflicting with its objective of improving economic performance. Hereby, economic success and social/ecological engagement is seen as pure tradeoffs as every social or environmental protection initiative (e.g., adding environmental filtration technologies or wastewater treatment plants) is expected to reduce the firm's profit due to associated incremental cost up to a point where the company becomes unprofitable (below point F).

A different and more revisionist and progressive view tries to overcome this tradeoff thinking with regard to sustainability. It is based on the idea of creating and leveraging win-win situations between the economic and social/ecological performance. To create a so-called *business case for sustainability*, a company deliberately pursues activities that help to solve societal problems—either environmental or social—and thus, improves its economic performance. Business cases for sustainability may be driven by

- **cost** (e.g., improved energy, water or waste management is leading to cost reduction)
- **revenue, price, margin** (e.g., enhanced customer value through sustainably produced products leads to an increased demand or the ability to charge a higher price)
- **risk** (e.g., reduced business risk reduces unforeseen cost or negative revenue impacts related to potential ecological or social issues)
- **reputation, intangible values, brand value** (e.g., improved positive company image or avoidance of negative image through voluntary ecological and social

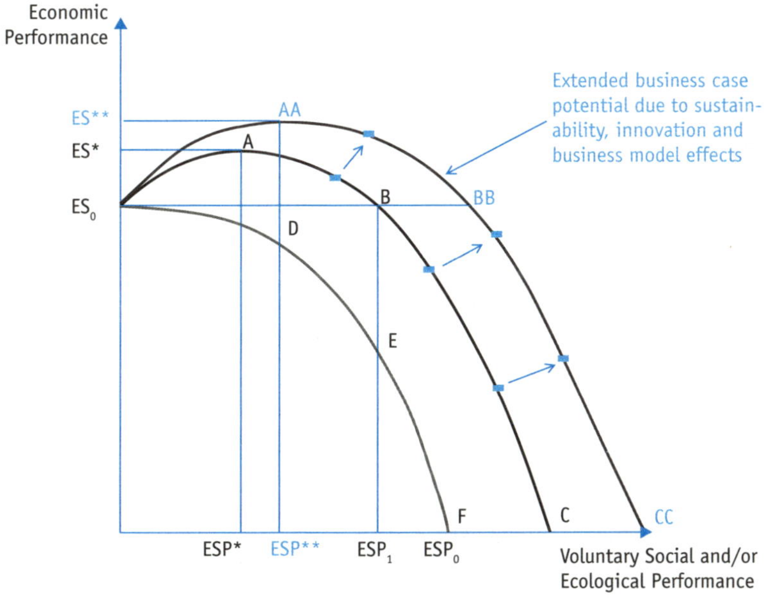

Source: Lüdeke-Freund 2013 adapted from Schaltegger/Burritt 2005: 197 and Schaltegger/Synnestvedt 2002: 8; adapted

Fig. 3-44: Illustrative relations between economic performance and voluntary social and/or ecological performance

Continued on next page

Continued from previous page

engagement enhances customer and employee relationships and retention)
- **product/service or process innovation** (e. g., new profitable products or services that contribute to solving societal problems)
- **business model innovation** (e. g., adjusting existing or creating entirely new business models that leverage social and/or ecological solutions for creating economic success).

In a business case for sustainability, economic success is driven *by* the voluntary, intentional and effective contribution of a company to a sustainable development of the society. This does *not* include accidental and random economic effects or an opportunistic short-term capture of "sustainable" market opportunities (e. g., by adding a "green" product to an otherwise conventional product portfolio). Business cases for sustainability are created and managed based on an entrepreneurial rationale that tries to leverage the company's core business for creating positive ecological or social impact, and thus, directly or indirectly drives economic success. This is shown in figure 3-44 when moving along the half-circle ES_0-A-B. Within this half-circle above line ES_0-B, a company's voluntary ecological and/or social engagement is considered to have a beneficial effect on its economic performance (based on a starting point ES_0). It thus represents an area of various business cases for sustainability with point A representing an optimal economic *business case* (ES*). This is based on the assumption that up to this point every well thought out voluntary social or environmental initiative is improving the company's economic performance. After reaching that optimum, additional ecological/social efforts are assumed to reduce marginal economic success contributions showing that even the revisionist and progressive view on sustainability considers tradeoffs, though at a different stage. Slightly above point B would be a business case for sustainability with optimal *societal case* (i. e. highest environmental or social performance without economic tradeoff compared to baseline ES_0). Which sustainable business case aspiration a company pursues strongly depends on its strategic approach, key objectives, and core values. For example, even if economically neutral in the first place, a business case in point B may create competitive advantage through, for example, the raise of de-facto ecological or social standards in the market. This may increase the pressure to improve sustainability performance on competitors who may not be capable of reaching the same level, and thus, face a competitive disadvantage in the future.

A company's existing business model may impose some limitations for creating business cases for sustainability as shown in the "border" ES_0-A-B. To reach a higher performance level in these situations, firms need to enhance their scope of building business cases of sustainability. They may significantly adapt their core business, redirect their market activities to new segments, fundamentally alter their existing or create completely new business model(s) with sustainability innovation. In figure 3-44 this highly entrepreneurial approach means a change of the curve's slope, that is the dashed line ES_0-AA-BB-CC. The company seeks to extend its business case potential through enhanced own contributions for solving environmental or social problems of the society (see also table 3.2).

(Source: Lüdeke-Freund 2013, Schaltegger 2012, Schaltegger/Lüdeke-Freund/Hansen 2012), Schaltegger/Wagner 2006

Addressing sustainability in different business model building blocks

With respect to the generic business model framework described in chapter 3.4.4.2, each building block may capture elements that drive corporate sustainability. Examples are (Boons/Lüdeke-Freund 2013: 13; Wunder 2014a: 74):

- **Sustainable Profit:** Accounting for the company's ecological impact (Hengstmann/Seidel 2014) or social impact, improving the company's profitability through extended customer value as well as appropriately distributing economic costs and benefits among actors involved to make the business model happen.
- **Customer Value:** Co-creation of customer value that drives economic value as well as ecological and/or social value, for example, offering products and services that create customer value and, at the same time, positively contribute to solving ecological and social issues.

- **Customer Segments:** Focusing on customer groups that embrace the idea of sustainability such as consumers with a lifestyle of health and sustainability (LOHAS) or fulfilling unmet needs of entirely new consumer groups, for example, empowering people of the poorest socio-economic group in the world and raise their standards of living with a corresponding offering (bottom of the pyramid; see table 3-2).
- **Customer Relationships:** Encouraging customers to consider ecological and social aspects of their consumption along with increasing their consciousness and knowledge about how to address sustainability challenges.
- **Channels:** Distribution through sustainability-oriented wholesalers and retailers or own low-energy stores as well as sustainability-oriented marketing communication.
- **Key Processes:** Engaging in sustainable management along the entire value chain, such as sustainable new product development, sustainable sourcing, sustainable logistics, etc.
- **Key Resources/Partnerships:** Sustainable own assets and resources as well as collaboration with partners who take responsibility for the requirements of their own and their partner company's stakeholders (e. g., engaging suppliers in sustainable supply chain management).

Sustainable business model patterns

Table 3-2 provides an overview of 10 selected sustainable business model patterns. The first group of patterns is addressing environmental issues as a key component for creating competitive advantage based on extended customer value. The second group is focusing on social issues as a driver for the corporate sustainability case. Sustainable business model patterns are archetypes that can be used by business leaders as blueprints of how to address sustainability from a strategic perspective. They may utilize those patterns to refine or enhance their core business or the success logic in certain strategic business units. As part of a business model innovation process, those and other patterns can be leveraged for ideation to create a business case for sustainability by advancing existing or creating entirely new business models (see fig. 3-40 in chapter 3.4.4.3). The coming decades will demonstrate how far ecological and social problems of our society can be effectively tackled or even solved by companies that leverage sustainable business models. It will also provide more clarity about whether an integrated triple bottom line perspective will become mainstream in strategic management and business administration or remain a topic managed primarily by specialized environmental, sustainability or corporate social responsibility (CSR) experts, compliance and legal managers as well as corporate communication departments.

3.4.4.6 Benefits and Pitfalls for the Business Model Approach

Integrating concepts and insights from various management disciplines

Business modeling integrates different concepts, mechanisms and insights from various management disciplines and research areas such as strategic management, marketing, finance, operations, sales and distribution, etc. to a holistic approach of how to compete and succeed in the marketplace. A key reference point for the design of a business model is provided by the results of strategic analysis such as macroeconomic

Tab. 3-2

Selected sustainable business model patterns

TITLE	DESCRIPTION	COMPANY / INDUSTRY EXAMPLES
Green Business Model Patterns		
Dematerialization	Replacing physical products by digital offers saves scarce resources, reduces expensive infrastructure, and avoids pollution and waste. However, it may cut jobs raising concerns about social sustainability.	Newspapers like **Seattle Post-Intelligencer** or **The Yorkshire Times** are not available in print but online only. Various **grocery delivery** companies eliminate resource-intense brick and mortar stores.
Eco-supportive Products	Offer products or services that support customers in reducing environmental impact.	The cloud-based carpooling platform **SAP TwoGo** offers ecosensitive customers the opportunity to find somebody for a rideshare to work instead of driving separately.
Closed Loop	The linear "take-make-dispose" production model ("cradle-to-grave") is replaced by continous recycling of material used to creat a product in the production system ("cradle-to-cradle") avoiding outside inputs.	**Interface Inc.** makes carpet tiles from reclaimed and recycled material. It transferred its business to a floor-covering service leasing model providing a defined level of comfort instead of selling carpets.
Longevity	Products are designed, manufactured and repaired with respect to a maximized product life in order to minimize their environmental impact (e. g., reduce waste, use of scarce resources, and pollution).	**Vaude** focuses on best quality and designs of its products with due consideration of a long product life. An own repair service and special product usage education for customers is provided to extend product life.
Rematerialization	Value from waste is created by developing and marketing entirely new types of products from waste or from extracted and remanufactured materials derived from waste.	**TerraCycle** collects waste and repurposes the material into new products, e. g., bags out of drink pouches or ashtrays and plastic pallets out of cigarette waste.
Social Business Model Patterns		
Bottom (Base) of the Pyramid	Enabling and emporing people in the largest but poorest socio-economic section of the population—bottom of the pyramid (BOP)—to fulfill their needs by creating special offerings, enabling entrepreneurship, etc.	**Grameen Bank** offers microcredits to very poor people for projects that shall secure their long-term income. **Novartis** created a new marketplace by providing education and access to healthcare for poor rural communities.
Buy One and Do Good	Social impact is explicitely linked to sales by donating a profit portion to provide similar products to those in need. Firms charge a premium or accept lower profits but expect higher sales due to the social cause.	With every pair of shoes bought at **TOMS**, the company donates another pair to people in need. With every pair of eyewear bought, the company gives sight to a person in need.
Consumer Health Models	Fostering consumer health and well-being is at the core of the product offering. Sourcing and production is done in a way that avoids product elements with negative impact on consumers' health and well-being.	The products of the frozen food producer **Frosta** do not contain any artificial additives such as flavor enhancers, stabilizers or emulsifiers. Its business is based on self-imposed purity and quality standards.
Shared Value	Creating both economic value and societal value through a long-term development of suppliers in poor countries in order to improve their living standards and, at the same time, the quality of the supplied products.	**Walmart** and **Nestlé** create shared value in parts of their business by improving social conditions in rural areas where their product ingredients are grown, e. g., by training farmers, building schools and infrastructure, etc.
Cooperative Ownership	Business ownership and management by its members replaces the traditional shareholder (investor) orientation by a broader perspective resulting in the distribution of greater value to more stakeholders.	**Ocean Spray** is an agricultural cooperative owned solely by more than 600 cranberry and grapefruit farmers. These growers are paid as much as possible for their fruits and labor by the cooperative.

(Source: Bocken et al. 2014; Boons/Lüdeke-Freund 2013; Chandler/Werther 2013; Clinton/Whisnant 2014; Porter/Kramer 2011; Prahalad 2005; Prahalad/Hart 2002; Sommer 2012) trends, new customer requirements, market and competitor developments, new technologies or industry structures, etc. (see chapter 2.2). Based on those insights, some companies identify so-called "strategic challenges" and define more or less siloed strategic projects as a course of action for dealing with each individual strategic challenge. Business modeling follows a different, more holistic and integrated approach. Here, the key question is which business model as a whole might be best suited for dealing with anticipated pictures of the future instead of focusing on single strategic challenges.

Systematic and intuitively understandable strategic ideation framework

A business model approach needs to be simple and intuitively understandable, without simplifying the complexity of business success mechanisms too much. If this balancing act has been successful, the business model helps to systematically describe, and thus, better understand the own current business logic as well as success models of other companies as a basis for strategic ideation. Through its standardized structure, differences between various existing business models (business model patterns) and possibilities for differentiation can be identified more easily. Based on this, the method can be used for own business improvements and for identifying new sources of competitive advantage. With the different building blocks and their interaction, fundamental strategic decisions can be made from a comprehensive and integrative strategic perspective. Business model patterns shall facilitate the strategic idea generation process by serving as a source of inspiration and by forcing companies to break with the dominant logic of their business. This should help to think about completely new strategic ideas that might never come to mind when they do not fit into the industry's dominant logic.

Strategy articulation and link to strategy execution

Another benefit of using business models is the possibility to effectively communicate the somehow abstract strategic success mechanisms internally to decision makers at different management levels and employees. Furthermore, a business model can be used to clearly articulate strategic intent to supervisory bodies, investors, lenders, or cooperation partners. Representing a set of fundamental strategic decisions, a business model fulfils an important bridging function from the often rather abstract strategy formulation to the specific strategy execution system, for example, based on strategic goals and strategy maps, balanced scorecards and strategic action programs (see chapter 4.3). Hereby, a conceptual compliance of the selected business model framework and the tools and methods used for strategy execution is considered essential for acceptance and practical applicability in company practice. If the business model has fundamental conceptual differences from other strategy tools already used by the company there is the risk that the required "translation" from one concept to another might not be accepted. Moreover, it contradicts the pursued simplicity and intuitive understandability of the business model approach.

Continual adaption to a changing competitive arena

To benefit from using the described business model framework, it is crucial that strategy and business model management as well as strategy execution are interlinked and integrated effectively. Hereby, the business model approach has to be considered as an integral part of the applied strategic management framework (see chapter 1.3.3). As the company's environment is changing permanently, its business models need to be monitored continuously and, if necessary, adjusted. The better the understanding of the own business model, the easier it can be aligned with new opportunities. This can lead to an incremental or to a revolutionary business model innovation as a response to new requirements or as a lever to create completely new markets. In any case, business model innovation seems to be a promising approach for dealing with those challenges. More and more companies are addressing the increased relevance of sustainability in their business models. They are driving positive social and ecological transformation with innovative business models as a pillar for sustainable growth and profits. This includes business models for solving social problems or ecological issues with so-called *green business models* (see table 3-2).

Watch out: multi-level and interlinked business models

In general, a business model framework can be applied at various levels in the company. However, there are currently hardly any recommendations regarding the interaction between those levels. For example, the desired success mechanism in a business model may be strongly based on leveraging synergies with other businesses in the company, and thus, need to be reflected in all relevant business models involved. Furthermore, in large diversified companies there is a risk that business models at the levels above the business units (e. g., corporate or divisions) include only general and vague statements and therefore provide little direction and value. In such a case, basic strategic guideposts might be more appropriate at the corporate or divisional level (see chapter 3.1).

Watch out: missing business model patterns

Many available business model patterns are based on commercial business activities in the internet and differ from each other based on the business model framework they are described in. Leveraging business model patterns for strategic ideation can be best supported when patterns are described in the same business model framework that is applied by the company. Various authors provide corresponding patterns with their frameworks (e. g., Gassmann/Frankenberger/Csik 2014; Osterwalder/Pigneur 2010: 56–119). Other patterns are described more anecdotally without using a particular descriptive model, and some potentially valuable business models in certain industries may not have been described at all yet. To facilitate business model innovation effectively, it can be useful to "translate" anecdotal patterns or describe undocumented business models in the corresponding framework that is applied by the firm. This process may already foster creativity and ideation for a firm's own business model.

Watch out: comprehensive business model evaluation

An essential pitfall of using the business model approach is the fragmented and mostly qualitative representation of factors that are supposed to influence the business success. Although there is a financial element embedded in a business model (sustainable profit), a more comprehensive quantitative analysis of correlations between various elements is important for a deeper understanding of success mechanisms of a business, and thus, for strategic decisions. Also, an explicit risk assessment regarding the different business model options is missing. To avoid those pitfalls, additional quantitative and qualitative evaluations of various design options in the business model are necessary. Essentially, these business model design alternatives represent strategy options. Along with other strategy alternatives they need to be evaluated and the best options need to be selected for execution. This will be described in the next chapter.

3.5 Strategy Option Evaluation and Choice

Selecting and combining the best strategy options

It is likely that strategic ideation and strategy formulation including business modeling results in a set of strategy options that a firm generally considers to follow on a corporate and/or business level. Given the specific company situation, options may significantly differ regarding their performance impact (e. g., contribution to achieving commercial or non-commercial objectives) or questions like plausibility with the company's strategic context, attractiveness to key stakeholders, feasibility, consistency, business risk, etc. Some strategy options may reflect alternative approaches for creating competitive advantage whereas others might be compatible but cannot be realized simultaneously due to limited resources. Also, some strategy options may be more in line with personal values of strategic decision makers than others. In any case, strategic choice requires tradeoffs (see chapter 1.1.2.3). It needs to be decided which of the identified strategy options the company wants to realize and combine or integrate into a consistent overall strategic direction. This requires an evaluation of the pros and cons for each strategy option and, based on this, a selection of options for execution.

3.5.1 Key Criteria for Evaluating Strategy Options

Strategy options need to be suitable, acceptable and feasible.

Executive judgments on proposed strategy options can easily result in comments like "plausible ... but not feasible" or "strong performance impact ... but not very likely" or "profitable ... but not sustainable". To validate those statements and provide solid decision support, it is suggested to systematically appraise each strategy option based on pre-defined criteria. Such a systematic approach for evaluating strategic alternatives assumes that the more rational and fact-based such an appraisal is conducted, the more likely a selected strategy option will be successful. Whereas in business practice companies have typically defined their own specific evaluation approaches and principles, the following six generic criteria are identified that can be used for evaluating strategy options (see fig. 3-45). They are grouped into the three evaluation categories of suitability, acceptability and feasibility (Johnson/Whittington/Scholes 2012: 261–265):

- First, it has to be checked how **suitable** (i. e. plausible and consistent) a strategy option is for dealing with a certain strategic situation. Does it address key strategic issues identified in strategic analysis and does it fit in the company's strategic set up?
- If the general suitability is given, it needs to be evaluated whether the strategy option is **acceptable** in terms of its expected contribution to the company's performance based on the company's specific measurement as well as the related overall business risk to the organization.
- Finally, assuming the strategy option being both suitable and acceptable, it needs to be assessed whether it is **feasible** related to meeting interests and perceptions of different relevant stakeholder groups (external feasibility) as well as to the internal practicability for execution if it comes to organizational resources and capabilities (internal readiness).

Fig. 3-45

Six criteria for evaluating strategy options

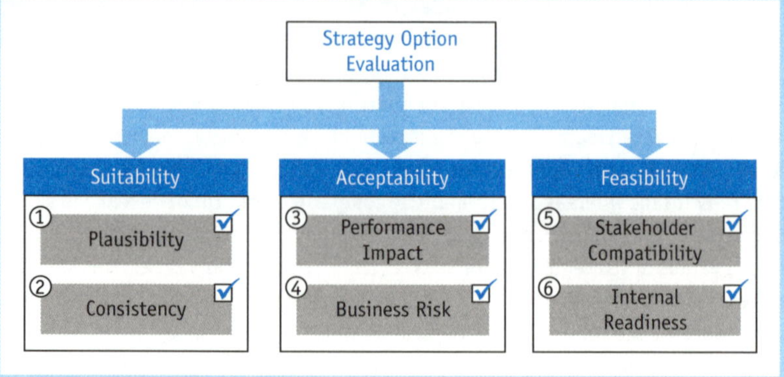

All those different considerations provide an orientation of what to take into account for decision making but may need refinement based on the specific company or industry situation as well as the personal value set of the executive leadership team and company owners. Independent of how the different strategic alternatives have been generated, it is suggested that each strategy option is rigorously evaluated based on the set of financial and non-financial criteria within the three categories. Each of those criteria will be explained in the following (Johnson/Whittington/Scholes 2012: 261–265; Lynch 2012: 379–414; Müller-Stewens/Lechner 2011: 322–330; Welge/Al-Laham 2012: 730–776; Wheelen/Hunger 2010: 251–265):

3.5.1.1 Plausibility

Convincing logic in dealing with a strategic situation

One of the most important evaluation criteria is the capability of the proposed strategy option to deal with the specific strategic situation reflected in the results of strategic analysis. Does the option follow a convincing logic to deal with the strategic chances and challenges identified? Plausibility means to be appropriate for dealing with the company's external environment given the organization's internal resources and capabilities. Hereby, strategy options can be examined with reference to the most relevant elements of strategic analysis (see chapter 2.1.2). Corresponding questions to ask for each suggested strategy option are:

- Will the strategy option take advantage of promising opportunities that have been identified in the company's environment?
- Does the strategy option build on relevant strengths and core competencies of the organization?
- Is the strategy option appropriate to deal with critical threats in the company's environment?
- Does the strategy option avoid, lead away or even overcome the key weaknesses of the organization?

- Are underlying assumptions about the future and analogies based on which the strategy option has been developed suitable?
- Does the strategy option meet key expectations of the most relevant stakeholder groups?
- Are key competitor reactions likely and is their anticipated response profile reasonable?
- How robust is the strategy option if some of the underlying assumptions for the company's external environment are changing?

To support the evaluation whether a strategy option is plausible or not, portfolio analysis and the corresponding normative strategy recommendations provided by different portfolio techniques may be leveraged (see chapter 2.4.2 and Strategic Snapshot 3.4 in chapter 3.2.2.1). Also, the method of network thinking (Gomez/Probst 1999 and 1987) can be applied to illustrate the interactions of different factors identified in strategic analysis, and thus, better understand whether a strategy option is plausible in the context of such a system of interdependent strategic factors. In company practice, strategy options may be suggested more opportunity-driven without a prior comprehensive strategic analysis. In such cases, checking for plausibility may reveal gaps that need to be subject for further analytical investigation before a decision can be made. Such cases lead to iterations in the strategy processes where activities of strategy formulation alternate with tasks of strategic analysis.

3.5.1.2 Consistency

Consistency with strategic guideposts

Even in case a strategy option is plausible for a particular strategic situation of a company or business, it needs to fit in the overall strategic context of the organization. The consistency check is typically done based on qualitative considerations as the underlying complexity given all the different variables is hard to compute with mathematical models. Based on the understanding of strategy as being a "path to a destination" (see chapter 1.1.2.4), it has to be evaluated whether the proposed strategy option ("path") is directly or indirectly contributing to making the vision come true or to achieving the defined overall performance aspiration ("destination") in the context of a pre-defined purpose of the business ("mission"). Consequently, a strategy option needs to be consistent with the strategic guideposts of the firm, that is its vision, mission and values (see chapter 3.1). For example, if a proposed strategy option deals well with a particular strategic challenge derived from the external environment but is not in line with the vision and main purpose (mission) of the organization, it is either rejected or it is triggering a discussion about the validity of the strategic guideposts. If vision, mission and values have been carefully defined and agreed on by the executive leadership team beforehand, a rejection of the strategy option is the most likely course if it is not consistent with those guideposts.

Intra- and inter-strategy consistency: vertically and horizontally

Another consideration for checking consistency of a strategy option is intra- and inter-strategy consistency. First, based on the understanding that strategy is an integrated system of different interconnected activities or elements (Porter 1996 and chapter 3.4.4.1), a strategy option needs to be evaluated regarding its *intra-strategy consistency*. Hereby, it is assessed whether the different relevant building blocks of a proposed strategy or business model are complete, logically interrelated and linked without contradictions. Evaluating intra-strategy consistency may also include the question whether strategic action programs or projects that may already be proposed at this point are clearly directed toward making the corresponding strategy option happen. Second, *inter-strategy consistency* has to be evaluated as well. This works in two ways. First, it needs to be assessed whether the strategy option supports superior strategic goals of the organization sufficiently in cases where the option is related to the managerial level of downstream organizational areas. Depending on the company's steering logic, a business strategy is supposed to support corporate strategy and a functional or regional strategy is typically required to contribute to business and corporate strategies, etc. This may relate to both financial and non-financial elements of the higher-level strategy. *Vertical consistency* means that higher-level strategies are supported by the strategy option. Second, it needs to be evaluated whether the strategy option fits to strategic goals of organizational units at the same management level in the organization. If it is not complementary or even conflicting with strategic goals of other organizational units, it will either be rejected, or an adjustment of the option or the other goals is required. *Horizontal consistency* means that the strategy option is synchronized with strategies of different organizational units at the same management level (e. g., businesses, functions, regions). This is of particular relevance when the company wants to realize broader strategic benefits with the strategy option, such as economies of scope based on cross-unit synergies or value chain linkages (see also chapter 4.3.3).

3.5.1.3 Performance Impact

Clarity about the overall performance definition and aspiration

Whereas the focus of consistency is the general vertical and horizontal "fit" of a strategy option, assessing its performance impact is related to the expected quantified success contribution. To which degree does a proposed strategy option satisfy agreed on performance objectives of the company? For a better understanding whether some strategy options are more suitable than others, it needs to be clearly defined how a company measures its ultimate performance and what it wants to achieve relative to its competitors. There is a variety of approaches for measuring competitive advantage and company performance (see chapter 1.1.3). Firms can generate some sort of *profitability* metrics from accounting to measure performance. Publicly traded companies tend to evaluate their economic performance by looking at the *shareholder value* they generate. In recent years, some companies have started to evaluate their performance more broadly by following a triple bottom line perspective and try to capture their *corporate sustainability performance*. Hereby, economic results remain one crucial performance dimension. However, these firms try to also account for the societal impact—that is the ecological and social performance—of their business activities when making strategic decisions. This may actually result in an adjustment of their economic performance results. Although there are currently no sufficient and commonly accepted ways to measure particularly the social performance, companies like PUMA have enhanced their performance measurement methodologies by implementing new approaches like the environmental profit and loss (EP&L) accounting

(Hengstmann/Seidel 2014). Other methods for measuring and reporting a corporation's sustainability performance include social profit and loss accounting (SP&L) and total impact measurement and management (TIMM). A sustainability-oriented transformation of overall accounting principles toward an accounting for negative and positive externalities are currently being developed (Elkington/Zeitz 2014: 55–85; Sawczyn/Fink/Werner 2014). Whatever key performance measurement is used in a company, it is a critical checkpoint for evaluating strategy options and making a decision (see Strategic Snapshot 3.7).

Financial analysis based on pro forma financial statements and ratios

Most business methods for evaluating the performance impact of strategy options today are related to financial analysis. Hereby, a strategy option is essentially seen as an investment in the company or strategic business unit, which is supposed to deliver a certain expected financial outcome. What is of interest is the specific impact caused exclusively by the proposed strategy option. There are a variety of methods that can be applied for this purpose. Some of them are based on simple spreadsheet-based simulation models and others are utilizing more comprehensive financial calculations such as the real options approach or Monte Carlo simulations. A starting point for analyzing the financial impact of strategy options is to develop common-size pro forma balance sheets and profit and loss statements. Based on these statements, top and bottom line impact of the strategy option as well as corresponding financial profitability ratios can be calculated (e. g., return on sales, return on equity, etc.). Profitability ratios represent the profit (EBIT, operating profit, etc.) proportionately to a measure like revenue, equity, capital, net assets, etc. (see chapter 1.1.3.2). They show the forecasted financial effect of a strategy option on the corresponding financial performance of the company or strategic business unit over a certain period of time such as the next five years. Pro forma financial statements are derived from the strategy option's estimated levels of sales, expenses and other categories for future years. If expected future top and bottom line changes caused by a strategy option are too hard to develop, working with a rough order of magnitude (ROM) may be useful. Hereby, a broader range of potential sales or profit contribution is estimated for each strategy option to get a basic idea of the expected magnitude of their performance impact.

Return on capital employed (ROCE)

A widely used method for assessing strategy options is to measure the return on capital employed (ROCE). Its definition is based on accounting data and typically looks back at a project's past performance. For the purpose of assessing strategy options, the *expected* operating profit—usually defined before interest and tax—after the strategy has been in place for an agreed number of years is divided by the specific capital employed to make the option happen. Due to the tendency of varying capital requirements during the course of a year, capital employed is commonly averaged across 12 months. Many companies apply a so-called "hurdle rate" for ROCE that are often related to their cost of capital (e. g., cheap capital means a low hurdle rate). If the strategy option does not earn at least the hurdle rate, it is very likely to be rejected. Apart from determining this hurdle rate, one of the major challenges of using ROCE for assessing strategy options is the definition of the incremental capital required only for making this specific strategy option happen. In practice, strategy options are frequently based on utilizing existing plants, equipments or services where the capital involved for the sole purpose of executing the proposed strategy is

Strategic Snapshot 3.7

Different Performance Metrics Result in Different Strategic Decisions

Figure 3-46 illustrates in a simple way how three different approaches for measuring performance can lead to different evaluations of a strategy option. The starting point is a company's actual performance provided in the first column of the table. Then a strategy option is suggested that needs to be decided on, which impacts the financial indicators (second column). By looking only at *profit*, the strategy option leads to an increase and would thus be evaluated positively regarding its performance impact. Also, if the company is emphasizing *revenue* growth for all areas of the business, the strategy option seems to be supportive. However, is the growth sufficiently profitable? It actually reduces the overall *profitability* of the company based on metrics like *return on equity (ROE)* and *return on sales (ROS)* as compared to the actual (base) performance of the firm. If the firm's top performance goal is achieving superior profitability measured with these types of indicators, the strategy option will probably not be selected as it drives corresponding performance metrics down. From a *shareholder value* perspective, though, the strategy option is evaluated positively as the costs of capital are treated as a kind of hurdle rate. Although investing in the new strategy will decrease the ROE and ROS, and thus, the company performance based on accounting profitability, the proposed strategy option will improve the company's performance from a shareholder value perspective. Its additional profit is higher than the cost of capital, and thus, it creates economic value. Finally, from the perspective of *corporate sustainability performance*, (see chapter 1.1.3.4) the strategy option's financial profit contribution is generated at the sake of significant environmental or social cost and a corresponding destruction of "total" value. This hypothetical example assumes that shareholder value accounts for the economic sustainability dimension and is supplemented by the environmental and social performance dimension. The strategy option is likely to be

Impact of strategy option on ...		actual performance	+	strategy option	=	new performance	evaluation
	financial profit	200 €		150 €		350 €	↑ ☺
Accounting Performance (Profitability) ☹	capital invested	1,000 €		1,000 €		2,000 €	
	revenue	2,000 €		2,000 €		4,000 €	↑ ☺
	return on equity (ROE)	20 %		15 %		17.5 %	↓ ☹
	return on sales (ROS)	10 %		7.5 %		8.75 %	↓ ☹
Economic Performance (Shareholder Value) ☺	WACC	8 %		8 %		8 %	
	cost of capital	80 €		80 €		160 €	
	shareholder value added	120 €		70 €		190 €	↑ ☺
Corporate Sustainability Performance (Value Co-Creation) ☹	environmental/social value*	-60 €		-90 €		-150 €	☹
	adjusted profit	140 €		60 €		200 €	↑ ☺
	adjusted ROE	14 %		6 %		10 %	↓ ☹
	adjusted ROS	7 %		3 %		5 %	↓ ☹
	total value added	60 €		-20 €		40 €	↓ ☹

* Note: financially assessed or estimated negative or positive environmental and social impact of the business activity

Fig. 3-46: Company performance criteria determine the evaluation of strategy options (illustrative example)

Continued on next page

Continued from previous page

rejected by the company applying such a performance philosophy and mindset as it destroys total value (negative total value added). Note that an alternative strategy option achieving the same economic results as the one in the example and, at the same time, adding ecological or social value will increase total value, and thus, outperform the previous option regarding corporate sustainability performance. In this case, the co-creation of economic and ecological/social value that can be achieved with, for example, a sustainable business model (see chapter 3.4.4.5) is reflected in a positive evaluation of the corresponding strategy option.

hard to separate out. Furthermore, when using accounting based profitability metrics, it has to be considered that a sinking remaining book value of investments corresponds with increasing profitability.

Payback period or break-even point

Instead of looking at profitability ratios like the ROCE, strategy options may also be assessed by looking at two other numbers—one related to time and the other to sales volume—indicating when investments are recouped:

- **Payback period:** The time it takes to recover initial capital investments of a strategy option is called payback period. Given all the uncertainties in the company's environment it may be very difficult to estimate particularly future sales accurately for calculating profitability ratios when strategy options are related to long time spans up to 10 years or more. Especially in case of strategy options requiring significant incremental capital investments, it might be more appropriate to focus on the payback period and simply add cash flows in payback whenever they occur and calculate how long it takes to recover the invested capital. In such situations, looking at *cash flow* payback periods is considered more effective for strategy evaluation than looking at accounting performance. Companies may report or anticipate satisfactory levels of profitability such as ROCE and—at the same time—are at risk of going bankrupt due to a significant lack of cash.
- **Break-even point:** Another way of evaluating strategy options is to explore the incremental sales volume of the option required to recoup its initial investment. This is a way to explore whether the anticipated future number of product units to be sold driven by a new strategy is reasonable. A strategy option's break-even point is reached when the total incremental revenue caused by the strategy equals the total costs of undertaking it. Since it may be hard to model future sales, determining the break-even point is an effective method for a sanity check. It allows assessing on a high level if the strategy option is realistic or not.

Discounted cash flow (DCF)

Another widely used method for evaluating the financial impact of strategy options is the discounted cash flow (DCF) approach. Unlike the payback period for evaluating strategy options explained before, DCF is considering the time value of money, that is the fact that cash in five years is worth less than cash today taking into account the cost of capital. There are different ways for DCF calculations that will not be elaborated on here. Essentially, DCF assesses the net cash flow delivered by a strategy option for each year of its lifespan. Net cash flow is the profit *before* depreciation, *less* taxation to be paid to the government and *less* period investments in fixed assets and working capital. Each resulting annual cash amount as well as the residual value of a strategy option at the end of its lifespan—assuming its continuation—is then discounted back to the present using the cost of capital. Discounting is usually done by using the weighted average cost of capital (WACC). WACC is the a sum of the *cost of equity*—considering a risk-free interest rate and the market risk premium adjusted by a company-specific risk discount—and the after-tax *cost of debt*. A strategy option is ultimately evaluated by its net present values (NPV), that is the sum of all future discounted cash flows today.

3.5.1.4 Business Risk

Avoiding unnecessary hazards or danger to the business

Assessing the performance impact of strategy options is a prerequisite for evaluating whether they are acceptable regarding the results they are supposed to deliver either isolated or in a broader strategic context within an organization. However, the attractiveness of a strategy option is also a function of the amount of risk it entails given the fact that different managers are likely to have different attitudes toward risk. In this context, risk means the degree of unnecessary hazards or danger that an entire company or a company's business will be exposed to by pursuing a strategy option. It is composed by elements such as the

- probability that the strategy option will be successful given the company's environment including potential competitor reactions
- legal concerns (e. g., something to draw a lawsuit or headline risk that damages the company reputation and brand)
- amount of assets and resources a company has to allocate to the strategy option
- lengths of time the assets and resources are dedicated to make the strategy option happen
- availability of contingencies to mitigate the negative impact of things that might go wrong.

In general, the greater the assets and resources involved and the longer they are committed, the more a strategy option will be required to have a high probability of success in order to be selected. Some of the proposed financial analysis methods for evaluating the performance impact of strategy options capture risk to a certain degree, such as the payback period (cash flows and time length indicate risk regarding liquidity) or DCF (risk-adjusted cost of capital is used as discount factor). However, one of the major challenges for evaluating strategy options is the prediction of future developments based on which the numbers of those calculations are estimated. With a high level of uncertainty in the environment given macroeconomical global developments, market dynamics, technology changes, government interventions, etc., it is very difficult to estimate financials such as future sales numbers and corresponding cash flows accurately. However, strategic decisions have to be made today in order to capture the strategic opportunities of the future. In such a situation, scenario and sensitivity analysis can provide managers with a useful estimate of the risk involved in a strategy option they have to decide on:

- **Scenario analysis:** Scenarios are used to assess the potential impact of future changes in the company's external environment (see chapter 2.2.1) on the effectiveness of the proposed strategy option. By elaborating on different "what if" scenarios, it can be evaluated how robust the strategy will be in case basic planning assumptions underlying the corresponding option are changing. For example: What would happen to the proposed strategy and its estimated future financial contribution to company performance in case of changing key economic factors, emerging government regulations, NGO activities, technological disruptions, competitor moves, changes in customer preferences, etc? What impact would be there if our key suppliers or customers go bankrupt? What would happen in case of a certain unexpected response of our key competitors to our strategy option? To support managerial decision making, companies tend to develop three types of scenarios: optimistic, pessimistic, and realistic (most likely). This can be done for both assessing the impact of changing assumptions on a strategy option's plausibility (see 3.5.1.1) and—more quantitative—its financial performance impact (3.5.1.3). The assessment does not have to be a onetime event for strategic choice only but needs to be part of a continuous process of environmental monitoring for the purpose of strategic control (see chapter 4.3.7).

- **Sensitivity analysis:** With a sensitivity analysis the impact of individual variables on the strategy option's financial performance contribution can be assessed. Hereby it is analyzed how arbitrary variations of certain variables influence performance indicators such as the return on capital employed, discounted cash flow or other metrics used for performance measurement. For example, when applying a DCF-approach, different variables such as the expected market share can be moved up or down based on different scenarios to investigate what this means for the net cash flows in the corresponding periods and ultimately for the strategy option's discounted cash flow. Does it change proportionally or deviate significantly either up or down? The ultimate goal of sensitivity analysis is to identify the variables that have the strongest impact on the success of a strategy option. Consequently, a special emphasis of investigation and high management attention is put on those crucial items for the purpose strategy evaluation and choice.
- **Monte Carlo simulation:** A sensitivity analysis is conducted by changing only one variable per calculation sequence and keeping the remaining variables of the model fixed. However, for evaluating strategy options under different scenarios, it might be more appropriate to change a number of variables at the same time and understand the corresponding performance impact. A Monte Carlo simulation is based on a mathematical model that captures the dependencies between different variables. It allows the change of a limited number of variables at the same time and computes the different cash flows by applying corresponding probabilities. Sensitivity analyses and, in particular, the Monte Carlo simulation allow a more comprehensive evaluation of strategy options given the dynamics and uncertainties of the future. However, they are highly complex approaches that do not only require well-trained users but also decision makers that are willing to base their risk assessment and strategic choice on data derived from such comprehensive mathematical models.
- **Real options:** Another way of dealing with strategy options under conditions of high environmental uncertainty is to keep a broad range of options open. The strategy evaluation approaches explained before boil everything down to one

most appropriate scenario and do not provide flexibility when circumstances change. They are even more difficult to apply when it is not possible to anticipate future financial effects today due to a high level of uncertainty. The real options approach helps to evaluate alternative options under such conditions by breaking down a strategy option in a set of multiple strategic projects at different stages. Based on this, the funds required for making a strategy option happen is broken down according to those stages and allocated to initiate various strategic projects. Projects that turn out to be not successful will be canceled and funds for projects that are doing well are continued.

3.5.1.5 Stakeholder Compatibility

The likelihood of a strategy option to be successful depends on its compatibility with expectations of key stakeholders. This is why stakeholder analysis is an integral element of strategic analysis (see chapter 2.1.1). Hereby, an effective stakeholder dialog early in the strategy process may not only ensure compatibility but is an important source for strategic ideas and priorities itself. If a strategy option proposed for evaluation and choice is not sufficiently appealing to those stakeholders that are most affected and that have a strong impact on the company's performance, it is likely to fail. A more general assessment whether expectations of key stakeholders are met by the strategy option have already been considered when checking for plausibility. Now, an explicit and more thorough analysis is required. There are two key challenges when assessing stakeholder compatibility.

Prioritizing stakeholder interests

Interests and perspectives of different stakeholder groups may not be the same or even conflicting. For example, a strategy option might suggest significant investments to improve company performance and enhance employment security for *employees* in the long term but does not increase wealth of *shareholders* in the mid or short term. The *society*, *governments* and *non-governmental organizations* demand corporate social responsibility whereas *consumers* may want cheap products and *shareholders* expect an attractive return for their capital (see Strategic Snapshot 3.8). As a first step for prioritizing stakeholder interests, internal and external stakeholders need to be listed. Then it needs to be identified which stakeholder groups such as shareholders, customers, competitors, suppliers, regulatory authorities, non-governmental organizations (NGOs), etc. are of particular relevance for the particular strategy option proposed. This can be done by categorizing stakeholders according to the following questions:

- What is the level of interest the stakeholder has in the company?
- What is the stakeholder's power to influence the company's activities and performance?
- What is the general tendency of high-power stakeholders to actually use their influence (e.g., public campaigns by NGOs, unions initiating and organizing strikes, etc.)?
- How much can the company influence and control the stakeholder?

For identifying and prioritizing stakeholders relevant for the strategy option, it is important to focus not only on external groups but also internal parties at different management levels such as employee groups, individual managers, supervisory boards, or other business units or departments that are affected by the strategy option.

Assessing attractiveness to key stakeholders

As different stakeholder groups may not find proposed strategy options equally attractive, the compatibility of strategy with expectations of particularly *high priority* stakeholders is of utmost importance and requires further and more comprehensive evaluation. It needs to be assessed whether the company's strategic ambition that is articulated with the strategy option fits to key expectations of relevant stakeholders. Those expectations are either anticipated by the company, systematically collected in stakeholder dialogs and surveys, or both. The key question is how much of what each relevant stakeholder wants from the company is delivered by the corresponding strategy option. For this purpose, different proposed strategy options may be plotted in a matrix showing each strategy options' degree of attractiveness to stakeholders (e. g., in terms of met expectations) on one axis and to the company (e. g., in terms of business impact) on the other. The top right quadrant of the matrix chart contains strategy options that are not only highly attractive for the company but also appealing to relevant stakeholders. In case there are strategy options that are supposed to deliver important business impact for the company but are not in line of the expectations of high-priority stakeholders, the potential stakeholder responses as well as possibilities to resolve the issue need to be evaluated:

- Are there ways to bring the stakeholder's expectations in agreement with the proposed strategy?
- How can the stakeholder respond if the company pursues the proposed strategy despite the missing fit?
- How likely are the different stakeholder responses?
- What can be done to mitigate the risk of harmful stakeholder reactions (e. g., political influence and government affairs departments, change management programs, etc.)?

Strategic Snapshot 3.8

Former CEO and Honorary President of Nestlé, Helmut Maucher, on Social Responsibility and Business

"Entrepreneurs today are in a dilemma: On the one hand, they are confronted with the ever-increasing demands of society with regard to ethics, ecology, and social conduct (including sponsorships and support of charitable foundations); on the other hand, they face increasing financial pressures with regard to short-term results, shareholder-value orientation, and profit maximization. Finding and maintaining a sound balance in the midst of these conflicting requirements is not easy. (...) To make one thing clear: The most important social and ethical responsibility of entrepreneurs is to be successful in a competitive market over the long term, thus ensuring lasting profitability for their companies. In this way they make a major contribution to the prosperity of the economy from which everyone benefits at the end. (...) Apart from that, I believe that a company has to generally strive for responsible and ethical conduct. Abiding by the law is not enough: Responsible entrepreneurs pursue policies that serve the long-term development and optimization of revenues, rather than short-term profit maximization. As a result, many measures driven by social, ethical or even ecological motives which would be rated negatively under short-term profitability aspects, will turn out to make perfect sense (even economical sense) for the company from a long-term perspective."
(Source: Maucher, H. 2007: 50–52)

In case key expectations of high-priority stakeholders are not met by a proposed strategy option and cannot be brought into agreement as well as negative stakeholder responses are likely and cannot be mitigated, the strategy option is considered not feasible. Usually, it will be rejected even though it may fulfill the other evaluation criteria. However, if strategy options have been developed based on a thorough strategic analysis including a comprehensive stakeholder assessment and dialog (see chapter 2.1.1), serious misfits should not be the case as key stakeholders' expectations would have been already considered for strategy formulation.

3.5.1.6 Internal Readiness

Identifying internal constraints of making a strategy happen

The final evaluation criterion is internal readiness or internal feasibility. Once the other criteria are met and a general external feasibility is given, it needs to be checked whether a company is actually capable of making the proposed strategy option come true. This is an area that has gained significant attention in company practice over the last decades due to the fact that only a very limited portion of supposedly well-developed and selected strategies are actually executed successfully (see chapter 4.1.1). To evaluate the internal feasibility of a strategy option, it is important to get a basic idea of how it will be realized. Based on a tentative execution plan, internal challenges the organization is likely to face when trying to execute the strategy option are anticipated. Internal constraints that may limit the feasibility of a proposed strategy option may be related to organizational culture, organizational structure, business processes, and managerial systems. For example, required skills and capabilities (human capital readiness), necessary financial and other resources, and sufficient commitment from management and employees need to be in place to make a strategy option come true. One of the major challenges companies are facing when it comes to internal feasibility is that the management of key initiatives tend to fall on the same group of people who have only a limited bandwidth. If too many projects are pursued at the same time the results do not materialize or it takes forever and a day to implement. Accessing external resources to solve this issue can be expensive, will not develop the buy-in and require the company to go up a learning curve.

In cases where there are some serious constraints in one or more of these areas, the management has to decide if it should reject the strategy option due to a lack of practicability, adjust the strategy option to better fit with internal circumstances, try to reduce or overcome the internal constraints for a better fit with the strategy option, manage around the limitations, or take a chance and ignore them.

Creating internal readiness with excellent strategy execution

On the one hand, trying to pursue a strategy option with significant internal constraints can be a risky and dangerous undertaking. Some limitations such as a lack of funding may leave the management with no other chance than rejecting the strategy option. On the other hand, restricting a company to only those strategy options that are completely feasible might eliminate alternatives that are most critical for driving the performance and long-term survival of the company. To capture those strategic opportunities, firms may decide for a strategy option even though it has a limited "fit" with the current internal set-up but room for matching strategy, structure and culture in a transformational process later on. In such a case, internal readiness can be created with corresponding organizational design activities (see chap-

ter 4.2) as well as an effective strategy execution system covering elements such as (see chapter 4.3):
- systematic strategy refinement and strategy alignment in the organization
- effective development, prioritization and management of strategic action programs
- alignment of financial resources, people and information (e.g. key performance metrics)
- effective strategic control.

3.5.2 Strategic Choice—Selecting Strategy Options

Strategic choice based on constructive debate and critical thinking

Strategic choice is the selection of the best alternative following a strategy option evaluation with the six key criteria explained. Unfortunately, there is no such method that automatically computes the best strategy option. One way to consolidate all the different results and suggest an "optimal" solution is a use-value analysis in which the various evaluation criteria and their subcomponents are weighted and rated for each strategy option. The option with the highest use-value or utility will be selected (Welge/Al-Laham 2012: 736 f.). However, for this method to be effective, it is of utmost importance to ensure a productive and sophisticated weighting and rating process with a competent team. Hereby the best strategic decisions may not be arrived when everyone agrees on a certain rating for a strategy option too quickly. To improve the quality of decisions, use-value analysis may actually be leveraged to facilitate the presentation and discussion of diverse opinions and drive a constructive debate in a task-oriented environment legitimizing critical thinking and evaluation (Wheelen/Hunger 2010: 257 f.).

Watch out: Why an analytical evaluation process might fail.

Having said this, there are some additional words of caution to note regarding the suggested rational strategic decision making process for evaluating strategy options. Drawing on the remarks and explanations about strategic decision making in chapter 1.2, there are some evident difficulties with this type of mechanistic appraisal of strategy options. First, as strategic decision situations are often highly complex, management leadership teams might be well advised to enhance their analytical approach with a great portion of intuition and not only rely on the supposedly mathematical rational of a use-value analysis or other analytical methods. The right combination of analytics and intuition in a particular situation is probably one of the secrets of good decision making (see chapter 1.2.2.2).

Second, cognitive biases might interfere with the idea of optimal decision making. For example, managers might unconsciously select and consider only information that confirms their strategic assumptions and neglect contradictory information (confirmation bias). They might opt for a strategy option because of past investments in this strategic theme (sunk cost effect) or, for some reason, have an emotional link to a certain strategy option (inappropriate attachment). Also, top managers that have to decide on a strategy might be driven by their personal needs and desires, either consciously or unconsciously (egocentric bias). For example, "narcissistic (self-absorbed and arrogant) CEOs tend to favor bold actions that attract attention, like many large acquisitions—resulting in either big wins or big losses. A person's ego may be tied to a particular proposal to the extent that all other alternatives are strongly lobbied against. As a result, the person may have unfavorable forecasts altered so that they are more in agreement with the desired alternative." (Wheelen/ Hunger 2010: 256) More cognitive biases as potential pitfalls for decision making as well as some ideas of how to counter them are explained in chapter 1.2.2.3.

Third, decision making and choice is likely to be strongly influenced by personal values of the decision makers and strategic leaders (see chapter 1.3.3.3) which may be different from the values of relevant stakeholders. If unambiguous decision criteria are not in place and there are tradeoffs between different evaluation categories, the ultimate choice comes down to the personal values of decision makers. For example, should limited funds be spent on the strategy option with outstanding economic but moderate ecological/social performance or on the one with moderate economic but good ecological/social performance?

Fourth, a systematic and analytical strategy evaluation process requires collecting and processing detailed information for all the different criteria for each proposed strategy option. This may require significant time and effort. Some companies may hesitate to invest time and resources for developing a great level of detail for each potential strategy option during strategy formulation or for the purpose of strategy evaluation. They rely on a more general assessment of the different criteria and follow up with incremental evaluation during execution. First, strategic leaders make an initial—sometimes highly intuitive—decision for one or more options based on a rough assessment. Second, once they move forward with an option, those firms refine, calibrate and sometimes even terminate the strategy as part of their ongoing execution process based on continuous individual and organizational learnings (see chapter 4.3.7).

3.6 Summary

The starting point for strategy formulation is the clarification, refinement or development of strategic guideposts that are supposed to provide a normative frame for all subsequent strategy development activities. These guideposts include the definition of a company's purpose (*mission*), that is why it is in business. The mission often finds its expression in a compelling and easy to understand (brand) claim. Another important guidepost is the company's *vision*, which is supposed to guide strategy formulation toward what the company wants to become in the future. Strategy in turn is supposed to define the path from today to that desired picture of the company's future. A vision should provide orientation, create enthusiasm, foster stimulation, and offer security to employees in the organization. The (proclaimed) values are the third element of strategic guideposts. They are supposed to articulate behavioral principles for interaction that have to be embraced in the organization for defining and executing strategies.

In multi-business companies, strategy formulation on a corporate level has to be distinguished from the business level. *Corporate strategy* is covering the following key strategic themes:

- basic *corporate parenting* approaches that the corporate center or headquarters wants to leverage for creating value with its portfolio of strategic business units
- its general *strategic orientation toward growth*
- the corporation's *scope of business*, that is *which businesses* it wants to compete in (diversification), *which stages in the value chain* it wants to capture (vertical integration) and *which international markets* it wants to address (geographic coverage).

In the field of *international strategy*, a corporation or business needs to decide *whether* it wants to go international, *where* it wants to go, and *how* it wants to enter international markets (i.e. the specific go-to-market strategy which is typically crafted on a business level). Additional, multinational corporations can leverage their "internationality" for gaining competitive advantage in the context of global integration and local responsiveness. Corresponding *strategy postures* are global strategy, international strategy, multinational strategy, and the transnational strategy. Multinational corporations following a transnational strategy are trying to create different layers of competitive advantage leveraging global efficiency, multinational flexibility as well as worldwide learning.

Strategy on a business level includes both competitive approaches such as *intensive growth strategies* and *generic competitive strategies* as well as approaches to restrain or bypass competition such as *cooperative strategies* and *blue ocean strategies*. Three strategic approaches for growing a business are *market penetration*, *market development*, and *product development*. To gain competitive advantage over its rivals, a company may pursue *best product*, *lock-in* and *complete customer solution* strategies with a variety of strategic approaches in each type. The best product approach is based on competitively priced (low-cost) or differentiated products and services, or a combination of both. The lock-in strategies are trying to prevent customers from switching to competitor offerings through establishing de-facto standards, complementary products, or exclusive customer intimacy. Complete customer solution approaches are targeted at establishing intense long-term customer relationships through a complete and broad offering, compelling product and service packages, as well as tailor-made solutions based on profound customer insights.

The *business model* approach can facilitate the development and refinement of business strategies effectively. A business model is a simplified expression of the anticipated mechanisms how the business is supposed to be successful. It can be considered as a refinement of strategy on a business level. Hereby, the firm describes the fundamental principle of how it creates and delivers value for certain target customers, and thus, typically realizes profits by capturing some of this value. The seven building blocks of a business model are *sustainable profit*, *customer value*, *customer segments*, *channels*, *customer relationships*, *key processes*, and *key resources/partnerships*. Effective business modeling drives strategic ideation by leveraging *business model patterns* for creating *business model innovation*. It translates high-level strategic intent into competitive advantage and provides a linkage for strategy formulation with strategy execution. With a unique business model, a firm differentiates not only with products or services but its entire business model that may be much harder for rivals to imitate. A *sustainable business model* is designed to co-create superior customer value and societal value by voluntarily integrating environmental, social and economic entrepreneurial business activities. In today's era of sustainability, providing a substantial long-term contribution for solving environmental and social problems of our society is increasingly considered a strategic success factor that contributes to the economic performance and long-term survival of a company.

Strategy formulation on both the corporate- and business level results in a number of *strategy options* that need to be *evaluated* and decided on. A strategy option has to be *suitable*, *acceptable* and *feasible* to be selected for execution. Hereby, six key criteria can be used to assess proposed strategy options more in detail. Those are *plausibility*, *consistency*, *performance impact*, *business risk*, *stakeholder compatibility*, and *internal readiness*. *Strategic choice* is the selection of the best alternatives based on a rigor evaluation of each strategy option according to those six criteria. This, however, does not have to be seen as a mechanistic analytical procedure that will ultimately lead to an optimal result. Given the complexity of a strategic decision situation, the quality of strategic decisions do not only depend on a systematic and thought through strategic analysis and strategy formulation process but also on constructive debate among competent strategic decision makers supplemented by a great deal of intuition. The right *combination of analytics and intuition* will strongly determine the effectiveness of strategy formulation, which makes this type of managerial task—the art of crafting a strategy—both exciting and challenging.

PART II

Bringing it into Reality: From Strategy to Action and Control

4 Strategy Execution

Learning Objectives

After studying this chapter, you should be able to

- understand why strategies might fail and the importance of excellent execution.
- explain the barriers to successful strategy execution.
- describe the drivers for effective strategy execution and their relevance.
- understand the role of organizational design for strategy execution.
- explain organizational culture and its key dimensions.
- explain the three elements of managerial infrastructure.
- describe how to integrate an organization's culture and structure with strategy.
- explain origins and key principles of the balanced scorecard approach.
- describe key components and closed-loop process for effective strategy execution.
- explain and apply the six steps toward developing a strategy execution system.
- describe how strategy becomes a continuous process.

4.1 Making Strategies Happen: Barriers and Drivers

Strategy x execution = results

There are a variety of reasons why strategies might fail. First, the *strategy itself can be wrong*. It does not lead to the desired market success because key assumptions do not prove to be valid or they are outdated due to changes in the competitive environment. In such cases, excellent strategy execution will pull down the company's performance even faster (Kaplan/Norton 2008: 253). Second, the strategy might well be right for delivering results, but for some reason *does not get the corresponding funding*. In such a case it fails because it is not supported by key stakeholders such as investors or lenders. Following the strategy framework outlined in this book, both cases should ideally not make it into the phase of strategy execution because they are filtered out during the strategy option evaluation and choice. A third reason why strategies fail is *bad execution*. A traditional business formula says that business results equal strategy times execution (Covey 2008: 20):

S x E = R

Excellent strategy execution as a key success factor

The best-laid strategy is useless if it is not executed properly. Moreover, excellent strategy execution is a key success factor for any organization. Companies that use a formal system for executing strategy (see chapter 4.3) or consider execution as a strength (see Strategic Snapshot 4.1) have been reported to outperform their peers consistently (Wunder/Bausch 2014: 59 f.; The Economist Intelligence Unit 2013: 7; Kaplan/Norton 2008: 4–6; Horváth & Partner GmbH 2008: 7). Although it is not possible to relate outstanding company performance *only* to excellence in strategy execution it still indicates that successful companies tend to be also successful strategy executors. Strategy execution becomes even more important in situations where strategies of companies in the same competitive arena are very similar. Although strategy is about being different (see chapter 1.1.2.3), truly innovative strategic ideas and unique business models are rare. The reality for many companies is a head to head competition. In a number of studies, about 70 % of companies have reported that they pursue similar strategies as their competitors (Wunder/Bausch 2014b: 58; Horváth & Partner GmbH 2008: 9). Also, the half-life period of strategies has been shortened making it necessary to challenge the current pillars of competitive advantage continuously (see chapter 1.1.2.1). In such an environment excellence in strategy execution has become a competency that is increasingly important for winning in the marketplace. "Execution is not just something that does or doesn't get done. Execution is a specific set of behaviors and techniques that companies need to master in order to have competitive advantage." (Bossidy/Charan 2002: 7)

Strategy execution—the big challenge in strategic management

The bridge from formulating a strategy to making it happen is actually one of the biggest challenges in strategic management. For many years, companies have been repeatedly reporting that they experienced significant problems particularly in trying to execute their strategies (Sull/Homkes/Sull 2015; Wunder/Bausch 2014b: 59 f., Horváth & Partner GmbH 2008: 29, Kaplan/Norton: 2008: 3 f. and 2001: 1 f., Huber 2006: 45, Niven 2002: 9–11, Al-Laham 1997: 461). "In the majority of cases—we estimate 70 %—the real problem isn't [bad strategy] ... It's bad execution. As simple as that: not getting things done, being indecisive, not delivering on commitments." (Charan/Colvin 1999: 70) It is no surprise that improving strategy execution has

Strategic Snapshot 4.1

Food Companies with Excellent Strategy Execution Outperform Their Peers

A survey of 104 companies from the food industry in German-speaking countries revealed some fundamental differences between great and average companies and their state of strategy execution. In the survey, 12 reference companies emerged that outperformed their competitors clearly. They succeeded to increase both their sales volume and profit more than their peers for three consecutive years. These companies were labeled "champions". All of those champions state that the execution of defined strategies is one of the strengths in their company. Among the rest of study participants only 43 % confirm this, which means that more than half of the rest of firms consider their strategy execution rather weak. The majority of study participants point out that their employees understand the strategy only insufficiently. In such cases strategies rapidly become a "paper tiger". As for the champions, 82 % of them stated that their employees had a clear shared understanding of strategy and the strategic direction is developed collectively in interdisciplinary teams. (Wunder/Bausch 2014: 59 f.) The survey results confirm several empirical studies over the past decades which reveal that excellent strategy execution is a key success factor in a firm's performance (Kaplan/Norton 2008: 3–6).

become number one priority of many senior executives (Kaplan/Norton 2008: 3). But why is strategy execution such a challenge? Obviously, companies seem to find it naturally difficult to orchestrate all the various interrelated areas and elements that drive the organization toward strategy. But what are the key barriers to effective strategy execution and what are the organizational elements and areas to elaborate on for overcoming those barriers? This will be explained in the following.

4.1.1 Barriers to Strategy Execution

Nine out of 10 companies fail to execute strategy.

Only a small portion of effectively formulated strategies are executed successfully. Various studies indicated that 60 % to 90 % of organizations fall short of realizing the success expressed in their strategic plans (Kaplan/Norton 2008: 3 and 2004: xi). Management tends to emphasize strategic analysis and strategy formulation but struggles to bridge the gap to its day-to-day execution. Key issues that companies are facing when executing their strategies are summarized with the six barriers to strategy execution that are illustrated in figure 4-1. The percentages are adapted from selected literature and management surveys on strategy execution of the past decades and meant to illustrate the relative importance of the corresponding barrier (Kaplan/Norton 2008: 3–6 and 1996: 193–198, Niven 2002: 9–11).

Fig. 4-1

Six barriers to strategy execution

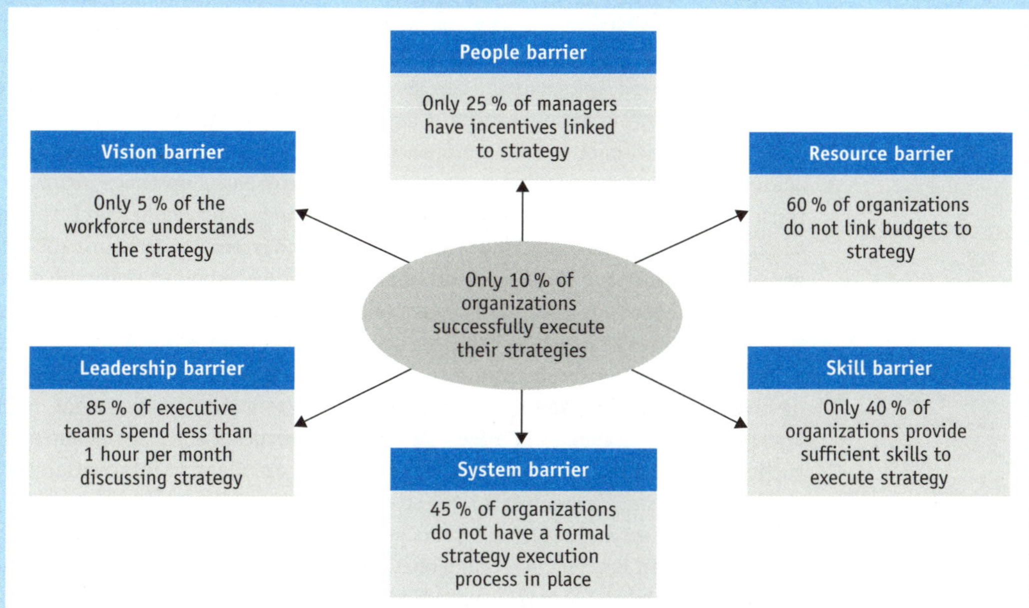

A first barrier occurs when the vision and strategy is too vague and generic (**vision barrier**). In such cases, employees do not understand the company's strategy and cannot link their daily activities to it. Due to a lack of clarity and missing strategic guideposts there is much room for interpretation which makes it difficult to create a shared understanding and commitment for strategic priorities. A clear orientation is missing. Another barrier is the missing link between goals of organizational units, departments, teams or individual people and related incentives with the overarching strategy (**people barrier**). Employees that are expected to drive strategy execution do not understand what they can do in their specific job environment to make the strategy come true. Furthermore, they are rewarded based on achieving short term financial targets but not on their individual contribution to the long-term corporate or business strategy. The consequence is a lack of commitment regarding strategic priorities. Missing or insufficient strategy alignment caused by failures to coordinate across units is a key factor that undermines execution (Sull/Homkes/Sull 2015; see also chapter 4.3.3). Another barrier is related to operational planning and budgeting (**resource barrier**). A missing link of resource allocation to strategy means that people who are in charge of executing strategic initiatives and who are expected to achieve strategic goals do not get the funds, people or time to excel. In such a case, annual budgets are frequently based only on short-term considerations and are not synchronized with long-term strategic planning. The fourth barrier to execution is a lack of comprehensive strategy reviews by management teams (**management barrier**). Executive teams do not spend enough time discussing the strategy, and strategy execution as a whole does not receive appropriate senior leadership attention. Bossidy and Charan wrote: "Typically the CEO and his senior leadership team allot less than half a day each year to review the plans—people, strategy, and operations. Typically too, the reviews are not particularly interactive." (Bossidy/Charan 2002: 22) Many management systems facilitate short-term operational performance reviews based on accounting numbers rather than long-term strategic performance reviews covering the discussion of assumptions, the correlation of actions and strategic goal achievements and ultimately the adaption of strategy based on those insights. In those cases, management reports do not reflect strategy. Another barrier is the lack of a formal framework for making strategies happen (**system barrier**). In a 2006 survey of 143 performance management professionals, 46 % reported that they do not have a formal strategy execution system (Kaplan/Norton 2008: 4). In those cases, firms are missing, for example, explicitly designed processes regarding strategy articulation and communication, strategy review meetings, strategy alignment, strategy-oriented budgeting and rewarding, strategic performance measurement, or the management of strategic initiatives. As a result, execution activities might not be sufficiently coordinated or ineffective, combined with a lack of information about their status. A final barrier exists if the strategic competencies, that is the knowledge as well as business and leadership skills people should have to support the strategy, are missing (**skill barrier**). A recent global survey with 587 senior executives revealed that approximately 60 % of the companies do not provide sufficiently skilled personnel needed for strategy execution (The Economist Intelligence Unit 2013: 13).

Other issues that may interfere with effective strategy execution are a lack of trust (Covey 2008) as well as an unexpected company crisis (Alexander 1985: 92) or an endless need to solve newly arising operational problems immediately. The latter divert management and employee's attention away from the strategic priorities and may lead companies into a destructive cycle of overreacting. Another issue might appear, if the time of implementing an effective execution system and making the strategy happen takes longer than expected and is neither perceived nor used as a continuous process (see chapter 4.3.7).

4.1.2 Strategy Execution Excellence

Strategic projects are not sufficient.

In some companies, strategy execution means nothing more than developing strategic projects as "answers" to the strategic challenges that have been identified earlier and measuring results with a set of mostly financial metrics. Companies that execute strategies effectively, however, use a more sophisticated approach. They integrate strategic projects (actions) with various other elements to establish alignment, focus and impact systematically. This way, they overcome the barriers to successful strategy execution, which have been explained before, and produce results. After having explained what can go wrong in strategy execution, the following will provide key drivers for getting it right.

Driving behavior toward execution

Figure 4-2 shows seven generic elements that are supposed to drive strategy-oriented behavior of people. Although strategy processes differ from one company to another, a variety of management concepts and methods developed for executing and managing strategy are essentially based on those drivers. Here is a corresponding rationale:

- **Goals** and **actions** articulate what people *should* do (strategic priorities of the company).
- **Resources** and **skills** are necessary that they *can* do (ability to act and contribute to strategic priorities of the company).
- **Incentives** and **attitudes** are foundations so people actually *want* to do (motivation to help make the strategy happen).

Information is vital for all those mechanisms. It helps to understand the aspiration, provides "as is" versus "to be" comparisons, and shows whether executed actions deliver the expected results. Based on those types of analysis, information helps to understand whether additional resources or skills are required to make the strategy happen. Finally, without the described transparency it will be difficult to influence attitudes and it is not possible to see whether commitments are fulfilled as a base for allocating the corresponding incentives. In cases where some of those elements are missing, the strategic priorities may not be reflected in the actual day-to-day behavior of employees.

Leadership ensures accountability.

Having the seven generic elements in place and integrating them effectively is supposed to establish buy-in for results. It ensures that people are committed to the goals and action plans they have helped to create, that they follow through and get

Fig. 4-2

Drivers for strategy-oriented behavior

things done, and that the best performers get rewarded. Making this happen strongly depends on strategic leaders that practice what they preach, engage personally and deeply in the strategy execution process, ensure that the organization is aligned around strategy and hold people and themselves accountable for results. Hereby, leadership and accountability go hand in hand.

Measuring execution excellence

Each of those seven drivers is supposed to facilitate a certain role in the strategy execution process. The bullet points in the following checklist illustrate the benefits that an organization can expect for its strategy execution from having those drivers in place. They can be used as a practical guide for achieving and measuring execution excellence. A low level of agreement to the statements provided for the seven drivers below can be seen as an indicator for areas of improvement regarding the organization's strategy execution capability. The corresponding strategy execution excellence profile can be illustrated with a radar chart (see fig. 4-3). The more layers of concentric rings are filled out in each driver or their elements the better a company is performing in strategy execution.

Goals—creating a shared and aligned understanding of the strategic priorities

- Executive leadership teams have a shared understanding of the corporate strategy.
- Individual strategies of businesses, regions and functions are aligned with each other and with corporate strategy.
- The strategy is communicated clearly within the entire organization, so employees are aware of it and understand their specific contribution.

Fig. 4-3

Strategy execution excellence radar

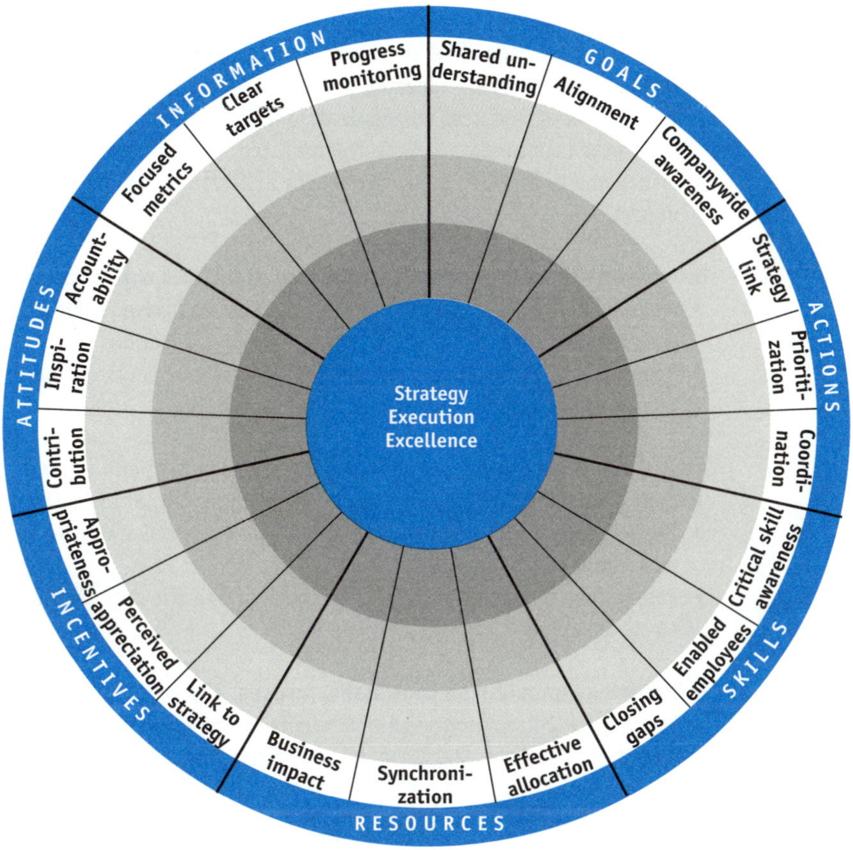

Actions—making the strategy happen most effectively

- Current strategic actions are derived from the corresponding organizational strategies (corporate/business/region/function).
- Strategic action programs are effectively prioritized, agreed upon, committed to and regularly updated.
- Cross-organizational actions are well coordinated to make the strategy happen most effectively.

Skills—ensuring that qualifications necessary for execution are in place

- There is a clear understanding which skills (know-how, competencies, talent) are critical for strategy execution.

- Employees are enabled and empowered to deliver their expected strategy contribution.
- In case of skill gaps there is a process in place for closing those gaps through, e. g., employee training and development programs as well as recruiting.

Resources—providing time and funds required for high impact initiatives

- Company resources are allocated to the highest value strategic projects.
- It is clear what resources are required from which area to execute the key strategic actions and those resources are synchronized.
- Management attention is focused on the achievement of the committed business impact.

Incentives—linking personal performance objectives with strategy and rewards

- Personal performance objectives of employees are derived from strategic goals of the organization and aligned with incentive compensation.
- People feel appreciated and valued for their strategic performance.
- The incentive system rewards results appropriately.

Attitudes—inspiring people to make the strategy happen

- People are energized by the vision and strategy and want to contribute.
- People understand how their individual work can make a difference in strategy execution.
- People feel accountable for delivering their specific contribution to make the strategy happen.

Information—providing a consistent, focused, and reliable performance view

- Strategy execution progress and strategic performance is adequately reflected in a limited number of vital financial and non-financial key performance indicators (KPIs).
- For each KPI, there are defined targets and thresholds for achievement as well as reliable actual data available.
- There is transparency about progress in executing the defined strategic action programs and how they relate to achieving the corresponding targets.

Missing a driver has serious consequences.

What can happen in an organization if one of those drivers is missing or implemented insufficiently (see fig. 4-4)? In case the *goals* are missing, there is a risk of **actionism** due to a lack of guiding strategic direction for systematically deriving and coordinating initiatives. On the other hand, a situation in which strategy execution is only driven by goals and the *actions* are missing can lead to **chaos**. The identification and alignment of effective action programs is one of the key levers for successful strategy execution (Wunder 2001: 137 f.). If the direction is clear but the *skills* are not in place, people may feel **fear** as they are supposed to deliver something they are not capable of. If the *resources* are missing, there is a danger of creating **frustration**. People know how they could contribute to achieving the defined goals or executing actions, but they do not have the time or funds to excel. Missing *incentives* for performance could

Fig. 4-4

Relevance of the seven drivers for successful strategy execution

"should"	Goals		Goals	Goals	Goals	Goals	Goals	Goals
"should"	Actions	Actions		Actions	Actions	Actions	Actions	Actions
"can"	Skills	Skills	Skills		Skills	Skills	Skills	Skills
"can"	Resources	Resources	Resources	Resources		Resources	Resources	Resources
"want"	Incentives	Incentives	Incentives	Incentives	Incentives		Incentives	Incentives
"want"	Attitudes	Attitudes	Attitudes	Attitudes	Attitudes	Attitudes		Attitudes
	Information	Information	Information	Information	Information	Information	Information	
	= *Effective Execution*	= *Actionism*	= *Chaos*	= *Fear*	= *Frustration*	= *Disengagement*	= *Apathy*	= *Confusion*

Source: following Motorola University and Horváth & Partners; adapted

lead to **disengagement** as people are not allowed to share in the value they are expected to create. They are particularly unlikely to invest their attention and energy if they are rewarded for other items than reaching the strategic goals and executing the required actions. If strategy-oriented **attitudes** are not present, there is a risk of **apathy** because people do not buy-in the priorities intrinsically that they are expected to drive. Finally, if important *information* is missing, there is a risk of **confusion**. Questions may arise like: What exactly do the goals mean? Is there a shared understanding about the desired aspiration level? Where do we stand as compared to where we want to be? Do well executed actions actually have an impact on the strategic performance? How is our progress in strategy execution and how does our performance compare with competitors? Do we need additional funds or people? Are there any skill gaps we have to close? Are the incentives based on valid information?

Successful strategy execution requires both a supportive organization and an effective strategy execution system that captures all the elements explained above. To align the organization with strategy, it is important to understand essential components and key considerations of organizational design first. This will be covered in the next chapter. Afterwards it will be explained in detail how to build an effective strategy execution system.

4.2 Organizational Design

Reciprocal relationship

The effectiveness of strategy execution strongly depends on the existing architecture of an organization. If certain building blocks are not supportive of the strategy, they have to be altered or the strategy is not very likely to happen. On the other hand, the current organizational set-up influences strategic analysis and strategy formulation, and thus, the strategy itself. This is comparable to an individual who has set a certain aspiration (vision) and knows the basic paths to achieve it (strategy) but needs to make sure that the corresponding psychological and physiological requirements to make it happen are in place. At the same time, the individual's physical and psychological set-up is likely to have an impact on setting the aspiration and formulating the strategy.

Matching strategy, culture and structure

There are many interdependencies among the three components strategy, culture and structure. This makes an optimal organizational design for a certain configuration of the environment very difficult. However, regardless of the complex and reciprocal relationships between strategy, structure and culture, an organization needs to be designed in a way that provides the stability to realize the current competitive advantages and the flexibility to develop future ones. In other words, there needs to be an appropriate match of strategy, culture and structure. Figure 4-5 illustrates this interrelationship that is important for effective strategy execution, given a certain environment of the company (Bleicher 1991: 777; Hax/Majluf 1996: 39; Rühli 1990 and 1992: 1167 f.).

Organizational design: hard and soft factors

The key components of organizational design are *structure* und *culture* (Hill/Jones 2010: 381; Rothaermel 2013: 303). Structure—or managerial infrastructure—includes the organizational structure, business processes as well as the managerial systems.

Fig. 4-5

Fit between a company's strategy, culture and structure and the environment

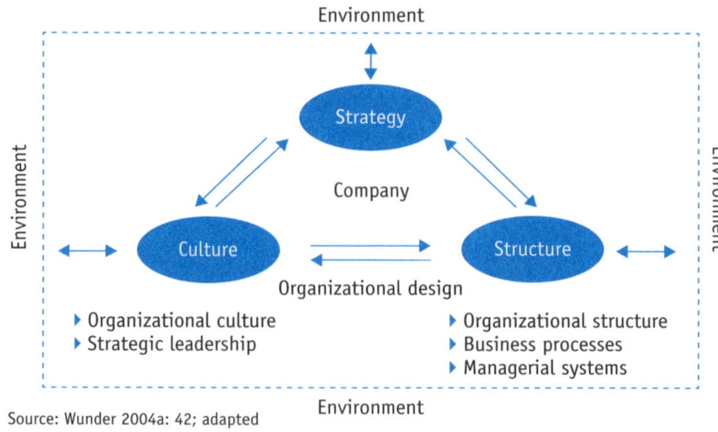

Source: Wunder 2004a: 42; adapted

They can be considered "hard" factors for execution and can typically be adapted more quickly to a changing competitive environment. "Soft" factors like the organizational culture and strategic leadership supplement the hard factors (Lombrieser/Abplanalp 2004: 328 f.). Typically it takes longer to change them, and it might be harder for competitors to imitate them, which makes these factors interesting levers to consider in the quest for competitive advantage (Scheuss 2012: 177). The strategy-structure-culture trilogy has similarities to McKinsey's 7-S-model (Peters/Waterman 1982), which shows seven variables that determine the company's success (see chapter 2.3.1.3). For a strategy to be effective, it is important to take a closer look at the culture and structure of a company and see whether there is a match of those elements or, if not, identify what needs to be done.

Watch out: organizational inertia

Although the idea of an organizational design as a means for matching strategy and structure as well as creating a strategy-supportive culture sounds compelling, company practice might well reveal a different picture. Leaders and employees usually do not like unpredictable or ambigious situations and do all they can to stabilize their work environment. Therefore, any prospective change of structure or culture might launch huge amounts of anxiety along with a resistance to change. Having this in mind, strategies might well follow a company's existing structure and culture without even considering altering these two elements. Executive teams may only decide on strategy options that neither require changes of the current organizational structure nor depend upon lengthy corporate culture development efforts. The company's resistance to change its structural and cultural status quo can be referred to as organizational inertia and should be considered as a crucial point for strategy execution as it can set the stage for a company's subsequent failure (Rothaermel 2013: 303 f.).

4.2.1 Culture—How Things are Done Around Here

Organization's "personality" fulfilling four key functions

Whether planned or unplanned, successful or unsuccessful, favored or rejected, any company or organization has a specific corporate or organizational culture with more or less clear subcultures in, for example, certain departments, business units or locations. This must not be confused with work and office climate or working atmosphere, terms that are often used for describing the staff's current morale and how well they get along. What corporate culture does for a company can be compared to what the personality does for a human being (see Strategic Snapshot 4.2). It serves as a guide for setting priorities and for deciding on what action and behavior is appropriate or inappropriate. This way it fulfills four key functions in an organization that are important for day-to-day business activities: It reduces complexity, helps to coordinate actions, provides sense and a source of meaning to organizational members, and establishes continuity (Sackmann/Bertelsmann Stiftung 2006: 29).

Collectively shared basic beliefs and values

Organizational culture captures the subtle and largely unconscious forces in an organization that determine how "things are done around here". It describes the collectively shared basic beliefs and values, norms and expectations of the organization's members. At its core are the basic beliefs collectively held by a group of people that

Strategic Snapshot 4.2

Organizational Culture Compares to Personal Psychology of a Human Being

"The *culture of any organization* is the corporate equivalent of the personal psychology of a human being. An individual's psychological makeup defines his personality, how he behaves, what he looks like to others and what he does and does not do in relation to any set of circumstances. The culture of an organization operates in a similar way and defines the kind of things the organization can and cannot do. It becomes a predictor of what changes are likely to succeed and those which will not. The culture of an organization defines most aspects of its potential in the same way that an individual's psychology largely controls his success and failures in life." (Levicki 1996: 111)

determine their thinking, actions, feelings, and attitudes, and that are typical for that group, and hence, for the company's management and employees. "These basic beliefs can manifest themselves in verbal and nonverbal behavior, such as how people interact or how they address each other. They can appear in a company's jargon, its dress code, its products, tools and services." (Sackmann/Bertelsmann Stiftung 2006: 26).

How culture begins: founders imprinting

In entrepreneurial new ventures, founders typically imprint an initial strategy, culture and structure on their firms in order to make their visions come true. Thus, corporate culture emerges when the company is founded and develops more or less distinctively as the company evolves. First, it is created when founders attract, recruit, promote and retain only those employees who seem to share the same basic beliefs, assumptions and values as they do. Second, founders impose their values and assumptions, and thus, socialize those employees to their way of thinking and doing things. And third, the founders' own behavior and way of thinking set examples for employees who may identify with them and internalize their basic beliefs. Although elements of the founders imprinting will evolve over time based on corporate development stages and changes in the environment, it can persist and shape the organization for many decades (Nelson 2003; Robbins/Judge 2013: 553; Schein 2009: 123–131). In companies like Apple, Henkel, Hilti, Hipp, Microsoft, W. L. Gore & Associates, or Würth, the imprint of the founding people seems to be still present, particularly when it comes to organizational culture.

Two layers of culture: executives and employees

Culture is more of a group than an individual phenomenon. Hereby it is helpful to distinguish between two groups that relate to different levels in the organization. On the one hand, there are the collectively held basic beliefs of employees and, on the other hand, the ones of senior executives (Davis 1984: 21 f.). Executive behavior has a significant impact on shaping, preserving and modifying corporate culture. Leaders and managers tend to serve as role models for employees and are likely to set examples for what is considered right or wrong. From the perspective of employees, the verbal or nonverbal behavior of their leaders is a living expression of corporate culture, and thus, might well impact their basic beliefs (Sackmann/Bertelsmann Stiftung 2006: 196–200). For example, leaders impose their basic beliefs, assumptions and values through what they pay attention to, what and how they measure and control, how they inspire, reward or punish their employees, how they react to critical situations and so forth (Schein 2009: 131). With respect to the strategic management process, the concept of *strategic leadership* has been introduced and discussed at the beginning of this book as the core of the applied strategy framework and is considered crucial for making a strategy process effective (see chapter 1.3.3.3).

Visible and invisible elements of organizational culture

Organizational culture consists of various interdependent elements in different layers that are more or less visible (Sackmann/Bertelsmann Stiftung 2006: 226–29; Schein 2009: 21–27; Rothaermel 2013: 318): In the outer layer, organizational culture finds its expression in **artifacts** which are visible manifestations, symbols, processes, and systems. Artifacts include the layout or design of office space, company reception desks and buildings, the official or unofficial dress code, company-specific vocabulary as well as stories, myths and anecdotes that are told in the firm (see Strategic Snapshot 4.3). The next layer addresses commonly known rules and standards for attitudes and behavior—the **norms.** Third are the openly **proclaimed values** that define whether a specific mode of conduct is important and preferred to another. Proclaimed values might be not necessarily expressed in people's nonverbal behavior. In the core of organizational culture are **basic beliefs** or assumptions. These basic, collectively held convictions guide thinking, perception, feeling, and behavior. They can be characterized as hidden, unconscious, based on experience, habituated, and attached with emotions (Sackmann 2002: 27). Figure 4-6 shows the described elements. The most important but least visible ones are the basic beliefs at the core. The more organizational culture is moving from proclaimed values to norms to artifacts, the more observable it becomes.

Strong versus weak organizational cultures

Organizational cultures can be distinguished regarding the intensity of how strongly members of an organization are committed to their organization's core values and the degree of how widely those values are shared among the entire group. In strong organizational cultures, the attitudes and behavior of employees is strongly aligned with the company's core values. The more people of an organization buy in and internalize the core values, the greater an organization's culture influences their behavior. It promotes consistency, shared expectations and self-control. When there is little alignment, meaning that core values are only moderately held by individuals and shared only by a small portion of the organizational members, the company needs to rely more on other means to control behavior of its employees, such as extensive

Strategic Snapshot 4.3

Artifacts—the Easiest Level of Organizational Culture to Observe

"The easiest level to observe when you go into an organization is that of artifacts—what you see, hear, and feel as you hang around. Think about restaurants, hotels, stores, banks, or automobile dealerships. Note your observations and emotional reactions to the architecture, the décor, and the climate, based on how people behave toward you and toward each other. You can sense immediately that different organizations do things differently. (...) As a customer or new employee, you may like or dislike one or the other of these organizations; (...) But you have to be careful. All you know for sure is that they have different ways of presenting themselves and different norms of how to deal with each other. What you don't know is what this all means. In other words, at the level of artifacts, culture is very clear and has immediate emotional impact. But you don't really know why the members of the organization are behaving as they do and why each organization is constructed as it is." (Schein 2009: 22)

Fig. 4-6

Four elements of organizational culture

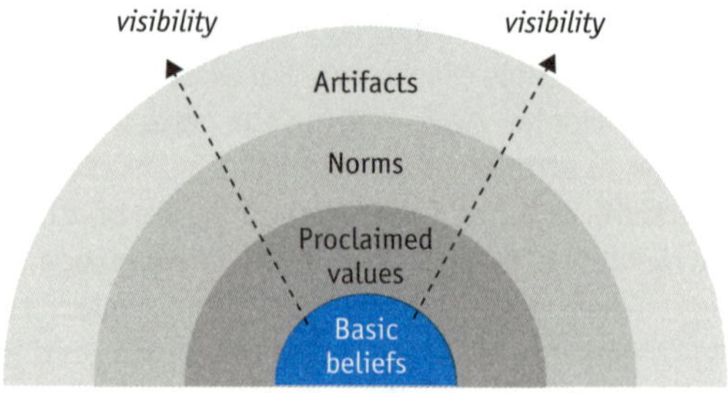

Source: following Schein 2009: 21; Sackmann/Bertelsmann Stiftung 2006: 28; Rothaermel 2013: 318; adapted

rules and procedures. In weak organizational cultures there is a lack of shared purpose that could well harm employee cohesiveness, loyalty and commitment (Robbins/Judge 2013: 548 f.). Depending on the core values, strong cultures can be very powerful but also difficult to adjust in case of major changes in the environment and a corresponding strategic reorientation. This might be less of a risk if core values include elements of agility and adaptability. Such strong and adaptive organizational cultures can be characterized by promoting autonomy, entrepreneurship and initiative across all managerial levels (Hill/Jones 2010: 397 f.) and are strong in creating change, customer focus and organizational learning (Denison/Haaland/Goelzer 2003).

Typologies of organizational culture

In general, organizational culture can be considered as a highly unique phenomenon. Its characteristics are strongly linked to a certain company with its specific heritage and its past and present leaders and employees. Although organizational cultures tend to differ from one company to another, there are a variety of typologies that try to categorize organizational culture by defining cultural archetypes. Some of those typologies are listed below. They distinguish corporate culture archetypes based on different dimensions:

- Organizational culture typology linking organizational structure to culture (Handy 1976; Harrison 1972): *power* (centralized leadership and decisions), *role* (control through job descriptions and competency profiles), *task* (project- and team-oriented), and *person* (focus on individuals and quest for self-glorification) *cultures*.
- Culture types are based on the dimensions flexibility versus stability and internal versus external focus (Cameron/Quinn 1999): *clan* (internal focus and flexible: friendly workplace with caring leaders), *adhocracy* (external focus and flexible: dynamic workplace fostering innovation), *market* (external focus and controlled: competitive workplace with demanding leaders), *hierarchy* (internal focus and controlled: formal and structured workplace) *cultures*.

Strategy Practice Example 4.1

The BMW Group's Corporate Culture Summary

"The BMW Group's corporate culture is particularly characterized by a pronounced sense of social and environmental responsibility, as the numerous activities of the company's foundations and the initiative of employees themselves demonstrate impressively. Great emphasis is placed on securing jobs, as awards (...) show. An additional strength of the company lies in its clear commitment to shared and aligned goals and its employees' exceedingly strong identification with their employer. This shows not only in employee surveys, but also in the entrepreneurial experience as it is actually lived. The vibrant corporate culture guarantees employees the freedom to take initiative and exercise an entrepreneurial spirit in the context of agreed goals, which ultimately strengthens the innovative power of the BMW Group." (Source: Sackmann/Bertelsmann Stiftung 2006: 69)

- Archetypes of organizational culture are defined based on the speed of feedback that the organization receives, the way members are rewarded, and the level of risks taken (Deal/Kennedy 1982): *work-hard/play-hard* (rapid feedback/reward and low risk: consistency), *tough-guy macho* (rapid feedback/reward and high risk: gambling), *process* (slow feedback/reward and low risk: zero failure), *bet-the-company* (slow feedback/reward and high risk: entrepreneurship) *cultures*.

Almost all of these typologies refer to human relations within the organization and in the interrelation with the environment. "When culture change is proposed it is almost always in relation to more teamwork, employee involvement, reducing the layers of supervision in the organization, creating lateral communication, building loyalty and commitment in the organization, empowering employees, and becoming more customer oriented" (Schein 2009: 37). To understand organizational culture, it is important to look at and deeply understand its key dimensions.

4.2.1.1 Key Dimensions for Analyzing Organizational Culture

Culture audit

A pragmatic way for understanding an existing organizational culture is a culture audit in which senior managers are elaborating on different subjects and what effect they have had on the present "way we do things around here". It can be used to better understand the present and to think about a desired future organizational culture. Key questions and elements of a culture audit include (Levicki 1996: 112–130):

- What are important **historical events,** and which of those left patterns still affect the way people do things today or prevent people from doing things in a more effective way? For example, companies may set up ways of conduct based on past failures which can stop or prevent certain legitimate actions. Instead of creating learning approaches from these events they tend to hinder as organizations avoid these at all costs.
- How can the **mental environment** and atmosphere in the organization be characterized (e. g., openness, fairness, fear and anxiety, (dis)honesty, respect, intellectual ambience, etc.)?

- What is the **physical environment** of the head office as compared to the physical environment of the rest of the organization?
- What are the most memorable **past leaders** and what behavioral elements are still attributable to them today?
- What are key **present leaders** and what particular qualities, attitudes and styles can be associated with them?
- What are the best known **stories** and **anecdotes**, **myths** (probably untrue stories that are told anyway) and **legends** (source long in the past but carried forward) that still affect how people behave?

Such a pragmatic approach for auditing corporate culture can be used as an attempt for getting a first read on cultural elements that drive behavior in an organization. However, it might be supplemented with a more comprehensive analysis to better understand the various dimensions of organizational culture that are likely to drive performance in a specific company situation.

Assessing culture dimensions that drive company success

There are a variety of more academic approaches and methods for a multi-dimensional assessment of organizational culture. In some approaches, internally developed company-specific cultural dimensions are used for the assessment reflecting the specific organizational context. Other approaches apply a set of predefined generic dimensions that are supposed to influence company performance and, through its standardization, allow for cross-company and cross-industry comparisons. And finally, there are approaches for culture assessment that start with defining the requirements of a "to be" culture based on the specific organizational context such as a strategic reorientation, and then assess to which degree those particular requirements are fulfilled (Bertelsmann Stiftung 2006). Currently available methods are frequently assessing corporate culture along the dimensions of performance orientation (efficiency, effectiveness, deliverables, profitability, cost), employee orientation/development, leadership, human relations, innovation/corporate development, openness/adaptability/willingness to change, task-/job-oriented aspects, team orientation, customer orientation, communication, value orientation, and strategy/vision/goal orientation (Sackmann 2006a: 6).

Adaptability, mission, consistency, involvement

A comprehensive approach for analyzing corporate culture with predefined dimensions is based on an empirically supported general framework with the four performance-related traits *mission* ("Do we know where we are going?"), *consistency* ("Does our system create leverage?"), *involvement* ("Are our people aligned and engaged?"), and *adaptability* ("Are we listening to the marketplace?"). According to the model, high-performing organizations can be characterized by a culture that is strong along those four traits, and thus, are able to resolve some of the underlying contradictions between providing stability and flexibility as well as internal and external focus without simple tradeoffs. Each of those four traits can be measured and evaluated with three criteria. The organizational culture profile resulting from the evaluation of those 12 criteria can be visualized with a radar chart (see fig. 4-7). The assessment is conducted with a standardized 60-item questionnaire. Strengths and weaknesses and corresponding culture development needs can be derived by comparing the

Fig. 4-7

The Denison organizational culture model

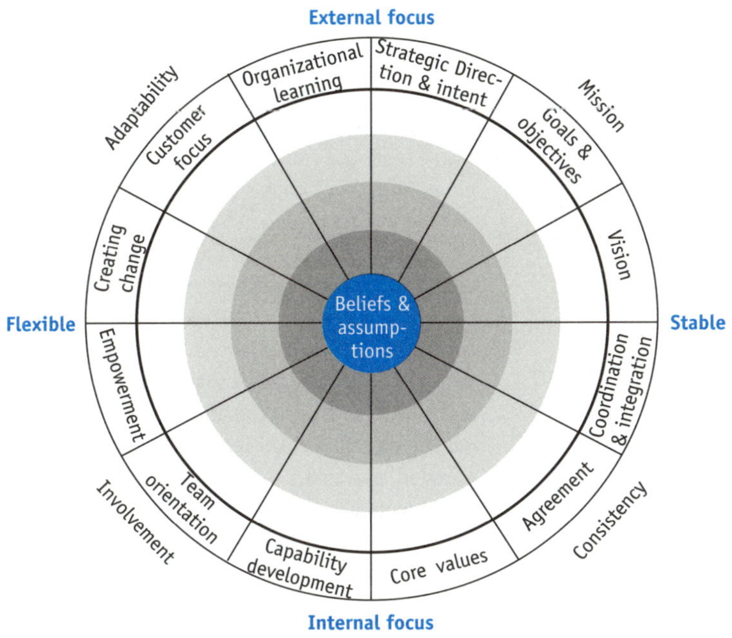

Source: Denison/Haaland/Goelzer 2003: 209

results with a database of organizational culture assessments of other higher and lower-performing organizations (Denison 2014).

A practical example of results derived from a similar multi-dimensional corporate culture assessment is provided in Strategy Practice Example 4.2 and figure 4-8. The ten dimensions and their corresponding evaluation criteria applied in this example have been developed by an international working commission composed of experts from both the academic and the business world (Booz Allen Hamilton 2003 and Bertelsmann Stiftung 2003 quoted in Sackmann/Bertelsmann Stiftung 2006: 44).

Gaining competitive advantage with organizational culture

The resource-based view of strategy suggests that a company may achieve a sustainable competitive advantage when its combination of resources and capabilities meet several conditions (see chapter 2.3.2): They must be valuable, rare, difficult to imitate, and non-substitutable. In addition, the organization has to be capable of exploiting those resources and capabilities and capturing the value created (VRIO principles). Once an organizational culture meets those criteria, it is not only something that needs to be matched with strategy to facilitate effective strategy execution but becomes a source of competitive advantage, and thus, a part of strategy formulation itself (Rothaermel 2013: 319–322; Sackmann/Bertelsmann Stiftung

Strategy Practice Example 4.2

Hilti's Corporate Culture Core Characteristics and Summary

Hilti is a provider of innovative products, systems and services for professional customers in the construction and energy sectors. The company's headquarters are located in Schaan, Principality of Liechtenstein. With its approximately 22,000 employees worldwide the Hilti Group achieved sales of more than $16 bn in 2014. "Five dimensions of Hilti's corporate culture stand out: customer orientation, fundamental convictions, beliefs and values and their consistent implementation; employee responsibility; corporate governance; and continuity in leadership.
Customer orientation is the top corporate goal, underpinning operations throughout the company at all levels. It is achieved particularly through the direct sales strategy and the company's commitment to integrating customers in the development process. Two-thirds of Hilti employees work in sales.
The fundamental convictions, beliefs and values are coherently lived by managers and employees. Hilti supports this consistently with coordinated measures and instruments. Employees are actively involved, given responsibility and expected to act responsibly. Accordingly, organizational structures are flat. Communication and interaction flow freely and openly across hierarchical levels and departments.
Hilti's model of corporate governance sets an example not only for family-run companies. At Hilti, the focus is clearly on securing the company's long-term course. The members of its Board of Directors are more involved in the company than is usual for such bodies and are expected to make an active contribution. The company's continuity of leadership and culturally aware management are exemplary. With self-criticism anchored in the corporate culture and opportunities for improvement actively pursued, the company's leaders have joined its Culture Journey, confidently traveling the road to a secure future." (Source: Sackmann/Bertelsmann Stiftung 2006: 134 and 148)

Fig. 4-8

Evaluation of the core characteristics of Hilti's corporate culture

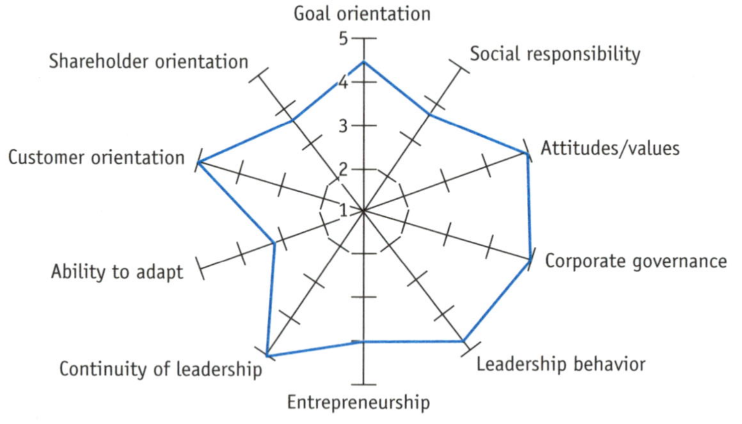

Source: Booz Allen Hamilton 2003 quoted in Sackmann/Bertelsmann Stiftung 2006: 134

2006: 32 f.). Competitive advantage can be achieved when companies are able to produce their products and services in an efficient way (operational effectiveness), when they come up with proximal or radical innovation regarding their product and service offerings (see chapter 3.4.1.2), and when they find new ways of creating customer value with innovative business models (see chapter 3.4.4.3) where products and services are only one of the various elements. The challenge, however, is that those sources of competitive advantage can often be imitated in a more or less short period of time. Best practices in efficiency enhancements, like principles of the Toyota Production System (TPS) or Lean Management and Six Sigma methods applied in operations and administration, can be acquired and applied by all companies and implemented in a fairly short period of time. They are neither rare nor difficult to imitate, and thus, create only parity with key competitors but no advantage for a prolonged period of time. The same holds true for product renovation and innovation that are imitated quickly nowadays unless they are protected by law for a certain period of time. Business model innovation is proclaimed to make imitation more difficult, and thus, creating temporary or even sustainable competitive advantage. However, extensive research and popularized publications that describe different generic business model patterns and their mechanisms are widely available, and hence, are an open source for imitation as well. So what is left for creating sustainable competitive advantage? With an effective configuration and orchestration of performance-relevant dimensions it might be organizational culture (see fig. 4-9).

Fig. 4-9

Corporate culture development and competitive advantage

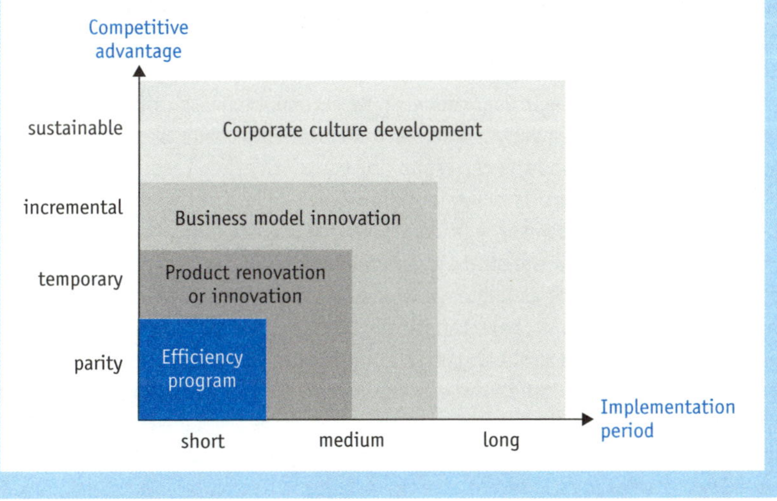

Performance-relevant dimensions of organizational culture

What are the dimensions of organizational culture that could be a source of gaining such a sustainable competitive advantage? By analyzing different approaches that try to measure organizational culture, empirically supported key dimensions of organizational culture that have an impact on company performance have been identified and are listed below (Sackmann 2006a: 7–18):

1. Clear and communicated identity
2. Consistent strategic (goal-)orientation
3. Customer orientation
4. Ability to learn, adapt and innovate
5. Leveraging employee potential
6. Participative and culture-compliant leadership—open communication
7. Performance orientation/performance commitment and capability
8. Balanced stakeholder orientation

Although those dimensions have an impact on company performance, they need to be aligned with strategy and structure (see fig. 4-5). They can be used as starting point for assessing and creating a fit between strategy and organizational culture.

4.2.1.2 Integrating Strategy and Culture

Culture shapes strategy and vice versa.

Creating a strategy-supportive culture is a difficult task. Organizational culture cannot be thought of as a simple variable in the strategy process that can be managed and controlled at will. Instead, it is a multi-dimensional, multi-layer and dynamic phenomenon that strongly drives behavior of people at all levels in a firm. The relationship between strategy and culture is twofold: On the one hand, organizational culture shapes strategy through, for example, selecting or neglecting certain strategy options. On the other hand, organizational culture is being shaped by strategy through, for example, cultural changes imposed by a strategic reorientation. One way to grasp this twofold mechanism is by separating corporate culture into "guiding beliefs" and "daily beliefs" (see Strategic Snapshot 4.4). Any attempt to match strategy and culture requires a comprehensive but yet practical understanding of the "as is" and the "to be" organizational culture and the corresponding transformational requirements.

Assessing strategy-culture "fit"

Organizational culture is a key building block of organizational design. When diagnosing a culture, however, it is important that it is not done for its own sake or as a standard procedure in the strategy process but tied to an organizational issue or "business problem" such as implementing a new strategy or business model. When doing so, the challenge is characteristically not "the culture" as a whole but certain elements or dimensions in the organizational culture that may either hinder or support the strategy. Those critical dimensions need to be identified and either enhanced and leveraged or altered to make a strategy come true (Schein 2009: 77–102). For conducting an assessment of the "fit" between organizational culture and strategy the following steps are suggested (see fig. 4-10):

1. **Decomposing strategy in its essential elements (strategic goals):** First, strategy has to be decomposed in its essential elements such as strategic goals or strategic action programs (Lombriser/Abplanalp 2004: 350). In chapter 4.3.2, it is suggested that strategy should be translated into a limited and balanced number

Strategic Snapshot 4.4

Guiding Beliefs and Daily Beliefs

"Capturing the essence of guiding beliefs and distinguishing them from daily beliefs are important because of the link between beliefs and strategy. Strategy proceeds from guiding beliefs; they are the roots from which strategies grow. If strategy is a statement of *what* a company wants to accomplish, and organization is the vehicle of *how* the company will accomplish it, then guiding beliefs are the statements of *why* the company wants to accomplish the strategy. And whether the strategy will succeed or fail depends on the match or mismatch of all of these with the daily beliefs, embedded in the people, structures, and systems within an organization. (...) Guiding beliefs are precepts upon which strategies get formulated, while daily beliefs affect whether strategies get implemented. When the corporate culture is healthy, the daily beliefs flow from the guiding beliefs. They are translations and enactments of the basic tenets. The more unlinked the daily beliefs are to the guiding ones, the more unhealthy is the corporate environment. Guiding beliefs, therefore, are about the way things ought to be, and daily beliefs are about the way things are on a daily basis (...)." (Davis 1984: 4 f.)

of strategic goals which are derived following some key principles to ensure an effective execution. Following those guidelines, strategic goals articulate the desired outcome ("what"), reflect desired changes in the business model and are selected only if they have a high competitive impact and a high necessity to act. If those principles are applied, strategic goals might be the best "check-points" for a culture-fit assessment as they reflect both a high importance for making the strategy happen as well as a need to do something about it. This way they represent a *strategic change agenda* that needs to match the organizational culture for realizing the desired transformation.

2. **Identification of relevant dimensions and definition of desired culture ("to be"):** In the second step, dimensions of organizational culture that are most relevant for achieving the strategic goals are identified. Hereby, culture dimensions from previously explained organizational culture frameworks can be used as a guideline. For each of the relevant dimensions it needs to be defined, what characteristics are required to make the strategy come true in terms of the strategic goals. Those dimensions might only cover a portion of the culture assessment frameworks explained previously in this chapter. The idea is not to change "the

Fig. 4-10

Assessing strategy-culture "fit" in four steps

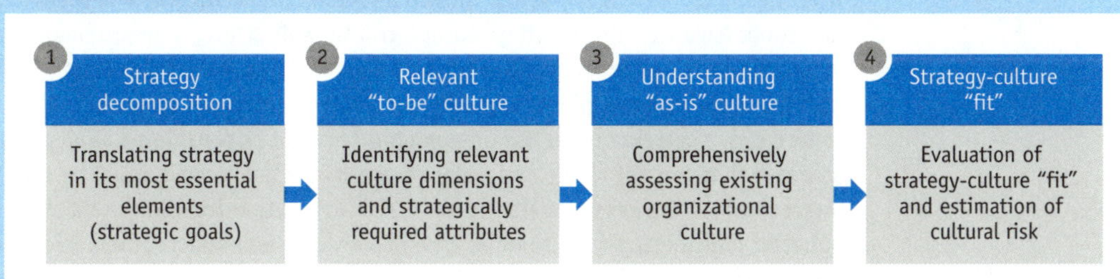

culture" as a whole but to focus only on those concrete cultural areas with the highest relevance while ensuring consistency with other strategically less relevant culture dimensions. A "to be" culture can be defined based on the identified strategic priorities (step 1), the relevant culture dimensions, and their strategically required characteristics.

3. **Assessment of existing organizational culture ("as is"):** To understand the current organizational culture comprehensively, it is recommended to rely not only on one method of data collection such as a questionnaire. Survey-based culture assessments have a variety of limitations and the response may reflect more of a cultural artifact itself than a valid and reliable indicator of the deeper shared basic beliefs and assumptions (Schein 2009: 79 and 101). It seems to be more appropriate to apply a so-called triangulation approach, which collects different types of data from different sources (e. g., survey, interview, behavioral observations, documents, etc.) and evaluates its consistency. When drawing a conclusion of the existing culture profile, it should be considered that culture is a multi-dimensional phenomenon and cannot be explained with only one element, and that there need to be consistency between proclaimed aspirations and actual behavior (Sackmann 2006a: 19). A comprehensive culture assessment approach bares the risk that it results in a wealth of partially inconsistent information that is hard to digest and draw valid conclusions from. For this reason, some authors suggest more pragmatic checklist-based approaches knowing that those do not comply with solid empirical research methods but are simple and quickly applicable. They can be used for getting a high-level understanding and descriptive representation of an organizational culture (Hinterhuber 1997: 238 f.; Lombriser/Abplanalp 2004: 330–332).
4. **Evaluation of the strategy-culture "fit" and cultural alignment risk:** A final step in assessing the strategy fit of organizational culture is a comparison of what is strategically required versus how things are currently done. Hereby, the focus is primarily on the relevant culture dimensions identified earlier. Characteristics of the existing organizational culture can either be strengths or constraints on what the company is trying to do strategically. In company practice, it is generally easier to draw on strengths of the current organizational culture than to overcome the constraints by a long-term culture change effort (Schein 2009: 77–102). Therefore, companies should focus on preserving, leveraging and building upon aspects of an existing organizational culture to support proposed strategic goals. What needs to be done, however, in case relevant culture dimensions seem to be not supportive of strategy? Depending on how well strategic requirements are matched in the existing culture, a level of risk for strategy execution can be estimated in case culture is not aligned with strategy (cultural alignment risk). In some cases, a certain degree of non-compliance might be considered as an acceptable risk that will not break a strategy, whereas in other cases it might be seen as unacceptable. The result of this cultural risk evaluation functions as a starting point for considering what to do in case of a serious and unacceptable mismatch (Schwartz/Davis 1981: 36).

Fig. 4-11

Strategic culture alignment (SCA) matrix

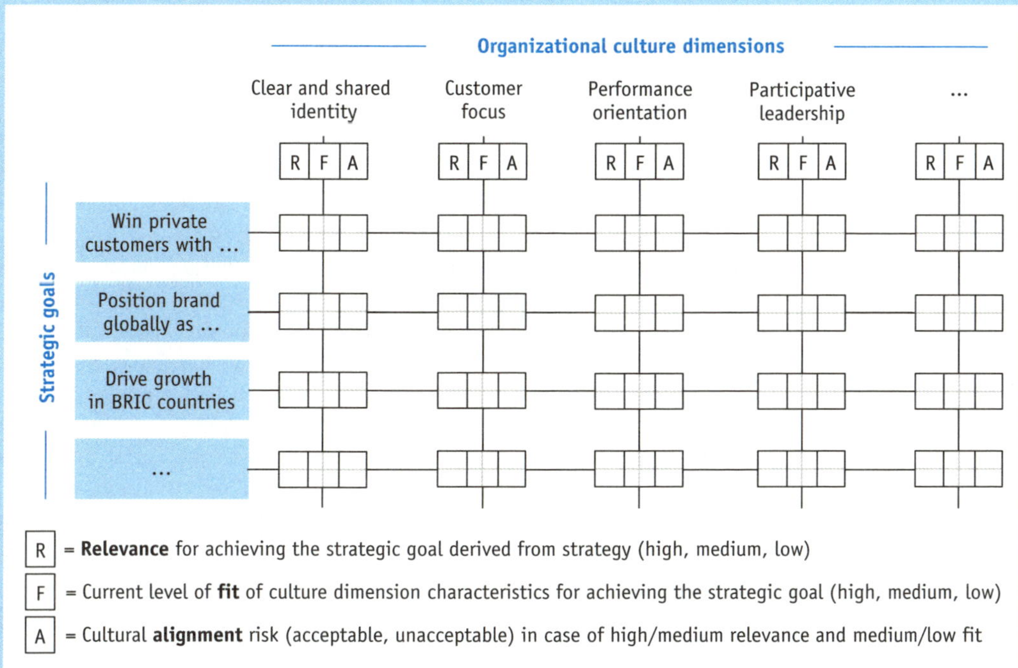

The strategic culture alignment matrix (SCA) presented in figure 4-11 shows a framework for a culture-strategy fit assessment following the same approach as the strategic process alignment matrix that is introduced in chapter 4.2.2.4 (see fig. 4-31).

Options for dealing with a strategy-culture-mismatch

Once the "fit" of strategy and culture has been assessed and a mismatch has been identified, a suitable course of action needs to be defined. There are several ways of dealing with a mismatch between strategy and organizational culture (Schwartz/Davis 1981):

- First, culture could be ignored for strategy execution. Managers might rely on cultural self-adjusting mechanisms that are caused by pursuing the corresponding strategic goals. This can be a dangerous undertaking, though, particularly in cases where there is a serious mismatch between strategy and culture.
- Second, strategic action programs derived for achieving strategic goals (see chapter 4.3.5) are defined in a way that is compliant with the existing organizational culture. This means that strategic goals remain unchanged but concrete strategic initiatives and programs might be different from what they had looked like in case of a better strategy-culture fit (e. g., less aggressive; different schedules, budgets or responsibilities; other actions; etc.).

- Third, strategic goals are changed in a way that the strategy better fits with the existing culture through, e. g., reducing the aspiration level or refining the basic strategic course.
- Fourth, organizational culture is systematically developed toward a better match with strategy. Although this might be difficult and take long, it could be an effective way for establishing a more sustainable competitive advantage as explained earlier (see fig. 4-9). In such a case, the culture transformation effort might well become a strategic goal itself, and thus, be managed as an element in the strategy execution process. For achieving such a strategic goal, it is particularly important to follow change management principles such as effective communication, training, and explaining and to build this into the execution plans. This approach is reflected in strategy map templates where "culture" is defined as an own generic topic in the potentials perspective (see fig. 4-47 in chapter 4.3.2.2). Corresponding strategic goals are typically addressing only certain dimensions of corporate culture that need to be developed for making a strategy happen, such as driving performance, enhancing customer orientation, establishing a market leadership culture, improving cross-functional collaboration, living sustainability principles, or improving communication.

Changing and matching corporate culture with strategy

Developing organizational culture for a better fit with strategy can be facilitated with a variety of means (Lombriser/Abplanalp 2004: 351 f.): Leaders can set examples of the desired behavior (e. g., way of decision making, design of meetings, etc.) for their employees. Furthermore, symbolic actions or events such as celebrations, ceremonies, conferring distinctions (awards, honors) may contribute to driving culture toward strategy. Also, visualization of corporate identity as well as the physical environment of the organization can be changed to better reflect a desired new culture (e. g., architecture, office space, etc.). Another effective way to influence organizational culture and match it with strategy is the design of corresponding organizational structure, business processes and managerial systems such as planning and budgeting, human resource as well as management reporting. This relates to the managerial infrastructure, which will be explained next.

4.2.2 Structure—the Managerial Underpinning

Managerial infrastructure

Structure is the second building block in designing a strategy-supportive organization. In analogy to a human being, culture relates more to the psychology of an organization whereas structure is related to the physiology. Hereby, the term "structure" relates to the managerial infrastructure of the firm, which can be constituted by three elements:

- organizational structure
- business processes
- managerial systems.

When designing an effective managerial infrastructure, it has to be determined how to divide the work and tasks (differentiation) best, and how to coordinate the different tasks in an optimal way (integration). People need to be assigned to certain activities, and their work efforts in the organization need to be connected and orchestrated. Here are some key questions that have to be answered (Hill/Jones 2010: 381–386; Lombriser/Abplanalp 2004: 330–332; Robbins/Judge 2013: 514–520; Rothaermel 2013: 304–309):

Differentiation and integration

- To what degree are formal rules and regulations supposed to direct behavior of employees and managers (formalization)?
- At which management level is the decision making authority positioned (centralization)?
- How distinct is labor division by cutting a task into separate jobs and grouping it together in units such as departments (specialization)?
- How many employees should directly report to a manager and who should formally report to whom (hierarchy)?

Formalization

Formal organizational structure

Formalization captures the degree to which activities in an organization are standardized by explicit rules and procedures that are often codified in employees' handbooks. Companies tend to promote formal structures to achieve the desired business outcomes. A comprehensive organizational design including well-structured organizational charts, formal processes and procedures as well as well-designed systems and tools are supposed to foster results.

Informal structure

The elements of formal structure are supplemented by an informal structure at all levels in the organization. This refers to the host of informal interactions that exist among members of the organization and that mirrors or interferes with the laid out design of the organization. This could mean, for example, that communication or decision making does not follow but cuts across formal lines along which the business is supposed to work. The informal structure might be based on personal relationships between members of the organization that get formed when people work together at challenging tasks and move around the organization during their career. These personal bonds may be stronger than any laid out organizational design and encourage interaction in a way the formal structure could not explain (Levicki 1996: 163 f.). Companies may explicitly develop an environment that fosters informal structures by means like abandoning formal job titles and job descriptions, operating without formal communication channels and organizational charts and with no or little written rules and operating procedures (Rothaermel: 2013: 306). Other means are limiting the number of employees per subsidiary, creating an office space with social meeting areas or innovative ride-sharing concepts (see Strategy Practice Example 4.3) that facilitate informal communication processes. The desired informality is supposed to foster employee satisfaction, personal initiative and creativity as well as innovation.

Centralization

Allocation of decision making authority

Another element that needs to be decided when designing an organizational structure is the level of centralization, that is how to concentrate decision making at a

Strategy Practice Example 4.3

SAP's TwoGo Ride-sharing App Drives Sustainability and Informal Networking

"Organizations committed to sustainable business practices are concerned with creating value while mitigating their environmental impact. TwoGo helps companies meet that goal simply and effectively. For example, the solution offers cost-efficiency opportunities to companies with corporate fleets, such as how to minimize maintenance costs and preserve vehicle resale value. It also supports companies looking for a way to make vanpooling more affordable and effective by seeking riders to find and regularly fill the van. Finally, TwoGo helps organizations demonstrate their commitment to sustainability by reducing carbon emissions and engaging employees in that mission. For individuals, there are also significant benefits. In very large U. S. urban areas, the average commuter spends $16 and 52 minutes going 32 miles round trip to work each day. TwoGo can help people save money and time. By carpooling, employees may recoup some of this lost productivity by encouraging networking and brainstorming on the road. In addition, TwoGo may create a more networked culture, helping employees to feel better informed and involved with their peers. (…)
Deployed internally at SAP since July 2011, SAP estimates TwoGo has generated more than $5 million in value for the company, specifically by helping to:

- Save costs related to commuting to work; e. g., reducing fuel consumption and maintenance costs and increasing the resell value of company cars
- Avoid expenses related to employee travel such as reimbursements for fuel, limo services or parking
- Increase networking among employees by matching employees into carpools more than 36,000 times and adding 2,200 additional days of networking time among employees
- Reduce greenhouse gas emissions by helping to eliminate 400,000 miles of driving and avoid 88 tons of greenhouse gases." (Source: SAP 2013)

central point in the organization. A company has centralized decision making when only managers at the top or upper levels of its hierarchy have the formal authority to make important decisions. Lower-level managers are supposed to merely carry out the directives derived from those decisions. Centralization gives top managers greater control over the company's activities. Strategic decision making is likely to become simpler and faster. It can be beneficial, for example, in times of crisis where strong leadership is required to make the entire organization respond quickly and in a highly coordinated way. Decentralization, on the other hand, pushes down decision making authority to lower-level employees who are closer to the sources of issues and have more detailed knowledge to handle them. This leads to fast response times in solving problems at the point where they occur, which can be particularly important when dealing directly with customers. It promotes flexibility and reduces an information overload of top managers allowing them to spend more time on strategic issues. Also, empowered lower-level employees are likely to feel more motivated and responsible as they are in charge of their local contribution to strategy and less alienated from those who make all the important decisions that affect their local work environment.

Democracy of opinions but not decisions

Centralized decision making does not automatically mean that people provide little input. This can be illustrated by the anecdotal statement of the division CEO of a Fortune 500 company: "We are a democracy of opinions but not a democracy of decisions". Typically, top-down strategic planning is applied by centralized organizations whereas decentralized companies are more likely to use a bottom-up approach and leave room for emergent strategies. Given its specific context, some organizations benefit from centralized authority as others from decentralized authority or from a combination of both. Some excellent companies combine centralization and decentralization in a way that they push autonomy down in the organization and empower people for the most part but concentrate decision making authority for a few issues critical to strategy at the top (Peters/Waterman 1982: 15).

Specialization

Division of labor

Specialization refers to the division of labor, that is the breakdown of jobs into narrow repetitive tasks or activities. The key idea of specialization is that jobs are divided into a number of steps, each completed by different employees who are specialized at their particular step. Large firms tend to have their tasks divided very clearly into separate jobs that are conducted by highly specialized employees. In smaller companies, there usually is a lower degree of specialization. This means that people need to be more of a generalist as they are required to take on many different tasks. Employees, who are used to work in an environment with a high degree of specialization, such as in Fortune 100 companies, may struggle when they move into a small or medium-sized enterprise (SME) where people tend to perform a variety of different tasks themselves. The same may happen the other way round when people who used to solve a variety of different problems in a SME move on to a specialized work environment of a large corporation. The economic goal of specialization is to increase productivity by enhancing people's skills and know-how, and thus, saving time in doing the job. A pitfall, however, is that a high degree of specialization can lead to reduced job satisfaction as employees tend to repeat the same tasks over and over again.

Departmentalization

Departmentalization captures the extent to which common jobs or tasks, and thus, people are grouped to be managed better (see also chapter 4.2.2.1). Essentially, a company can departmentalize jobs in grouping them by:

- The type of *functional activity performed*; that is functional specialists are put together to gain efficiencies. Examples of organizations that might be organized by a functional departmentalization are a manufacturing plant (e. g., engineering, manufacturing, assembly, logistics, quality control, accounting, HR, etc.) or a hospital (e. g., surgery, inpatient care, research, administration, equipment, etc.).
- the type of *products* or *services* (or brands) offered by the organization, that is all activities related to a particular product, service or brand are under the control of a single manager with P&L responsibility to enhance accountability for performance. An automotive company might be organized by its different product categories such as passenger cars, commercial vehicles, motorcycles, financial services, etc. or by the different brands they have in their portfolio (e. g., VW, Audi, Seat, Bentley, Bugatti, Porsche, etc. in the case of the Volkswagen Group).
- *geography* (region or country) that can be beneficial when the company needs to be highly responsive to different local market environments in order to be successful (e. g., the company's customers at particular geographic locations have similar

needs). In these cases performance might be managed best by putting accountability for a territory into the hands of a regional or country general manager.

- the type of *customer* that might have different kind of needs and wants that have to be served in different ways. An example is the departmentalization into private customers and business customers of a bank.

Hierarchy

Span of control

Hierarchy answers the question who reports directly to whom based on the formal organizational structure and the manager's formal position. The number of employees that a manager officially directs and controls is called span of control. In tall organizations with many management levels, the span of control is narrow as employees are spread on several organizational levels with managers in charge at each level. This allows managers to maintain close control of their subordinates. However, it is costly due to the high amount of management salaries, it may slow down decision making due to more complex vertical communication lines, and it can increase the distance of the upper management from the operational level. In the worst case, top-level decision makers lose their grip on the organization's business reality (see chapter 1.3.3.3). In flat structures, that is with only few management levels, the span of control is wide. This means that a relatively high number of employees directly report to a manager. Cost cutting initiatives or mergers and acquisitions, in which particularly middle managers are laid off, tend to increase the span of control as the remaining managers have to supervise and monitor more direct reports. Although this puts a higher pressure on those managers, flat organizational structures with a wide span of control support companies in their goals to maintain competitive cost structures, get closer to the customer and accelerate decision making as well as empower employees and increase flexibility. The challenge is to determine the span of control at which a manager can direct the organization's employees in the most efficient and effective way.

Chain of command

Adding up the direct vertical reporting lines of different organizational levels leads to the chain of command. It defines the unambiguous and unbroken line or hierarchy of authority that extends from the top management of the organization through middle and first-line management to the non-managerial employees. A clear chain of command assumes that certain management positions have the right to give orders and can expect them to be followed (authority). Each management position has its place in the chain of command with the degree of authority that is necessary for a manager to meet his or her corresponding responsibilities at the specific level in the organization. Furthermore, a clear and unambiguous chain of command requires the principle of a unit of command, which means that an individual should have only one superior to report to directly. A broken unit of command means that employees have to cope with potentially conflicting demands coming from several superiors that follow different priorities. A clear chain of command with clear responsibilities tends to reduce friction among managers and helps to make things happen effectively. Nowadays, the chain of command seems to be far less important than it used to be. This is caused by the popularity of decentralized decision making with empowered employees, self-controlled and cross-functional teams or task forces as well as new organizational structures like the matrix or tensor organization, which are supposed to better deal with today's highly complex and dynamic management challenges.

4.2.2.1 Organizational Structure

Several basic types of organizational structure can be distinguished based on a combination of some of the described design elements. They range from simple, functional or divisional to more complex types like matrix or tensor. Before suggesting how to deal with the issue of matching structure to strategy (or vice versa), main types of organizational structures are briefly explained (David 2013: 250–260; Homburg/Kuester/Krohmer 2013: 466–470; Nicolai 2009: 111–130 and 262–264; Wheelen/Hunger 2010: 280–288).

Simple Structure

An entrepreneur makes all important decisions.

The simple structure (see fig. 4-12) is typically found at a start-up or small business enterprise where an entrepreneur tends to make all important decisions personally, has a direct individual relationship with the employees and is personally involved in key managerial functions like planning, staffing, controlling, etc., which are usually performed to a very limited degree. This "structure" is typically found in an early corporate development stage. It has the advantages of rapid decision making and a high degree of flexibility. However, it totally relies on the decisions of the entrepreneur who may struggle to have all in one hand once the company has grown.

Fig. 4-12

Example of a simple structure

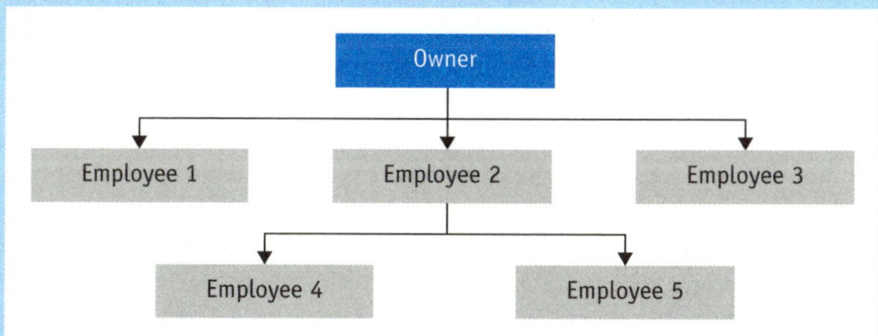

Functional Structure

People are grouped based on domain expertise.

In the functional structure (see fig. 4-13), the entrepreneur is supplemented by a team of managers with functional specializations to run the company. Tasks and activities are grouped to functions like Finance, Marketing, HR, Production, Research and Development, etc., which are required to manage the organization. On the one hand, the specialization of business activities improves efficiency; on the other hand, it may cause problems in coordination, such as misalignments of functions like Marketing and R&D when planning for products and markets. This could lead to a situa-

Fig. 4-13

Example of a functional structure

tion where requirements of certain customer groups may not be considered sufficiently, which is a major disadvantage. A functional structure becomes particularly problematic once the company expands and diversifies its product lines and services and operates in different markets and industries. At a certain degree of diversification or size, an organization would be very difficult to manage and control with all those different functions that would be required then.

Divisional Structure

Distinct units with P&L responsibility

Diversified growth typically leads to a divisional structure (see fig. 4-14) that can be characterized by a business-oriented specialization instead of a focus on types of tasks. The decision making authority is usually decentralized with each division or business unit having its own profit-and-loss (P&L) responsibility and functional structure. The divisions can be established by *product or services* as well as *brands*, by *customers* or by *geographic areas*. Key advantages are the clear accountability, entrepreneurship and control for each division. The divisional structure is an effective structure for strategies that strongly emphasize specific products or services, customer groups or geographies. Major disadvantages are the duplication of functional activities and an opportunistic behavior of the divisional management to a degree that the overall corporation cannot optimize its performance. Competition between divisions may become so intense that ideas and resources are purposely not shared. Optimizing the division's own sales and profits becomes the one and only focus of the division management, which can be inconsistent with the overall strategic intent that may want to leverage certain interdivisional relationships and synergies for realizing competitive advantage. A special type of divisional structure is the management holding (see Strategic Snapshot 4.5).

Strategic Business Unit (SBU) Structure

Bundling of divisions within a discrete competitive arena

Particularly when the number, size and diversity of divisions increase, they are organized into strategic business units (SBU) to better reflect independent product-market combinations (see fig. 4-15). Typically, a SBU is an internal organizational unit that is formed to address an external strategic business field (see also chapter 2.1 and 2.2.2.1). It serves a certain market with a certain product or service offering and dis-

Fig. 4-14

Examples of divisional structures

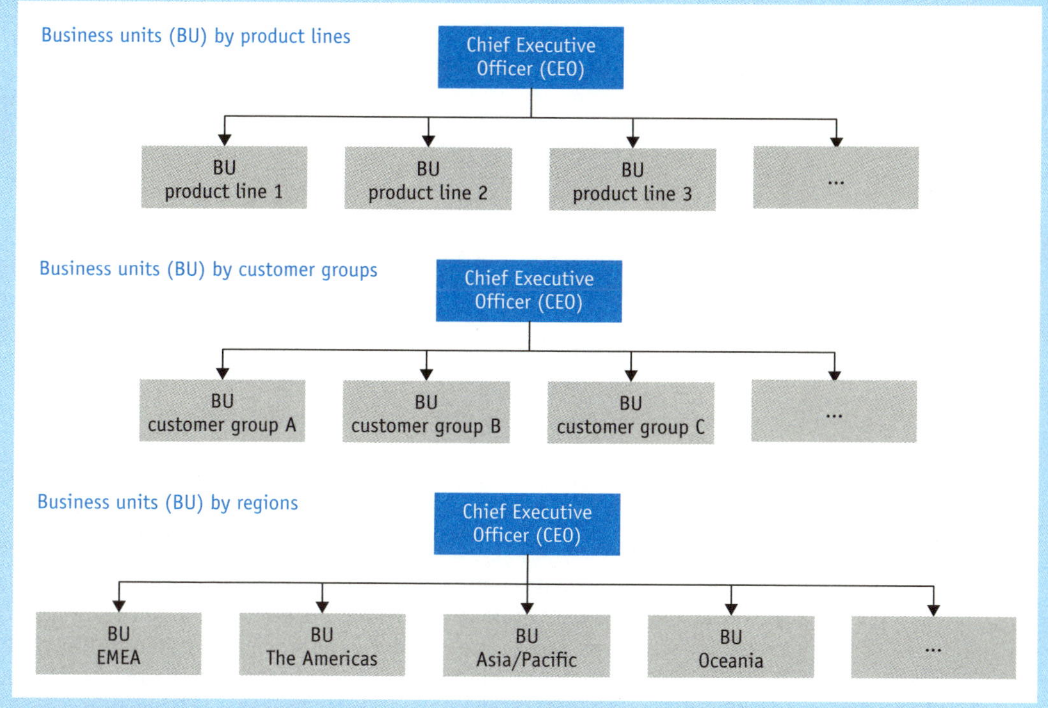

Strategic Snapshot 4.5

Management Holding

A special form of the divisional or strategic business unit (SBU) structure is the conglomerate structure (management holding company). This describes an assembly of legally independent firms (subsidiaries) under one corporate umbrella. The key difference to pure divisional or SBU structures is that the different divisions or strategic business units are legally independent companies. Whereas pure financial holdings only focus on the return of their investments (subsidiaries), management holdings are additionally leading their divisions or businesses strategically (strategic management holding) or operationally (operational management holding). In company practice there also exist corporations with divisional or SBU structures that are a mixture of legally independent subsidiaries and non-independently operating divisions and strategic business units. (Macharzina/Wolf 2010: 491–493; Wheelen/Hunger 2010: 148 f.)

tinct competitors independent of other SBU. A SBU can be seen as a sort of "company within a company" that has its own strategic plan and independently contributes to the overall success of the corporation. Classically each SBU has its own CEO with profit-and-loss (P&L) responsibility as well as a functional and/or divisional structure. Whereas one SBU might design its organization with a basic functional structure,

Fig. 4-15

Strategic business unit structure with functional/divisional and matrix structure

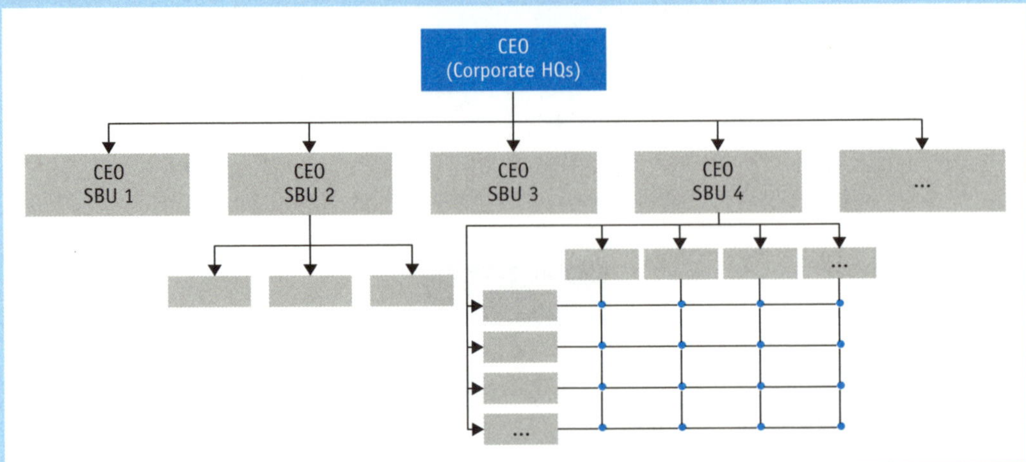

another one might be organized in a matrix with different product lines and regions (see fig. 4-15). With a SBU structure the coordination between similar divisions can be improved and accountability can be channeled. The performance of the divisions or business units can be managed more effectively and efficiently within an SBU. Ideally, certain resources (e. g., functions) can be shared. On the other hand, an additional layer of management is added with a SBU structure. For example, a conglomerate with 100 divisions may be grouped into 10 strategic business units with its corresponding SBU leader. This will increase bureaucracy and cost (e. g., salaries of the SBU leaders) and may enhance the dysfunctional competition for limited resources in the corporation that might already have been an issue within the divisional structure.

Multidimensional Structures (Matrix and Tensor)

Mixture of two or three organizational forms on one level

Particularly in large companies, several types of organizational structures may be combined. If this is done across different hierarchical levels, the question arises as to which type should be at the top and which at the subsequent levels. For example, the first level after corporate management could be based on strategic business units with each having its own functional units. Within certain functions there might be a regional structure applied such as different sales units for different regions in the sales function or a certain SBU. The combination of two forms of specialization on one level, that is the divisions or strategic business units, on the one hand, and the functions or regions, on the other hand, results in a matrix structure. It was designed for combining the advantages of different organizational forms to deal with very complex and dynamic environments and encouraging interdepartmental collaboration. Decision making is split between different structural dimensions. This is supposed to lead to better results particularly in situations with equally important but potentially conflicting strategic priorities. A clear hierarchical steering logic is replaced by supposedly constructive problem solving mechanisms of managers from different organizational units. In company practice, there is typically a solid and dotted line reporting implemented with a multidimensional organizational structure. Within a global matrix structure (see fig. 4-16), for example, the benefits of a global SBU striving for economies of a global scale and a region striving for location economies through a high degree of responsiveness to local market requirements are pursued simultaneously. In such a case, a company can choose to have solid reporting on the dimension it wishes to emphasize the most (e. g., the SBU for scale economies and global coordination or the geography for more of a local market and customer emphasis). A manager working for the global SBU in a specific geography might report directly to the global SBU CEO (i. e. solid line) and have a dotted reporting line the general manager of the region (e. g., country manager). This allows the company to feed regional learning back to other SBUs, and thus, drive company-wide innovation. A matrix also evolves when people from functional units are assigned to product divisions or projects. The advantage is the combined expertise of cross-functional teams that is leveraged across products and projects (project organization).

Tensor structure

Adding another organizational structural dimension to the matrix leads to the three-dimensional organizational structure called tensor organization (see fig. 4-17). It can be applied in multinational enterprises with heterogeneous markets and regions requiring, for example, a tensor of SBUs (with product lines or global brands), global functions and regions. An example from the professional service industry is a management consulting company with a tensor of regional units, solution centers

Fig. 4-16

Example of a matrix structure with SBU and regional divisions

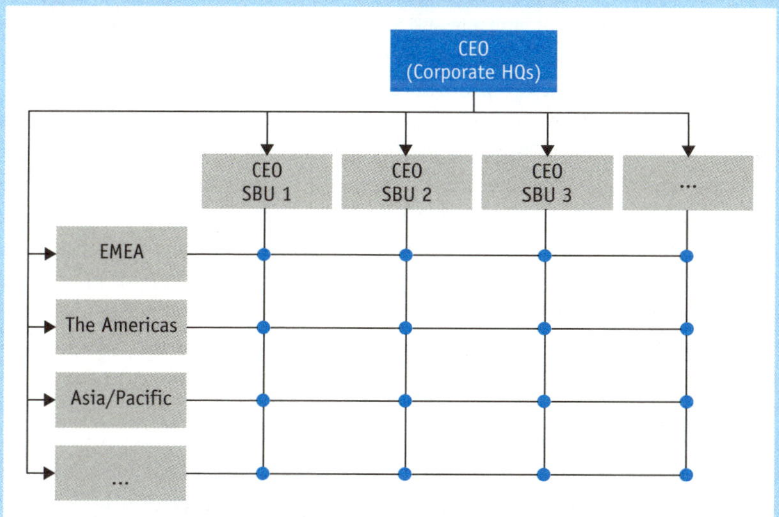

Fig. 4-17

Example of a tensor structure

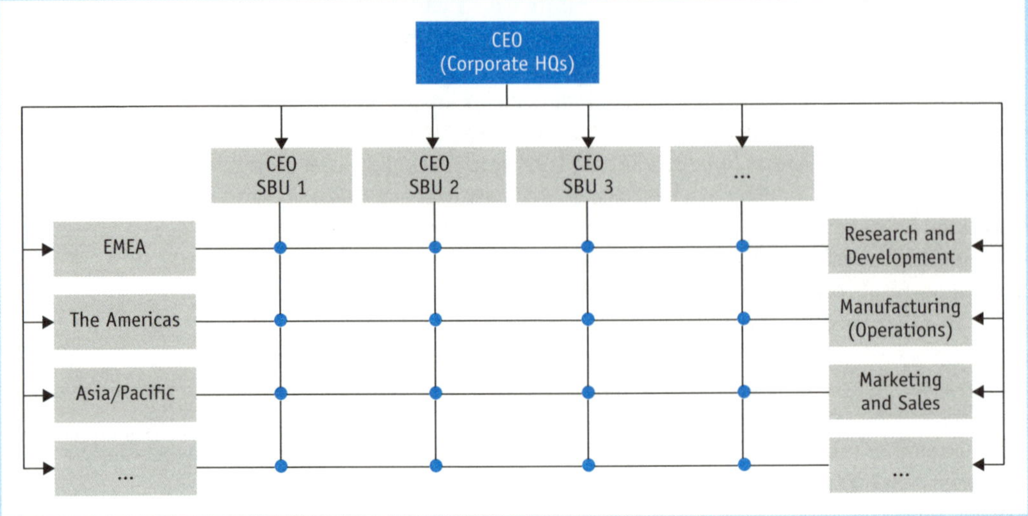

and industry centers at the second management level. For project-based firms this can be an effective structure as they constantly create temporary project teams for solving client problems that require different types of expertise. The advantages of matrix and tensor structures are their high degree of versatility and the enhanced interorganizational collaboration. Effectively implemented, they can support the alignment and execution of multiple and partly conflicting organizational strategies. Another advantage is a more efficient use of human resources since those structures may reduce idle time. Their key disadvantage lies in the increased effort for coordination and conflict resolution, particularly if they are not well installed. In such cases, managers are balancing multiple strategic roles and employees end up with two or three bosses and may not be sure whom they should be reporting to at any given time. Ambiguous reporting and leadership structures can undermine accountability, slow down decision making and increase bureaucracy. It can place a significant amount of stress on individuals, particularly if they do not like ambiguous situations and desire security. They may find themselves in the middle of power struggles between functional managers, SBU or product managers, and regional managers who fight over getting resources and the best talent assigned to their initiatives.

New Types of Organizational Structure

Network structure

Apart from the basic types, a number of new organizational structures have been developed that tend to have fewer layers of hierarchy and little to no in-house departments. One of those new design options is the network structure, a kind of "virtual" organization in which most or all primary functions of a business, such as R&D, Manufacturing, Marketing & Sales, etc., are outsourced and coordinated by a small group of executives. Managers in these types of organization spend most of their time managing the remaining in-house core activities and orchestrating external relationships or project groups that are not located in a single building or area but that are scattered around the world. A key benefit of network organizations is their increased flexibility and adaptability in coping with a high degree of dynamics, such as rapid technological changes, in the competitive arena. The organization focuses on those activities it can best perform based on its core competencies. Everything else might be managed with contractors for a limited period of time or service level agreements with affiliated but legally independent firms. A network structure allows individuals or smaller companies with innovative ideas but limited funds to start entrepreneurial ventures or compete with larger, more established companies. Firms belonging to a network organization can leverage resources and know-how of their affiliated network partners to better serve their own customers (see Strategy Practice Example 4.4). Furthermore, they might get access to markets and customers they would not be able to serve when acting alone. Disadvantages of the network structure are the difficult process of building trust, aligning potentially conflicting goals, and thus, avoiding tension among the various affiliated partners, a higher need for coordination and communication, the risk of losing know-how, and the danger of neglecting the firm's own strategy for the sake of the network.

Strategy Practice Example 4.4

Network Structure: Global Management Consulting Based on an Agile Partnership Model

"Cordence Worldwide is the only major global management consulting partnership made up of independent Member Firms, each of which is a leader in the regions it serves. By maintaining their independence, each Member Firm maintains focused accountability to clients; firsthand knowledge of relevant business and cultural context; the autonomy to do what's right for each client; and the ability to access the global expertise, resources, and infrastructure of Cordence Worldwide whenever and wherever they're needed. It's a uniquely agile consulting model that creates strong, responsive partnerships with clients and produces powerful business solutions all over the world.

How this model benefits clients

- Expertise: The combined global skills of Cordence Worldwide's Member Firms give clients access to cutting-edge knowledge that cuts across industries, geographies and business functions.
- Accountability: Each Member Firm is independently owned and operated, and answers directly to its clients—not a remote board of directors.
- Diversity: The complementary regional expertise and consulting capabilities of Cordence Worldwide's Member Firms put a diverse set of cutting-edge skills and knowledge at every client's disposal. And our consultants are part of their local cultures and communities. We speak our client's languages, figuratively and literally." (Cordence Worldwide 2014)

4.2.2.2 Business Processes

Value creating activities beyond departmental boundaries

Looking at business practice today, the dominant perspective on the structure of a company still seems to be primarily driven by hierarchical organizational charts with departments and vertical reporting lines. Companies tend to build the organizational structure first and supplement it with procedures and processes afterwards (Osterloh/Frost 2006: 31 f.). Whereas basic types of organizational structure relate to the configuration of hierarchy, the process organization of a firm refers to its value creating activities beyond departmental boundaries (see fig. 4-18). Fundamental principles and considerations when defining a process organization within a company will be explained in the following section.

Customer-driven view on the firm's activities

In general, a process is a set of interrelated activities that transform a defined input into a defined output. Processes are "structure for action" (Davenport 1993: 5). In this generic understanding, the goal of a process, its trigger, scope and content as well as the desired output and its recipient are left open (Schmelzer/Sesselmann 2013: 51). In a business context, a process can be defined as "a set of activities that, taken together, produce a result of value to a customer (...)." (Hammer/Champy 1993: 3) The key idea of (business) process orientation in a company is a horizontal customer-/market-driven perspective on the firm's business activities instead of a vertical view based on functions, products, regions, or projects (see fig. 4-18).

End-to-end perspective

A business process starts at a customer's desire and demand and ends with products and services that the customer values and pays for. In this context, an organization can be seen as a "value-producing machine" (Bergsmann 2012; Rummler/Ramias/Rummler 2010: 27 f.). A business process can be characterized by the following criteria (Bergsmann 2012: 17–21; Horváth 2011: 90; Mayer 2005: 3; Schmelzer/Sesselmann 2013: 52–57):

- Business processes have external or internal **customers** for whom value is created.
- The **customer needs, wants, demands and requirements** (see chapter 2.2.3.1) is the raison d'être of a business process. It starts at the customer and ends at the customer (see Strategic Snapshot 4.6). This has to be distinguished from the *event* that triggers a process on an operational level such as a request, order or complaint that comes in via phone, mail, e-mail, etc. Also, it does not have to be confused with the *input* of a process such as different types of resources or information.
- Business processes have defined tangible or intangible **outcomes** (results) in terms of value that is derived from the needs, demands and requirements of the external or internal customers. Outcomes might be products, services, information, or a combination of these. There might be additional *outputs* of a process on an operational level such as information, documentation, waste, etc. that are not part of the desired business process outcome in terms of customer value.
- A business process is a set or chain or **sequence of interrelated cross-functional activities** that are required to deliver the desired outcome for the customer.
- Business processes have defined tangible or intangible **inputs** such as human, financial and technical resources (material, parts, IT, tools, etc.), services, and information that are required to conduct the process and that are utilized to deliver the desired outcome. They are provided by external and internal suppliers at the beginning or at certain points within the process.

- A business process has certain requirements in terms of **process performance indicators and targets** regarding quality, cost and time that are tight to the desired outcome and are typically derived from the pursued business strategies or from benchmarks.
- Business processes have **procedures, rules and responsibilities** that are supposed to ensure an efficient and effective transformation of the inputs into the desired outcome.

Process categories

Processes within a company have different functions. Some are directly related to the market and the external customers whereas others have more of a supporting role. To take this into account, processes are typically categorized into primary and second-

Fig. 4-18

The business process idea

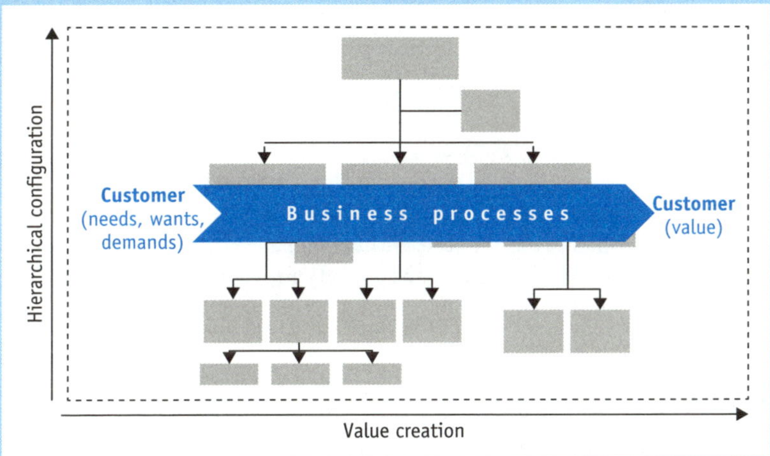

Strategic Snapshot 4.6

Michael Hammer on Customer-driven Business Processes

"Process is the way in which the abstract goal of putting customers first gets turned into its practical consequences. Without process, companies decay into a spiral of chaos and internal conflict. (...) From the customer's point of view, a company exists only to create value for them, to provide them with results. Yet in far too many companies, the actual creation and delivery of customer value is not the responsibility of any particular individual. One searches in vain for people who are focused on and responsible for the end-to-end work of filling customer orders; on seeing new products through from conception to realization; or on resolving customer problems. Instead, the work that creates results for customers is broken into pieces and scattered across numerous departments and units. In these companies, workers, managers, and departments focus on each of the steps that lead to creating results for customers, but no one focuses on all the steps together in a unit. One person takes the customer call, another gathers needed information, a third decides what is to be done, a fourth takes action, and no one oversees the whole thing. These companies (...) suffer from a crisis of processes." (Hammer 2001: 52f.)

ary processes. Primary processes are business or core processes that are directly creating value for external customers. Hereby, the term "business process" is used especially for this category whereas—depending on various process typologies from different authors and firms—the term may also be related to all processes of a company. Secondary processes include contributing or support processes as well as control or management processes (see fig. 4-19) and are in place to enable efficient and effective primary processes. Secondary processes typically address internal customers. The different categories of primary and secondary processes can be described as follows (Bergsmann 2012: 49–81; Schmelzer/Sesselmann 2013: 65–68; Seidenschwarz 2008: 61–64; Vahs 2012: 241 f.):

- A **business (core) process** creates value for external customers. It is usually something the company does exceedingly well. Those processes might also be labeled core processes as they represent the core of the business, that is value creation. An example is the process "sale of a machine", which is initiated by the external customer's need, want and demand and ends when the customer pays for the value delivered. In company practice, value creation within a business process might be further distinguished into processes such as product life cycle management, supply chain management and customer relationship management, although each of those "business processes" alone does not represent the complete end-to-end perspective.
- Business processes are enabled by **contributing (support) processes**. Frequently, the term "support processes" is used although this might imply that they are of lower priority, which is not the case. Without proper contributing processes, business processes as well as control (management) processes are not effective. Examples are processes that contribute financial resources, personnel, infrastructure, legal advice, etc.

Fig. 4-19

Key categories of processes and their relationships

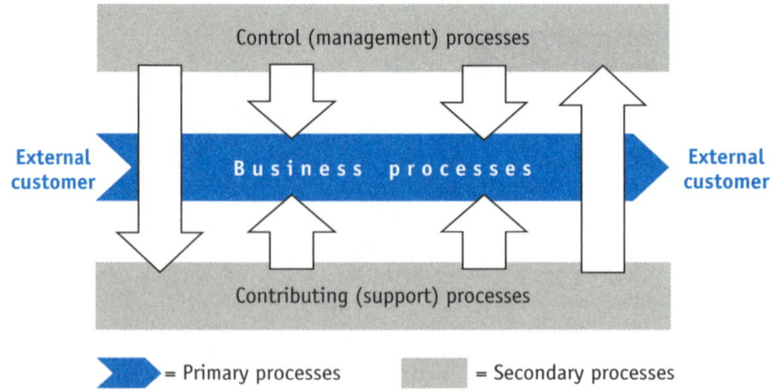

Source: Vahs 2012: 242; adapted

- **Control (management) processes** include strategic and operational planning as well as process performance measurement and management. Control processes are also called "management processes" and are required to continuously manage the primary and contributing processes regarding efficiency and effectiveness.

Process levels

Particularly in larger organizations with many business, contributing and control processes, the process organization can become fairly complex. In this case, it is advisable to describe and structure processes comprehensively across different hierarchical levels. Each process in the three process categories consists of a set of main processes (level 1), each main process of a number of subprocesses (level 2), and each subprocess of a set of activities (level 3) that are the smallest pieces that still fit the criteria of a "process". The corresponding levels and their application to the three different process categories are illustrated in figure 4-20. As an example, the business process of selling a customer-specific machine might have main processes such as sales, order processing, engineering, production, and service on level 1. The main process "sales" may consist of level 2 subprocesses like acquisition, handling a customer request, preparation of an offer, etc. Level 3 covers the different activities required for preparing an offer (the level 2 subprocess). The number of levels as well as the specific term for each level varies depending on the company or the applied reference model. Ideally, processes at all three levels should have a responsible process owner with clear tasks and authority.

Fig. 4-20

Process categories and process levels

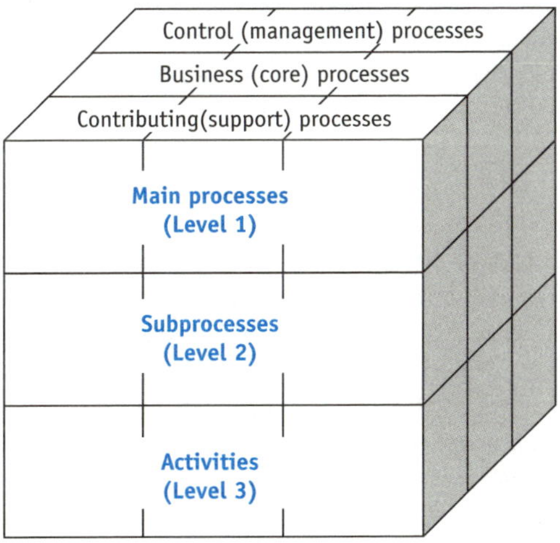

Source: Vahs 2012: 259; adapted

Processes of the different levels can be developed either top-down, bottom-up, or in a combination of both approaches (Schmelzer/Sesselmann: 140–143; Vahs 2012: 254 f.):

Top-down process definition

In a pure top-down approach, there are basically two ways of identifying business, contributing and control processes, that is based on the strategy or based on generic process reference models. In both cases, processes are decomposed top-down across the different levels. When processes are derived from the corresponding corporate or business **strategy**, the customer or market requirements and strategic goals are used as a guideline. Hereby, processes for achieving those requirements and company goals in the most effective and efficient ways are defined independently from the existing structure and its constraints. An example for such an approach is the development of the value chain in a way that enables generic strategies such as cost leadership or differentiation, which has been explained in chapter 3.4.2.1. Second, processes can also be identified and defined based on so-called **process reference models** (Schmelzer/Sesselmann 2013: 240–261). This is another deductive way where business processes are derived top-down based on available generic process frameworks, which typically provide not only top processes but also their decomposition across various levels. There are many different reference models available for specific industries, process categories (e. g., supply chain, value chain, operations, IT governance, finance, etc.), or software solutions. Some examples are:

- Supply Chain Operations Reference model (SCOR®), which is a cross-industry framework for supply chains provided by the Supply Chain Council (SCC 2012)
- IT Infrastructure Library (ITIL®), which is an IT service management framework (ITIL-IT Infrastructure Library 2014)
- Controlling Process Model of the International Group of Controlling, which is a hierarchical process model covering main processes for corporate performance management (IGC 2011)
- Value Creation Hierarchy (VCH) for organizing, performing and managing value-adding work in the entire company (Rummler/Ramias/Rummler 2010: 39–70).

Companies can use reference models as templates from which firm-specific process models are derived (see Strategy Practice Example 4.5).

Bottom-up process identification

Contrary to the strategy- or reference model-based process definition, a bottom-up approach follows an inductive logic and identifies processes within the existing organizational structure. It typically starts by analyzing relevant employee activities in various departments or cost centers. For practical reasons and to avoid unnecessary complexity, it might be useful to focus on repetitive and mainly formalized activities with certain minimum time thresholds (e. g., a certain minimum percentage of the employee's total working time). Activities might be analyzed based on existing documents, interviews with employees or their supervisors, self-recordings of employees, workshops, or observations. In the next step, activities are consolidated to subprocesses based on, for example, the requirements of internal or external customers on the subprocess, efficient resource utilization, and the avoidance of unnecessary interfaces. The same principle can be applied for consolidating various subprocesses into a main process.

Top-down/bottom-up combination

With a bottom-up process identification and consolidation, the "as is" process model of the entire organization can be developed. Compared to a top-down definition and de-composition, however, the bottom-up process identification typically requires much more effort in terms of time and cost. It can also lead to a suboptimal process model as it might be oriented too strongly on the existing organizational structure and not representing a true end-to-end customer orientation. As a result, it might be disconnected from the business and the strategy. Furthermore, it may not be useful for benchmarking as a means of process performance measurement and management as it is highly company-specific and not standardized. Therefore, a rough blueprint of main processes is typically defined top-down as an orientation for the bottom-up activity analysis and process consolidation. An application of such an approach is the introduction of an activity-based costing model as a base for analyzing product or customer profitability and for process cost based calculations (Mayer 2005, 1998).

Company-specific process model and architecture

All processes and activities within a company are captured in a so-called process model, which should be geared toward meeting external customers' needs, demands and requirements. A process model consists of an individually (i. e. company-specific) defined number of hierarchy levels up to the point where the effort to add another layer of detail does not justify the benefits. In company practice up to four hierarchy levels are frequently found. The different process categories and levels and their interrelations within this hierarchy constitutes the process architecture.

A process map provides a high-level overview or master plan over all relevant processes and their basic relationships. It represents the highest level in the process hierarchy. Each process of the process map can be drilled down several layers in the process hierarchy. For designing a process architecture it is critical to ensure vertical consistency regarding this drill-down from rough to detailed processes and activities. Figure 4-21 shows the basic principle of a process architecture with the analogy of a world map that is zoomed in more and more at each level.

Process variants

When defining business processes, it might be necessary to design process variants (Bergsmann 2012: 91; Schmelzer/Sesselmann 2013: 161 f.): This is the case if the standard procedure within a business process for delivering a desired outcome needs to be altered depending on the type or category of customer, product and service, order, sales channel, etc. A process "deliver product", for example, might require all subprocesses and steps for customer category A, a different combination for customer category B, and another different combination of subprocesses and steps for customer category C. In this example, the process "deliver product" would have three process variants. Other examples for process variants are:

- preparing routine offers for standard products that are sold in high quantities versus specific offers for some technically comprehensive products
- sale of a product to key accounts versus to small customers
- development of standard products versus customer-specific development of a product
- differences regarding make or buy for certain products in the production process.

Fig. 4-21

Multi-level process architecture

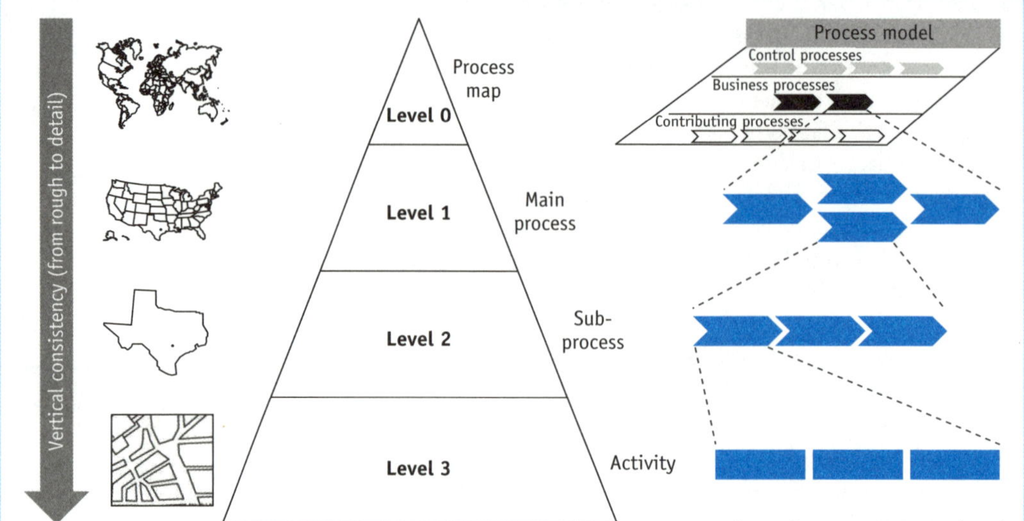

Source: Bergsmann 2012: 93; adapted

When defining process variants, it needs to be decided at which level in the process hierarchy the standard procedure is altered. A typical approach is to differentiate at the level of subprocesses. By defining the differing procedures clearly for each process variant, complex exception clauses for standard processes can be avoided. On the one hand, this enhances transparency of the entire process model, and thus, simplifies process management. On the other hand, increasing the number of process variants requires additional documentation and makes it more difficult to coordinate the different processes.

Process standardization

A key challenge when designing a process organization is the optimal balance between process standardization and customization. On the one hand, standardization drives process efficiency and enables specific strategic goals. From an operational point of view, standardization of processes can lead to improvements in cost, time and quality particularly at highly repetitive processes. From a strategic point of view, process standardization might be a vital lever for achieving strategic goals. For example, it fosters a strong business portfolio with standardized management processes or innovation and differentiation based on standardized technology platforms as part of a product life cycle process. Furthermore, operational excellence can be driven with, for example, a standardized supply chain management process. Process standardization in multinational companies helps to realize company-wide synergy potentials, benchmark processes internally and externally, multiply best-practices and reduces the risk for reinventing the wheel at each international subsidiary (Seidenschwarz 2008: 53–63).

Process customization

On the other hand, process standardization might limit flexibility and customer orientation, and thus, hinder the realization of competitive advantage. As one principle of the underlying strategy concept is about being different (see chapter 1.1.2.3) the standardization of processes might be counter-productive as it does not reflect customer and business specific requirements. Furthermore, it can be imitated easily by competitors. One example might be the renowned Toyota Production System (TPS). It used to be a unique source of competitive advantage for Toyota by realizing enormous production efficiencies. Over time, however, many other companies including competitors adapted this model as an industry standard, and thus, competitive advantage based on efficiency has diminished over the course of time. In times of buyer markets, companies need to be highly responsive to the specific wants and requirements of their target groups. If customers need to be addressed highly individually and not as an anonymous mass, companies are forced to meet customer requirements in the best possible way through process customization.

Combining process standardization and customization

When defining a company-wide process model (see Strategy Practice Example 4.5), the optimal balance between process standardization and customization needs to be considered. Multinational companies which manage their international subsidiaries with a high degree of freedom and decentralized decision power typically struggle when they want to exploit the advantages of process standardization across their organizational boundaries. One of the key arguments found in company practice is that "things are different" on a local subsidiary level, and thus, require customized processes. On the other hand, there might be pressure for standardized and documented processes coming also from a control perspective. For example, auditors of financial processes typically see a risk that misguided (or malicious) employees may cause a problem in looser and more customized processes. Standardizing business processes that reach across different subsidiaries or strategic business units is particularly challenging when the multinational enterprise grew through acquiring companies all over the world that operate with different IT applications (legacy systems) and different related process definitions and architectures. Particularly for organizations that are managed like strategic or operational management holdings firms (see Strategic Snapshot 4.5), globally highly heterogeneous process models make communication and collaboration between the different organizational units difficult and might lead to strategic misalignment and unexploited potential for operational efficiency. For example, harmonizing the global IT-tool landscape with standardized processes can be a huge lever for cost savings as up to 90% of the IT budget is usually necessary just for maintaining and managing the functionality of a complex tool infrastructure. A way to tackle this issue is by defining binding process standards for the higher levels in the process architecture and to leave freedom for local customization at the lower level subprocesses (Schmelzer/Sesselmann 2013: 255f.). To avoid over-standardization, this might be best achieved by a top-down, bottom-up counterflow procedure where high-level processes are standardized based on a reference model or strategic goals, best practices are identified at customized lower-level processes and fed back to become a company-wide standard in the process model (Seidenschwarz 2008: 75–77).

Strategy Practice Example 4.5

Process Model of the Siemens AG

The Siemens Process Framework (SPF) provides all organizational units and affiliated companies within the Siemens AG with a common framework for the design and management of business processes. It is based on the three elements Siemens Reference Process House, Process Modeling Methods and Process Management (see fig. 4-22).
Reference processes are defined and described in the *Reference Process House*. Those business as well as the management and support processes—the primary and secondary processes—are the binding foundation for process management in the different divisions, business units, regions, and the headquarters (corporate) of Siemens. They are used for leveraging best practices and for benchmarking as a base for process optimization. Three types of primary processes are distinguished focusing on customers, orders and products: customer relationship management (CRM), supply chain management (SCM), and product life cycle management (PLM).
Part of the element *Process Modeling Methods* is a level concept. The different process levels in the architecture of the Reference House are derived from the SCOR® reference framework. Level 0 to 3 are standardized and compulsory for all organizational units. Starting at level 4, processes might be customized to reflect unit-specific or regional requirements. This way both benefits of process standardization and customization are pursued. Furthermore, predefined conventions and methods ensure a common modeling and documentation of processes (convention manual and modeling manual). This ensures a

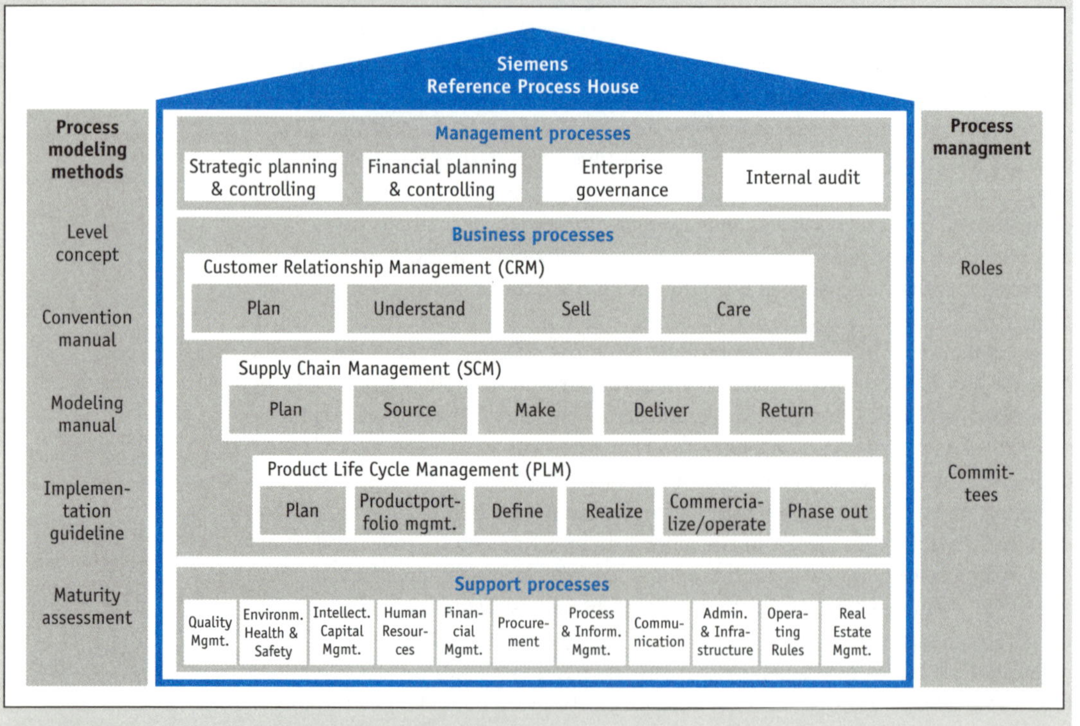

Fig. 4-22: The Siemens Process Framework (SPF)

common group-wide process language and understanding. Based on an implementation guide and a corresponding intranet-based knowledge hub for best practices, divisions and regions are enabled to establish and align their corresponding processes. To control the implementation of the principles of the Siemens Process Framework, a process management maturity model is applied where the maturity of processes can be assessed based on the five maturity stages initial, managed, defined, quantitatively managed, and optimizing with clearly defined characteristics for each stage.
A clear and company-wide applied definition of common roles and committees for *Process Management* is the third pillar of SPF. Those roles are process sponsor, process executive and process framework executive on a strategic level, and process owner and process manager on an operational level. With the defined roles and committees, clear responsibilities are specified for designing and anchoring process models, modifying them within in the given scope, and continuously optimizing processes according the SPF principles. This can be seen as a vital foundation for making process management successful.

(Source: Schmelzer/Sesselmann 2013: 256-260; Seidenschwarz 2008: 67–70)

Primary and secondary organization

Managing processes requires clearly defined process responsibilities that need to be aligned with the corresponding organizational structure (see 4.2.2.1). Hereby, the key design question is how much emphasis the company puts on its value creation activities (i. e. process orientation) versus its hierarchical configuration. How much process orientation should be integrated in the organizational structure and which element comes first? In business practice there is hardly any company with a structure purely based on processes (Schmelzer/Sesselmann 2010: 182; Jochem/Landgraf 2010: 7). Two basic design options are (Fink 2003: 58 f.; Gaitanides 2012: 26–46; Nicolai 2009: 185 f.; Vahs 2012: 251 f.):

- In a *process-oriented organization* or *secondary process organization (SPO)*, processes are a secondary dimension that enhances or overlays the organizational structure as a primary dimension. This leads to a process-oriented matrix structure or, in case the primary organizational structure is already a matrix, to a process-oriented tensor structure. This type of structure can be frequently found in companies who want to combine advantages of the corresponding primary organizational structure with the benefits of a process orientation. Process-oriented organizations can be facilitated with projects, own staff for processes, or by additional responsibilities of line managers.
- In a *pure process organization* or *primary process organization (PPO)*, the entire company is viewed as a wealth of interdependent processes. The smallest organizational unit is not a job but a process that needs to fulfill its task on its own authority. This type of structure requires entrepreneurial process managers and employees who are able to give up traditional thinking regarding organizational structures, flexible and highly customer-oriented.

Process management as critical success factor

During the last decades, the process perspective on an organization and the effective management of processes have become a critical success factors. Changing market conditions such as increased international competition, changing customer preferences, and technological developments force companies to achieve competitive cost structures, gain flexibility, drive innovation as well as ensure responsiveness and deliver maximum value to its customers. Old rules for organizational design that go back to Adam Smith (1723–1790) and Frederick W. Taylor (1856–1915) and that are used to drive efficiency gains through specialization seem to be outdated. Specialization creates interfaces causing organizational problems that interfere with today's requirements for competitiveness and profitability, such as being lean, flexible, innovative, responsive, and highly customer-focused (Hammer/Champy 1993: 7–30; Osterloh/Frost 2006: 17–25).

Figure 4-23 shows a comprehensive framework for process management that will be used for explaining relevant elements that need to be considered when designing and managing a process organization (Mayer/Brenner 2005: 19–23):

Strategic process alignment and benchmarking

On a highly aggregated firm level, the following five process areas need to be elaborated on: management process, innovation process, operations and supply chain process, market and customer relationship process, and the internal services process. Requirements for the design of those processes as well as their performance level are derived in two ways. First, the corporate or business strategy determines which processes are particularly relevant and where there is need for action. Key processes need to be aligned with the strategy (see fig. 4-31 in chapter 4.2.2.4). Second, best-in-class process designs and performance derived from benchmarking provides important information of requirements that need to be fulfilled for getting a leading position in the market.

Fig. 4-23

Horváth & Partners process management framework

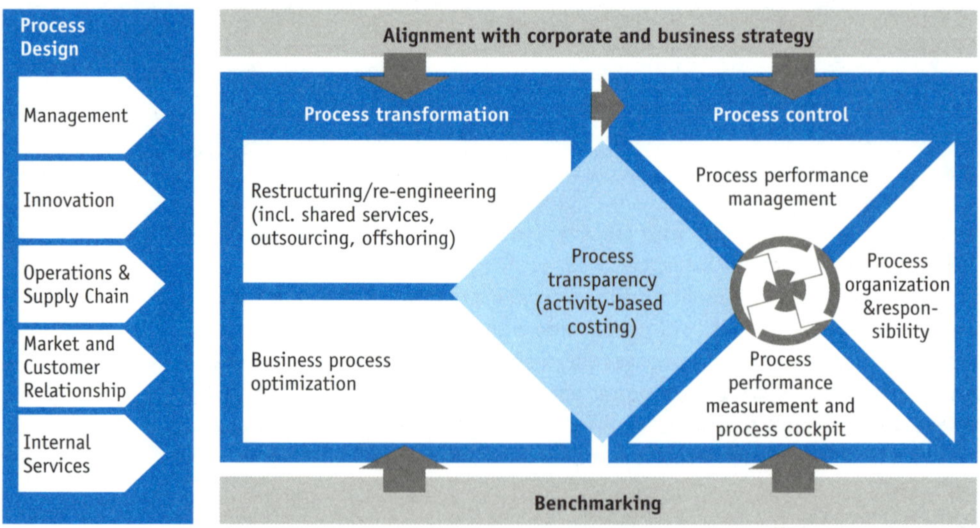

Source: Mayer/Brenner 2005: 19; adapted

At the core of the process management framework are three building blocks: process transformation, process control, and process transparency. Process transparency is a prerequisite for optimizing and controlling processes. It covers methods and tools for modeling and analyzing processes to provide information about cost (e. g., activity based costing), time and quality as well as corresponding areas for improvement. Process transformation refers to the design and change of processes, which can be achieved in two basic ways: through restructuring or re-engineering, and through business process optimization.

Restructuring and re-engineering

Restructuring or re-engineering is a radical redesign of business processes to achieve major cost, quality and speed gains. It covers a redesign of existing processes, the modeling of new processes, and includes also concepts like outsourcing (i. e. contracting of processes to another external or internal party), Shared Service Centers (i. e. entity responsible for specialized processes such as accounting, payroll, recruiting, purchasing, etc. as a service shared by all divisions and units of an organization) and offshoring (relocation of processes to another country). Instead of fine-tuning existing processes, re-engineering calls for designing or redesigning processes in a completely new way following—in its most radical version—a greenfield approach based on top-down company goals (Hammer/Champy 1993). Although there are a number of firms that were able to strongly improve their cost position and competitiveness with re-engineering, other organizations report mixed results when looking at the performance gains caused by their company-specific re-engineering programs (Hess/Schuller 2005; Wheelen/Hunger 2010: 289). Nowadays, the emphasis of process management tends to be less on the radical approach but more on an optimization and continuous improvement of existing processes (Hammer 2001: 52).

Business process optimization

Other than with restructuring and re-engineering, business process optimization starts at existing processes and tends to be less radical. The goal is to systematically identify areas for improvements based on strategic requirements, benchmarks, or current issues such as failure rates, service reliability, quality issue, long lead times, or cost running off course. The approach is team-oriented and consensus-driven. People from different areas that are involved in the process jointly develop ideas and implement measures to optimize current processes regarding time, cost and quality in so-called process optimization teams (POTs). This way, the detailed process know-how of employees is leveraged and their buy-in for the necessary changes is fostered. Based on this approach, 10 % to over 50 % of cost savings or lead time improvements regarding the defined base lines have been reported (Kieninger 1998: 44). There are a variety of effective production oriented efficiency methods such as Six Sigma or Lean Management that are increasingly applied to optimize business processes in the entire organization. (Wunder/Bausch 2015)

Process performance measurement and management

A onetime process transformation needs to be supplemented by process control to continuously ensure and improve process performance. This requires process performance measurement and process performance management. The generation of financial and non-financial indicators and their consolidation in, for example, a process cockpit should provide a fast and reliable view on the process performance. Based on those indicators, targets are set and corresponding initiatives to achieve the targets are defined and managed. The performance is regularly monitored and, in case of deviations, corrective actions are defined. Furthermore, resources that are required for running the processes need to be planned and allocated.

4.2.2.3 Managerial Systems

Internal governance mechanisms

The third and final key element when designing an organization that is supposed to enable a strategy are the managerial systems. Those are internal governance mechanisms covering an integrated set of processes, procedures and tools that are utilized for specific management tasks. They are strongly related to some of the control (management) and contributing (support) processes explained in the previous chapter. Managerial systems need to be linked to strategy to direct people's attention on strategic priorities, drive people's actions in the desired strategic direction and ultimately focus the entire organization on strategy. There are many different managerial systems in company practice. Three of them tend to be particularly important when it comes to strategy execution:

- planning and budgeting system
- human resource system
- management reporting system.

Practical key considerations for those managerial systems will be explained subsequently but not discussed in detail, given that such methods are well established and adequately covered in corresponding publications. Approaches for the challenging task of aligning those systems to strategy will be explained later in chapter 4.3.6.

Planning and Budgeting System

Strategic planning

A planning system can be distinguished into strategic planning and operational planning and budgeting. The strategic planning system deals with defining long-range plans for where the company strives to be in three, five, ten or even more years including the corresponding targets and strategic programs. In many companies it operates on an annual cycle where senior executives go off-site sometime in the middle of the fiscal year and engage in strategic discussions for several days. Usually facilitated by senior strategic planning or corporate development managers and, sometimes, with the support of external consultants, they develop a strategic document that is supposed to provide strategic guidance for the company and direct behavior as well as resources toward strategic priorities. Usually separate from the high-level strategic planning system run by strategy or corporate development staff, there is an operational planning and budgeting system operating throughout the year that is run by finance staff to set the primarily financial operational targets for the next annual planning period. Strategic planning and operational planning and budgeting are frequently treated independently (Kaplan/Norton 1996: 247; Mintzberg 2000). To make a strategy come true, though, those systems need to be strongly integrated along with other important management systems (see chapter 4.3.6.1).

Operational planning & budgeting funnels limited resources

A key system that is involved in the allocation of resources is the operational planning and budgeting system. It has a long tradition starting at the beginning of the 20th century (see Strategic Snapshot 4.7) and still is a standard managerial system in most companies around the world. With the introduction of budgets, business leaders wanted to make their increasingly diversified and complex large corporations "manageable" (Weber/Schäffer 2011: 285). Budgeting is one of the most critical systems in terms of strategy execution as it funnels limited resources (financial funds, manpower, material, etc.) into those activities that a company expects to have the strongest impact on its long- and short-term business results (e.g., revenue, profitability) and that it needs for running its operations. In other words, it tells "who gets which resources" and "who is expected to generate which revenues". In company practice, the term "budget" is often related to primarily financial targets and sometimes referred to only to fix annual operational funds that a company expects to be required for hitting the corresponding annual sales budget. "Traditionally," say Kaplan and Norton, "the annual budgeting process establishes detailed short-term targets for financial measures, such as sales, operating expenses, gross margin, general and administrative expenses, operating margin, net profit, cash flow, and return on investment. It also establishes and authorizes spending levels for capital investments, research and development, and for marketing and promotional activities." (Kaplan/Norton 1996: 248) In general, though, budgets are either fixed or flexible (flex budget) and can be derived for (Horváth 2011: 204)

- functions, processes, products, regions, or projects as well as hierarchical levels
- time periods or intervals such as months, quarters, years, or multiple years
- expenses, cost, contribution margins, revenues (sales).

Furthermore, companies usually distinguish their budgeting in two main categories: capital budgeting and operational budgeting.

Capital and operating budgets

The capital budgeting system relates to the planning and managing of a firm's long-term investments. The type of capital budgets that typically need to be considered depends in part on the nature of a company's business. It may include funding of research and development programs, new product development, new ventures or physical assets. For example, the decisions of a large food discounter whether to open another store, an automaker whether to invest in a new machine, a software company whether to develop a new type of enterprise resource planning (ERP) system or a large

Strategic Snapshot 4.7

An Early Plea for Budgeting from McKinsey Founder James O. McKinsey

"Modern business administration tends more and more to become a standardized routine. In a large organization such standardization is essential to the maintenance of a unified business policy and to the coordination of the activities of the several departments; and coordination means subordination to a common head. Business men are gradually coming to realize that this can best be accomplished by the formulation of plans submitted for approval in black and white, if indeed coordination can ever be accomplished in any other way for a great length of time. (...) All of which goes to show that there should be some systematic method of gathering information from the past and formulating on this basis plans for the future, and of subsequently reporting how these plans have been carried out. Such an accounting and statistical organization we may call a budget system." (Source: McKinsey 1922: 10f. quoted in Weber/Schäffer 2011: 285) pharmaceutical company whether to conduct research to find a new molecule for curing a certain disease would be capital budgeting decisions. For the most part, capital budgets are discretionary funds derived from the company's strategy that are typically evaluated in terms of size, timing and risk of future cash flows. Operating budgets, however, are related to funding the day-to-day operations. They specify the ongoing expenses (e.g., labor, material, outside services, travel, fees, taxes, etc.) that are expected to be incurred for the goods and services to be produced and delivered to customers as well as a forecast of the revenues that are expected from sales of those goods and services. (Bragg 2002: 258–292; Rainey: 2010: 464 f.; Ross/Westerfield/Jordan 2010: 2 f.)

Annual budgeting process and control

During the budgeting process, managers are generally requesting an amount of financial funds and human resources that are required to execute their desired operational activities and to make investments. Operational budgeting is usually done on an annual basis. It might include requests for acquiring equipment and material, hiring personnel, advertisement and promotion, traveling and entertainment, assigning manufacturing capacities to certain products or rooms of training facilities for employee trainings, etc. in the following fiscal year (annual budget). Those requirements regarding financials, headcount and physical resources are outlined in budget plans. Then, once the budget is approved, the plans will usually be reviewed on a regular basis (e.g., monthly or bi-monthly) by the manager in charge in order to ensure that the actually required amount for what was planned is not exceeding the budget. Allocating resources is one of the key roles of any business leader. Comparing the results of a business or organizational unit to their budgets is a common tool for determining the performance of the corresponding managers in charge.

Master budget (pro forma statement)

A total view of the various key budget types and their logical flow is provided by the master budget. "A master budget summarizes the planned activities of all subunits of an organization—sales, production, distribution, and finance. The master budget quantifies targets for sales, cost-driver activity, purchases, production, net income, and cash position, and any other objective that management specifies. Thus, the master budget is a periodic business plan that includes a coordinated set of detailed operating schedules and financial statements. It includes forecasts of sales, expenses, cash receipts and disbursements, and balance sheets. Sometimes master budgets are also called pro forma statements, another term for forecasted financial statements." (Horngren/Sundem/Selto 1993: 223) The master budget distinguishes between the operating budget or profit plan and the financial budget. Whereas the operating budget focuses on different plans that lead to a budgeted profit and loss statement, the financial budget focuses on cash and how it is impacted by the operating budget's effect as well as other plans such as capital budgets (see fig. 4-24).

Shortcomings of traditional budgeting

While the basic principles of budgeting seem simple and clear, in business reality it is a complicated system that leads to a complex web of spreadsheets capturing activities of virtually all areas within a company. In recent years, traditional budgeting has been widely criticized as being inefficient, disconnected from the reality of dynamic and highly competitive marketplaces, and leading rather to dysfunctional instead of entrepreneurial behavior of managers and employees (see Strategic Snapshot 4.8). Some of the criticism regarding budgeting in its traditional form, as

Fig. 4-24

Master budget of a manufacturing company

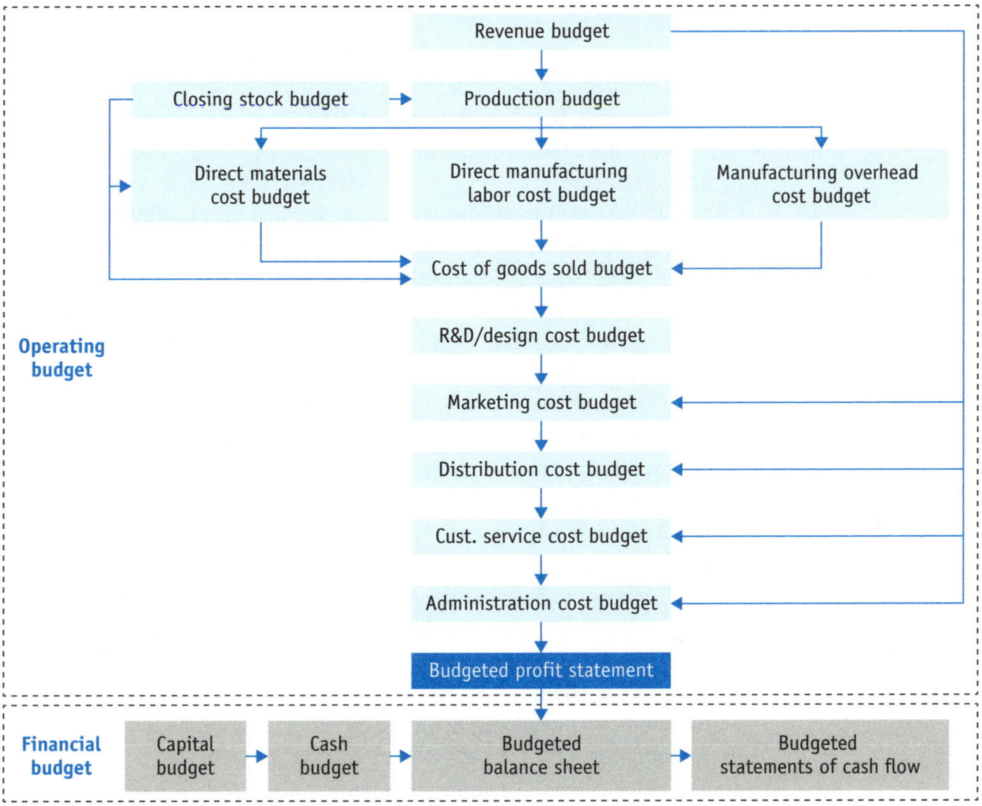

Source: Bhimani et al. 2008: 474 modified and quoted in Weber/Schäffer 2011: 474; adapted

explained before, are (Gleich/Kopp/Leyk 2003: 315 f.; Hope/Fraser 2003: 3–15; Horváth/Sauter 2004: 3):

- **Disconnect from strategic planning:** Approved budgets do not seriously take into account the amount of funds required for achieving mid- or long-term strategic goals as defined in the strategic plan. As a result, business managers might have difficulties in executing strategic priorities as they do not have the resources to excel.
- **Fixed time period:** Budgets are typically set and approved for a fixed time period such as one year. Even when budgets are adjusted based on forecasts, the numbers are usually forecasted toward the end of the fiscal year and not beyond. This means that from approving the plan toward year-end, the control period, that is the time for corrective actions in case of deviations from the plan, gets shorter.

- **Outdated assumptions:** Annual budgets do not reflect changes in the competitive environment between the plan approval and the end of the planned period. For a fiscal year starting in January and ending in December, e. g., budgets are generally developed and approved in the second half of the previous year. In company practice this could mean that financial revenue or cost targets are set up to 18 months prior the time when the business activities actually occur. During the fiscal year, companies may notice that planned numbers are not appropriate anymore due to changes in the company's environment.
- **Focus on financials:** Traditional operational planning systems are commonly focused on financial targets (i. e. budgets) only. Non-financial performance indicators such as the number of new customers or customer loyalty, on-time delivery, process cycle time, speed to market new products, or employee skill levels are often not reflected in the planning system. If this is the case, it is a severe shortfall from a managerial point of view as these elements are often leading indicators for the lagging financial results. Anecdotally spoken, you cannot properly steer a car if you only look in the rear mirror.
- **Internal perspective:** Whereas strategic planning is typically based on internal and external factors such as market developments and competitor behavior, the perspective of operational planning and budgeting tends to be purely internal. Budgets are frequently derived as a forward projection of previous year budgets with a lump sum adjustment. This approach is particularly applied when cost budgets are planned. Those are often based on the previous year plan or actual numbers instead of deriving them with an outside-in perspective based on comparisons with other companies (benchmarks).
- **Motivation of the wrong behavior:** Traditional budgeting fosters bureaucracy and prevents people from acting in their company's best interest. Instead of elaborating on how to best execute strategic priorities and allocating funds accordingly, budgeting tends to reflect more of a gambling exercise. This could lead to managers focusing on "making their numbers" instead of "beating competitors", deliberately lowball financial targets (budget slack) to increase the probability of getting rewarded for "over-performance", and completely spending their approved funds during the fiscal year (budget wasting) to avoid getting budget cuts in the next planning period (see Strategic Snapshot 4.8).

The described issues with many traditional budgeting systems are serious challenges for establishing an effective internal governance system. On the one hand, budgeting is one of the most critical systems for allocating limited resources, and thus, enabling or preventing important activities in an organization. On the other hand, it is historically burdened with dysfunctional mechanisms that have to be overcome. Different types of advancements have been developed as a reaction to the shortfalls of traditional budgeting systems and are supposed to overcome some if not all of its limitations: beyond budgeting, better budgeting, and advanced budgeting (Gleich/Kopp/Leyk 2003: 316–326; Weber/Schäffer 2011: 294–297).

Beyond budgeting

The beyond budgeting approach is based on the idea of radically abandoning budgets as an element of planning and control. Traditional planning processes are

Strategic Snapshot 4.8

Jack Welch on Shortcomings of Budgeting: Disconnect from Reality

"The budgeting process as it currently stands at most companies does exactly what you'd never want. It hides growth opportunities. It promotes bad behavior—especially when market conditions change midstream and people still try to 'make the number'. And it has an uncanny way of sucking the energy and fun out of an organization.

Why? Because most budgeting is disconnected from reality. It (budgeting) begins in the early fall when people in the field start the long slog of constructing the next year's highly detailed financial plans to make their case to the company brass. The goal of the people in the field, of course, is unstated and laser-like: Come up with targets that they absolutely, positively think they can hit. After all, that's how they are rewarded. So they construct plans with layer upon layer of conservative thinking.

Meanwhile, back in the headquarters, executives are also preparing for the budget review but exactly the opposite agenda. They're rewarded for big increase in sales and earnings, so they want targets that push limits.

(After heated negotiations, both sides compromise in the middle) ... What's wrong with this picture? First, what you see: an orchestrated compromise. More important what you don't (see): a rich, expansive conversation about growth opportunities, especially high risk ones. ... So why in the world would they ever dream big?

They won't unless a new reward system is put in place. One in which bonuses are not based on a negotiated number but on real-world measures: how the business performed compared with the previous year and how it did compare with the competition." (Source: Welch/Welch 2006 quoted in Zenie 2011: 51)

replaced with highly flexible and adaptive managerial processes and structures that are based on different principles such as relative instead of absolute targets, continuous rolling forecasts, highly decentralized organizational structures with self-governance processes based on internal market mechanisms as well as relative, team-based reward and incentive systems. To make this happen, beyond budgeting is based on core management tools such as balanced scorecard (see chapter 4.3), activity based budgeting, benchmarking, and rolling planning. One of the main criticisms regarding beyond budgeting is its limited practical feasibility given the radical change required to transform the structure and culture of an organization. Some managers that tried to abandon budgets based on beyond budgeting principles have reported that they reintroduced them after a while (Boesen 2005: 24).

Better budgeting and advanced budgeting

Whereas beyond budgeting is a radical approach of overcoming the shortfalls of traditional budgeting, better or advanced budgeting concepts are following an evolutionary approach. These management methods do not fundamentally question the application of budgets at all but suggest a step-by-step simplification and optimization of traditional planning and budgeting processes. Guiding key principles for improving both effectiveness and efficiency of operational planning and budgeting are (Gleich/Kopp 2001; Horváth/Sauter 2004: 4 f.):

- **Align budgeting with the company's strategic priorities** through integrating strategic and operational planning instead of operating with highly autonomous managerial systems. Strategy-oriented target settings should be supported by internal and external benchmarks.
- **Focus on the relevant, not the detail** through the use of aggregated budgets. Hereby, the planning effort is adjusted based on business relevance (cost/benefits) instead of budgeting everything at the same level of detail.

- **Apply a moving constant planning horizon** such as rolling quarterly forecast instead of fixed year-end budgets. This way, budgets are less outdated and better reflect the dynamics of the business environment.
- **Link relevant non-financial performance indicators** to budgeting as those are performance drivers and leading indicators for financial results.
- **Use relative, self-adjusting targets** for rewarding people. This fosters an outside-in perspective on performance and drives behavior toward winning in the competitive arena instead of making the defined budget numbers.

Based on those principles, management attention is shifted from detailed planning of annual budgets to elaborating on relevant forecast and action planning. Apart from an improved effectiveness, those means are also supposed to reduce the duration of operational planning from, for example, 3–6 months to 6–12 weeks. How to implement these principles through separating operational from strategic expenditures and applying different planning mechanisms for each will be explained more in detail in chapter 4.3.6.1.

Human Resource System

Managing the type and behavior of human capital

The human resource (HR) system relates to the type of individuals a company decides to employ as well as the policies, practices, and systems that have an impact on employees' behavior, attitudes, and performance. It considers the organization's employees—its human capital—as crucial resources of the company as their skills and competencies, experience, judgment, intelligence, relationships, and insights significantly contribute to the firm's economic success. Important practices for managing human resources include (Noe et al. 2009: 1–3)

- job design
- HR planning
- recruiting and retention (attracting potential and keeping current employees)
- selection (choosing employees)
- training and personal development for the future
- individual performance appraisal and management
- rewarding and incentivizing people (compensation and incentives)
- creating a positive work environment (employee relations).

Of particular importance for managing human resources to best support an organization's strategy is creating awareness of strategic priorities through effective communication, aligning the performance management system, and ensuring that HR and personal development plans for leaders and employees are in line with strategic requirements (see chapter 4.3.6.2).

Creating effective internal communication

Effective internal communication is crucial for conveying an organization's vision, mission and values, its strategies and explicit culture as well as for staying on top of employees concerns and suggestions. Clear, compelling and honest internal communication can be a motivational tool, and thus, influences employees' behavior, attitudes and, ultimately, performance. Keeping people informed of what is going on within the organization and how the company they are working for is performing is likely to create mutual trust and appreciation and may well affect employee retention and satisfaction in a positive way. The formal internal communication system can take many forms including e-mails, intranet, newsletters, messaging boards as well as different types of meetings. From a strategic point of view, employees have to be kept up with the strategic priorities of their company or business area as it is them who execute plans and strategies. Thus, managing human resources is not only about teaching employees how to perform their jobs but also about engaging them with effective and inspiring communication.

Performance management system

A performance management system covers formal means of "establishing performance standards and evaluating performance in order to arrive at objective human resource decisions—such as pay increases and training needs—as well as to provide documentation to support any personnel action" (Robbins/DeCenzo 2008: 171 f.). Whereas enterprise performance management (EPM) or corporate performance management (CPM) relates to managing performance of the entire company and includes elements of strategic management, business planning and financial management, employee performance management is a subsystem within the human resource system. Its primary goal is to encourage employees to achieve performance levels that the organization considers desirable and to help correct poor performance. In company practice, performance management is typically a recurring cycle of defining expectations, developing critical skills and capabilities to meet those expectations, reviewing whether defined expectations are fulfilled and skill development plans are successful and, finally, rewarding achievement of the predefined outcomes, which is a kind of incentive-based performance feedback (see fig. 4-25).

Performance management cycle

The cycle starts by clarifying which specific goals of the company or business area an individual or group is expected to contribute to. Those business priorities are the foundation for deriving specific measurable goals for an employee that may combine strategic (e. g., contributing to a cross-functional strategic project for expand-

Fig. 4-25

Performance management cycle with four stages

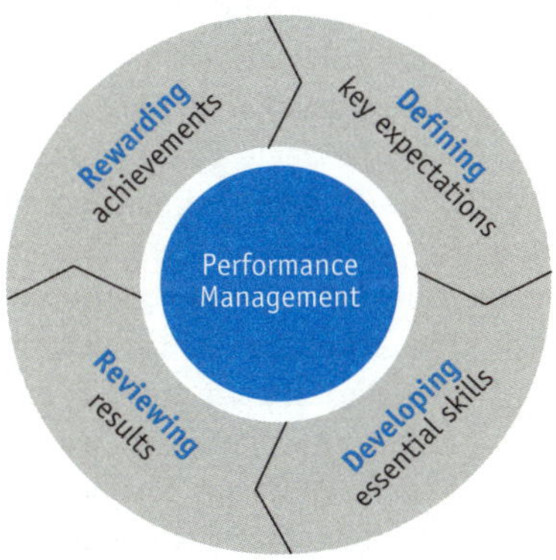

ing market share in Asia), operational (e. g., performance improvement relative to the employees job description) and individual (e. g., improving foreign language skills such as Chinese) elements. The second stage is related to developing the essential skills that the employee needs to perform successfully in the short or long term, based on the defined expectations. Corresponding goals in order to close a skill gap may be defined within the individual component of stage one. Development involves acquiring knowledge, capabilities and behavior, and occurs on the job through work assignments, ongoing feedback, formal and informal reviews as well as planned education, training or development programs. The final stage is rewarding the achievement of predefined expectations by incentive pay, promotion or other means of rewards such as awards. Criteria for effective performance management include (Noe et al. 2009: 215–218)

- fit with the company and business area strategy, that is performance measures are linked to the corresponding strategic goals, targets or strategic initiatives (see also chapter 4.3.6.2)
- validity (actually measuring what is intended to be measured) and reliability (consistency of results, for example, when more than one person is measuring or when being retested over time)
- acceptability by the people who are using the measures (if employees believe it is unfair, they will not accept feedback and might become frustrated in case of neglected incentive pay)
- clarity of what is expected as otherwise employees will neither know how to help the organization to achieve its goals specifically nor how to improve individually.

Performance appraisal and rewards/incentives

Performance appraisal methods and the corresponding rewards and incentives have an effect on people. For example, the impact of bonuses that are tied to monthly or quarterly sales figures is evident in the behavior of sales people in certain industries such as the automotive trade. Toward the end of the month or quarter, those sales people either want to "make their numbers", that is achieve their sales objectives to get their sales commission or they shift their sales pitches to the next period if they already hit the required targets. Appropriate behavior is not only encouraged through financial compensation or bonuses but also through promotions or symbolic appreciations such as public praise based on awards. The desire to get the same rewards as the high performing peers in an organization tends to encourage imitative behavior and is a means for business leaders to create a performance oriented culture (Johnson/Whittington/Scholes 2012: 242). A reward and incentive system needs to be designed very carefully to achieve the desired behavioral results. Managers who apply such control mechanisms need to watch out that they do not leave a great portion of their employees frustrated at the end of the year because they do not get what they expected. In company practice, performance-related pay or incentive pay is a widely used mechanism. Depending on the type of business, company and the managerial level, financial bonuses could be a significant portion of the total annual salary. Incentive pay means that various forms of incentives are linked to an employee's performance as an individual (e. g., pay increase linked to performance appraisal, performance bonuses, commissions), group member (e. g., group bonuses, team awards), or organization member (e. g., profit sharing, stock options) to recognize the person's achievement with regard to the previously defined performance indicators and targets (Noe et al 2009: 339–351).

Skill gap analysis and skill development

Great business models may be doomed to failure unless there are the required skills available. As an essential part of performance management, companies are proactively managing both its leadership and human capital development. They are continuously updating the skills and knowledge of the organization's employees based on anticipated business needs in a changing environment. This can be achieved through training and development of internal employees as well as through external acquisition of talent. An important stage in the performance management cycle usually deals with identifying skill gaps and defining specific individual training and development activities for both managers and employees. Training is traditionally focused on employee performance improvements at their current jobs. In contrast, personnel development tends to be more future-oriented. Some of those skill requirements might be built into job descriptions but most tend not to. Many companies have well-developed systems in place for their personnel development that typically combine formal education at the workplace or off-site, job experiences, interpersonal relationships (e. g., mentoring, coaching), and different types of assessments with feedback and follow-up activities (Noe et al. 2009: 246–262). Some of those development programs are based on predefined key skills for different levels in the organization that are supposed to have a clear influence on the performance and that are systematically assessed. Examples of such essential skills are leadership and management capabilities, interpersonal and team competencies, communication skills, decision making capabilities, customer orientation, professional knowledge, etc. Other companies follow a more strategic approach to workforce management and systematically identify skill gaps based on their specific strategic requirements with less emphasis on the hierarchical level (see chapter 4.3.6.2).

Management by objectives (MBO)

Today, many companies are applying a company-wide performance management system in which people at each level of the organization set logically related goals in

Strategic Snapshot 4.9

Peter Drucker on Management by Objectives

"Every job in the company must be directed toward the objectives of the whole organization if the overall goals are to be achieved. In particular, each manager's job must be focused on the success of the whole. The performance that is expected of managers must be directed toward the performance goals of the business. Results are measured by the contribution they make to the success of the enterprise. Managers must know and understand what the business goals demand of them in terms of performance, and their superiors must know what contribution to demand and expect. If these requirements are not met, managers are misdirected and their efforts are wasted. Management by objectives requires major effort and special techniques. In a business enterprise managers are not automatically directed toward a common goal. On the contrary, organization, by its very nature, contains four factors that tend to misdirect: the specialized work of most managers, the hierarchical structure of management, the differences in vision and work and the resultant isolation of various levels of management, and the compensation structure of the management group. To overcome these obstacles requires more than good intentions. It requires policy and structure. It requires that management by objectives be purposefully organized and be made the living law of the entire management group." (Source: Drucker 2008: 258)

Fig. 4-26

Cascading goals in a management by objectives (MBO) system

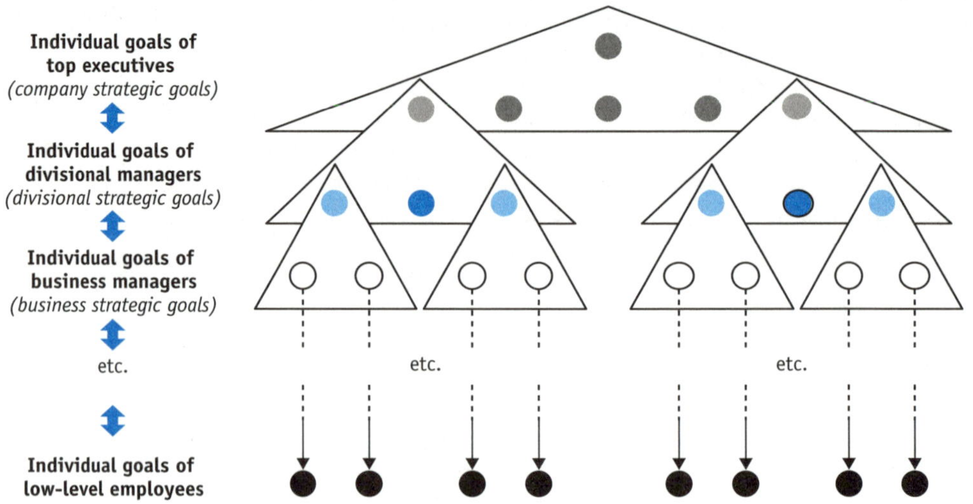

Source: Robbins/DeCenzo 2008: 77 and Likert 1967: 50; adapted

a systematic process and based on certain principles flowing from top to bottom. Such a system is called management by objectives (MBO) or target agreement. The concept goes back about 70 years (Drucker 1954) and tends to remain appealing and effective for driving an organization's performance (see Strategic Snapshot 4.9). Management by objectives is "a system in which specific performance objectives are jointly determined by subordinates and their supervisors, progress toward objectives is periodically reviewed, and rewards are allocated on the basis of that progress (Robbins/DeCenzo 2008: 77).

In a management by objectives or target agreement system the organization's overall objectives are cascaded to each succeeding level where they are translated into individual objectives for each employee (see fig. 4-26). This way, managers and employees on all levels get specific personal performance targets that are consistently linked throughout the entire organizational structure. If all people achieve their goals, the goals of the corresponding department or business area will be attained, and thus, the organization's overall goals are achieved. Grounded in goal-setting theory, MBO systems are typically designed around some key principles (Drucker 2008: 258–268; Eyer/Haussmann 2014; Latham/Locke 1991; Robbins/DeCenzo 2008: 77–80; Robbins/Judge 2013: 246–249):

"SMART" and challenging

- **Goals are specific, difficult, and objectively measurable.** To develop commitment to goals, people have to understand what is expected of them and why (e. g., corresponding business goals), what and how they will be precisely measured against as well as the explicit time period to accomplish the objectives. This can be achieved, for example, with a concise goal statement, a metric with which the goal can be measured, a target value that articulates the desired accomplishment as well as a time frame in which the target value should be reached. In company practice different variations of "SMART" guidelines are applied for goal formulation. For example, goals should be **s**pecific, **m**easurable, **a**chievable, **r**ealistic, and **t**imely. A certain level of difficulty related to the goal accomplishment ("stretch targets") tends to be motivating, particularly for people who are generally seeking challenge and success rather than avoiding failure, and if it is assured that the challenging goal is generally achievable and realistic (i.e. the person wants to achieve the goal and beliefs he or she is able to achieve it).

- **Managers and their direct reports work together when setting the goals.** Collaboration in goal setting and participation in decision making is supposed to ensure that agreed targets are accepted by the people who will be measured against those targets. Superiors and their subordinates jointly define and agree on the goals. Just providing a sense of participation when goals are ultimately set by the boss and assigned to employees is not sufficient. Acceptance can be further encouraged by involving particularly managers in the formulation of superior company or business area goals they are supposed to contribute to. To think through what are the objectives of the higher unit, contributing managers should participate responsibly in defining them. In general, goals that are agreed upon can be purely individual, shared or a combination of both. Shared goals means that achieving the departmental or company objectives is the base for individual rewards. This way, team, departmental or company performance is personally appraised in addition to individual performance (e.g., group goals in case of highly interdependent contributions to goal achievement).

Acceptance through sincere participation

- **Continual employee performance feedback and periodic formal employee performance reviews.** For allowing individuals to monitor and correct their own action, the progress toward objectives is periodically reviewed. Feedback guides behavior and can be supervisor-generated or self-generated. Frequent and recurrent feedback mechanisms are built in the MBO-system, so managers or employees know whether their level of effort is sufficient to accomplish their goals and what overall performance can be expected for the corresponding organizational unit or the entire company. Companies that run their MBO-process on an annual cycle typically conduct formal performance review meetings between superiors and their subordinates halfway (i.e. after six months) and toward the end of the time frame. Those performance reviews are documented in a corresponding agreement. Goal achievements that are discussed and signed-off in the final annual performance review are the base for making decisions on incentives and rewards. Figure 4-27 provides an overview of different types of incentives and rewards in terms of time horizons and organizational elements.

Link of goal achievement and rewards

Management by objectives is not about command and control but about making the company's interest the aim of every manager and employee through collaboratively creating a shared understanding of expectations, clarifying people's contributions at all levels, and creating commitment to objectives. A key principle of MBO is self-con-

Self-control

Fig. 4-27

Examples of incentives and rewards

	Immediate	Longer term
Strategic leaders Managers	▸ Enhanced total compensation ▸ Variable performance compensation ▸ Better benefits ▸ Awards	▸ Long-term performance / one-time bonuses ▸ Profit sharing / stock options ▸ Promotion ▸ More autonomy ▸ Recognition
Professional staff Employees	▸ Improved base pay ▸ Commissions ▸ Short-term performance bonuses ▸ Better benefits ▸ Awards / vouchers / prices	▸ Performance pay / bonuses / flexible benefits ▸ Enhanced work environment ▸ Promotion ▸ Recognition / commendation / respect ▸ Learning / training / development
Team- / group-based	▸ Provided group performance pay such as gainsharing ▸ Short-term team bonuses ▸ Team awards / vouchers / prices	▸ Team recognition / commendation ▸ More decision making input ▸ Self-directed leadership ▸ Enhanced roles / work quality

Source: Rainey 2010: 478; adapted

trol of managers through agreed upon performance measurements. "The greatest advantage of management by objectives", states Peter Drucker, "is perhaps that it makes it possible for managers to control their own performance. Self-control means stronger motivation: a desire to do the best rather than do just enough to get by. (...) All managers should have the information they need to measure their own performance, and they should receive it soon enough to make any changes necessary for the desired results. This information should go to the managers themselves, and to their superiors. It should be the means of self-control, not a tool of control from above." (Drucker 2008: 266) This, however, assumes that people want to take responsibility and to contribute and that they are motivated by challenging aspirations.

Pitfalls of management by objectives

Does an MBO-system have a downside? What are the crucial points in designing an effective MBO-system? Here are six important pitfalls of management by objectives and some practical suggestions to overcome them (Eyer/Haussmann 2014; Kaplan/Norton 2001: 232–271; Noe et al. 2009: 227; Robbins/DeCenzo 2008: 80; Robbins/Judge 2013: 247–249):

- First, set goals might become outdated at the time of appraisal. Goal achievements might not relate to the manager's performance but are caused by unforeseen significant developments in the environment. This relates to both goal accomplishments (windfall profits) and failures. This issue might be overcome by defining relative self-adjusting targets, that is performance levels in relation to peers that are impacted by the same developments. Also, it might be necessary to update goals when warranted during the regular feedback and review process.

- Second, a high specificity of goals may get people too focused on elements by which they will be judged, and thus, neglect other important performance contributions they could provide but that are not part of their target agreement. Some companies try to tackle this issue by linking rewards to both individual and overall company performance. Furthermore, there might be situations that require more generic goal statements. For example, too specified goals might not be desirable at the early ideation and decision stages in highly complex business situations as they may limit creativity.
- Third, people may lose energy and engagement once a set goal has been achieved prior to the end of the defined time period. This discourages continuous improvement and might even limit somebody in exploiting his or her full upside potential. Furthermore, subordinates may deliberately lowball their targets to increase the likelihood of a positive performance review and the corresponding rewards. Managers should treat the MBO-system with its target agreement dialogs and feedback mechanisms as an ongoing process with a regular review and update of goals as necessary. It should encourage ambitious goals and a spirit of beating competition and not beating the agreed targets.
- Fourth, the principle of setting goals that are difficult to reach can lead to unrealistic expectations that might be biased with overconfidence or positive outcome at both the superior and the subordinate (see also chapter 1.2.2.3). Setting stretch targets might be applied consciously by superiors to increase the pressure at their subordinates for extraordinary performance and to bring about unconventional solutions for sometimes unrealistic aspirations. This can be avoided by conducting sincerely participative and fact-based target agreement meetings and by adhering to the "SMART"-guideline of goals being achievable and realistic.
- Fifth, managers may tend to set and agree on too many goals in order to capture each and every strategic, operational as well as personal development need. If the number of goals become a burden to the subordinate and is not manageable anymore, target agreements and the entire MBO-mechanism might well lose effectiveness. Being achievable and realistic should not only be applied to single goals but to the entire number of agreed goals as well. In company practice, three or four to seven single goals are typically set in a target agreement.
- Finally, the question of how specific and difficult a goal should be, whether to apply individual or rather team goals as well as the level of participation when setting goals may have different effects on goal acceptance and goal commitment in different national cultures (e. g., collectivists versus individualist and high versus low power distance cultures). This needs to be considered carefully in multinational enterprises that may apply a standardized MBO-process in their global organizations.

Management Reporting System

Information relevant for planning and control

Information ties all business functions and areas together and helps managers to integrate and coordinate different tasks. The management information system in an organization collects, stores, synthesizes, and presents relevant internal or external information in a way that is useful for the manifold internal and external recipients. Management reporting is the particular part of the organizational management information system that assists managers at different functions and levels of an organization to fulfill their managerial tasks of planning and control. The following sections will provide an overview of key considerations for designing a management reporting system (Horváth 2011: 534–549; Weber/Schäffer 2011: 242–250) and are based on the following questions:

- Why do we report (purpose)?
- What do we report (objects, content)?
- To whom do we report (recipients)?
- How often do we report (frequency)?
- How do we report (types, design)?
- Who is reporting (originator and sender)?

Documentation, planning, and control

Companies develop and use a management reporting system for different purposes such as *documentation* (e. g., historical planning and actual data, minutes of internal management meetings, etc.), *planning* (e. g., an overview of political stability of served country markets is used for target setting and decisions on how to mitigate risk), and *control* (e. g., the analysis of deviations from defined targets trigger corrective action). The system's key role is to provide managers with the information relevant for effective and efficient decision making and evaluation. Effective management reporting allows managers to prepare decisions, monitor and evaluate whether their decisions and plans are working as intended and how they should be changed or improved in case they are not working. Those feedback mechanisms are critical for managers to evaluate organizational progress and the extent of organizational efficiency and effectiveness as well as for initiating action. Ultimately, the purpose of management reports strongly determines what is reported to whom and how. Hereby, the management reporting system typically follows the managerial control and steering logic of the company and reflects its organizational structure and processes.

Reporting objects, recipients, and frequency

The objects of management reporting define what information is gathered and reported. To each reporting object, financial and non-financial value figures are linked and reported in a certain frequency such as on an annual, quarterly, monthly, weekly, and daily basis or by demand. The definition of reporting objects is strongly linked to the question for who reports are prepared and what type of information those recipients regularly need for decision making. For example, in order to support a global key account manager, the reporting object "global key customer (account)" is defined and gets the value figure "revenue" which is reported on a quarterly basis. The purpose is to identify needs for initiating additional customer relationship management and sales effort in case defined revenue targets are not met. Reporting objects that are relevant for internal planning and control are the company itself with its different organizational units such as corporate, divisions and strategic business units, regions and countries, functional units including departments and manufacturing sites as well as processes. Typically, legal entities are also reporting objects not only for external reporting but also for the purpose of internal control. Other reporting objects that can be found in company practice are individual customers (e. g., key accounts), customer groups or markets as well as individual products and services or product/service areas. The different reporting objects and the corresponding recipients result in a reporting pyramid with consistent information (one source of truth), which is illustrated in figure 4-28. Some information may be relevant to specific managers only, some is aggregated for higher management levels or other recipients, and some typically relate to managerial controls relevant for reporting across all levels (see also chapter 4.3.6.3).

Three types of management reports

In company practice, different types of management reports can be distinguished. With regard to the planning and managerial control process, standard reports, deviation reports and on-demand reports are of particular importance:

- **Standard reports** are based on predetermined information requirements of a defined group of recipients such as individual executives or management teams. They tend to cover a broad and comprehensive range of information that is prepared and reported on a regular basis. Content, layout and schedule are predetermined and cannot be modified by the managers for whom the reports are prepared. In standard reporting with its fixed information flow, recipients generally get a wealth of information and need to identify and select for themselves which specific elements are relevant for their decisions or evaluation. An advantage of standard reports is the efficient preparation and broad utilization as they can be used by many different recipients who all get the same wealth of information. A main disadvantage is the fact that they do not fulfill specific or just current relevant information needs of certain recipients.

Fig. 4-28

Reporting pyramid

Report recepients	Reporting objects
Top management, division directors, SBU leaders	Corporation/entire company, divisions (sectors), strategic business units (SBU)
Regional directors, key account manager, global product manager	Regions, global key accounts, product areas
BU-leaders, country directors, product managers, account managers	Business units, countries, products, customers
Functional leaders, sales manager, plant managers, process manager	Functions, sales areas, plants, processes

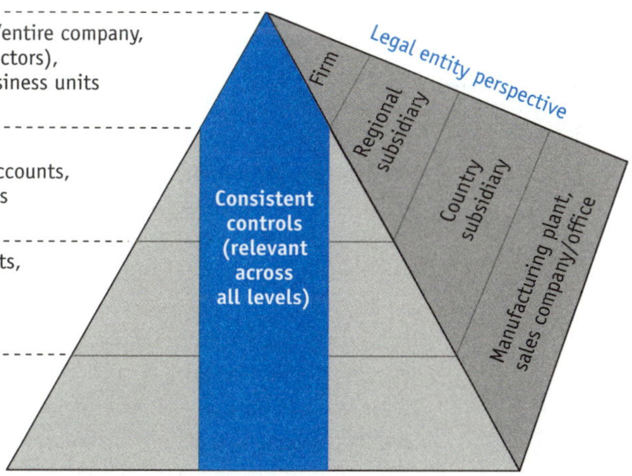

Source: Gräf/Nase 2008: 51; adapted

- **Deviation reports** are meant to direct management attention to specific issues that require decisions or action. This way they only cover information that is particular relevant for supporting current managerial control efforts. The preparation of deviation reports is usually triggered when certain tolerance values in plan versus "as is" comparisons are exceeded. Deviation reports can be used for practicing the principle of management by exception. To best utilize precious managerial time and effort, management attention is drawn only toward exceptional matters, which is a main advantage of deviation reports. A key challenge is the question of setting the right trigger. Tolerance values must be set in a way that they direct attention, and thus, trigger vital corrective action in case of deviations without creating a burden of getting too many deviation reports.
- **On-demand reports** are triggered by specific actual information requirements of their recipients. They neither have a regular schedule nor a predefined content. Specific on-demand reports may provide deep dives into issues that have been identified through deviation or standard reporting. Their major advantage is the focus of management attention and the level of insights provided for a particular issue. A disadvantage is the required effort, which may make on-demand reporting very costly. With advancements in information technology, the relevance of ad-hoc reporting has been increasing. Effective IT-based management information systems provide managers direct access to reliable and most up-to-date information that is relevant for tackling their particular management issue and for conducting their tasks. Drill down reporting is common to identify the root cause of an issue that may only be vaguely highlighted at a high level standard report.

Never before there have been such technological opportunities to assist management reporting. Nowadays, many types of user-friendly IT applications, networked mobile devices, and cloud computing are available to assist managers in fulfilling their specific information requirements in a highly efficient way. This leads to an increased individualization of management reports as managers have the opportunity to customize their reports and drill-down to more detailed information depending on their specific needs.

Principles of effective management reports

In general, management reports for internal managerial control have to answer the following key questions (Axson 2010: 145):

- What has happened?
- What are the consequences?
- Why did it happen?
- What has to be done about it?

For answering those questions, management reports typically cover different types of numbers: plan ("to be"), actual ("as is"), and expectation ("outlook"). By comparing planned with actual values, deviations can be calculated. Effective management reports need to be designed in a way that reflects the requirements of the recipients. The information needs to be provided in a way that it can be best utilized by the corresponding managers for their particular purposes. This can be achieved by following basic principles when designing a management report:

1. The reported information needs to be *relevant* to the corresponding recipient. Hereby, the appropriate scope and depth of information need to be determined by looking at the purpose for which the report is used. Information regarding the same reporting object may be reported to different recipients in the company with different levels of detail.
2. Relevant information should not only be reported continuously but also *explained* with commentaries. Some companies do this only in case of issues; others follow the principle of "no number without commentary".
3. In case of unwanted deviations it should be made clear how severely those deviations *impact* performance. This may require weighting or prioritization of deviations.
4. Effective management reports do not only include the diagnosis of an issue but also corresponding *risk* levels as well as therapeutic elements, that is proposals of *corrective or mitigating actions* to solve the issue.
5. The structure and contents of the report should be straightforward and *intuitively understandable* for the recipient. This can be supported by a comprehensive form that follows a logic sequence and includes easy and quick references and management summaries. Depending on the recipient, visual aids such as figures, diagrams and charts may be considered although some people prefer to work with tables or lists of numbers.
6. Information reported should be *accurate* and *consistent*. There needs to be a "single source of truth" for data in the organization, which requires that the owner or originator and the source of information are defined clearly and traceable. This can be a significant challenge if the company applies various reporting processes and uses multiple reporting tools in different organizational units across the company. Some firms supplement their management reports with a visible watermark as a kind of quality seal indicating that the information is accurate and approved.

Figure 4-29 provides an overview of key processes and activities of management reporting that captures some of the presented principles. Based on the report preparation and its related data management activities, information relevant for managerial control is derived from calculating and analyzing deviations and elaborating on corresponding root causes. Prior to the actual control activities of management (phase 4), key insights are explained, decision needs are identified, and suggestions for action are developed (Gleich/Temmel 2008: 74–77).

Improve business impact

A great portion of management resources can be unleashed with effective management reporting that provides consistent, reliable, focused, and action-oriented management information. The corresponding reporting process is typically facilitated by people from Finance, Corporate Performance Management, or Management Control (Controlling) departments. When following the illustrated principles and activities, employees in charge of management reporting should spend most of their time on business analysis, utilizing information to generate insights and decision support. As far as possible, data collection and report generation should be highly automated and efficient, thus capturing only a small portion of time and cost. This way, Finance departments act as an internal business partner and can significantly

Fig. 4-29

Key processes and activities of management reporting

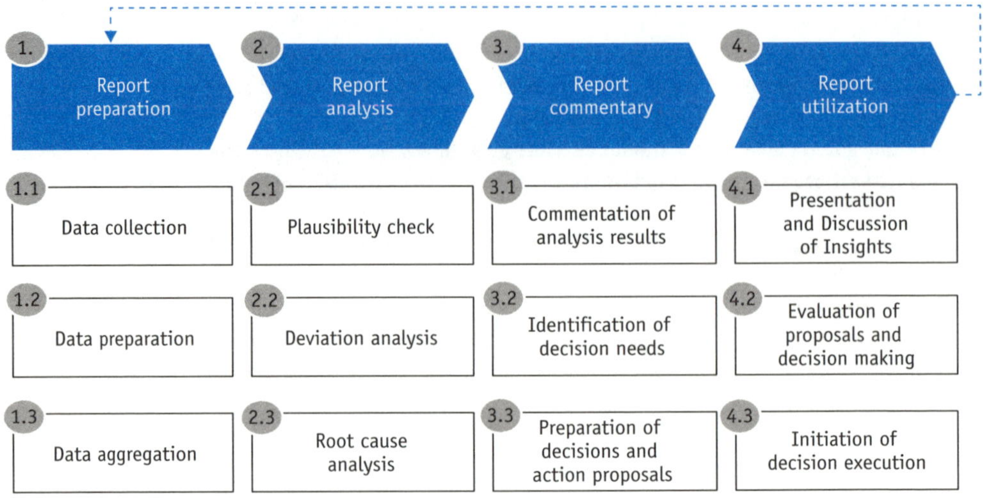

Source: Gleich/Temmel 2008: 77; adapted

improve the effectiveness of management in the operating business areas (Wunder/Müller 2008: 42). The same is true for strategic management. Effective strategic performance reporting is a key element of strategy execution. Many of the principles discussed can also be applied to supporting strategic managerial control. The specific information requirements, however, tend to be different and can hardly be fulfilled with a highly automated standard reporting. Key considerations for effective strategy-oriented management reporting are presented in chapter 4.3.6.3.

4.2.2.4 Integrating Strategy and Structure

Fit between strategy and organizational architecture

Following the same rationale that has been applied to organizational culture (see chapter 4.2.1), there also needs to be a "fit" between strategy and the structure of the company. In order to execute strategies effectively, companies need to integrate strategy and organizational structure, business processes as well as managerial systems. Many interdependencies exist among the different elements, which makes it very difficult to suggest a specific "optimal" constellation for strategy and structure. However, there needs to be a basic match—just like in culture—to make a strategy come true effectively. Key considerations for integrating strategy and structure will be outlined in the following section.

Organizational Structure

Structure follows strategy

How does organizational structure relate to the success of strategy execution? For many decades, the answer to this question seemed to be clear: structure has to follow strategy (Chandler 1962). The underlying rationale is straightforward: Because structure is a means to achieve strategic goals that are derived from the overall strategy, it is only logical that organizational structure has to follow strategy. As explained earlier on, the organization (resources) and its environment (markets) have to be integrated (see chapter 1.1.2.2). This can mean that the structure has to vary based on the specific company context. In general, changes in the environment identified during strategic analysis need to be reflected in the organization's strategy, which, in turn, may lead to changes in its structure. Therefore, it is no surprise that firms within one industry that follow similar general strategies (e. g., automotive or branded consumer goods) tend to have similar structures (Wheelen/Hunger 2010: 280).

Stages in corporate development

Companies tend to follow certain structural patterns when they grow and expand. In those cases, structure is not following strategy but reflects evolutionary stages in corporate development. Entrepreneurial new ventures typically start with a simple structure. When sales increase and they get larger, firms generally adopt a functional structure with departments such as Marketing and Sales, R&D, and Manufacturing. They may also decide to obtain a larger geographic reach to achieve the growth aspirations. With continued success, companies tend to expand their vertical integration and add new product lines in different related or unrelated businesses. As a result, they organize themselves with interconnected business units in a (multi-)divisional or matrix structure (Rothaermel 2013: 309 f.; Wheelen/Hunger 2010: 279–285).

Pitfalls of "strategy first—structure afterwards"

If the existing organizational structure does not easily support a chosen strategy, the executive leadership team has to decide whether the structure should be changed. This decision, however, needs to be considered very carefully, as reorganizing the company can lead to a loss of stability and confusion (see Strategic Snapshot 4.10). Managers may only choose strategic options that do not require changing the existing organization, which means in those cases that strategy follows structure. Furthermore, there might be situations where developing organizational structure by following the general strategy might be suboptimal for a firm. Some considerations are (Lynch 2012: 461 f.):

- Rapid and radical changes in the competitive arena may require strategy and structure to emerge over time in an experimental way as the organization has to learn how to adapt. Simple strategic and structural solutions may not be available.
- An organizational design enabling emergent strategies and adaptive strategies needs to be less formal and more open. Rigid, hierarchical and bureaucratic organizational structures might be inappropriate for the organizational learning and experimentation processes that are required.
- The organizational structure itself may be a cornerstone of the strategy. The source for competitive advantage might lie in a unique structure, and the strategy is developed to leverage the existing structure for realizing the competitive advantage.

Matching organizational structure and strategy

With all the innovation in new structures, there is no answer to the optimal organizational design for a given strategy or company situation (Wheelen/Hunger 2010: 280). "We found that strategy rarely seemed to dictate unique structural solutions." (Peters/Waterman 1995: 4) With regard to the strategy process it might make sense that structure does not "follow" strategy if its execution can be facilitated with the

Strategic Snapshot 4.10

"Structure Follows Strategy" versus What Happens in Practice

"Academic studies on organizational design recommend that structure should be designed after the leaders have decided the long-term strategy. Theoretically, the advice is sound. (...) However, in practice, most executives behave differently and tend to sort out structure and design quite early when appointed to a new position. (...) First of all, structure and design change in organizations makes for very high levels of political instability, demotivation, and a loss of work impetus. Put simply, it is the basic nature of human beings to need structure and stability. (...) It is, therefore, difficult to understand the underlying arguments leading modern theorists (and some executives) to assert that it is advisable to continuously change structure to adapt to the faster changing conditions which prevail in the business world. I contend that frequent change only causes chaos, which is never useful when trying to manage an organization or business for long-term profitability. Similarly, classic strategy theory (...) suggested that one should develop a complete strategy for the long term and only then design a structure to enable that strategy to be achieved. Unfortunately, the real world is rarely as straightforward as that. For example, in reality, most managers coming into a new post tend to restructure the department or business first and only afterwards get down to analyzing their strategic purpose or create a plan. (...)." (Levicki 1996: 151 f.)

existing organizational form. The key point is, however, that the chosen organizational structure has to match the strategy. Here are some examples (Hitt/Ireland/Hoskisson 2011: 324–338; Lynch 2012: 481; Rothaermel 2013: 307–317):

Corporate strategy and structure

On a corporate level, different organizational structures can be supportive of different corporate strategies. If the company is pursuing a high degree of related or unrelated diversification (see chapter 3.2.2.2), for example, a divisional or SBU-structure is recommended. If the businesses are operating in separate markets and are not linked, a SBU-structure with decentralized decision making and a low level of integration at corporate headquarters might be appropriate. If the company wants to realize parenting advantage (see chapter 3.2.1) by integrating its different divisions and strongly link them, a more centralized decision making with a highly coordinated divisional structure or a matrix structure might be best suited. Also, the different strategy postures for multinational enterprises (see chapter 3.3.2.1) are best supported by different organizational forms. Whereas a transnational strategy requires a global matrix structure, a multidivisional structure is appropriate to realize a multinational strategy (division by geographic areas) and a global strategy (division by products). The best match for an international strategy tends to be a functional organizational structure allowing the company to leverage its home-based core competencies and related processes in different countries.

Business strategy and structure

Different types of business-level strategies, such as cost leadership, differentiation or outpacing (see chapter 3.4.2.1), can be effectively supported with functional organizational structures in different variations that need to be carefully defined on a case-by-case basis. The functional structure seems to be particularly relevant when the firm focuses only on few products or services with a limited geographic scope. Matching strategy and structure particularly on a business level can be illustrated with reference to the models of a mechanistic and organic organizational structure (Burns/Stalker 1961). Key characteristics of these two organizational forms tend to be in favor of different types of business strategies (Lombriser/Abplanalp 2004: 331; Robbins/Judge 2013: 531 f.; Rothaermel 2013: 306–309).

Strategy Practice Example 4.6

W. L. Gore & Associates: Organic Organization for Driving Innovation

W. L. Gore & Associates, the inventor of breakthrough products like the breathable Gore-Tex fabrics, is an example of an organic organization. The company operates with very flat structures to enable its employees (called "associates") utilizing their full entrepreneurial and creative potential, and thus, realizing the company's successful product-leadership strategy. Gore's strategy process is emphasizing sponsor-led project-based teamwork, entrepreneurial spirit and individual initiative instead of a hierarchical chain of command with centralized decision making. The various company sites are limited in size to foster informal communication in flat structures even though the firm has been growing to approximately 10,000 employees and about $3 billion in revenues since founding the company in the 1950s. People are encouraged to try out new ideas, strategize in an experimental way and make their own commitments as long as their decisions are above the so-called "waterline", that is not exceeding a certain risk-level that could cause serious damage for the company ("drilling holes in a boat below the waterline"). Successful product strategies emerge based on the many ideas, decisions and personal commitments from individuals throughout the organization. (Source: Berlet-Maiweg/Oppermann 2014; Rothaermel 2013: 306)

Mechanistic organizations and cost leadership

A cost leadership strategy in a stable market, for example, tends to match with more mechanistic organizational forms. Those can be characterized with a strong division of labor and rigid departmentalization, extensive formalization, a tall hierarchy with narrow spans of control and a clear chain of command as well as centralized decision making. The fast food chain McDonald's somewhat follows this type of structure. The specific combination of organizational design elements enable tight cost control, avoid unnecessary expenses, and ensures a high degree of efficiency. It provides an organizational environment supportive of standardization and economies of scale.

Organic organizations and differentiation

Otherwise, an organic organizational model operates with cross-functional and cross-hierarchical teams, a low degree of formalization, a flat structure with wide spans of control and free flow of information as well as decentralized decision making. This type of structure tends to be supportive of a differentiation strategy at the business level as it typically correlates with a higher rate of entrepreneurial behaviors and innovation. Therefore, it seems to be appropriate for companies like Apple, 3M, or W. L. Gore & Associates, who are renowned for their innovativeness (see Strategy Practice Example 4.6). Pitfalls of an organic structure are extensive coordination and alignment processes, a lack of orientation along with a significant amount of stress on individuals, particularly if they do not like ambiguous situations.

Business Processes

Figure 4-30 shows two ways of matching strategy and processes. One view is that strategy is preceding processes. In other words, you first define the strategies of the company and afterwards you design the structure including the business processes to facilitate strategy execution (Hax/Majluf 1996: 38 f.; Stöger 2011: 60 f.). However, there is also a reverse relationship. Here, the processes are a key source for competitive advantage, and thus, a cornerstone of strategy. The processes are defined first

Fig. 4-30

Strategy-process relationship

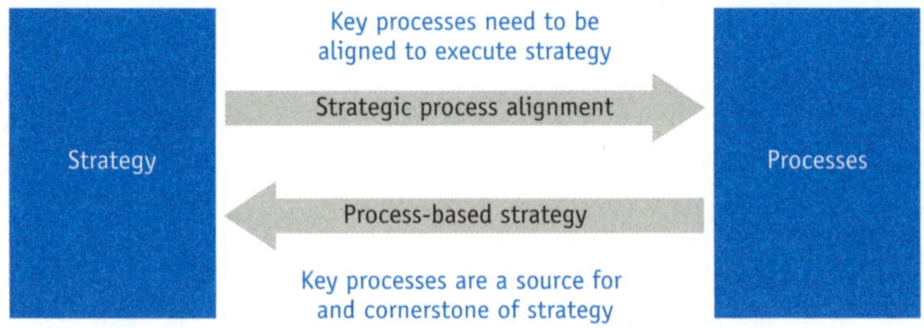

Source: Horváth & Partners 2008: 5; adapted

and the strategy or business model is build around them. For example, an analysis of how to best configure, compress or break down a company's value chain as compared to its competitors can be an important source for strategic ideas (Slywotzky et al. 1999: 96–123). In company practice it might well be an interdependent relationship of strategy and processes resulting in a combination of these two perspectives.

Processes follow strategy

Business processes are a key driver for achieving competitive advantage. Whether it is about producing superior quality for a premium automaker, speed in processing a credit application for a bank, reliability and service provided by an online-dealer or offering competitive prices for a food company—customer value is created through business processes. This is truly where the "rubber meets the road". Therefore, key processes of the company need to be tailored in a way that they help to make strategy happen. There are basically two ways for matching processes to strategy. In a pure deductive approach, the "to be" processes are defined top-down, based on the strategic priorities of the company independently from the existing structure and its constraints. In an inductive approach, the existing process model with its actual processes is compared with the strategic priorities and aligned in case of significant deviations (Fischermanns 2012: 20–22). Although the second approach does not provide the opportunity for a fundamentally new or even revolutionary process organization as suggested in the concept of re-engineering (see chapter 4.2.2.2), it tends to be a more realistic and applicable approach for company practice (Hammer 2001: 52).

Strategic process alignment

An effective alignment of a company's existing process model to strategy requires an assessment, which processes are particularly relevant from a strategic point of view. In case those processes are currently not supportive of strategy, they need to be aligned. Key processes for making the desired strategy happen are articulated in the process perspective of a strategy map, and thus, become a key priority in the strategic execution framework (see chapter 4.3.1). This ensures that they get the required management attention and resources for effective alignment and execution. Figure 4-31 illustrates the idea of strategic process alignment. Hereby, main- and subprocesses in the different process categories within the firm's process model are compared with strategic goals. Strategic process alignment is following three guiding questions:

1. How relevant is the main- or the subprocess for achieving the strategic goal?
2. Does a highly relevant process with its current design and process priorities fulfill the requirements for achieving the strategic goal, e.g., regarding time, quality and cost?
3. If not, what are the process alignment needs or, in other words, what has to be changed regarding the current process design and management?

Apart from a critical business process that might require strategic alignment, there are also alignment needs regarding control (management) processes and contributing (support) processes to make the strategy happen (see fig. 4-31).

Strategy follows processes.

Processes can also be an important pillar for strategy formulation itself. This is particularly relevant in cases where core competencies of a firm as a source of competitive advantage are strongly based on specific process capabilities, for example, innovation, logistics, production, sales, brand management, etc. As described earlier, a business model can be seen as a refinement of the company's strategy on a business level. "Key processes" are essential for value creation, and thus, are typically an integral part of business model frameworks. For "process-based strategies" (see fig. 4-30), key processes are not only an element in the business model of a company, but they are *the* dominant element from which the entire business model is originating and on

Fig. 4-31

Strategic process alignment (SPA) matrix

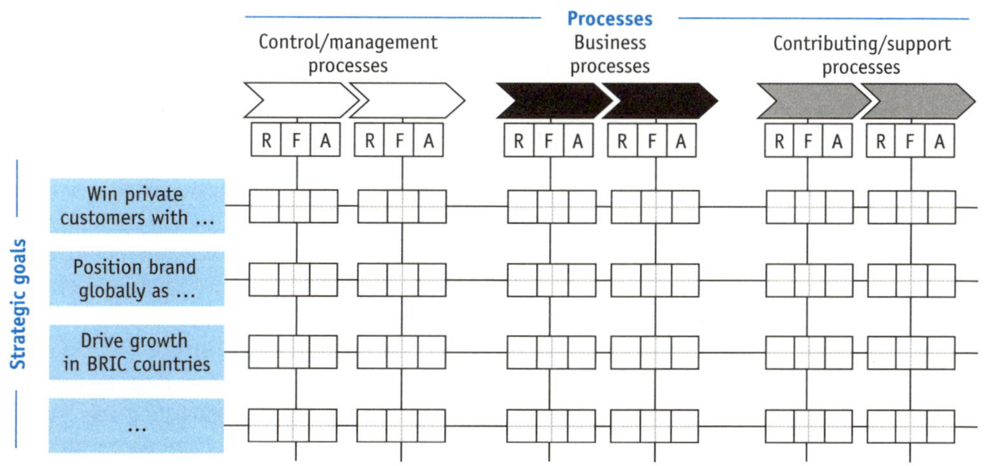

R = Process **relevance** for achieving the strategic goal derived from strategy (high, medium, low)

F = Current level of **fulfillment** of process requirements for achieving the strategic goal (high, medium, low)

A = Process **alignment** need in case of high/medium relevance and medium/low fulfillment

Source: Horváth & Partners 2008: 5; adapted

which it is based (see "Process Patterns" in chapter 3.4.4.4). Here, processes do not follow strategy, but they are a source of strategy.

Continuous improvement is not strategy.

Strategically designing key processes must not be confused with continual improvement programs. Process improvement activities driven by tools such as Six Sigma or Lean Management typically target at ongoing improvements in process performance. Strategy-oriented process management, however, should be more strongly related to the identification, selection, management, measurement, and further development of business processes that are critical for delivering value, and thus, making the strategy come true (Bergsmann 2012: 7 f.). Although continuous improvement activities are a lever for operational excellence, and thus, are driving performance and results, they alone are usually not sufficient for winning in the marketplace (see chapter 1.1.2.3). Outstanding companies are well positioned regarding both operational effectiveness and strategy (Wunder/Bausch 2014b: 55 f.).

Managerial systems

Alignment of key managerial systems

One of the success factors for effective strategy execution is the link of key managerial systems with strategy. It needs to be ensured that resources, people and management information are aligned with strategy (see chapter 4.3.6). Whoever is in charge of executing strategic priorities needs to have the necessary resources to excel in terms of funds, staff and time, which relates to the operational planning and budgeting system. The organization has to communicate strategies and provide its staff at all levels with the corresponding incentives and rewards. This relates to the performance management and management by objectives systems. Within the personnel development system, it needs to be ensured that corresponding skills for executing the desired strategy are in place. Another important interface is the information and management reporting system, which is a crucial facilitator of a strategic management process. It has to provide the leadership team with a reliable, balanced and focused view on the strategic performance of the company in order to enable them to make more informed decisions. In company practice, aligning all those managerial systems represents a significant challenge as they have emerged historically and are run in different departments and areas of the company. Linking them to an effective strategy execution system may require changes in established procedures. Processes in different areas of responsibility such as the Human Resource, Finance and IT departments need to be orchestrated and focused on the strategy.

Strategy execution system tunes structure and culture to strategy.

According to Kaplan and Norton a new organizational chart is not necessarily critical for making the strategy happen (Kaplan/Norton 2001: 12). They de-emphasize the prominent role of organizational structure for strategy execution referring to McKinsey's 7-S-model (see chapter 2.3.1.3), where strategy and structure are only two of seven S's: "The insights from our work with hundreds of organizations is that organizations should not search for the perfect [organizational] structure for their strategy. Instead, they should choose a structure that is reasonable and seems to work without major conflicts, and then design a customized, cascaded *system* of linked Strategy Maps and Balanced Scorecards to tune the *structure*—the corporation and its collection of centralized functions and decentralized product groups and geographical units—to the *strategy*." (Kaplan/Norton 2006: 38) Furthermore, effective strategy execution has a lot to do with effective communication, which shapes culture. Communicating the vision of what the company aspires to become as well as corresponding strategic priorities along with concerns about how things are currently done addresses the existing organizational culture. Corresponding cultural messages and development efforts in company practice include an emphasis on commitment to performance and accountability (performance culture), collective strategic dialog and learning processes, focus on customers or on emotionalizing brands, and on engaging in creative ideation and innovation (Kaplan/Norton 2008: 141). The approach to strategy execution a company applies represents a managerial system on its own. Although its development is influenced by the existing structure and culture of the company, an effective strategy execution system also facilitates the tuning of structure and culture to strategy with respect to the strategy-structure-culture trilogy explained earlier on (see fig. 4-5). It is considered one of the core elements of the "Wheel of Strategy" (WOS) framework applied in this book. The following provides a detailed outline of how to develop such a system in a way that encompasses the generic elements for a successful execution and overcomes the barriers to strategy execution explained at the beginning of this chapter.

4.3 Strategy Execution System

Balanced scorecard (BSC) approach and its four perspectives

Probably the best known and most widely used management concept for strategy execution is the balanced scorecard (BSC) approach. First published in 1992, there is hardly any other management concept that has been so extensively applied for strategy execution by organizations all over the world during the last 20 years (see Strategic Snapshot 4.11). The classic balanced scorecard approach covers specific components and a corresponding management process for strategy execution that can be strongly related to the seven generic drivers described earlier (see chapter 4.1). Key components of the approach are strategy maps with strategic goals, key performance indicators with target values, and strategic action programs with budgets, responsibilities and time frames. Four distinct balanced scorecard perspectives are used to structure all three components—goals, metrics and actions—and provide a comprehensive way for articulating a strategy (following Kaplan/Norton 1996: 25–29 and Horváth & Partners 2007: 41 f.):

- The **financials** perspective reflects the shareholder expectations and the desired economic consequences of strategy.
- The **customer/market** perspective covers targeted markets, customer segments and value proposition that are pursued for achieving the financials.
- The **internal processes** perspective addresses the most critical processes that need to be performed excellently in order to achieve the customer/market and financial goals.
- The **potentials** perspective (originally called "innovation & learning" or "learning and growth" perspective) is related to human capital, information capital, and organization capital that is required as a base for achieving all other goals and for long-term success.

Strategic Snapshot 4.11

Origin and Advancement of the Balanced Scorecard Approach

The origin of the balanced scorecard was a one year management study called "Measuring Performance in the Organization of the Future". Study leader was David P. Norton, managing director of the Nolan Norton institute, a research unit of KPMG that supported the study. Academic advisor was Robert S. Kaplan, a professor of Leadership Development at the Harvard Business School. Goal of the study was to overcome the limitations of the so far highly financially focused performance measurement systems applied in company practice. The study was conducted with 12 large companies—Advanced Micro Devices, American Standard, Apple Computer, Bell South, Cigna, Conner Peripherals, Cray Research, DuPont, Electronic Data Systems, General Electric, Hewlett-Packard and Shell Canada—and supplemented by innovative performance measurement systems of other firms from outside the research group. The outcome of the iterative research and learning process was the balanced scorecard (BSC) framework with its four distinct perspectives: financial, customer, internal, and innovation & learning. The key principle was to *balance* short- and long-term objectives, financial and non-financial performance measurements, lagging and leading indicators as well as external and internal performance views (Kaplan/Norton 1996: vii–xi). According to Kaplan and Norton, 80 percent of performance indicators on a balanced scorecard should be non-financial (Kaplan/Norton 2001: 376).

In the first publication of the approach, the balanced scorecard was presented as a new type of **performance measurement** system covering a set of different financial and non-financial key performance indicators (Kaplan/Norton 1992). "At that time, we thought the Balanced Scorecard was about measurement, not about strategy." (Kaplan/Norton 2001: 3). Since then the concept has evolved to an approach for clarifying strategies and for **strategy communication** (Kaplan/Norton 1993), to a comprehensive **strategy management** system (Kaplan/Norton 1996), and to a holistic approach for **organizational alignment**—the "strategy-focused organization" (Kaplan/Norton 2001). "What we could not have anticipated was that it was also a solution for a much bigger problem: Organization's inability to implement new strategies and to move in new directions (...)" (Calabro 2001: 73). The rather mechanistic cause-and-effect chains of the initial balanced scorecard approach evolved to the concept of **strategy maps** (Kaplan/Norton 2004). Strategy mapping has not only proven to be an effective tool for strategy articulation but also for strategy alignment of different areas within an organization (Kaplan/Norton 2006). Companies which are capable of **linking strategy to operations** and executing them more effectively than their peers can leverage this capability as an additional source of competitive advantage—the "execution premium" (Kaplan/Norton 2008). Today, **strategy execution excellence** is generally perceived as an important pillar for achieving outstanding performance results. Figure 4-32 shows key advancements of the original balanced scorecard approach by Kaplan and Norton since 1992.

Fig. 4-32: Evolution of BSC-based strategy execution methodology

Continued on next page

Continued from previous page

The balanced scorecard was labeled as one of the most influential management ideas of the past 75 years (Sibbet 1997). The different books by Kaplan and Norton received numerous recognitions and their original publication *The Balanced Scorecard: Translating Strategy into Action* has been translated into 24 languages (Harvard Business School 2015). In German-speaking countries the approach was strongly driven and advanced by Horváth & Partners with a vast number of publications and various management studies (e.g., Horváth & Partner GmbH 2008; Horváth & Partners 2007). The balanced scorecard approach has been applied to a broad range of managerial problems including questions of how to implement corporate sustainability (Hansen/Schaltegger 2014), how to effectively align and manage international subsidiaries and value chain activities in multinational corporations (Wunder 2005 and 2004a), and how to support the implementation of business model innovation (Horváth & Partner GmbH 2008; Wunder 2013; see also fig. 4-41 and fig. 4-44). Today, balanced scorecards are applied in companies all over the world (Rigby/Bilodeau 2015). Despite its manifold applications, is has to be noted that the balanced scorecard approach and its effectiveness is not without skepticism and pitfalls (Brown 2007; Nørreklit 2000; Rieg/Esslinger 2012; Wunder 2004b). Hereby, it must be considered that organizations and authors have different interpretations of the approach which makes an evaluation and generalization of its effectiveness very difficult. It goes without saying that there is no automatism which guarantees that the use of the concept will be beneficial. Ultimately, the effectiveness of the concept comes down to how it is applied for a specific managerial situation. As with most management methods and tools, they are only as good as the skills of the managers who use them for tackling a particular organizational problem.

Traditionally, the term "balanced scorecard (BSC)" conceptually referred to all three elements together as well as to the corresponding management process (Kaplan/Norton 1996; Wunder 2004a: 113–157). Nowadays, however, the "balanced scorecard" term is typically used to describe the performance measurement component only. This contemporary understanding of a "balanced scorecard" as being a set of key performance indicators with targets including actual/target comparisons in different perspectives is applied in this book (Kaplan/Norton 2004: 53; Speckbacher/Bischof/Pfeifer 2003: 372 f.).

The different components and perspectives of the balanced scorecard approach and the corresponding management process are the foundational cornerstones for building an effective strategy execution system. The next chapter provides a brief overview of key components and the process of developing a strategy execution system. This is followed by a detailed description and discussion of its essential components as well as of how to embed them in the organization for effective ongoing strategy management.

4.3.1 Developing a Strategy Execution System

Four perspectives provide a comprehensive view of the strategy.

Following the balanced scorecard approach introduced at the beginning of this chapter, it is recommended to view strategy from different perspectives to effectively drive its execution. The perspectives are supposed to capture financial and nonfinancial factors, results and enablers, leading and lagging elements as well as tangibles and intangibles. Figure 4-33 illustrates such a "balanced" view of the strategy that follows a top-down logic. The approach of viewing strategy from those different angles is derived from empirical experiences (see Strategic Snapshot 4.11). The four perspec-

Fig. 4-33

Four perspectives for looking at strategy and their underlying logic

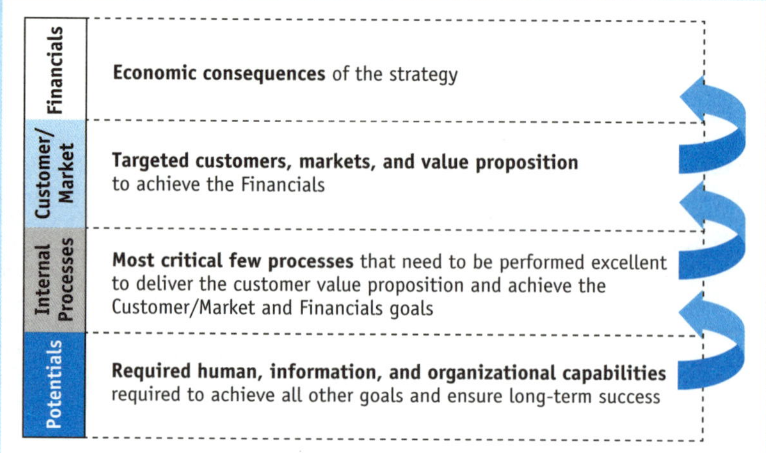

tives are considered of equal importance ("balanced") for describing strategy and making it happen. Essentially, they reflect and integrate different views of strategy (see Strategic Snapshot 4.12) and provide a standard structure that can be applied in most firms but can also be adjusted to the specific industry or company. In non-profit organizations, for example, the financials perspective is typically replaced or supplemented by a perspective reflecting the mission results or purpose of the organization.

Trilogy for execution: strategy map-balanced scorecard-strategic actions

Figure 4-35 shows three key questions relevant for strategy execution and the corresponding components of the strategy execution system that address those questions. The traffic lights are used for strategic performance reviews (see chapter 4.3.7.1). On the strategy map (see fig. 4-36) they illustrate the status of goal achievement that is triggered by actual/target comparisons of key performance indicators in the balanced scorecard (see fig. 4-37). Green or "+" means that the strategic goal achievement is well on track, yellow or "~" means that the achievement of the goal is somewhat off track and at risk, and red or "X" means off track, that is the desired targets will not be achieved. Traffic lights in the strategic action program reflect the progress of executing strategic actions (see fig. 4-38).

To provide an initial overview of the methodology, the three components are briefly described next. Considerations and guidelines for their development are explained in detail in the subsequent chapters.

Strategy map: Strategy cannot be executed if it cannot be described.

"Strategy cannot be executed if it cannot be understood, however, and it cannot be understood if it cannot be described. If we are going to create a management process to implement strategy, we must first construct a reliable and consistent framework for describing strategy." (Kaplan/Norton 2001: 65 f.) A strategy map gives an answer to the question what the strategy is trying to achieve. Strategic goals are arranged on *one* page and visually linked across the four perspectives to describe and

Strategic Snapshot 4.12

Integrating Different Strategy Views with Different Perspectives

From a strategic management point of view, the four perspectives of the classical balanced scorecard approach reflect and integrate different views of strategy (see 1.1.2.2), which is illustrated in figure 4-34. The *market-based view* of strategy can be related to the customer/market perspective whereas the potentials perspective somehow reflects a *resource-based view* of strategy. If companies follow a (shareholder) *value-based view* of strategy, this is typically reflected in the financials perspective and in the internal processes an *activity-based view* of strategy can be addressed. A *society-based view* of strategy can be related to the customer/market perspective if it is a key element of the value proposition and market positioning as well as to the internal processes or potentials perspectives if corresponding strategic goals regarding ecological or social elements are addressed here. (Wunder 2004a: 118 f.; Servatius 2002: 184)

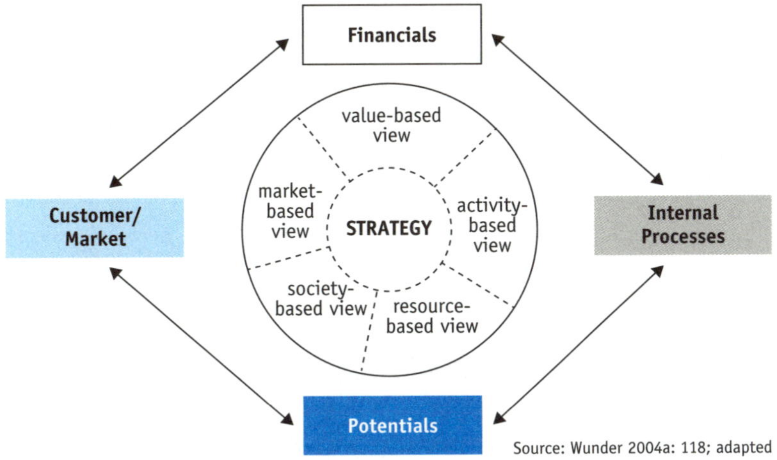

Source: Wunder 2004a: 118; adapted

Fig. 4-34: Integrating different views of strategy through the balanced scorecard perspectives

visualize the key rationale of strategy (see fig. 4-36). The logic underlying those linkages can practically be derived from the discussions and sometimes controversial dialogs that occur during strategic goal development. Strategy maps can be used for aligning the entire organization around strategy creating a shared understanding and buy-in to strategic priorities throughout the company.

Balanced scorecard (BSC): You cannot manage what you cannot measure.

The balanced scorecard includes strategic measurements or key performance indicators (KPI) used to monitor success or failure against the strategic goals on the strategy map. Each key performance indicator has a target value that reflects the desired aspiration level. The term "key" refers to the idea that out of all performance measurements in a company those are the few vital ("key") indicators that determine the strategic performance based on a limited set of strategic goals that has been defined before. "The Balanced Scorecard is not a replacement for an organization's day-to-day measurement system. The scorecard measures are chosen to direct the attention of managers and employees to those factors expected to lead to competitive breakthroughs for an organization" (Kaplan/Norton 1996: 164). The BSC should provide a focused, reliable, and balanced view on the strategic performance of the company.

Fig. 4-35

Key questions for strategy execution and corresponding components

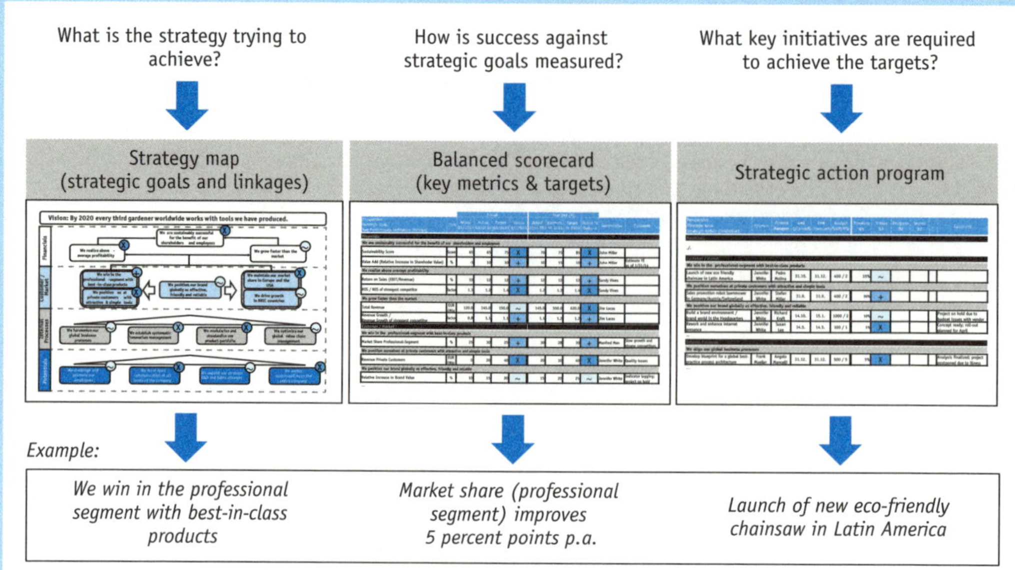

Source: following Horváth & Partners 2007 and Kaplan/Norton 2004: 53

Strategic action programs: Goals without the actions to realize them are daydreams.

Ultimately, strategy execution is about "translating strategy into action" (Kaplan/Norton 1996) and making those actions happen. The strategic action program covers key initiatives with corresponding budgets, responsibilities and time frames that are pursued to achieve the defined targets. Figure 4-38 shows an excerpt of the illustrative strategic action program developed for achieving the defined targets, and thus, ultimately realizing the strategic goals. The color codes and status symbols are related to the execution progress of strategic actions, for example, on time, on budget and on quality. Detailed strategic action programs are the linkage between strategy and budgets as well as target agreement.

Integrated closed-loop process for strategy execution

Apart from the different components, the strategy execution system also includes a management process for executing strategy. The process draws on the different components explained above and is designed to overcome the different barriers to strategy execution mentioned earlier (see chapter 4.1.1). Figure 4-39 shows this high-level process for strategy execution that links strategy and operations. Different management systems that typically exist in different areas of a company, such as strategic and operational planning and budgeting, the target agreement and incentive system as well as management reporting, are integrated in a closed-loop process. The vision and strategy needs to be translated into strategic goals and strategy maps, balanced scorecards and strategic action programs. Strategic priorities of people, teams and departments in different organizational units of the company need to be aligned

Fig. 4-36

Illustrative strategy map of a producer of landscaping power equipment

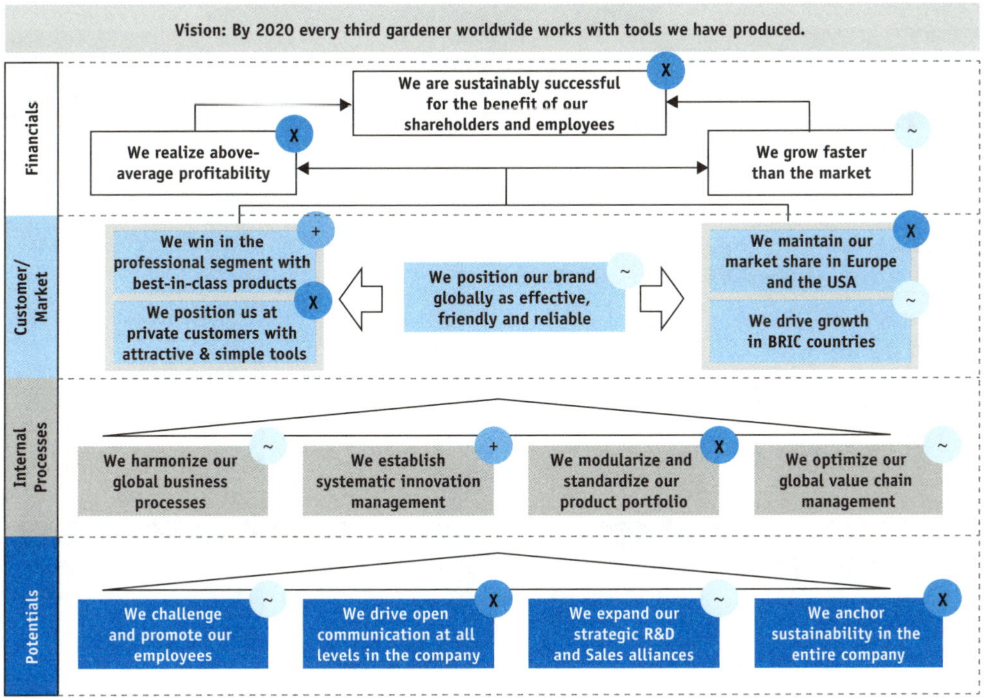

Source: Wunder 2012: 274; adapted

around strategy. Also, budgets, incentives, reports, etc. need to reflect the strategy, which means that key processes, structures and systems in the organization need to be aligned. This is followed by the actual execution of strategic initiatives, which impacts operational activities and is supposed to deliver improvements regarding the achievement of strategic goals. Finally, strategic learning is driven by a continuous analysis of progress and the underlying assumptions, which may lead to an adaption of strategy. In essence, this process is following the classic "plan-do-check-act" logic from the quality management movement (Deming 1982: 101–104) but is enhanced by the strong strategy focus, explicit strategy and organizational alignment activities as well as environmental scanning and monitoring processes (see chapters 4.3.3, 4.3.6 and 4.3.7). Strong executive leadership is vital for successfully applying this process and driving a strategic change agenda (Kaplan/Norton 2006: 259 f.), which has been described in chapter 1.3.3.3.

Designing a strategy execution system: components and process

Figure 4-40 shows six steps of how to design a strategy execution system. This system is based on the assumption that important strategic analyses have been conducted and fundamental strategic decisions are already made in advance according to

Fig. 4-37

Illustrative balanced scorecard (excerpt) of a landscaping tool producer

Perspective Strategic Goal Key Performance Indicators (Metrics)	Unit	Actual: Actual Q1/ 2013	Actual: Actual Q1/ 2014	Actual: Target Q1/ 2014	Actual: Status Q1/ 2014	Year End (YE): Actual 2014 YTD	Year End (YE): Estimate YE 2014	Year End (YE): Target YE 2014	Year End (YE): Status YE Outlook	Responsible	Comment
Financials											
We are sustainably successful for the benefit of our shareholders and employees											
Sustainability Score	Score	65	65	75	X	70	75	85	X	John Miller	
Value Add (Relative Increase in Sharehoder Value)	%	-5	10	10	+	10	15	10	+	John Miller	Estimate YE as of 1/31/14
We realize above-average profitability											
Return on Sales (EBIT/Revenue)	%	10	12	12	+	12	12	12	+	Sandy Vines	
ROS/ROS of strongest competitor	factor	1.1	1.2	1.4	X	1.2	1.2	1.4	X	Sandy Vines	
We grow faster than the market											
Total Revenue	EUR (Mio	120.0	145.0	150.0	~	145.0	550.0	620.0	X	Jim Lucas	
Revenue Growth/Revenue Growth of strongest competitor	factor	0.8	1.1	1.1	+	1.1	1.2	1.2	+	Jim Lucas	
Customer/Market											
We win in the professional-segment with best-in-class products											
Market Share Professional-Segment	%	25	30	25	+	30	28	30	+	Manfred Man	Slow growth and strong competitors
We position ourselves at private customers with attractive and simple tools											
Revenue Private Customers	EUR (Mio	0	20	40	X	20	30	40	X	Jennifer White	Quality issues
We position our brand globally as effective, friendly and reliable											
Relative Increase in Brand Value	%	10	15	20	~	15	20	25	~	Jennifer White	Indicator lagging; project on hold

...

Source: following Horváth & Partners 2007

the "Wheel of Strategy" (WOS) framework (chapters 2 and 3). The execution system covers both the development of different *components* for strategy execution as well as the design and set-up of the corresponding management *process*. The components can be characterized as being holistic, integrative, balanced, strategy-oriented, focused, precise, measurable, and action-oriented. The corresponding management process captures principles of how strategies are translated into the three key components, communicated in the organization, embedded in managerial systems, continuously tested and adapted, and ultimately leveraged for organizational learning. (Wunder 2004a: 115–152) Note that certain steps in the presented strategy execution process (see fig. 4-39) are split in the development of the framework (see fig. 4-40) for practi-

Fig. 4-38

Illustrative strategic action program (excerpt) of a landscaping tool producer

Perspective Strategic Goal Strategic Action (Initiative)	Sponsor	Project Manager	End (planned)	End (fore-cast)	Budget TEUR/ FTE	Progress Q1	Status Q1	Progress Q2	Status Q2	...	Comment
Financials											
./.											
Customer/Market											
We win in the professional-segment with best-in-class products											
Launch of new eco friendly chainsaw in Latin America	Jennifer White	Pedro Molina	31.10.	31.12.	400 / 2	35%	~				
We position ourselves at private customers with attractive and simple tools											
Sales promotion robot lawnmower in Germany/ Austria/Switzerland	Jennifer White	Stefan Miller	31.9.	31.9.	400 / 2	90%	+				
We position our brand globally as effective, friendly and reliable											
Build a brand environment/ brand world in the Headquarters	Jennifer White	Richard Kraft	14.10.	15.1.	1000 / 2	10%	~				Project on hold due to budget issues with vendor
Rework and enhance internet presence	Jennifer White	Susan Lee	14.5.	14.5.	100 / 1	5%	X				Concept ready; roll-out planned for April
...											
Internal Processes											
We align our global business processes											
Develop blueprint for a global best-practice process architecture	Frank Mueller	Angelo Mancelli	31.12.	31.12.	500 / 5	5%	X				Analysis finalized; project postponed due to illness
...											

Source: following Horváth & Partners 2007

cal reasons, which will be explained in the coming chapters. Translating the vision and strategy is reflected in the development steps 1, 3 and 4, aligning people and the organization is addressed in steps 2 and 5. Execution and improving as well as analyzing and adaption of strategy are related to step 6 and the loop to previous steps.

Strategy needs to be precise and actionable.

In a first step, strategies are described by translating them into strategic goals. Then a strategy map is developed by logically integrating the strategic goals, and thus, articulating a "story of strategy" in a simple but compelling way. In the second step, strategy maps are used to cascade strategy in the organization and align strategic priorities of different organizational units, and thus, create a shared understanding and commitment for strategy in the entire company. In case only one strategy map has been developed for the purpose of strategy clarification and articulation (e. g. in a start-up or small business enterprise), step two may not be required. Third, the strategy is made measurable by developing key performance indicators (KPIs) for each strategic goal of the strategy map and by clarifying the desired aspiration level

Fig. 4-39

Strategy execution process

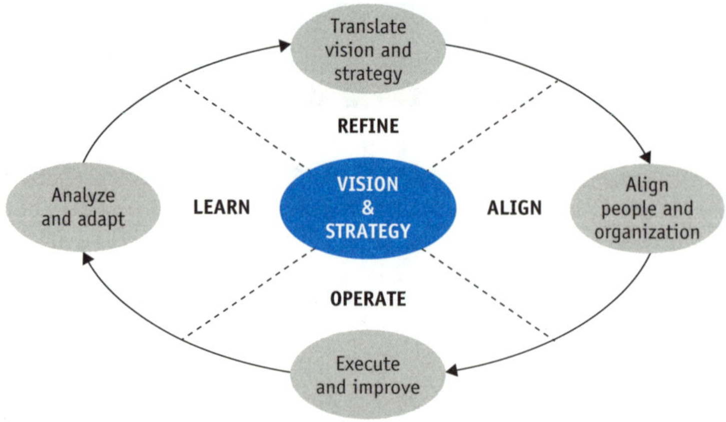

Source: following Kaplan/Norton 1996: 77; 2001: 9; 2006: 260; 2008: 70; adapted

by setting targets (balanced scorecard). For planning how to actually make the strategy happen, strategic action programs are developed and prioritized in step four along with the definition of responsibilities, milestones, timeframes, and budgets. In step five, key management processes of the organization such as budgeting (resources), target agreements and skill development (people) and reporting (management information) will be aligned to strategy. The final step, execution and strategic control, ensures that strategy becomes a continuous process. This requires regular attention in management meetings, a dynamic perspective on the strategic action portfolio, monitoring of the environment and the assumptions as well as a regular test and adaption of strategy. It leads to a review and update of strategy maps, key performance indicators, actions, budgets, etc. based on new strategic insights (Kaplan/Norton 2008; Horváth & Partners 2007).

With reference to the seven generic elements for successful strategy execution, which are explained in chapter 4.1.2, the expected results of this process are:

- Individual strategic perspectives and priorities are aligned across the organization based on well articulated strategic *goals* to optimize the company's overall results.
- The executive leadership team gets the relevant *information* for managing the strategy, meaning a focused, balanced, consistent, and reliable view on strategic performance.
- Time and funds are allocated to the highest value strategic *actions* to improve the overall business impact.
- People are committed to individually contribute to strategy execution and have the corresponding *skills*, *resources*, *rewards*, and *attitudes* (motivation) to excel.

Fig. 4-40

Developing a strategy execution system in six steps

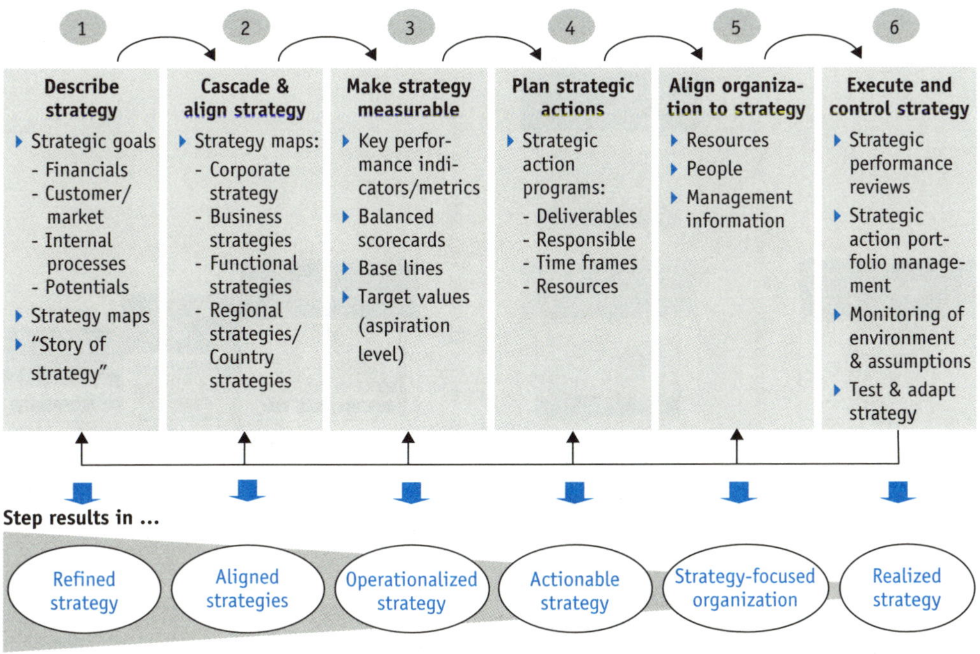

Source: following Kaplan/Norton 2008 and Horváth & Partners 2007; adapted

Practical guidelines for conducting the six steps illustrated in figure 4-40, foundations of the corresponding methods and processes as well as expected benefits, watch-outs and pitfalls of each step will be explained more in detail in the following chapters.

4.3.2 Describe Strategy

Strategic goals and strategy maps

Employees are key addressees of the strategy. They are looking for meaning and motivation in their work. The first step in strategy execution is to further sharpen the strategic direction by describing strategic priorities clearly as a base for aligning them systematically across the different organizational units. This can be facilitated effectively by the use of strategic goals and the corresponding strategy maps. The objective is to create a shared understanding and commitment for what should be achieved in the organization.

Bridge between strategy formulation and execution

Clear strategy description creates a bridge between strategy formulation, on one side, and performance measurement as well as the execution of strategic initiatives, on the other side (see fig. 4-41). Combined with balanced scorecards (measurements with

Fig. 4-41

Strategy map as a linking pin between strategy formulation and strategy execution

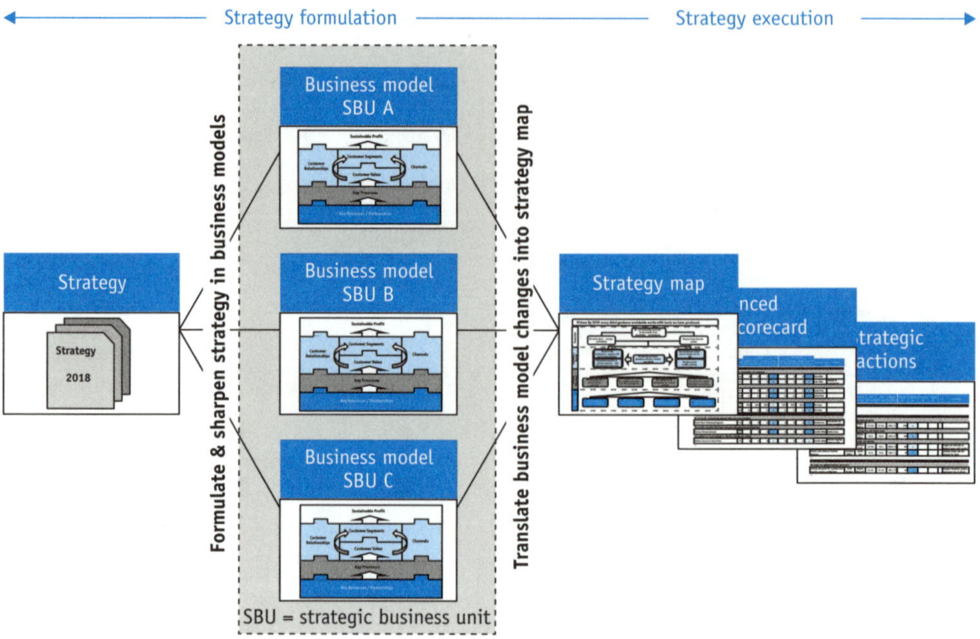

Source: Wunder 2013: 358; adapted

target values) and strategic action programs, strategy maps are powerful vehicles for a successful strategy execution (Kaplan/Norton 2004, Wunder 2012). In a strategy map, only those strategic priorities that require special management attention and special resources to provide results in the future are represented. These strategic priorities are expressed in the form of so-called strategic goals, which will be explained next.

4.3.2.1 Deriving Strategic Goals

Strategic goals require special attention and resources.

The first step for refining the strategy is the development of a limited number of strategic goals. These are action-oriented statements or objectives, which put the strategy in concrete terms. Note that this is not strategy development. Deriving strategic goals assumes that key strategic decisions are already made in advance. They represent elements of the strategy or business model that require special management attention and resources. In this sense, strategic goals refine and sharpen the strategy by identifying and precisely articulating the objectives that matter most. Strategic goals cover a variety of topics and are derived by asking guiding questions along the four perspectives introduced earlier (following Horváth & Partners 2007: 41–43). This is illustrated in figure 4-42 and will be explained in the following section.

Fig. 4-42

Translating strategy into strategic goals

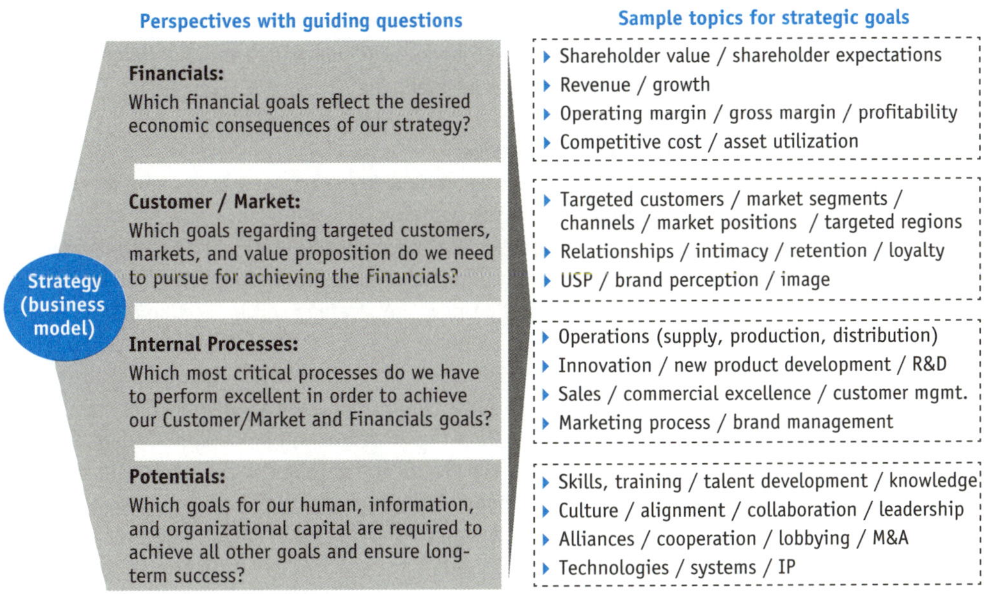

Source: Horváth & Partners 2007: 44; adapted

Financials Perspective:

Which financial goals reflect the desired economic consequences of our strategy?

- In profit-oriented private companies the financials perspective typically addresses the capture of shareholder value, growth aspirations, profitability objectives, and goals regarding the cost structure. Companies following a triple bottom line philosophy may want to adapt and rephrase that top perspective in order to reflect a more sustainable and stakeholder-oriented approach (e. g., "corporate sustainability" instead of "financials"). In such a case, the ultimate result of the strategy—the company's performance—would be articulated in terms of financial, social and ecological goals in the top perspective, and the goals in the subsequent three perspectives would logically drive those results (see also chapter 1.1.3.4). Also, in non-profit organizations the financial perspective is typically replaced with a "performance mandate" or "mission" perspective at the top that shows how the organization defines its non-financial purpose and success.

Customer/Market Perspective:

Which goals regarding targeted customers, markets and value proposition do we need to pursue for achieving the financials?

- In the customer/market perspective, organizations articulate which value they want to create for which customer segments. It covers goals related to the focused market segments, channels and regions that a firm targets in its strategy. Also, elements related to the desired relationship with certain customer groups, customer retention, unique selling propositions, brand perception or brand development as well as the desired image and customer awareness might be addressed with strategic goals in this perspective.

Internal Processes Perspective:

Which most critical processes do we have to perform excellently in order to achieve our customer/market and financials goals?

- Once an organization has articulated a clear picture of its targeted value proposition and the corresponding financial goals, the strategic goals in the internal processes perspective describe how the strategy will be accomplished. To find the vital few processes that are most relevant for making the strategy happen, the existing company's process model is compared with the defined strategic goals, for example, by establishing a strategic process alignment (SPA) matrix (see fig. 4-31 in chapter 4.2.2.4). Once the most relevant processes are identified, it needs to be evaluated whether they fulfill the strategic requirements. Does the current process performance support the achievement of strategic goals regarding customers and financials sufficiently? Furthermore, are the current operational process priorities (e.g., time, quality or cost) in line with the strategic priorities? Unless they are, corresponding strategic goals are defined for those processes, which ensures that they get the corresponding management attention and resources in the subsequent execution process. (Bergsmann 2012: 209–211 and Kaplan/Norton 2004: 43)
- The internal process perspective does not represent the process model of the company or organizational unit but highlights only those few processes that are of utmost importance for achieving the customer/market and financial goals and that require special management attention and resources. "The art of strategy is to identify and excel at the critical few processes that are the most important to the customer value proposition" (Kaplan/Norton 2004: 47). These might be processes related to operations, innovation and R&D, customer management including marketing and sales, market research and brand management as well as regulatory and social processes (Kaplan/Norton 2004: 11 f.). The achievement of goals in the internal process perspective represents leading indicators for realizing the customer/market and financial goals.

Potentials Perspective:

Which goals for our human, information, and organizational capital are required to achieve all other goals and ensure long-term success?

In the potentials perspective—originally called "innovation and learning" (Kaplan/Norton 1992: 72) and later "learning and growth" (Kaplan/Norton 1996: 9)—critical enablers for the strategic goals of the other three perspectives are addressed. The performance of potentials is another leading indicator regarding the achievement of strategic goals in the other three perspectives. Potentials may cover goals related to employees, skills, and knowledge as well as organizational culture, collaboration, and leadership. Typical goals in this perspective are also related to alliances and cooperation as well as the technological infrastructure. By using the more generic term "potentials" instead of labels like "learning & growth", "employees" or "innovation" a one-sided emphasis on certain potentials is avoided.

Adjusting or extending the standard perspectives

The four perspectives explained before can be considered as an empirically applied and practical framework but not as a dogma. Most companies applying such an approach for their strategy execution use the four standard perspectives explained above but sometimes label them differently to emphasize certain themes which are strategically important to them (Horváth & Partner GmbH 2008: 17). Examples are

- "financials & growth" or "financials & investors" for the financials
- "customer orientation & competitiveness" or "market, customer & competition" for the customer/market perspective
- "process excellence" or "productivity, processes and quality" for the internal processes perspective
- "people", "know-how" or "resources" for the potentials perspective.

Companies may extend or change the four perspectives to reflect their specific business situation (see also Strategy Practice Example 4.8 in chapter 4.3.2.2). Examples are a "goods and procurement" perspective at a trading company, an "innovation and technology" perspective at a company in a high-tech industry, or an explicit "supplier" perspective for a company that has a high percentage of vendor parts. Furthermore, companies may want to explicitly address strategic goals that have a strong impact on their performance but are outside of traditional market mechanisms in an additional "non-market perspective" to link sustainability management to business strategy (Figge et al. 2002). This can also be achieved by integrating corresponding social and environmental elements relevant for strategy in the standard perspectives (Wunder 2014a: 75 f.) or by combining both approaches in various ways (Hansen/Schaltegger 2014).

Guidelines for developing strategic goals

Strategic goals are the core for an effective strategy execution framework. They are the building blocks of the strategy map and for the corresponding strategy alignment process and are the basis for developing key performance indicators and strategic actions. With this in mind, there has to be a special emphasis on developing the "right" strategic goals as they ultimately determine whether the strategy can be executed successfully. Strategic goals can be derived by analyzing strategy documents, conducting strategy interviews, and by ideation within a group of people in a strategic goal development or strategy mapping workshop. To ensure an effective process, it is advisable to combine the different approaches and prepare a list of strategic goal candidates based on documents and individual conversations as a starting point for a workshop. Figure 4-43 provides eight principles that can be used as a guideline for developing strategic goals. Some principles are more related to the selection of strategic goals (1–4 and 6), others refer to the question of how to phrase a strategic goal best in order to make it understandable and inspiring (5, 7, 8). The principles have been grouped to the four categories relevance, perspective, altitude, and articulation, and are explained in the following (Horváth & Partners 2007: 155–202; Wunder 2005 and 2001: 134–136):

Fig. 4-43

Eight principles for developing strategic goals

Source: Horváth & Partners 2007: 162; adapted

Relevance

Selecting and specifying strategic goals

The definition of strategic goals requires that fundamental strategic questions (e. g., targeted markets, focused customer segments, etc.) are clarified in advance. In case this is not done, it will make it very difficult to find the relevant strategic goals due to missing strategic orientation. But even if the strategic direction is somehow decided on, there are typically plenty of strategic goal candidates that are suggested by managers involved in this process. Each of these goals usually has a solid strategic rationale regarding the company or the business whereas others might be driven by the personal agenda of the people who suggested them. All these strategic goal proposals need to be prioritized and cut down to a manageable number of goals (see principle 3). If strategy is understood as a set of "yes/no"-decisions or "tradeoffs" (Porter 1996: 70; see also chapter 1.1.2.3), the process of selecting and specifying strategic goals is of utmost importance for strategy clarification and, later on, for strategy communication and execution.

Business model changes become strategic goals.

Important sources of identifying relevant strategic goals are changes in the business models that have been decided on during strategy formulation (see fig. 4-44). In case of a strategy map for a certain business unit, changes in the business model might be directly selected and put into the corresponding strategy map if they fulfill the criteria for strategic goal development. Strategic goals for the top strategy map of a multi-business company, on the other side, could be sourced from different business models in the organization following the same logic (Wunder 2013: 355–358, Horváth & Partners 2007: 162–176; see fig. 4-41).

High competitive impact and necessity to act

To support the process of strategic goal development, the 2x2 matrix illustrated in figure 4-45 can be applied. The framework with its two dimensions *competitive impact* and *necessity to act* (see principles 1 and 2 in fig. 4-43) can be used as an effective

Fig. 4-44

Translation of business model changes into strategic goals for the strategy map

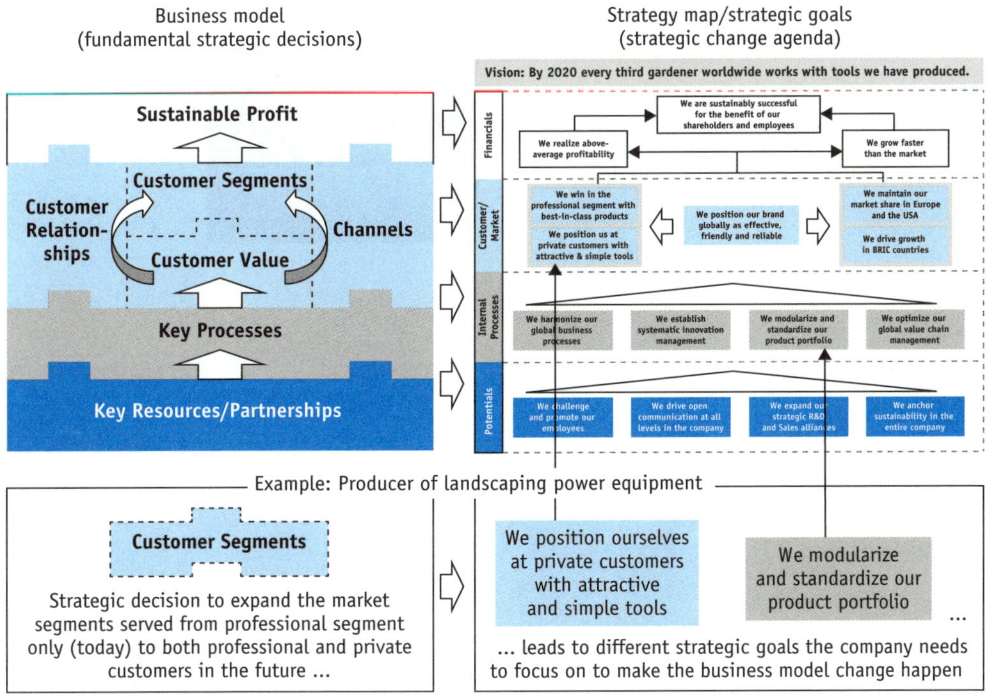

Source: Wunder 2013: 355; adapted

heuristic for prioritizing and selecting strategic goals (Horváth & Partners 2007: 165–168). The "filter" is based on the idea that only strategic goals that make a significant difference for the shareholders—or stakeholders depending on the company orientation—and help to win in the marketplace (competitive impact) AND that have a high need for action (necessity to act) are selected and put forward. These are the strategic priorities that require special management attention and resources to make the strategy come true. Identifying those few goals that are highly relevant for strategy execution is a balancing act and requires making tradeoffs.

"Must" priorities are not necessarily strategic goals.

An example of a goal proposal that would be considered as "low priority" for strategy execution is "improve efficiency in payroll processing". This might be an operational priority for the accounting department, but it is not likely to have a strong impact on the company's competitive situation in the market. Furthermore, there seems to be no urgent need, as even with an inefficient process, the payroll can be processed and people get paid. Also goals that require only selective attention due to a high need for action are filtered out if they do not have a strong competitive

Fig. 4-45

Horváth & Partners filter for prioritizing and selecting strategic goals

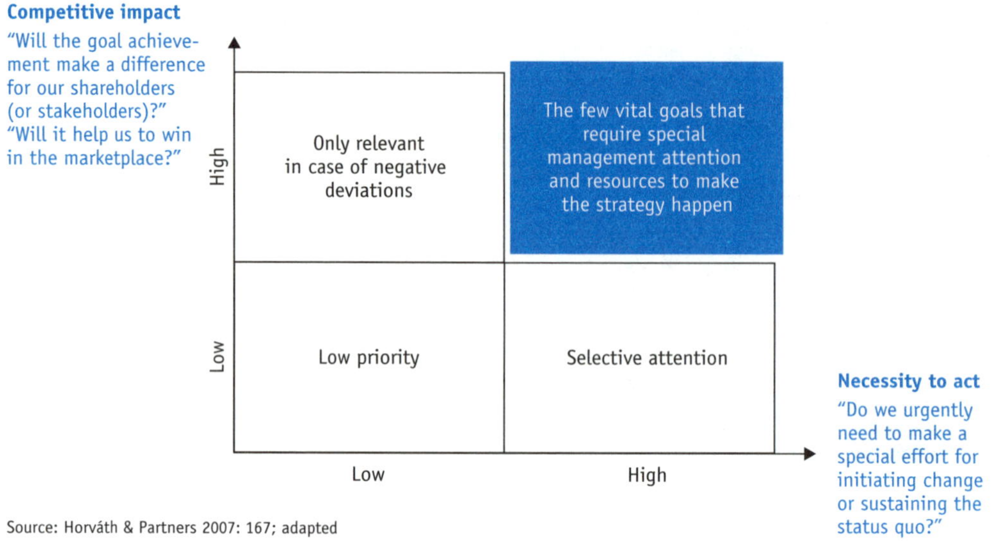

Source: Horváth & Partners 2007: 167; adapted

impact. An example was the so-called year 2000 problem or millennium bug. For several reasons, companies had to prepare their IT systems for the millennium change to avoid interruption of their operations. At this time, this was a severe problem with a high necessity to act and a deadline at December 31, 1999. However, all companies had to tackle that issue as a "must" without having an impact on how to win in the market. Other examples are required ISO or other certifications, the fulfillment of compliance requirements, the expansion of office space due to intense growth, or the breakdown of a machine in a manufacturing line that has to be solved quickly. The latter would only make a long-term strategic difference, if the manufacturing line was continuously facing those issues and the company was losing important customers.

Watch out for deviations from the status quo.

There are also goals with high competitive impact but little necessity to act. These are probably the most challenging objectives to discuss. Consumers in Europe, for example, expect a high level of food safety covering hygiene and quality standards when they buy products in the supermarket. There is no doubt that food safety is of high competitive relevance in the food and beverage industry. In general, however, the expectation is fulfilled by food producing companies offering trusted brands in this market. If those companies feel that there is and will be no issue with food safety as they have the corresponding processes in place and well managed, there is no necessity to act, that is no special attention and resources are required to maintain this standard. In those cases, this goal would not be considered as a strategic goal for

Strategic Snapshot 4.13

Strategic Goals and Body Temperature

Nobody will deny that body temperature is one of the most important indicators for well-being and health. Typically, this is one of the first things medical people look at or ask for if their patients do not feel well. However, there will be hardly any person who measures its body temperature regularly, say once a week. Why do we not focus our attention and resources continuously on this indicator if it is such an important factor for our well-being? Because it only becomes relevant in case of deviations, that is fever. It would be a waste of energy to focus on such indicators if they are within a normal range. The same philosophy applies to goals that have a high competitive impact because they are important factors for winning in the market place, but that are already well managed by the company. Because they are in good shape there is no high necessity to act. They only become *strategic* goals in case of deviations. If companies would not filter those items out but put everything forward, they would drown their executive leadership teams in strategic priorities, and they would not be able to focus attention and resources to the few vital goals for effective strategy execution. Those goals would have only diagnostic character. They can be easily monitored as part of the operative performance management system and raised to the strategic agenda only in case of severe deviations (management by exception).

the company (although it might be one in the food safety or quality department). It becomes strategically relevant only in case of deviations, for example, if the company has a difficult situation in this area, has been in the media for food safety issues and is losing market share due to this problem. As an analogy, people rarely measure fever if they do not feel ill (see Strategic Snapshot 4.13). Contrary to the illustrated example, the relevance of the same goal might be very different for a local food producing company that strategically plans to enter the European market if it is based in a country with moderate food safety standards. In such a case, a goal like "improve food hygiene and quality" might well be a strategic goal that needs special attention and resources in order to be well positioned for the new customer expectations, competition and regulations in the targeted regional market.

Strategy relates to critical performance variables.

This philosophy for identifying strategic goals strongly corresponds to the principle that strategy is about "being different and choosing what (not) to do" (see chapter 1.1.2.3). In essence, it is about distinguishing critical performance variables from diagnostic performance indicators (Simons 1995: 63–70), which can be illustrated with reference to the competition on quality that happened in the 1980s between U.S. and Japanese companies (see Strategic Snapshot 4.14). Executing strategy requires the identification of critical performance variables which correspond to the "right" strategic goals.

Both running a company and providing direction is important.

At the end of the prioritization and selection process there should be some vital few strategic goals that provide direction and determine the success of strategy. They should have a high competitive impact and a high necessity to act. Low priority goals, issues in need of action but without competitive impact, and purely diagnostic competitive aspects with little necessity to act should be filtered out of strategy. This does not mean that those items are not important for running the company. Sometimes people feel less relevant if "their" goal proposal does not make it onto the strategy map. This issue can be addressed by using an analogy: A boat on the ocean needs both an engine and a direction. None of these two elements can be skipped. Without the

Strategic Snapshot 4.14

Critical Performance Variables versus Diagnostic Performance Indicators

"After years of hard and diligent work, many U.S. companies have achieved excellent quality and are now at parity with their foreign competitors. At this point, quality has been neutralized as a competitive factor. Of course, companies need to maintain existing quality and continue to make continuous improvements, but quality is no longer a critical factor for determining strategic success. In such a situation, quality is monitored diagnostically, and the company needs to find other dimensions in its value proposition to distinguish itself from competitors. These other dimensions are the core of the balanced scorecard." (Simons 2000: 202)

engine there is no moving forward, without the direction there is no arrival. Goals with high competitive impact and high necessity to act represent the direction. Other important goals are related to the engine (Horváth & Partners 2007: 175).

Perspective

"Twenty is plenty"

Key principles for an effective strategy execution are to focus attention and resources on a *limited and balanced (principle 3)* set of strategic priorities and to distinguish strategic goals and actions clearly. First, it is important that goals are not only identified according to their strategic relevance but also limited to a manageable number (Nagel 1992: 2632). No more than 20 strategic goals are recommended for describing a strategy effectively and keeping execution feasible. The guiding principle "twenty is plenty" was shaped in the 1990s by ABB Switzerland, one of the early balanced scorecard adopters and has been applied in many companies since then (Horváth & Partners 2007: 163). This is in line with the recommendation of having no more than 20–25 measurements on a balanced scorecard (Kaplan/Norton 2001: 375), which implies no more than 1–2 metrics for each strategic goal (see 4.3.4). Experience has shown that less is more. However, it is difficult to get management teams to focus on "only" 20 or less strategic goals for their company or business unit in the first place. The same teams, though, tend to reduce complexity of their strategy execution framework by cutting down the number of strategic goals over time (Wunder 2004a: 223). Obviously complexity first needs to increase in order to be reduced subsequently later on.

Linking intangible assets to tangible results

Second, strategy is more than just profitability and revenue targets. It is about "linking intangible and tangible assets in value-creating activities" (Kaplan/Norton 2001: 67). Effective strategy execution requires that strategic goals are adequately *balanced*, that is they are financial and non-financial, tangible ("hard") and intangible ("soft"), externally and internally oriented, results (lagging) and drivers (leading) as well as short-/mid-term and mid-/long-term. These criteria are summarized in figure 4-46 and are applied to both strategic goals and measurements (KPIs) of a balanced scorecard. If strategic goals are developed and balanced according to these criteria, it leads to an equal spread across the four perspectives explained earlier (Kaplan/Norton 2001: 360 and 375 f.).

Strategic goals differ from strategic actions.

According to the forth principle (see fig. 4-43), strategic goals paraphrase the *desired outcome ("what")*, not the means (actions) for achieving this outcome (the "how"). Certain goals (drivers) foster the achievement of other goals (results) as illustrated in a strategy map. However, they all articulate a desired future state in their corresponding perspective derived from the strategic direction, and each requires certain initiatives to be achieved. The main reason for having strategic goals is to articulate the strategic intention. They need to be distinguished clearly from strategic actions or initiatives that serve the achievement of strategic goals. Although strategic goals may be realized with different time horizons (see fig. 4-46), they are in general gradually pursued over a longer period of time and not only temporarily. Their achievement strongly depends on the chosen metric and target value (see chapter 4.3.4). Strategic actions, on the other hand, are shorter term projects with clearly defined deliverables, budgets, responsibilities, and due dates. Actions have a finite duration and can be checked as completed/not completed at the end (see chapter 4.3.5), whereas goals are pursued gradually and might be adjusted over time due to changes in the competitive environment. Based on this differentiation, a statement like "implement activity-based costing" (ABC), for example, represents a strategic action and not a goal. If this had been suggested during strategy refinement, it would be necessary to analyze the underlying strategic relevance. If the company wants to implement ABC because they want to strategically focus on high-margin customers but have no understanding of their customer profitability, a corresponding strategic goal can be derived: "improve customer profitability" or "drive growth with key accounts". This guideline strongly relates to the next category, altitude, as strategic actions on a higher organizational level might function as strategic goals on a lower level (see Strategy Practice Example 4.7).

Fig. 4-46

Dimensions and criteria for balancing strategic goals and metrics

Dimension	Criteria		
Evaluation unit	financial	↔	non-financial
Measurability	tangible ("hard", "objective")	↔	intangible ("soft", "subjective")
View/perspective	external (Financials, Customer/Market)	↔	internal (Internal Processes, Potentials)
Past/future orientation	results ("lagging")	↔	drivers ("leading")
Duration	short-/mid-term	↔	mid-/long-term

Source: Wunder 2004a: 124; adapted

Strategy Practice Example 4.7

Volkswagen Building an Assembly Plant in Chattanooga: Goal or Action?

In 2008, Volkswagen announced its intention to build a new automobile assembly plant in Chattanooga, Tennessee (USA). The company started production in 2011 with a capacity of 150,000 cars per year, and operates with about 1,700 employees. Volkswagen invested more than $1 billion in the development of the site (Volkswagen 2014).
Is "building an assembly plant in the U. S." a strategic goal or a strategic action according to the principles for developing strategic goals explained (see fig. 4-43)? In other words, would it be expected to be put on a strategy map and communicated as one of the top 20 strategic priorities, or is it rather something for the strategic action program? The answer depends on its "altitude", that is the level in the organization. For the Volkswagen group (top level), it probably had more the character of a strategic action that supports the strategic goal of boosting sales and increasing market share primarily in the U. S., and as a lever for becoming the world's biggest carmaker. Building the Chattanooga plant was considered a multi-year project on a VW group level with a "project manager" who became CEO of the Volkswagen Group of America Chattanooga Operations LLC once the "project" was completed. From the regional perspective of the Volkswagen Group of America, however, such a huge local undertaking may well be considered a strategic goal that requires special management attention and resources on a regional level and is one of the top strategic priorities locally.

Altitude

Generic strategy statements should be avoided.

Strategic goals should be *meaningful (principle 5)* to the people that are expected to contribute to the goal achievement. In other words, the goals should articulate strategic priorities as clear and precise as possible. "Frequently we all get trapped into signing on to a strategy because it's written at a sufficiently general level so that it can mean whatever we personally want to read into it." (Kaplan 1994: 19). General statements or buzz phrases that are true for any company in any industry such as "outstanding customer satisfaction", "optimized supply chain", or "excellent employee qualification" should be avoided as otherwise people will not get a clear understanding of what the company wants to achieve specifically (Wunder 2001: 135 f.). Which customers are particularly important for our strategy and why? What does "optimize" mean: Are we too slow, expensive or do we have issues with quality? Which part of the supply chain are we talking about? Where are our skill gaps regarding the strategic aspiration? What particular qualification do we need to develop with which employee groups to make our strategy come true? The chief executive officer of a leading European oil corporation puts it this way referring to the process of refining strategic goals: "In this process, we became aware of having a number of lofty statements in our strategy ... Merely preaching that we want to be bigger, nicer, and stronger are not strategic objectives." (Wunder 2004a: 231 and 2005: 37) For finding the right level or precision it sometimes helps to ask how to know whether the goal is achieved, though, this should not lead to an early discussion of measurements and target values. As experience has shown, elaborating on strategic goals at the beginning and postponing the sometimes difficult discussion about KPIs and target values (see chapter 4.3.4) is one of the success factors for developing a strategy execution system. It fosters the discussion and consensus building about strategic content instead of nitpicking over metrics and numbers that might be driven by a strong political agenda of the people involved (Horváth & Partners 2007: 176).

Cascading means refining strategic goals.

To avoid "lofty statements" it should be worked out which particular aspects that are related to those motherhood statements such as "excellent customer retention" are strategically important to focus on in the specific company situation. This could be a strategic goal such as *improve customer retention of wealthy private customers in our home market* for a domestic bank that faces issues with customers switching to aggressively competing foreign online financial institutions. The degree of precision depends on the organizational or hierarchical level where the strategic goals are defined. On the corporate level of a multi-business company, for example, strategic goals are usually more generic than on the business level as they should provide guidance for all businesses, functions and regions. The further strategic goals are cascaded in an organization, the more precise they should get as they represent the specific contribution of the corresponding unit, department or team to achieving the overarching strategic goal. For example, the company-level goal of an airline to *have all flights on time* could lead to the more refined goal to *increase the percentage of available airplanes* in a certain technical department and another goal to *ensure appropriate staffing for each flight* that might be relevant in the HR or personnel planning area in the cascading process (see chapter 4.3.3).

People defining strategic goals must have an impact on their achievement.

Through refinement, strategic goals become both meaningful to the employees that are supposed to execute on that level and *influenceable (principle 6)*. As a general guideline, units, departments or teams at a certain level in the organization should only define strategic goals that they can influence. In some cases, strategic goals are defined even though they may only be partially influenced by the corresponding people. Two reasons for this are: They have a role to play in driving that goal or otherwise nothing will happen, or the importance of an indirect contribution should be highlighted and communicated to improve strategic awareness. For example, some companies put the overarching financial strategic goals of the company in the financials perspective of functional departments like HR or IT in order to highlight how the corresponding department-specific strategic goals contribute to the overall aspiration. This way, a traditional financial focus on staying in budget is replaced by a thinking of how the functional department can enable the overall strategic goals.

Articulation

Understandable and inspiring statements

Strategic goals should be expressed in a way that they are understood clearly by the people who are supposed to drive execution, and that they are attractive to them. One way to support this is by using *action-oriented (principle 7)* statements with phrases like *build, maintain, enhance, develop, grow, establish, ensure, drive, improve, achieve, expand, leverage,* etc. This emphasizes that there is something to do regarding the corresponding topic, and implies that the individual contribution is needed to make that desired strategic change happen. Furthermore, adding an adjective to the verb could emphasize the desired intensity or aspiration level for communication purposes, for example, "aggressively grow in emerging markets" instead of just "grow". When doing so, it is important that the strategic goals still appear realistic and not as wishful thinking of executives, which could have a negative impact on the acceptance of employees. Furthermore, strategic goals should be phrased in a *compelling (principle 8)* way. Employees as the key addressees of strategy should not only understand the strategic goals clearly, but they should also be inspired by them. This way, it is more likely that people buy-in to the strategic content and want to contribute individually to the achievement of the strategic goals. In order to achieve this commitment, it is important that—apart from other factors such as the person's individual needs as well as the probability of success—employees understand clearly the strategic goals and why they are important (Wunderer 2001: 130; Latham/Locke 1995: 2224 f.).

4.3.2.2 Developing a Strategy Map

Externalization of mental strategy models

A strategy map expresses and visualizes the cognitive strategy model of the people who elaborated it. It tells the "story of strategy" through the arrangement of strategic goals and the intended logical linkages on *one* page. A well prepared strategy map with clear company-specific and compelling strategic goals can help employees to see the big picture and recognize their own strategy contribution. It transmits sense and drives performance.

Generic strategy map templates

A strategy map is based on several principles, which will be explained next (Kaplan/Norton 2004: 10–15; Gaiser/Wunder 2004; Horváth & Partners 2007: 187–202). Figure 4-47 illustrates those principles and shows generic components and relationships that can be used as an orientation for developing an own strategy map that, however, always needs to be customized to the organization's strategy.

Top-down logic

Upward logical flow of linked strategic goals

The strategy map architecture has a top-down logic. In case of a private enterprise, for example, financial goals are usually at the top of a strategy map. They are followed by strategic goals related to customers and markets that are required for achieving the financial goals. Here, the targeted customer segments and the value proposition that the company intends to deliver to those customers, such as best product, complete customer solution, or system lock-in, should somehow become clear (see chapter 3.4.2). Value is created through internal processes that enable or drive the customer/market, and thereby, the financial results. The potentials (or learning and growth) perspective describes intangible assets related to, for example, human capital, information capital, or organizational capital that are foundational for achieving all the other strategic goals. As a consequence, an upward logical flow evolves when linking the goals (although some goals may also be linked within the same perspective).

Few goal relationships

When developing a strategy map, it is not the idea to show all possible relationships between strategic goals in the sense of cause-and-effect chains as it was prescribed in the initial balanced scorecard concept (Kaplan/Norton 1996: 30 f.) or earlier on in the context of network thinking (Probst/Gomez 1987). Strategy maps focus on and illustrate only a few intended goal relationships. They should reflect the strategic way of thinking in terms of logical flows between strategic goals that are intended by the

Fig. 4-47

A strategy map represents how the organization creates value

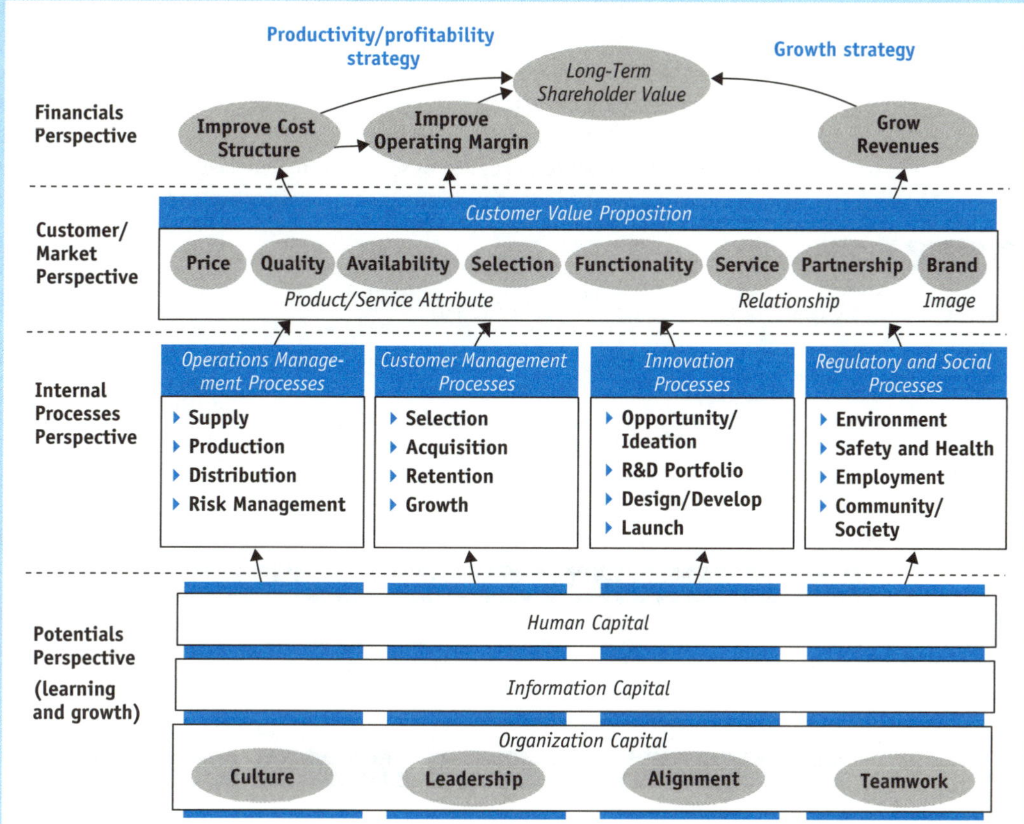

Source: Kaplan/Norton 2004: 11; adapted

corresponding management team, and support strategy communication instead of providing the illusion of analytically derived relationships or even quantified correlations. Using the analogy of a traffic network, the strategy map does not have to describe the entire road system of a country but to highlight the most important arterial roads or highways (Hahn/Kaufmann 2002: 535).

Strategic themes

Those logical vertical flows typically represent complementary strategic themes that the company is pursuing simultaneously (Kaplan/Norton 2008: 132–134 and 2006: 101–118). For example, by linking strategic goals vertically across the four perspectives, an efficiency-oriented theme for driving a competitive cost structure and profitability, on the one hand, and a growth-oriented theme with differentiation as a key

Linked strategic goals articulate strategic themes.

pillar, on the other hand, might evolve. Those resulting themes can be effectively applied for high-level strategy communication and alignment (see chapter 4.3.3.1). Strategic themes can also reflect the customer value proposition and how it is created. Companies such as Apple or Daimler following a differentiation approach with product leadership would typically emphasize strategic goals such as a high margin (financials) based on a market offering with innovative high-performance products as well as strong brands (customer/market) driven by excellent processes to capture customer insights, develop highly innovative product features and an exceptional speed to market (internal processes) as well as a company culture fostering innovation and creativity combined with strong technical expertise and a high level of cus-

Fig. 4-48

Generic strategy map template for a product leadership strategy

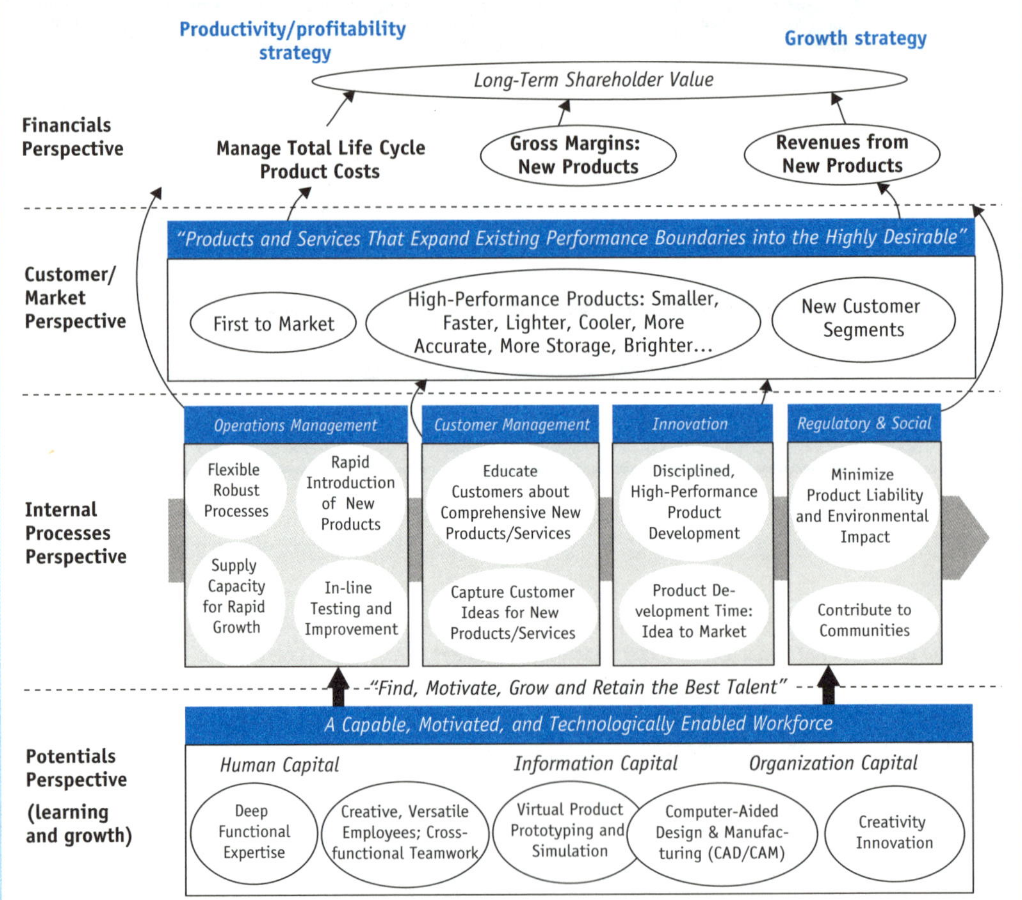

Source: Kaplan/Norton 2004: 327; adapted

tomer orientation (potentials). Those and other potential components are illustrated in figure 4-48, which shows a generic strategy map template for a product leadership strategy. Other templates are available for low total cost, complete customer solution, and (system) lock-in strategies (Kaplan/Norton 2004: 319–346). Each template illustrates potential components of the corresponding strategy pattern that are complementary across the four perspectives. They can be used for ideation or as a checklist when customizing an own company-specific strategy map. The four corresponding generic strategy patterns are explained in chapter 3.4.2.

Balance of contradictory forces

Contradicting strategic priorities become transparent.

Simultaneously pursuing strategic themes or the corresponding strategic goals of different perspectives can be quite challenging as there might exist contradictions (see Strategic Snapshot 4.15). The strategy map makes the contradiction of strategic priorities transparent, and thus, helps the management to "balance" strategic goals. It illustrates the relationships between financial and non-financial, tangible and intangible, externally and internally oriented as well as lagging (results) and leading (drivers) strategic goals of different time horizons constituting the organization's hypotheses that define its strategy. The strategy mapping process helps to clarify whether there is a shared understanding of those underlying strategic hypotheses in the management team.

Strategy maps can have a wide scope of designs, levels of detail, illustrative elements, and variety of contents (Kaplan/Norton 2004 and Horváth & Partner GmbH 2008: 13). Figure 4-49 shows a high-level strategy map that is designed to provide strategic guidance for all employees at the RECARO group. In this example, the four balanced scorecard standard perspectives explained earlier are adapted and rephrased to express the specific company situation and leadership philosophy (see Strategy Practice Example 4.8). The focus of this strategy map is put on simplicity with a few key messages, some graphical elements, and a few intuitively understandable linkages. The design and structure articulates clearly the strong branding focus of the group.

Figure 4-50, on the other hand, shows a more detailed version of a strategy map modified from a real-case example in the machinery and plant engineering industry with 16 strategic goals and 15 illustrated relationships (Wunder 2004a: 120). The different strategic goals and the illustrated relationships reflect a more analytical approach that tries to explain precisely the strategic rationale of the management team. The linkages are numbered for clarification and documentation purposes. Each relationship is described in a few sentences. By putting together those descriptions

Strategic Snapshot 4.15

Peter Drucker about Balancing Strategic Priorities

"There are few things that distinguish competent from incompetent management quite as sharply as performance in balancing objectives. There is no formula for doing the job. Each business requires its own balance—and it may require a different balance at different times." (Source: Drucker 2008: 111)

Strategy Practice Example 4.8

Strategy Map of the RECARO Group—the RECARO Brand in the Center

The RECARO Group is an international medium-sized family business with over 1,800 employees and more than € 350 million in revenue in 2012 (excluding license partner RECARO Automotive Seating). It looks back on a long corporate history with its origin in 1906 when Wilhelm Reutter established the "Stuttgarter Carosserie- u. Radfabrik" (Stuttgart Body and Wheel Factory). In 1963, Porsche acquired the body factory and REutter CAROsserie became RECARO focusing on its accumulated seating expertise since then. Today, it operates with two major divisions—Aircraft Seating and Child Safety—as well as the RECARO Holding. Additionally, customized seats for stadiums (team benches), for the office, and for control centers and 24/7 monitoring rooms as well as maritime seats for boats and yachts are offered under the RECARO brand today.

A few years ago, the company decided on a historic strategic change: The owner family sold its renowned automotive business covering 13 sites on three continents and approx. 4,800 employees as well as the biggest portion of the entire group revenue which was €1.26 billion in 2010. Since 2012, RECARO Automotive Seating has been a product business unit of the multinational U. S. company Johnson Controls, which licensed the RECARO brand from RECARO Holding. Customer references regarding RECARO

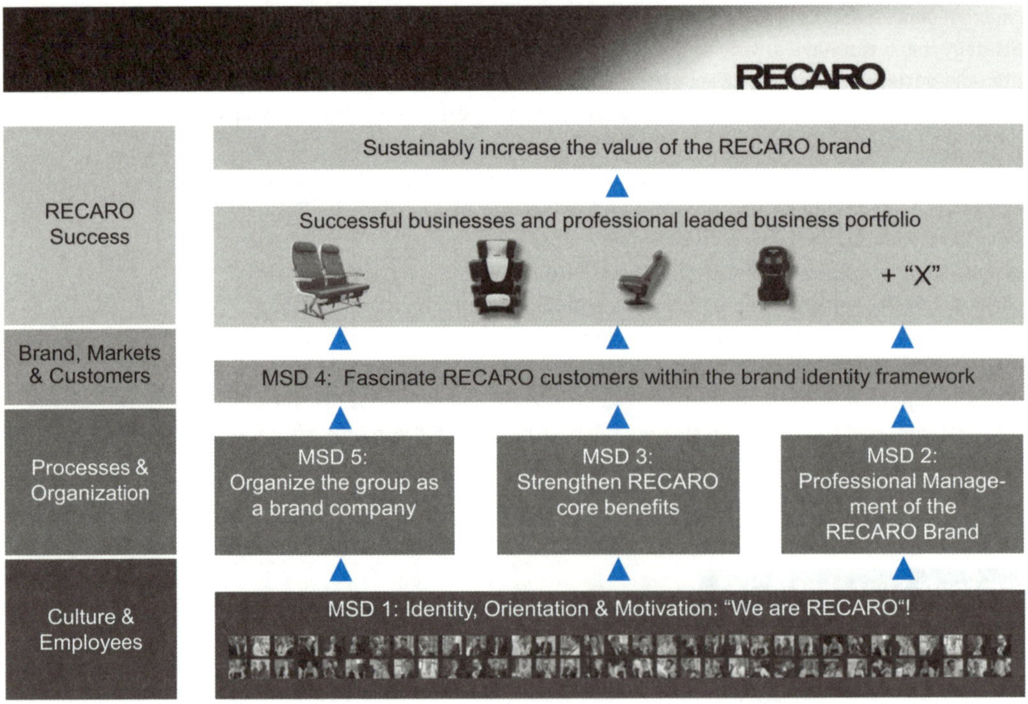

Source: Kellinghusen 2012: 122; adapted

Fig. 4-49: RECARO Group strategy map

seats include Aston Martin, BMW, Ford, Lamborghini, Mercedes-AMG, and the VW Group. Since then, the owner and his management team have been driving the evolutionary transition to a mid-sized branded family company. They put a professional management of the RECARO brand in the center of the strategy, which is articulated clearly in the group strategy map (see fig. 4-49). The ultimate strategic goal that determines the strategy's success is a sustainable increase in the RECARO brand value based on a well-defined business portfolio. The brand is considered a common "DNA" or umbrella that should be guiding all business activities. This is driven by four "main strategic directions (MSD)" in the three subsequent strategy map perspectives "brand, markets & customer", "processes & organization" as well as "culture & employees". All MSDs are related to the RECARO brand. The strategy map is used to summarize the results of a very rigid and systematic annual strategy process on one page that can be easily communicated to all employees. (Sources: RECARO 2014 and Kellinghusen 2012)

in a short essay, the so-called "story of strategy" is created, which describes the rationale of the company's strategy in one or two pages and can be used as a tool for further strategy clarification, communication and alignment (Horváth & Partners 2007: 195–202).

Intuitively understandable and meaningful

A combination of the more high-level approach and the more detailed version of a strategy map is illustrated at the beginning of chapter 4.3.1 (see fig. 4-36). This level of complexity turned out to be particularly useful for strategy alignment, which will be explained later (see chapter 4.3.3). Also, companies may use a more refined version to clarify strategy within the management team and use a more simplified version of the same strategy map for communication to employees and other stakeholders.

Not a company model but a strategic change agenda

Sometimes, a strategy map is confused with a business or operational model. The strategy map is not a description of the organization's day-to-day business activities. As explained earlier, strategic goals and their expected linkages are illustrated to direct attention and resources of managers and employees to those factors that are expected to lead to competitive breakthroughs (Kaplan/Norton 1996: 164; Gaiser/Wunder 2004: 458 f.). These breakthroughs can be achieved by different means such as the generic competitive strategies that have been explained earlier on in this book (see chapter 3.4.2). Also, strategy maps are linked to business models that are used to define more in detail how those breakthroughs are to be achieved and that are related to a specific business context.

4.3.2.3 Benefits and Pitfalls of Strategy Mapping

Strategy clarification, communication and alignment

Strategy maps provide a common framework for strategy clarification, communication and alignment through the externalization of mental strategy models. In practice, strategic action programs are often initiated too quickly, without prior creation of a shared understanding and commitment to the strategy among employees and managers who are supposed to execute it. The essential advantage of a strategy map is the intuitive, easy to understand and consistent logic for describing a strategy. Relevant tangible and intangible strategic aspects are clarified and integrated into a reasonable and understandable big picture. It can be implemented at different organizational levels and is applicable for large multinational corporations, small and

Fig. 4-50

Strategy map of a machinery and plant engineering company (modified)

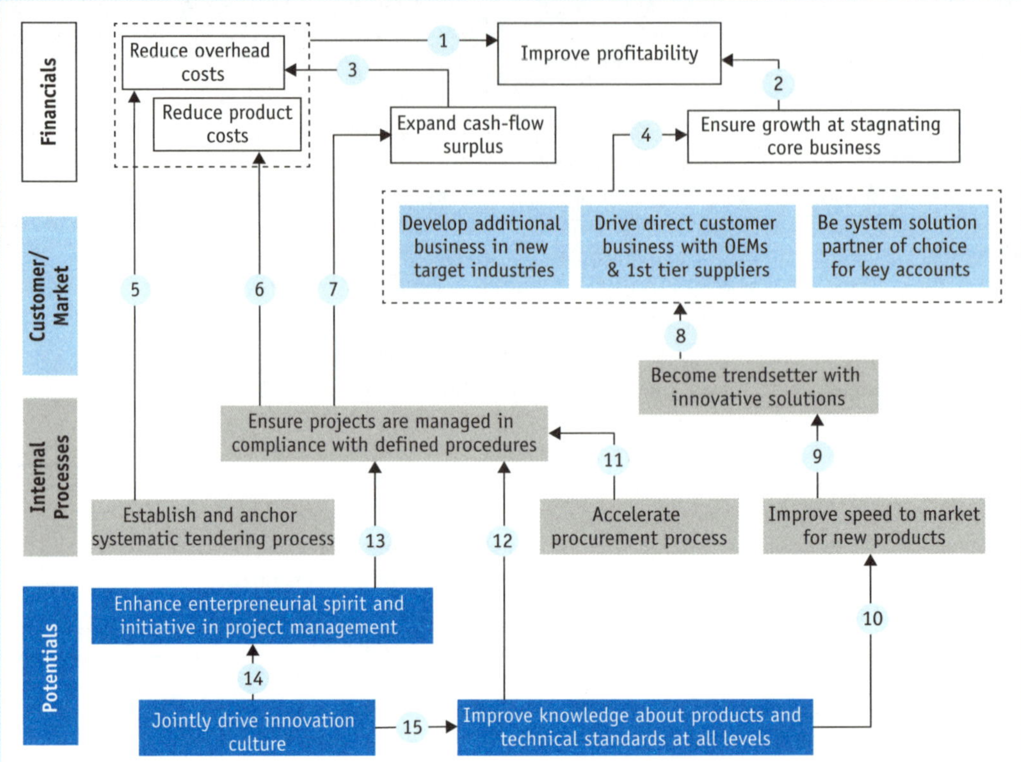

Source: Wunder 2004a: 120; adapted

medium-sized companies as well as departments, teams or individuals. Main benefits of applying a strategy map are:

- clarification of the strategy at the executive level
- evaluating plausibility and consistency of the strategy
- communicating the strategy and aligning employees and other stakeholders
- alignment of business, functional and regional/country strategies
- facilitation of strategic project prioritization ("filter").

Consistent strategy articulation with the "Story of Strategy"

Clarification of strategy at executive level

Experience has shown that aligning a group of managers around a list of unconnected strategic goals does not automatically mean that everybody will have the same understanding and tell the same strategic story related to those goals even though they may have developed these objectives together. In fact, unless the foundational strategic hypotheses are jointly discussed and clarified by illustrating and debating the intended linkages, there is a high risk for divergent mental strategy models from the people involved. This can lead to the articulation of inconsistent strategy messages in different areas of the organization later on when these leaders try to inspire their employees about the strategic priorities but each of them tells a different story. The result would be confusion and a lack of credibility. With the use of strategy maps a shared understanding not just for individual strategic priorities but also for the corresponding "story of strategy" is created. Such strategy stories help to layer out the overarching strategic themes of the company which can be more comprehensively articulated in narrative than in strategy map form. This can be particularly beneficial in situations where such a shared understanding is developed for the first time, for example, at newly formed management teams after a reorganization, mergers and acquisitions or personnel transitions, or in cases where systematic strategic team discussions had been missing in the past.

Strategy maps for key competitors

In addition to applying a strategy map framework for clarifying one's own strategy, it can also be used to describe the strategy of key competitors and gain a deeper understanding of their strategic intentions (Gaiser/Wunder 2004: 459). In this case, the strategy map becomes an additional tool for competitor analysis (see chapter 2.2.3.2).

Evaluating plausibility and consistency of the strategy

Validate if the strategy makes sense.

As part of the strategy clarification process described, strategy maps can also be used to check the plausibility and consistency of strategies (Gaiser/Wunder 2004: 460). This step is typically done during the strategy option evaluation and choice (see chapter 3.5). An additional validation cycle after having refined strategy with a set of integrated strategic goals might be necessary before moving on with the development of metrics and action programs. Key questions are:

- Does the strategy map fit well the results of our strategic analysis and our assumptions about the future?
- Are the strategic goals positioned logically and without contradiction?
- Does the strategy map describe the strategic priorities completely?
- Are the strategic themes reasonable?

Communication of strategy and aligning employees and other stakeholders

Employees understand their strategy contribution.

As stated earlier on, a lack of awareness for strategic priorities is a major barrier to a successful strategy execution. "Our research indicates", state Kaplan and Norton, "that less than 5 % of the typical workforce understands their organization's strategy" (Kaplan/Norton 2001: 215). Only if employees get a clear understanding of what the organization wants to achieve, they can adapt their work to contribute to effective strategy execution. With a well-designed strategy map—usually integrated in a broader strategy communication campaign (see 4.3.6.2)—this issue can be tackled. Companies using strategy maps report an improvement of strategy communication in their companies or departments (Horváth & Partners 2007: 15).

Stakeholders buy-in to strategic priorities

A strategy map can be used for effective and consistent strategy communication not only to employees but also to other stakeholders such as banks, shareholders, board of directors, key customers and suppliers, or alliance partners (Kaplan/Norton

Strategic Snapshot 4.16

Strategic Alignment for a Fundraising Campaign at the University of California

At a campus of the University of California the strategy map approach was used to align the university's leadership team and its foundation trustees for a major fundraising campaign. Strategic goals of the University regarding professional schools, enrollment, quality of life as well as teaching, learning and research were stated in the top perspective. Their achievement should be supported by the campaign, which was articulated with a set of 15 strategic goals across the subsequent three strategy map perspectives. Those strategic goals were related to strategic themes regarding the campaigns focus areas, donors, planning and messaging, financing, aligning campus groups, staff and infrastructure as well as best practice sharing. The strategy mapping process was based on several strategy interviews with the chancellor and vice chancellors of the University as well as with selected trustees of the university's foundation. The articulated strategic priorities were jointly discussed, prioritized, refined, and linked during the foundation's executive retreat. The resulting strategy map and a written three-page "story of strategy" was used as a baseline for further refining and focusing priorities and planning their execution.

2006: 193–245). All those groups should be able to understand quickly what the company or organizational unit strategically wants to achieve and buy-in to those strategic priorities. The process of establishing alignment by creating a shared understanding and commitment can be even more powerful when strategic goals and the corresponding strategy map are developed jointly by different stakeholder groups (see Strategic Snapshot 4.16).

Alignment of business, functional and regional/country strategies

Shared and aligned strategic orientation

In strategically-managed multi-business organizations developing strategy maps alone is not sufficient. Successful strategy execution substantially depends on an effective strategy alignment among the different organizational units. In and between these units individual and partly conflicting strategic priorities need to be identified, adjusted and coordinated to achieve the best overall result. Strategy maps can effectively support this process. They can be used for systematically cascading and synchronizing strategic goals of different businesses, support or functional units as well as regions within an organization. This will be explained in detail in chapter 4.3.3.

Facilitation of strategic project prioritization ("filter")

Some organizations are applying strategy maps and strategic goals for tackling the issue of disconnected initiatives mushrooming throughout their different organizational units. They want to identify which of the numerous siloed projects that evolved in their organization over time are still relevant and which can be stopped (see Strategy Practice Example 4.9).

Watch out: simplicity versus comprehensiveness

The advantage of simplicity by focusing only on a few selected strategic goals and their relations can also be a significant pitfall, because cause-effect relations are represented only on a fragmented basis. A sound analysis of correlations between strategic goals in the sense of calculable results and drivers is very difficult because the variables in the strategy map are not comprehensive enough. Also, practitioners using this tool tend to summarize as much strategic aspects as possible in one strat-

Strategy Practice Example 4.9

Strategy Execution at the American Red Cross

As part of a large strategy program, The American Red Cross Blood Services division conducted a rationalization exercise regarding its ongoing projects that evolved across its highly decentralized organization and in its different independent business units over time. Top-level strategic goals and strategy maps where developed to guide execution and to function as a strategic filter for prioritizing ongoing initiatives. American Red Cross leadership wanted to understand to which degree the different identified projects contributed to the organization's strategy. As a result of this exercise, the organization's project portfolio was reduced from over 200 siloed projects to about 50 strategic initiatives that were highly accountable and measurable for performance improvements. These key initiatives, in turn, helped to expand market share, develop new relationships with hospitals, and improve customer satisfaction. (Source: The North Highland Company 2014)

Strategic Snapshot 4.17

Recommendations for Strategy Map Development

- You have to distinguish strategic goals in the strategy map (rather long-term, being achieved gradually) from strategic actions (rather short-term, being evaluated as "done/not done"). Actions are derived after the goals are defined and are not placed on the strategy map.
- Strategy mapping can be more effective and efficient, if strategic goal candidates are defined in advance based on individual strategy interviews. There are often many ideas and perceptions about strategy in the organization. In a strategy map workshop the proposals are then discussed and sharpened among management team members, reduced to the maximum of 20 relevant goals, arranged and linked according to the strategy map perspectives.
- To support strategy clarification and refinement, so-called strategy map templates for different generic strategy types (product leadership, low total cost, lock-in, complete customer solution) can be used. They can be considered as a checklist or question catalogue for the strategic goal definition.
- Avoid discussions about measurements and target values when developing strategic goals or use it only if it is necessary to clarify the meaning of the goal. Otherwise there is a risk that the strategic goal development becomes a number-driven and rather political process instead of a constructive strategic discussion.
- Visualize the strategic goal relations in a way that it provides clear strategy messages and supports strategy communication. Here, creating a shared understanding and commitment is more important than analytical accuracy in the sense of "cause-effect".
- Document each strategic goal with a goal description. This provides clarity and prevents false interpretations of the goals. Based on the different goal descriptions it is easy to come up with the so-called „Story of Strategy", a one or two page essay describing the strategy map. This document supports strategy communication and strategy alignment between different areas.
- Define a "sponsor" for each strategic goal who drives and coordinates its execution while maintaining a shared responsibility of the management team for the entire strategy map.
- Provide your different areas the freedom of developing their own strategy map in the cascading process, based upon the one of the whole company. Only thus, you will get transparency about their mental strategy models and will be able to identify and solve possible goal conflicts between areas later on. In general, this process can also lead to new strategic ideas that, bottom-up, may well lead to the refinement of the corporate strategy map or others. If the areas are very homogenous, however, a standardized approach with limited freedom is more effective.

(Source: Wunder 2012: 276 and Horváth & Partners 2007: 157–177) egy map. This leads to the complexity not only of the strategy map but also of the subsequent system of performance indicators and of action programs as well as of the entire strategy execution system. In this case, the tool can easily lose its acceptance by employees. There is a risk that the company becomes overloaded and loses its orientation, and thus, ultimately fails in strategy execution. To avoid those pitfalls, recommendations for an effective strategy map development are summarized in Strategic Snapshot 4.17.

4.3.3 Cascade and Align Strategy

The whole must be more than the sum of its parts.

Many companies consist of multiple businesses, functional units and international subsidiaries that they strategically manage. In those organizations, corporate headquarters (the "parent") must ensure that the business units under its stewardship are collectively more successful than if they were acting individually (i.e. parenting advantage) and there is as little frictional loss as possible in the interaction of the different organizational units during strategy execution (see chapter 3.2.1). In those companies, cascading strategy as well as aligning and integrating strategic priorities of its different units such as businesses, functional departments as well as regions and international subsidiaries is a key factor for making the overall strategy happen effectively. Figure 4-51 shows the basic principle of aligning corporate strategy, business strategies, functional strategies, and regional as well as country strategies.

Risks of misalignment

To join forces toward a common strategic direction, the sometimes conflicting strategic priorities of different management teams must be aligned. The goal is to create shared understanding and commitment in all areas of the company for the overall strategy while ensuring a certain level of agility in the organization. Specific

Fig. 4-51

Basic principle of strategy alignment

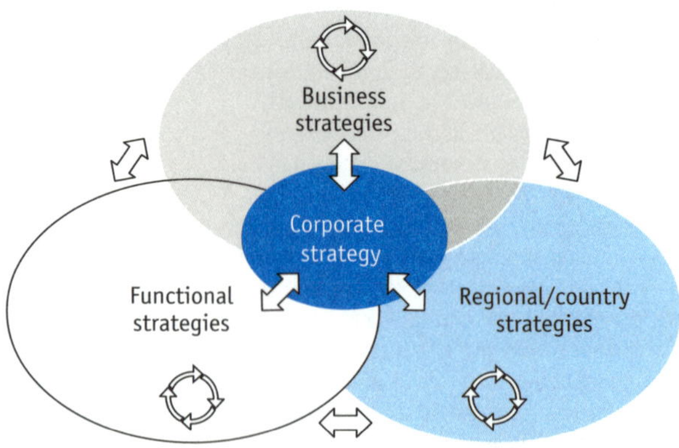

Source: Wunder 2014b: 408

strategy contributions of the organizational units are made transparent and responsibility as well as accountability for strategy execution is delegated. In case strategies are not aligned there is a risk of uncoordinated directions or even conflicting strategies among the different organizational units. Also, the senior executive team may be well aligned but the strategy can get morphed as it is cascaded which leaves the rest of the organization disconnected (see Strategic Snapshot 4.18).

Considerations for strategy alignment

As described in the previous chapters, strategic goals and strategy maps, balanced scorecards and strategic action programs are elements for effective strategy execution. In essence, cascading and aligning strategy can start at each of these three elements or at the strategy itself (Kaplan/Norton 2008: 125–156; Horváth & Partners 2007: 243–248). Key questions that need to be clarified for strategy alignment are listed and further explained in the following (Horváth & Partners 2007: 239–255):

- What are the strategic themes to be cascaded and who is expected to contribute?
- What are the specific strategic alignment needs within the organization?
- How and in which sequence are goals, metrics, and actions cascaded and aligned?
- Which organizational units should be involved in which sequence (cascading paths)?
- What methods are used for cascading: default versus freedom?
- In which way do strategic goals need to fit together (vertical and horizontal alignment)?

4.3.3.1 Applying Strategic Themes

Strategic guideposts for alignment

Strategy alignment usually starts with identifying overarching strategic priorities that have to be cascaded and embedded in the organization. As explained earlier, a corporate-level strategy map can be used to articulate the overarching strategy effectively. The four perspectives explained earlier have emerged as a useful taxonomy to identify and articulate how the corporate headquarters plans to create value with its different businesses (Kaplan/Norton 2008: 126–129). Some companies articulate only high-level strategic themes based on a combination of strategic goals that are logically linked, others use individual strategic goals as "themes". Examples of corporate synergies that can be articulated with such strategic themes and exploited through an effective strategy alignment among business units, countries and functions are (Kaplan/Norton 2008: 127 f.):

- leveraging brands across multiple business units and/or regions (e. g., common umbrella brand, product brands, ingredient brand, etc.)
- effectively acquiring and integrating other companies
- driving cross-selling activities with products or services from different business units at common customers
- leveraging customers in one region that also have a presence in other regions (global account management)
- achieving scale economies through sharing manufacturing, R&D, or marketing resources
- transferring best-practice capabilities from one business or country to another
- enhancing human capital through recruiting, training, and leadership development practices across multiple businesses and regions

Strategic Snapshot 4.18

Analogy: Company as Formation of Speedboats—The Alignment Matters!

Nowadays, many companies operate in an increasingly volatile environment. This means that both the frequency of economic boom and recession as well as its magnitude is increasing. Firms are facing unpredictable and massive fluctuations of market factors such as sales volumes, prices, or competitive positions where opportunities and risks lie closely together. In uncertain times it is important to create a common strategic orientation. A reliable long-term course provides employees with security, creates purpose, and fosters enthusiasm for the future. At the same time, a certain degree of strategic flexibility, that is the ability to react to short-term directional changes must be ensured. Given this context, figure 4-52 illustrates five different strategic company situations using the analogy of a speedboat formation that will be explained next.

Companies managed as large deep-sea tankers in the volatile environment are cumbersome and slow (1). They cannot avoid or bypass detected obstacles in the short term. Managing a formation of flexible and dynamic speedboats (= business units, functional areas, international subsidiaries) with fast reaction times seems more appropriate in this case. In corporate life this could be achieved through decentralized management structures and transfer of responsibility to different management teams. This alone, however, is not sufficient. The different speedboats must be coordinated in their respective course (business/function/country strategy) so that the formation could join forces and unleash its overall strength with aligned strategies (5). Without a shared vision and strategy, though, the strategic orientation will be missing (no common strategy) which can lead to chaos (2). But even if an overarching strategic framework is defined, the leaders of some speedboats may pursue their individual strategic course that collides with courses of other speedboats. This may cause frictional losses and, in the worst case, a

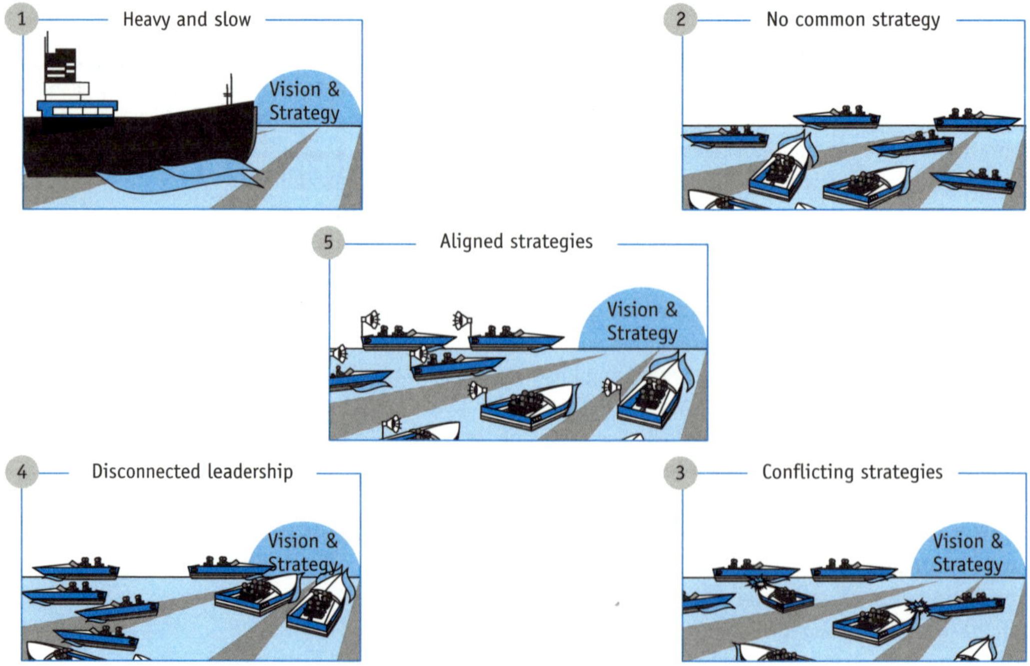

Source: Gaiser 2006: 162 f.; adapted

Fig. 4-52: Five strategic company situations illustrated with a speedboat analogy

Continued on next page

Continued from previous page

collision (conflicting strategies) within the fleet (3). Hereby, the supposedly false course can make a lot of sense from the perspective of the individual speedboat leader but hamper the flexibility, speed, and effectiveness of the entire formation. Furthermore, it may happen that the course is coordinated between only a few speedboats or leaders that drive ahead resulting in a disconnected leadership (4). In this case there is a risk of destroying the entire fleet. The leadership team disconnects from the rest of the organization and does not guide the other boats in their direction. (Source: following Gaiser 2006: 155–168, adapted)

Expected strategy contributions

Once the strategy is defined, it must be clarified, which organizational unit is supposed to contribute to the execution of which strategic theme (Kaplan/Norton 2006: 101–117). Figure 4-53 illustrates how common strategic themes derived from a strategy map are used to identify strategic support requirements of business, functional and regional/country units. Those strategic priorities (contributions) in the different organizational units need to be defined in the cascading process and finally aligned. Different methods of how this could be done will be explained in the follow-

Fig. 4-53

Identifying organizational strategy contribution requirements to strategic themes

Common strategic themes (e.g., from strategy map)	Corporation								
	Business units			Functional units			Regions/ countries		
Themes	A	B	…	R&D	MKT	…	CH	US	…
1. New/adapted business model	X		X	X	X		X		
2. Brand portfolio/ brand architecture	X	X	X		X			X	X
3. Customer mgmt./ global key accounts		X	X		X		X	X	X
4. Expanding product portfolio	X		X	X	X		X		X
5. Operational effectiveness and excellence	X		X			X			X
6. Innovation/learning	X	X		X	X		X	X	X
7. Sustainability/CSR	X	X	X	X	X	X	X	X	X

X = The organizational unit is expected to contribute to the execution of the strategic theme

Source: Wunder 2014b: 409 following Kaplan/Norton 2006: 108 and 1996: 245; adapted ing chapters. The resulting synergies and the focus of management attention and resources at all levels on the aligned set of strategic goals are a key for success in every company. Hereby not all organizational units may be of the same importance from the point of view of strategy execution. Given the heritage, structure and situation of the company, specific strategy alignment needs have to be identified in the organization first. Then the strategy alignment process can be designed and calibrated to address those specific requirements.

4.3.3.2 Identifying Strategy Alignment Needs

The company situation determines alignment needs.

Different company situations lead to different alignment needs. Some of those needs are vertical in nature—that is corporate to organizational units—others are horizontal in nature—that is organizational units among each other—(see chapter 4.3.3.6). Those have to be clarified prior to defining the alignment process and selecting the mechanisms and methods for synchronizing strategic priorities. For example, a multi-divisional corporation that has grown over the years by acquiring other companies worldwide might see the biggest need in aligning its international subsidiaries around certain strategic themes whereas another corporation needs to synchronize strategic priorities of their different businesses or functional units to realize corporate synergies and the desired parenting advantage. For a corporation with a global matrix structure all those mentioned alignment needs might be relevant.

Steering logic

Apart from the pursued corporate synergies, the company-specific alignment needs strongly depend on its applied organizational structure and steering logic:

- Which organizational units set strategic priorities as guideposts for others and which units are supposed to contribute to the execution of those priorities?
- Which comes first, the strategic business units or the regions and countries? Are products, brands or geographies the dominant steering dimension?
- What is the role of support units in the corporate strategy? Do global core functions provide strategic goals for realizing company-wide scale or scope efficiencies that have to be executed in the businesses and regions or—vice versa—are the functions supposed to enable the business or regional strategies? Etc.

Sometimes, those questions are not answered easily because strategy typically reflects both globally integrated elements for achieving corporate synergies as well as unit-specific elements. In the following, selected considerations for aligning business, country and functional strategies will be explained.

Aligning strategic business units: realizing parenting advantage

Strategy alignment needs for **strategic business units** strongly depend on the way and intensity the corporation wants to realize parenting advantage (see chapter 3.2.1). A key question is whether businesses need to be linked more strongly with other businesses to make corporate strategy happen, for example, improving efficiencies or market positions through corporate synergies. If the business unit strategies are developed independently without considering a corporate perspective, those linkages will not happen naturally, and thus, there is a strong need for strategy alignment. Those alignment needs are likely to vary depending on the type of corporation and are stronger in strategic or operational management holding companies as compared to pure financial holdings (see Strategic Snapshot 4.5 in chapter

4.2.2.1). Also, highly diversified conglomerates are likely to have less alignment needs among their different businesses as compared to corporations that want to add value by realizing corporate synergies such as leveraging global power brands across multiple businesses, driving cross-selling by combining products and services from different business units, or imposing a global best practice (e.g., customer or innovation management) identified in one business unit on others.

Aligning country/regional subsidiaries: exploiting "internationality"

Strategy alignment for **international subsidiaries** is particularly challenging due to several reasons: First, country subsidiaries usually have a physical and time distance from the headquarters. Second, their general strategic thinking patterns might differ from the corporate headquarters due to their specific sociocultural, political, economic and legal conditions. Third, unless the corporation is operating in a fairly global industry, the different country markets are likely to require different go-to-market strategies. The subsidiary management typically has more and deeper information about the local market, and thus, an information advantage over the headquarters. Fourth, experience shows that general managers of country subsidiaries tend to be strong personalities who want to determine strategy themselves instead of "being aligned". In addition, they might be in a strong position that allows them to act highly independently in case they cover comprehensive value creating activities, significant resources, revenue and profit contributions as well as growth opportunities locally. As a result "centrifugal forces" or opportunistic behavior of international subsidiaries may impede an optimal strategy execution from a corporate perspective (Wunder 2004a: 2–6). Otherwise, alignment is of utmost competitive relevance for multinational corporations that want to exploit their "internationality" as a source for competitive advantage (see Strategic Snapshot 4.19 and chapter 4.3.3.3). Here, effective strategy alignment can have a strong impact but requires special sensibility in selecting and applying the different mechanisms. For that purpose, it might be useful to assign to the various subsidiaries different strategic roles for the implementation of the corporate or business strategies considering their local resource strength as well as the importance of the local market and differentiate the alignment process in accordance with these roles (Wunder 2004a: 181–185).

Aligning functional units: supporting the business

Also core **functional units** such as R&D, Marketing and Production/Supply Chain Management as well as support functions like Finance/Controlling, IT, HR, or Law have to support corporate and business strategy effectively. These departments classically elaborate their own strategy maps and balanced scorecards. Figure 4-54 illustrates how corporate strategic goals can be cascaded to the Supply Chain Management area. The corresponding cascading method and options will be explained in chapter 4.3.3.5 more in detail. Some functional areas that are not directly related to the company's customers typically address strategic requirements of internal customers, such as the strategic business units, and align their employees in a way that they are aware of their critical strategy contribution in supporting the achievement of business priorities. The strategy map's customers/market perspective can be used for that purpose. Additionally, although the functional areas usually have no direct P&L responsibility except the cost budget, some companies integrate top- and bottom-line financial business goals in their financials perspective of functional maps to drive alignment with the business and highlight the own strategic importance. The inter-

Fig. 4-54

Cascading corporate strategic goals to Supply Chain Management (SCM) function

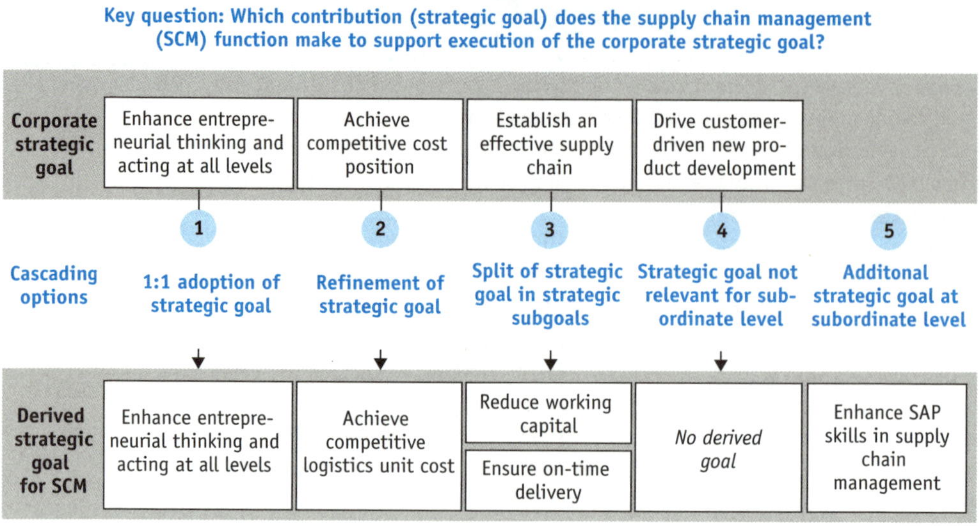

Source: Horváth & Partners 2007: 248; adapted

nal processes and potentials perspectives of functional units usually refer to procedures and competences within the department that are critical for enabling the business-related goals. Strategy alignment of functional units can also be enhanced through the development of service level agreements.

Aligning business process strategies

A company managed more strongly process-oriented and less with a classical hierarchical organizational structure might see the biggest need in aligning its key processes with strategy beyond the elements found in the internal processes perspective of an organizational unit's strategy map and scorecard. In such a case, business process strategies are derived from corporate strategy and strategic goals, key metrics (business process scorecards) and strategic actions are cascaded to those key processes (Fink 2003: 162–171).

Tackling ambivalent situations

As mentioned before, the cascading sequence usually follows the steering logic that is applied in the company. Sometimes, however, this might not be defined clearly or is not reflecting the strategic management reality or power structures anymore. For example, it is challenging to find the optimal balance between the strategic priorities of business unit heads with worldwide responsibility ("global entrepreneurs"), global functions, and strong regional or country managers in an organizational matrix structure (see chapter 4.2.2.1). One way to tackle this issue is by cascading corporate strategic goals to key countries, strategic business units and global functions simultaneously. Gaps or conflicts are identified afterwards and solved jointly in a global strategy alignment meeting (see Strategic Snapshot 4.20 in chapter 4.3.3.7). In ambivalent situations like the one described, the strategy alignment process itself can reveal gray areas, inconsistencies and unclear responsibilities, and thus, ultimately help to optimize the organizational structure or the steering logic in terms of structure follows strategy (Wunder 2004a: 229).

4.3.3.3 Sequencing Goals, Metrics and Actions

Aligning mental strategy models

In practice, two essential ways of cascading the three key components of strategy execution—strategy maps/strategic goals, metrics/balanced scorecards, strategic action—can be leveraged (see fig. 4-55). In a single-loop cascading all three elements are worked out one after another in a continuous sequence by downstream organizational units. This could be done in a series of workshops spread over a period of some weeks or in a sequence of some consecutive workshop days, for example, when cascading to international subsidiaries with on-site participation of senior managers from the headquarters. An advantage of this approach is that strategic goals, measurements and actions can be developed quickly and related strongly to each other. Nevertheless, there is a risk that in different organizational units individually developed strategic elements do not correspond later on, and the specific strategy maps, balanced scorecards and strategic action programs already developed need to be reworked. Those adjustment requirements identified later on in a company-wide alignment process usually causes frustration among the local management teams who developed those elements. Therefore, especially for companies with high strategy alignment needs among its different management teams, it is advisable to apply a double-loop cascading process. Hereby, the organizational units work out, at first, only strategic goals and strategy maps which will be jointly aligned (see Strategic Snapshot 4.20 in chapter 4.3.3.7). This way, mental strategy models of different

Strategic Snapshot 4.19

Competitive Advantage through Global Strategy Alignment in Transnational Organizations

In general, multinational enterprises (MNEs) that are managed with some kind of global matrix structure following a geocentric leadership perspective are facing a particularly challenging strategic alignment situation. They need to *optimally* balance strategic priorities of their (strategic) business units, functional units—that is core functions like R&D, Marketing, Manufacturing or Industrial Operations including Supply Chain Management as well as support functions like HR, IT, Finance or Legal—as well as regional units or international subsidiaries. Multinational enterprises that are capable of achieving this *optimal* balance are better positioned than others to win in the marketplace as they simultaneously exploit the different sources for a competitive advantage that lies in the internationality of their operations. International subsidiaries, business units as well as global functions should help the organization to develop different layers of worldwide competitive advantages: being responsive to local competitive environments (flexibility), on the one hand, and contribute to capturing scale and scope economies as well as to developing worldwide learning and innovation capabilities of the multinational enterprise, on the other hand. This means that their operations need to be integrated with other organizational units to simultaneously exploit those sources of competitive advantage (see chapters 3.3.2). (Source: Bartlett/Beamish 2011: 210–216; Bartlett/Goshal 2002; Wunder 2004a) Corresponding strategy maps need to reflect strategic goals regarding global integration, local responsiveness as well as cross-border knowledge management and worldwide innovation processes (Wunder 2004a: 176).

Fig. 4-55

Single- and double-loop cascading with vertical and horizontal strategy alignment

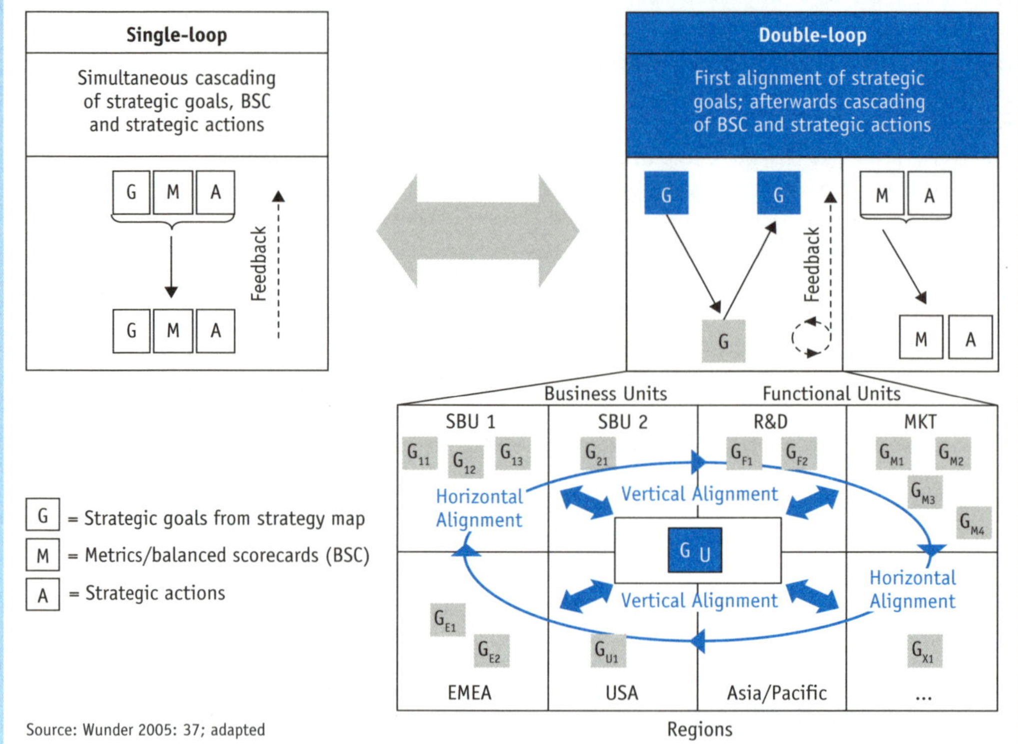

Source: Wunder 2005: 37; adapted

executives or management teams are externalized and systematically aligned before talking about how to measure and what initiatives to do. In a second step, balanced scorecards and strategic action programs are developed (Wunder 2005: 37–39).

4.3.3.4 Setting Cascading Paths

Based on the identified alignment needs, it has to be determined which cascading path—top-down, bottom-up or hybrid—is most effective and efficient for establishing alignment. "Most organizations, operating between the two polar cases of identical decentralized units and highly diversified corporations," outline Kaplan and Norton, "can choose from two implementation paths. One is a classic top-down process in which the initial Strategy Map and Balanced Scorecard are done at the corporate level, followed by scorecards at each business and support unit that implements the corporate-level strategy and scorecard measures. The second approach is more bottom-up, with pilot projects performed at the business unit level to develop local Strategy Maps and BSCs. These projects build knowledge and confidence about the management tool, and only subsequently is the project taken up to the corporate level for an explicit articulation of the corporate value proposition." (Kaplan/Norton 2006: 173)

Top-down cascading

Usually the cascading path follows the steering logic that is applied in the company based on the actual organizational structure and line of responsibilities. By leveraging the existing reporting lines, accountability for making the strategy happen can be cascaded most effectively. Based on this idea, companies typically follow a top-down approach when cascading strategy. The strategic direction defined at a higher level is cascaded into the organization. This does not automatically mean a top-down deployment of predefined strategic goals or metrics. "Top-down" indicates the direction but not necessarily the level of freedom that subordinate units or teams get for developing their strategic priorities. Often it is more a top-down communication and strategic guidance instead of a top-down deployment (see chapter 4.3.3.5).

Bottom-up consolidation

Does a multi-business company always have to start with a corporate strategy map and cascade top-down? Conceptually, the answer is "yes". Practically, some companies start strategy mapping and scorecarding at a business or regional level. They develop the corporate strategy map afterwards by consolidating the aligned bottom-up strategic insights and providing answers for the remaining strategic questions that can be addressed only at a corporate level (e. g., future portfolio and importance of businesses or regions).

Hybrid cascading

This can lead to a hybrid approach if the consolidated corporate map will be used in a second step to further cascade top-down to the remaining organizational units (e. g., functions).

Another way to combine the two paths and apply a hybrid approach is to define only a portion of the strategy map or scorecard top-down, the rest is developed bottom-up. This will be explained more in detail in the next chapter. Also, a company-wide rollout might as well be started at some pilot units to prove the effectiveness of the concept in a particular organization. If it is successful, it will continue as a top-down process in the second phase.

4.3.3.5 Choosing Cascading Methods

Strategic importance determines the level of freedom.

A variety of methods for cascading strategic goals, metrics and actions can be distinguished (Horváth & Partners 2007: 244–247). It needs to be decided how much freedom the different organizational units should have for developing their strategy maps, balanced scorecards and strategic action programs. The answer strongly depends on the type of organization. Key considerations are the size as well as the actual and future strategic importance of the organizational unit, the level of autonomy and the diversity of the business. In general, the bigger, the more strategically important, autonomous and different the organizational unit is, the more freedom it should get in the strategy cascading process (Horváth & Partners 2007: 247). An example would be a business unit in a highly diversified holding company with a significant revenue contribution. On the other hand, some organizations consist of very similar businesses like a franchise company. Here a more standardized approach with default strategic contents is more effective and efficient.

Metrics are more likely defined top-down.

In general, the question of setting defaults versus leaving freedom for individual development needs to be decided for strategic goals, metrics/scorecards, and partly also for strategic actions. In the following, four key methods will be explained more in detail. Method 1 and 2 are focused on strategic goals. They are typically applied in a double-loop cascading process emphasizing collaborative strategy alignment. Hereby, it is assumed that the corresponding balanced scorecards and action programs are developed in a second step. Particularly the metrics are often defined in a more top-down driven process later on to ensure efficiency and, as far as possible, comparability for performance measurement (see chapter 4.3.4). Method 3 and 4 are more applicable for a single-loop rollout.

Method 1: Guided autonomous strategy mapping

Top-down guidance instead of top-down deployment

At a guided autonomous strategy map development, the higher-level strategy map (e. g., corporate or division) provides a guideline for the subordinate organizational units (e. g., business or function) that develop their individual strategy maps with a high level of freedom. The relevance of each top-level strategic goal is discussed by the local management team with reference to the question whether and how the subordinate level needs to contribute to its execution (see fig. 4-56). If and how it intends to contribute is decided at the lower level and, along with other organizational units, later on discussed in the systematic alignment process.

Fig. 4-56

Five options for cascading strategic goals

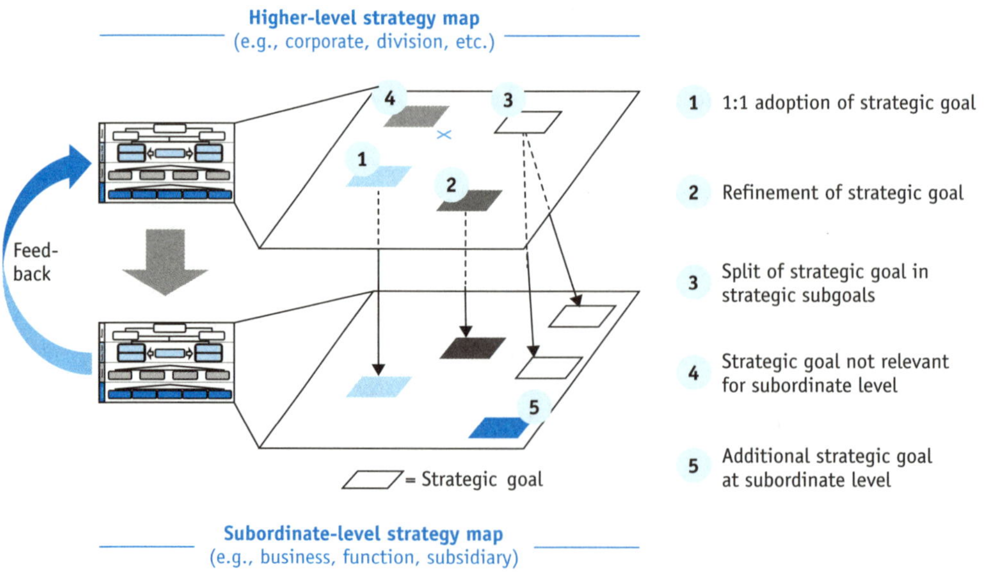

Source: Wunder 2012: 275; adapted

As shown in figure 4-56, some top-level goals may be taken 1:1 if they fit and cannot be further refined (1). Other strategic goals are adapted but refined to highlight the specific contribution of the organizational unit (2). This could also mean that more than one strategic goal is derived and required at the lower level to effectively support achievement of the higher-level goal (3). Some strategic goals of the higher-level strategy map might not be relevant at all (4), and some strategic goals may be necessary to make the local strategy happen but cannot be linked directly to a top-level goal (5). A practical example of how this method can be applied is shown in figure 4-54, with reference to a cascading process from corporate strategy into the supply chain management function.

Method 2: Selected direct goal deployment

Company-wide anchoring of selected strategic topics

In general, method 2 is following the same logic as method 1, however, selected strategic goals are set at a higher level and then deployed to all units in the organization. Those goals are not just a strategic guideline but predefined compulsory elements of the subordinate strategy maps. This hybrid method is used if the company wants to anchor specific strategic topics or messages in the entire organization but still leave freedom to define unit-specific strategic priorities bottom-up. In many cases the predefined strategic goals are related to financials (e. g., predefined revenue or profitability goals) or to potentials such as a certain company culture or entrepreneurial spirit that should be established in the entire organization (see Strategy Practice Example 4.10). Predefining not just one or two but many or all strategic goals leads to standardized strategy maps (method 3).

Method 3: Standardized strategy maps and scorecards

Standardized approach for homogeneous units

If the organizational units are very homogeneous, as it is the case with franchise outlets such as fast food restaurants or home improvement retail centers, it might be more efficient to standardize strategy maps as the value proposition among these units is similar and defined top-down (Kaplan/Norton 2006: 169–173). In those cases the key performance indicators are usually also standardized, and only the target values and specific strategic actions are defined at the local level based on its particular environment. Some companies standardize not all but a portion, such as 30 % of

Strategy Practice Example 4.10

Company-Wide Anchoring of HSEQ with Direct Goal Deployment

In the exploration and production (E&P) division of a leading oil company direct goal deployment was applied to anchor the topic HSEQ (health, safety, environment, quality) in the strategies of international subsidiaries and functional areas. During a strategy map cascading process a corresponding strategic goal was predefined at the top level and had to be adapted by all units at the next level down. Later on in the process, those default strategic goals were touch points for the HSEQ responsible who developed corresponding metrics and action programs together with the local management teams. Ultimately, strategy mapping could be leveraged to create a shared understanding and commitment for driving HSEQ in the entire organization.
(Wunder 2004a: 213–234 and 2014a: 75 f.)

goals in their strategy maps and metrics, to make the cascading process more efficient. This is done, for example, to accelerate the cascading process in an organizational structure with several semi-homogeneous international production sites, on the one hand, and fairly similar international sales units, on the other hand. The organizational units in each of these groups have certain commonalities that can be addressed top-down (e. g., the goals in the financials or customer/market perspectives of production sites) but also different specific local environments that rather need to be addressed with an approach like method 1 (Wunder 2004a: 239).

Method 4: Strategic actions only

Actions only for local implementers

In case opportunities to contribute to the overall strategy are very limited for a certain organizational unit that has just enough competence to fulfill its own specific tasks, the development of a local strategic action program might be sufficient. In those cases, strategy maps and balanced scorecards are not cascaded to that organizational unit at all. Local strategic actions are defined to implement the higher-level strategic goals. In general, companies apply this method because of resource constraints and accept that employees in those organizational units might lack strategic discussions and awareness compared to other areas in the company. Some firms chose this approach for certain functional units such as Legal or Regulatory Affairs whereas they use a more comprehensive cascading process with method 1 or 2 for other support functions such as HR, Finance or IT. Also, a multinational company with many international subsidiaries that cascades strategy to the country level may apply method 4 for certain international subsidiaries depending on their strategic role for the corporation to ensure an efficient process (see fig. 4-57).

Combination of cascading methods

Larger organizations typically apply a combination of cascading methods for their different organizational units based on, for example, their strategic role, resource situation or heritage (Wunder/Baur 2000: 551–553). In some businesses a strong discussion of the overarching strategic goals in the management team might be important to foster awareness and buy-in, thus calling for method 1. This can be particularly important for teambuilding in a newly formed business or post merger integration (Kaplan/Norton 2006: 238–242). In other areas it might be more effective to set clear strategic goals for the organizational units and apply methods 2 to 4.

Strategic roles of country subsidiaries and strategy cascading method

In multinational enterprises with a large number of international subsidiaries, it is advisable to select the cascading method based on the strategic role of the different country organizations. This role can be determined by the significance of the local market or environment, on the one hand, and the strengths of local resources and capabilities, on the other hand (Bartlett/Goshal 2002: 121–128). Figure 4-57 shows the explained cascading methods in relation to the various strategic roles of country subsidiaries (Wunder 2004a: 181–185):

- Country subsidiaries in the role of **strategic leaders** do not only sense local opportunities but develop their own strategies to capture those opportunities based on their own resources and capabilities. For them it makes sense to develop their strategy maps with a high degree of autonomy based on the corporate strategic goals. Their strategies might be used as reference for subsidiaries in other roles, e. g., the strategic implementers.

- Subsidiaries in the role of **strategic contributors** operate in country markets that are of low strategic significance for the multinational enterprise. Due to their high level of resources and capabilities they get strategic leadership roles that are important for the entire corporation, such as a center of excellence for a certain technology, but that are less relevant for the local market. Those company-wide mandates can be addressed by direct strategic goal deployment from the corporate level whereas the subsidiary develops strategic goals addressing its local market autonomously.
- Cascading (semi-)standardized strategy maps and scorecards is useful for international subsidiaries with a low level of local resources and competences that operate in a market with little strategic relevance. They have the role of **strategic implementers** and deliver the corporate value proposition to generate cash-flows that are important for the multinational enterprise to drive strategies and innovation in other regions. Those subsidiaries are supposed to execute predefined goals such as enabling scale efficiencies in a global production site or scope efficiencies through the exploitation of global brands in an international sales subsidiary.
- Country subsidiaries that are in the undesirable role of a **"black hole"** are not positioned to exploit the local market opportunities. They should be a strategic leader but lack the corresponding resources and capabilities. Due to that very limited ability to improve their strategic situation, it does not make sense to come up with an own strategy map. The corporation and the subsidiary management have

Fig. 4-57

Defining the cascading methods based on the strategic role of country subsidiaries

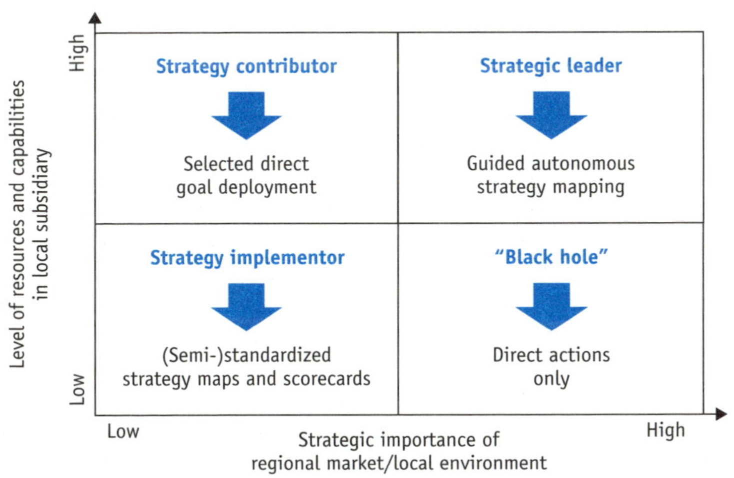

Source: Wunder 2004a: 185; adapted

to find ways of how to improve the situation, which can be addressed with a strategic action program. Once the subsidiary escapes its black hole position, other methods are more appropriate.

4.3.3.6 Aligning Vertically and Horizontally

Sufficient support for the strategy

Once strategic goals and strategy maps have been developed, they must be aligned in two directions (Kaplan/Norton 2008: 129 f.; Horváth & Partners 2007: 250–255; Wunder 2005: 39 f.). Within a *vertical strategy alignment* it will be checked whether the higher-level strategic goals are properly reflected and specified in strategy maps of downstream organizational units. Thus, it becomes transparent whether the corporate strategy is sufficiently supported in the entire organization. This process can also lead to a refinement of higher level strategic goals in case they are not understood at the next level in the organization, or bottom-up strategic insights require an adjustment of overarching strategic content. This way, vertical alignment provides feedback and new ideas for the corporate level.

Harmonizing conflicting strategic priorities

With *horizontal strategy alignment* strategic goals of organizational units at the same management level are synchronized. In case strategic goals in different units are not complementary or even conflicting regarding an effective overall strategy execution, they should be made transparent and harmonized in an open strategic dialog between the corresponding managers. Elaborating on those conflicting strategic priorities early in the strategy execution process will reduce friction later on. After the strategic goals of different units at the same level are synchronized, metrics and strategic actions will be elaborated (see fig. 4-55 in chapter 4.3.3.3). Some companies that are under high competitive pressure for improving their strategy execution postpone the development of metrics and define strategic action programs directly after the strategic goal alignment. They need to deliver results quickly, and thus, spend their management resources rather on making progress in achieving strategic goals than on the question of how to measure them immediately.

4.3.3.7 Benefits and Pitfalls of Strategy Alignment

Consistent strategic orientation

Most companies consist of multiple organizational units such as corporate, strategic business units, functional units as well as regional or country units. *Vertically aligning* the strategic priorities of management teams in those various units with an overall strategic direction and its higher-level strategic themes and goals is critical for a successful execution as it establishes a consistent strategic orientation throughout the company. Each organizational unit gets a clear understanding of how they contribute to the overarching strategic priorities. At the same time, those units have strategic goals that help them to execute their local strategies for winning in their corresponding competitive arena. This process is facilitated with strategy maps that help to articulate strategies clearly in a standardized format. If mental strategy models of managers or management teams from different areas are externalized and made transparent with the help of strategy maps, it will be easier to understand and align them. It also fosters identification and buy-in with strategy at all levels because individual strategy contributions to the "big picture" become clear and harmonized (see Strategy Practice Example 4.11).

Strategic Snapshot 4.20

Global Strategy Alignment—an Experiential Report

Various strategically managed multinational corporations or divisions have applied a global strategy alignment approach following a double-loop cascading process. They started off with the development of a corporate strategy map in a group of managers that did not only cover the senior leadership team but also the leaders of selected businesses, functions, and countries or regions. This way, all these managers became co-owners of the corporate-level strategic goals that had to be cascaded to their corresponding unit or department in the next step. Then the second level strategy maps were developed in the local management teams but with involvement of selected top-level executives that followed and engaged in the strategic discussions. A two-day global strategy alignment meeting was organized where all those 10–15 independently developed strategy maps were visualized with large posters and made transparent to everybody. After presenting each strategy map, a structured alignment process was conducted leveraging "bird's eye" strategic goal visualization (see fig. 4-58) of how top-level strategic goals were translated in the different local management teams. Clear conflicts or gaps identified prior to the meeting as well as key questions helped to reveal misalignments and synchronize strategic goals jointly in the group of top- and second-level executives. This process did not only foster alignment alignment but also awareness, shared understanding and buy-in regarding the global strategic priorities at different management levels.

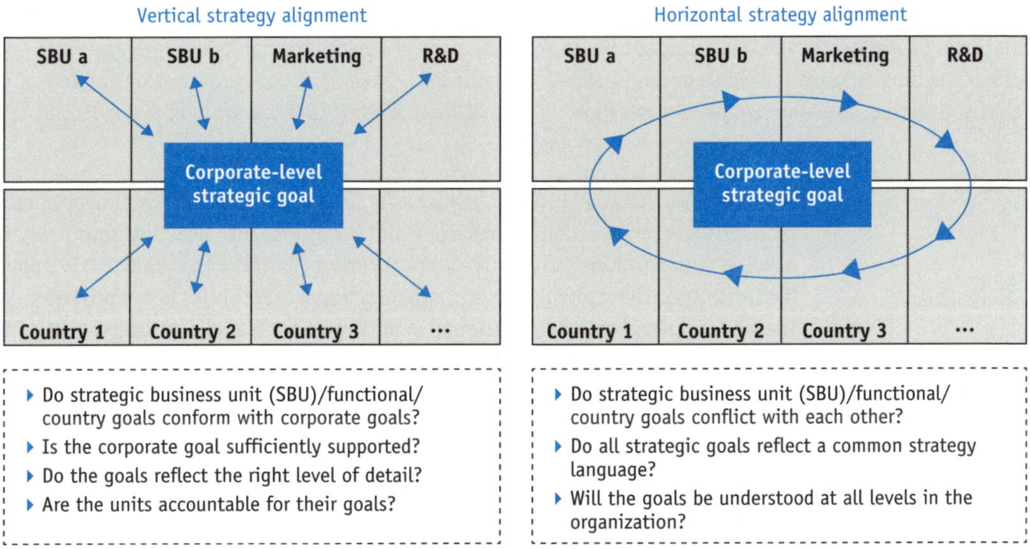

Source: Wunder 2005: 39; adapted

Fig. 4-58: "Bird's Eye" and key questions for strategy alignment

Realizing parenting advantage

The corporate headquarters has to ensure that its businesses are more successful collectively than if they were acting alone (see chapter 3.2.1). In addition to creating a consistent strategic orientation, *horizontal strategy alignment* enables the corporation to realize different types of synergies and ultimately parenting advantage. It supports management in realizing added value through the interaction of its organizational units. Here strategy maps act as "strategic language" that increases the awareness about possible goal conflicts or gaps between organizational units, which are then discussed and collectively solved in the management team. Ensuring that

Strategy Practice Example 4.11

Benefits of Global Strategy Alignment at Henkel

Henkel is a German group with revenue of € 16 billion, some 47,000 employees and worldwide operations in three business areas: laundry & home care, beauty care and adhesive technologies. Founded in 1876, Henkel holds globally leading market positions with well-known brands such as Persil®, Dial®, Purex®, Schwarzkopf®, Loctite®, and Pritt®. In 2004, the group's consumer adhesives division successfully implemented and synchronized strategy maps, balanced scorecards, and strategic action programs across the division level, three strategic business units, global functional units such as supply chain, operations, R&D and international sales, as well as nearly 30 regions or countries. Strategy maps helped to clarify, describe, and—most importantly—align strategic goals of these different units. The top strategy map was developed jointly by division managers, leaders of strategic business units, general managers of key regions, and heads of global functions. In a second step, these managers developed strategy maps for their organizational units together with their corresponding management teams. In a *global strategy alignment* meeting (see Strategic Snapshot 4.20) the different strategy maps were presented, discussed and aligned. Balanced scorecards and strategic action programs ensured strategic control and execution. After the implementation, Henkel's annual employee survey showed clearly the achieved degree of buy-in from the individual employee level. The adhesives division improved on the already excellent results of previous years with top marks in "organizational learning", "vision", "strategic direction and intent", and "agreement." (Source: Wunder 2005: 36–38)

strategic priorities at different managerial levels and units are well synchronized reduces performance-sapping frictions and disagreements, wasting resources, and missing opportunities during strategy execution (Sull/Homkes/Sull 2015, Kaplan/Norton 2006). The strategic focus of organizational units shifts from optimizing only their siloed results toward an aligned company-wide strategic perspective of collectively optimizing and creating the highest value strategy contribution for the firm.

Solving specific management issues

Finally, strategy alignment can help to solve specific management problems such as improved control and management of international country subsidiaries in multinational enterprises or the implementation of transnational strategies to realize different elements for creating competitive advantage (see chapter 3.3.2), better collaboration between specific departments like Marketing and R&D, the integration of businesses and departments along with required teambuilding activities of the new management in case of mergers and acquisitions, or an improved process orientation.

Watch out: risk of tighter control and high effort

A potential pitfall of strategy alignment from the perspective of the organizational units is the risk of being controlled tighter by their superior management team. Making mental strategy models transparent for discussion in a broad group of managers from different areas requires a high degree of trust as it might lead to increased aspirations and targets or to stronger supervision of their strategic action programs. This is contrary to the basic idea of strategy-oriented self-control based on open strategy dialogs and collective strategic learning underlying the described strategy execution framework. Furthermore, systematic strategy alignment is not an easy task. It requires additional effort for preparation and absorbs additional management capacities even though this might pay off with reduced friction and more effective deci-

Strategic Snapshot 4.21

Recommendations for Strategy Alignment

- Define strategy alignment as an explicit step in the strategy process and include it in the annual planning calendar. The success of strategy execution strongly depends on how well strategic priorities of different organizational units are aligned with respect to an overall strategic direction.
- Be aware of your explicit or implicit steering logic. This determines the sequence of strategy cascading and alignment. It needs to be clarified which organization units are providing strategic direction and which ones are contributing.
- Design the specific strategy alignment process and methodologies according to the management problem that needs to be solved and the strategic leadership culture. There is no silver bullet for cascading and aligning strategic priorities.
- Determine strategy cascading methods based on the strategic importance and role of organizational units. Depending on the business model, organizational structure or culture, some units may need more freedom than others.
- Strategy alignment should not only be done by few senior leaders but involve the entire management leadership team. Particularly the participation of next-level managers in developing higher-level strategic priorities creates awareness, acceptance and buy-in for what is strategically important. This makes the subsequent strategy cascading and the alignment process much easier.
- Keep the same perspectives in which strategic goals etc. are defined for all types of organizational units. There should be a common method for strategy articulation in the company. The challenging alignment of strategic priorities requires simple and intuitively understandable methods that are easy to apply and do not create confusion.
- Include financial business goals also in the financial perspectives of functional units such as HR or IT that typically have no direct impact on the financial performance other than staying in budget. This way, strategy contributions and their (indirect) business impact become very clear to the corresponding teams that capture strategic requirements of their internal customers in the customer/market perspective.

(Source: Wunder 2014b: 411)

sions and strategic control later on. Following a number of guidelines derived from company cases help to ensure that strategy alignment is both highly effective and efficient (see Strategic Snapshot 4.21).

4.3.4 Make Strategy Measurable

Clarity about strategic goal content and aspiration

If successful strategy execution means driving an agenda of strategic change and transformation, the definition and alignment of strategic goals and strategy maps alone is not sufficient. Effectively executing those goals requires clarity about what they exactly mean in terms of measurable values and what exactly should be achieved in terms of targets. Only then the aspiration becomes clear, progress can be measured unambiguously, and strategy execution can be managed. Defining metrics and corresponding targets is important for the development of strategic action programs and for the purpose of strategy communication as it provides clarity about strategic priorities and expectations with little room for interpretations: "Words are not sufficient to communicate the change initiatives. The same words mean different things to different people. It is only when word statements are translated into measures [measurements, metrics] that everyone understands clearly what the vision and strategy are about." (Kaplan/Norton 2001: 344)

4.3.4.1 Developing a Balanced Scorecard

Collection of quantifiable performance metrics

Executives use different types of data for measuring performance in their organization. They range from pure observational performance evaluations derived from "walking around" and experiencing it, to informal progress reports by asking people how certain things are going, to more formal investigations of specific performance aspects with studies, audits or assessments. A fourth type of looking at performance is provided by scorecards or dashboards that are a collection of quantifiable performance metrics. In general, they are derived by counting things such as financials (e. g. revenue, expenses), complaints, lost customers, patents, employees with a certain skill profile, etc. or making judgments by rating or ranking performance on a scale such as surveys on image, organizational culture, degree of collaboration, etc. (Brown 2007: 21–24).

Balanced strategic performance scores and targets

In the context of strategic management, a scorecard is a table of "scores" articulating how a company or organizational unit is strategically performing. Those scores are provided by metrics (measurements, indicators) that are supposed to quantify strategic goals that are qualitatively articulated. The score of how well a company is performing is derived from defined targets for each metric and a regular comparison with actual values (see chapter 4.3.6.3). Performance measurement with a balanced scorecard is considered "balanced" if it captures a mix of metrics measuring elements that are

- financial and non-financial
- tangible ("hard") and intangible ("soft")
- externally and internally oriented
- results (lagging indicators) and drivers (leading indicators)
- short-/mid-term and mid-/long-term.

Those are the same criteria that have been applied to the development of strategic goals explained in chapter 4.3.2.1. The effectiveness of a balanced scorecard for strategy execution strongly depends on the quality and applicability of its contents. Figure 4-59 provides six principles for developing an effective balanced scorecard, which will be explained next. They are derived from practical experience and relate to accuracy, impact, and practicability of key metrics (Horváth & Partners 2007: 202–210; Wunder 2001).

Accuracy

Measure strategic intent and strict goal achievement

There are plenty of handbooks and repositories for key performance indicators available and also companies usually have a wealth of different kind of metrics already established. However, performance indicators are only effective for strategy execution if they are strongly tied to the strategic intent, which means that they capture exactly what is articulated in the corresponding strategic goal. Finding the right metric can be facilitated by asking the following question:

How do we know we have made progress or have achieved the strategic goal?

If this principle is followed when deriving metrics for all strategic goals in a strategy map, it will automatically result in a "balanced" scorecard if the goals have been defined according to the guidelines presented earlier. Metrics that represent crucial strategic goals, and thus, are vital for measuring strategic performance are called key performance indicators (KPIs) or key metrics. From time to time, this KPI development process may require cycling back to strategic goal development. If it becomes clear that a goal statement is ambiguous and the discussed metric really articulates what is strategically intended, the corresponding strategic goal needs refinement. The more generic a strategic goal has been articulated, the more difficult it is to find the "right" metric, and the more likely it is that the corresponding strategic goal will have to be sharpened. An example is a generic strategic goal statement like "optimize processes". Looking for a metric leads to questions like: What are the key processes we need to focus our attention on for making our strategy happen? What do we mean with "optimize"—time, quality, or cost? The development process of key performance indicators should facilitate clarification of those questions. This way, performance measurement creates a common metric-based language in the company for communicating strategic priorities clearly (Schiemann/Lingle 2000: 23).

Measure non-financial goals with non-financial metrics

Intangible assets such as human capital, information capital or organization capital are crucial enablers for achieving business results (Kaplan/Norton 2004: 199–316). Corresponding metrics are typically non-financial and captured in the potentials perspective of a balanced scorecard. The corresponding strategy map shows and articulates how tangible and intangible elements for creating customer value and financial results are linked. Nearly 80 % of key performance indicators on a balanced scorecard tend to be non-financial (Kaplan/Norton 2001: 375 f.). Companies usually find it difficult to measure those elements that are traditionally not well established in existing performance measurement systems. As a guideline, corresponding non-financial indicators should not reflect indirect improvements of available financial numbers (e. g., sales) but capture the non-financial aspects directly (e. g., capability level of sales staff). Note that the financial impact is already captured with strategic goals and met-

Fig. 4-59

Six principles for developing key performance indicators

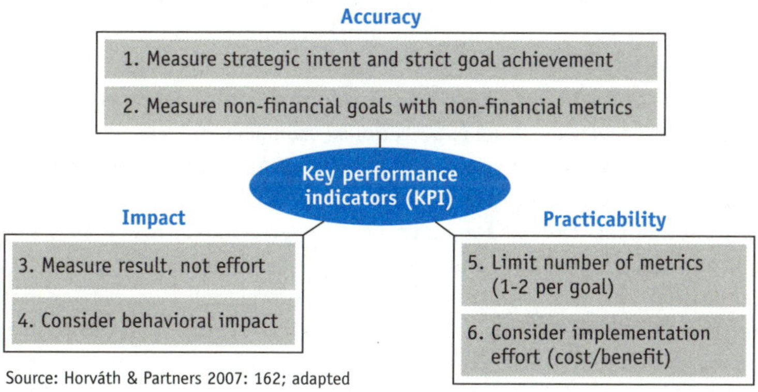

Source: Horváth & Partners 2007: 162; adapted

rics in the financials perspective of the balanced scorecard. "The scorecard does not attempt to 'value' an organization's intangible assets. It does measure these assets, but in units other than currency (dollars, yen, and euros)." (Kaplan/Norton 2001: 67 f.) Some new metrics such as the human capital readiness index have been developed particularly for this purpose (see chapter 4.3.6.2 and Wunder 2004b).

Impact

Measure the result, not the effort.

Another principle of key metrics is to measure preferably the output of what should be achieved and not the input. In general, the result of what is articulated with a goal is a better performance indicator than measuring the effort required for achieving this desired result. For example, strategic goals like "drive product innovation" or "improve innovation strength" may be measured by "R&D investment as a percentage of sales", which reflects an input. The company assumes that the investment will be well spent and drive innovativeness, but it does not capture the output directly with this metric. An alternative metric is "portion of revenue from products on the market for less than 3 years" (see also the KPI "Young Revenues" in Strategy Practice Example 4.12). The latter shows whether the company really improved its innovativeness reflected as a result (output) in the market, whereas the former measures only reflect what they invest in this topic without seeing if it pays off. For input versus output controls see also types of control in chapter 4.3.6.3 (see Strategic Snapshot 4.23).

Consider the behavioral impact.

Furthermore, the behavioral impact of metrics needs to be assessed carefully. Some performance indicators that were meant for driving behavior in a certain direction may actually lead to undesired behavior. Using the same example as before, imagine a strategic goal "improve innovation strength" has been defined based on the idea to implement a differentiation strategy and drive revenue growth with new products. A metric like "number of new patents per year" as a measure for improved innovativeness may direct employee behavior toward generating more and more patents with the risk of lacking a strong consideration whether those patents are actually leading to successful products in the market.

Practicability

Limit the number of metrics.

Keeping the number of key metrics at a manageable number is another crucial principle for balanced scorecards to be effective. This is a requirement for focusing management attention and resources on the "vital few" key performance indicators. As explained more in detail in chapter 4.3.6.3, additional information is provided only in case of strong deviations from expectations (management by exception). Hereby, the "twenty is plenty" guideline for developing strategic goals stated earlier can also be applied to metrics. Experience has shown that less is more particularly when it comes to the number of key performance indicators. Having too many metrics dilutes straight and unambiguous measurement of goal achievements, impedes strategy communication, and may lead to a situation of information overload—anecdotally referred to as "paralysis through analysis". In a company with strategy maps in different organization units, the number of key metrics can easily result in a few hundred indicators which all require target setting, regular generation of actual values, and calculation of deviations for feedback. Additionally, there is the exponential work of reconciling two or more metrics against one goal. Do you weight them? Is it double affirmative only? Should you blend the two results? (See also the "Watch out: benefit and risk of using analytics" at the end of this chapter). The strategy practice example 4.12 shows a balanced scorecard of a successful multinational corporation that captures only seven key performance indicators at the top group level.

Strategy Practice Example 4.12

Balanced Scorecard and Key Metrics Hierarchy at the Carl Zeiss Group

Founded in 1846 in Jena (Germany), the Carl Zeiss Group is an international leader in the fields of optics and optoelectronics with revenues of € 4.29 billion and 24,817 employees (FY 2013/14). The company provides solutions for the semiconductor, automotive and mechanical engineering industries, biomedical research and medical technology, as well as eyeglass lenses, camera lenses, binoculars and planetariums. The Carl Zeiss Group applies a highly consistent and comprehensive strategy process, which is based on a strong participation of employees and managers across different levels in the organization. An element in this process is the group balanced scorecard, which captures the vital few strategic key performance indicators across different perspectives:

- **People & leadership development: people & leadership index (PLI)**
  ... measures the leadership and corporate culture and how well proclaimed values are lived.
- **Innovation excellence: young revenues (YR)**
  ... measures the portion of revenue from products that are on the market for less than three years (innovation strength).
- **Customer excellence: net promoter score (NPS)**
  ... measures customer willingness to recommend the company and indicates customer loyalty.

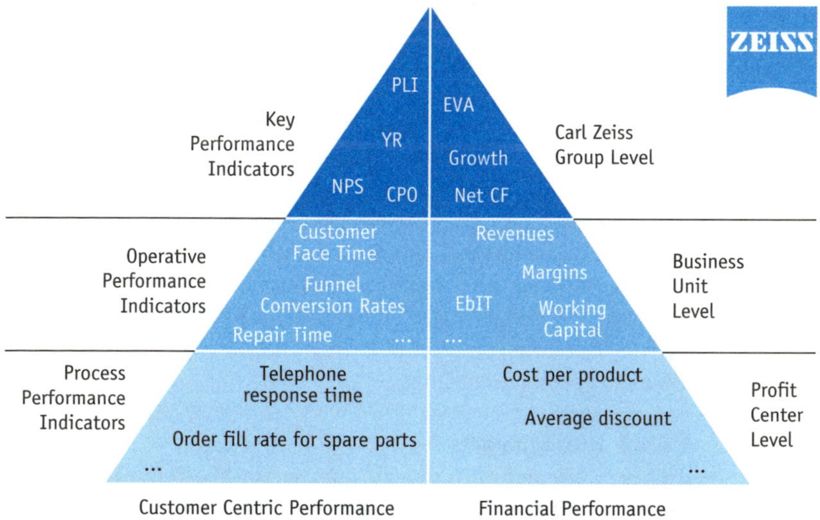

Source: Spitzenpfeil/Schneider 2011: 104 and Carl Zeiss Group 2013. Reprinted with permission of Carl Zeiss AG.

Fig. 4-60: Key metrics hierarchy at the Carl Zeiss Group 2010

- **Operational excellence: customer perfect order (CPO)**
  ... measures whether customer orders are fulfilled on time, complete, and with full product functionality.
- **Financial results: Economic Value Added® (EVA®), revenue growth, operating free cash flow**
  ... measure the financial group performance with EVA® as the top key indicator.

Each of those key metrics has been developed and selected together with the management, reflects the achievement of one strategic goal, is unambiguous, quantifiable, influenceable, and easy to measure in practice. Through focusing on only one key metric per strategic goal, the balanced scorecard is structured clearly and easy to understand. It is generated on a monthly basis for the entire Carl Zeiss Group as well as for the different business units. The group balanced scorecard is at the top of a metrics and reporting hierarchy in which additional performance indicators for operational control as well as for measuring process performance are applied (see also fig. 4.28—reporting pyramid—in chapter 4.2.2.3). (Source: Spitzenpfeil/Schneider 2011: 102–110)

Consider feasibility.

Generally speaking, every strategic goal can be measured—it is only a question of effort. For the final selection of key performance indicators, their implementation effort needs to be considered carefully. For some strategic goals on a strategy map, key metrics may already be established in the company. This is generally the case for financial strategic goals where values are available in existing databases or systems. The same might be true for a few selected goals in the other three perspectives. If not, additional key metrics can usually be established by, for example, leveraging or supplementing existing regular customer and employee surveys or other sources already available in the company. However, a great portion of metrics in a balanced scorecard development tend to be new to the company, particularly when it comes to non-financial indicators. Their development can be a time-consuming and challenging process, particularly in large organizations where the performance indicators need to be consistent across different management levels and "fit" each other. For limiting the effort, it is advisable to leverage what already exists and to harmonize KPIs across different organizational units as much as possible. Though, this needs to be done with due consideration of what was strategically intended in the first place in order to avoid losing strategic direction and clarity for the sake of getting enough metrics. If the expected cost/benefit-ratio for establishing a new performance indicator is not favorable, the company may alternatively start measuring goal achievement indirectly by measuring progress of strategic action programs that are supposed to contribute to goal achievement.

4.3.4.2 Setting Targets

Metrics or key performance indicators require the definition of targets, that is the desired values that have to be achieved after a certain time period. Those targets articulate precisely the desired performance level and are the base for corresponding strategic performance feedback processes. Key considerations for target setting will be explained in the following (Horváth & Partners 2007: 214–219):

Establishing reference values

Target setting should never be arbitrary. It requires reference values as a base for setting realistic aspirations. Those references can be derived from different sources such as

- historical data
- own baseline (that is current actual value such as the result of a customer survey)
- benchmarks from within the industry (e. g., the strongest competitors) or outside
- industry averages
- benchmarks from within the corporation (e. g., the own best performance).

Targets can be set in absolute values (e. g., 2 billion revenues in 2016), in relative terms to an internal reference value (e. g., 10 % revenue increase to the previous year), or to an external reference value (e. g., growing faster than the market). The latter is called relative self-adjusting targets as the aspiration level continuously adjusts relative to the peer group's performance (e. g., strongest competitors). This type of target tends to direct attention and action more toward beating the competition and less toward beating the "budget", which is one of the principles of a better budgeting or advanced budgeting approach (see also chapter 4.2.2.3). If there are no reliable reference values available, a small group of experts may estimate initial target values and update those targets once they have gained experience with the new numbers. Reaching consensus and starting with a best guess tends to be better than working without any target. Furthermore, progress of strategic goals can also be subjectively evaluated by the management team using, for example, a school grade scheme, or by directly setting a green ("on track"), yellow ("somewhat off track"), and red ("off track") status light. Just the discussion of those status lights in the management team may reveal diverse judgments and trigger important discussions about current strategic issues and need for action.

Setting realistic and stretch targets

Target values clarify the desired aspiration for each key performance indicator in a corresponding time frame. Generally speaking, they should be both realistic and ambitious without overtaxing the organization and its employees. The defined ambition needs to be in an appropriate relation to the capabilities and skills which are either already available in the organization or which can be developed or acquired. Furthermore, it needs to be realistic with reference to the company's ability for making investments that are required to achieve the target in the corresponding time period. Also, targets tied to metrics for different strategic goals on a strategy map need to "fit" each other. Targets linked to strategic goals articulating certain results (e. g., financial, customer/market goals) need to fit targets of "enabling" strategic goals (e. g., goals regarding internal processes or potentials).

Defining acceptable deviation ranges

The status that indicates the achievement for each target is usually determined by a certain range of deviation between the defined target value and the corresponding actual value. Hereby, it needs to be defined which range of deviation is acceptable ("green zone") and which not ("yellow zone" and/or "red zone"). Those status ranges are very important because they trigger attention, investigation and discussion about corrective action in performance review meetings conducted by the management leadership team (see chapter 4.3.7.1). Values that trigger a metric status light

Course of target values and hockey stick effects

turning from green to yellow or from yellow to red are generally defined as a percentage deviation from the target value.

Target values are set for the time period of strategy and may be adjusted over time. If the strategic plan's time horizon covers the next five years, target setting is typically done for each year starting at year five and working backwards. It starts by determining a quantified picture of what the strategy is supposed to achieve in the long run by setting all final year targets in the balanced scorecard. As those numbers are far in the future, they tend to be more generic than short-term targets and might be adjusted over time based on changes in the environment and internal feedback (see chapter 4.3.7). Target values for the upcoming year are usually planned in more detail and need to be synchronized with numbers from the operational planning and budgeting process. Once the different targets are set, the course of target values in the planning period can be analyzed and validated (see fig. 4-61). Targets rarely follow a linear development. A special emphasis needs to be put on so-called "hockey stick effects" or target closeness/target distance effects (Weber/Schäffer 2011: 50). In such a case, targets that are to be achieved sooner (that is they are "closer") are set more pessimistically, and targets that are further in the future (that is they are more "distant") are set more optimistically. Consequently, corresponding strategic initiatives are successful only because they deliver excellent results toward the end of the strategic planning period. This effect is illustrated with path 1 and 3 in figure 4-61, which both have the shape of a hockey stick. Those types of target value sequences may exist for different reasons. They may be caused by cognitive biases of the planners (see also chapter 1.2.2.3), set on purpose to increase the likelihood of rewards in

Fig. 4-61

Course of target values and hockey stick effect

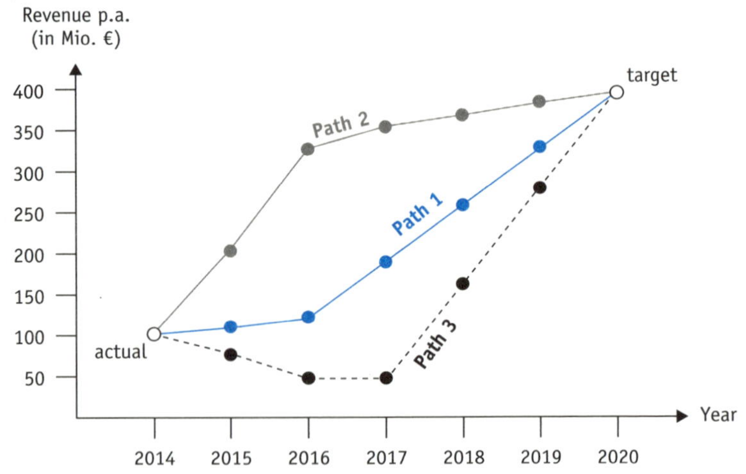

Source: Horváth & Partners 2007: 218; adapted

the short-run, or may reflect solid strategic thoughts such as a long introduction phase in the expected life cycle of a new product. Path 2 shows a flipped hockey stick that can be analyzed the same way. In any case, illustrating the course of target values can help to validate the rationale behind the numerical series, and thus, ensure more realistic aspiration levels.

4.3.4.3 Benefits and Pitfalls of Measuring Strategy

Four essential benefits of having metrics

Essentially, there are four benefits related to the definition of metrics that are tied to strategic goals (key performance indicators), and thus, operationalize a strategy (see also Strategic Snapshot 4.22):

- **Metrics help to understand what the strategic goal really means.** It replaces good guesswork and interpretation with certainty regarding the meaning of goals, which is a requirement for creating shared understanding and alignment about the strategic direction.
- **Metrics and targets clarify what exactly should be achieved.** This creates clarity about the desired aspiration level, which is essential for the development of effective strategic action programs.
- **Metrics help to evaluate the impact of (well executed) strategic actions.** They indicate whether the strategic initiative actually has an impact on the corresponding performance indicator and may disclose a need for more or another type of strategic actions.
- **Metrics help to get an outside-in perspective on the strategic performance.** Metrics that can be compared with other companies can be used to set relative self-adjusting targets in relation to peers in the competitive arena. This establishes a competitive performance culture as it drives behavior toward beating competitors instead of reaching the predefined (and maybe outdated) target.

Watch out: driving actions instead of developing metrics

Some practitioners question whether the benefit of developing performance indicators, particularly for non-financial strategic goals that are supposed to measure intangibles, is worth the effort. If the organization strongly struggles with execution, the company might be better off spending valuable time and resources on executing strategic action programs and getting things done instead of "wasting" time and resources for defining metrics that only help to clarify aspirations and measure results but not to accomplish initiatives. In such situations, executive teams might move directly to planning and initiating the execution of strategic action programs based on their experience and business acumen (conduct step 4 prior to step 3 in fig. 4.40, see chapter 4.3.1). Such companies start performance measurement by using relevant metrics that they already have in place and develop new and complex metrics simultaneously or postpone it to a later date in their transformation process of establishing a new strategy execution system. They do not wait with planning and executing strategic actions until their performance measurement system is completely established but get going.

Watch out: KPI development process

Developing and implementing new key metrics is not an easy task. It typically starts with developing KPI proposals based on internally available performance data, best practices and appropriate metrics from KPI repositories. Particularly for measur-

Strategic Snapshot 4.22

Purpose and Benefits of Metrics Illustrated with Sports

An easy way to understand the purpose and benefits of metrics can be provided by using a personal example related to sports. Imagine that you are using a strategy map for articulating your most important personal goals in life using four different perspectives. In this fictitious map there might be a perspective "personal well-being and health" and you have defined a goal "improve personal fitness". Most people would probably have a similar goal when following such an exercise. How would you know whether you made progress in achieving this goal? Let's say you want to use only one metric for telling you the answer. Which one would you look at—body weight, body mass index (BMI), resting pulse rate, blood pressure, muscular strengths, flexibility, fitness level derived from an oxygen consumption assessment tests (e. g., 12-minute swim test, 3-minute step test, 1-mile walk test, etc.), time to finish a certain distance (e. g., marathon), etc.? Although many people would probably embrace the goal "improve personal fitness", they may have different things in mind. Only the definition of a metric helps you to **understand what the goal really means**. Do you want to improve your cardio respiratory endurance or your muscular strengths or your flexibility? Let's further say you select the resting pulse rate, that is the number of times your heart beats each minute (bpm), as your key metric because you are focused on your cardio respiratory endurance. How much effort would you need to invest for improving it, say in six months? Well, that depends on your aspiration. Reducing it from 80 to 70 is different than from 80 to lower than 50. The latter would require a much tougher action program and might be unrealistic. The metric and target value for your goal "improve personal fitness" now helps you to **clarify what exactly should be achieved**. Next you set a target (e. g., reducing 10 bpm in 12 months) and you plan and initiate an action program of different sports activities per week for six months. Imagine you measure your resting pulse rate every four weeks for six months and you see no change, that is still 80 bpm. What does this tell you? Was your aspiration too high given a six month time period? Is it the right metric for measuring your goal? Did you execute your sports program as planned? Is your sports program effective? Answering those questions is triggering a learning process. You may have done everything right, but it does not change your key metric. You might have learned that your resting pulse rate is strongly influenced by your unhealthy diet and your high stress level at work. Therefore, you may want to change your action program and include those elements, that is change the daily diet composition and reduce stressors. After a further six months, you are at a resting pulse rate of 60 bpm and your action program was effective! Your metric helped you to **evaluate the impact of (well executed) strategic actions.** Finally, imagine you selected the "time to finish a marathon" as your key metric and your target is to finish in less than four hours as this was the last year's median time of all participants and you want to belong to the top 50 %. In that case, if you make it in 3:50 hrs you may have outperformed your target but not your underlying aspiration if your peers for some reason are performing better this year and the median time is 3:45. In such a case your aspiration might be better articulated with a relative self-adjusting target saying "finish within the top 50 % of participants". Hereby, the metrics will help you to **get an outside-in perspective on performance** and ultimately beat your competition instead of just making your number.

ing company-specific intangibles those KPI templates might not be available, which requires the development of new metrics. In such cases it needs to be ensured that those novel measurements are not overly subjective. From all metric suggestions, the most relevant performance indicators are then selected and refined following the explained guidelines (see fig. 4-59). To ensure consistency of metrics with strategic intent, the selection and refinement process should be conducted with the same team that developed strategic goals and the strategy map, whereas KPI proposals are typically developed outside that team and fed into the meeting to ensure efficiency. The decision which metrics should be implemented needs to be made considering a variety of criteria such as data availability, cost/benefit for measuring, acceptance of metrics, possibility to formalize and standardize the metric across different organizational units, frequency of data availability, etc. Following are 15 questions to ask when developing metrics (Brown 2007: 53):

1. What is the purpose of this metric—that is what behavior is this expected to drive?
2. What is the definition of this metric?
3. What type of metric is this (e. g., single measure, analytic/index, ratio, etc.)?
4. What is the unit of measure that will be used and what is the expected scale?
5. What is the formula for calculating this metric?
6. How often will data be collected on this metric?
7. What data collection method will be used?
8. Where does the data exist on this metric (data source)?
9. Does historical data exist on this metric?
10. Are there benchmarks available for this metric?
11. Who is the owner of this metric?
12. Who will be responsible for collecting the data?
13. What is the baseline and have targets been set on this metric?
14. What are the ranges for targets (e. g., green, yellow, red)?
15. What is the desired polarity of this metric (higher, lower, or in the middle)?

Watch out: benefit and risk of using analytics

In contrast to individual metrics, "an analytic is a metric made up of sub-metrics that are *different variables* and often units of measurement." (Brown2007: 29) Companies use analytics by analyzing an aggregation of different indicators if they want to evaluate and articulate how well they are doing regarding a specific strategic goal. This method results in a single strategic goal status (e. g., green, yellow, red) that is determined based on the statuses of the different submetrics, which have been aggregated following certain rules (e. g., weighting, status combinations, etc.). The advantage of this method is that analytics provide a more complete picture because strategic goal performance is rarely a function of only one element. For example, using analytics in the Strategic Snapshot 4.22 may lead to a "fitness index" derived from statuses of five submetrics: resting pulse, weight, average weekly stress score, average weekly calories intake, and daily diet composition. Note that each submetric needs a target, the definition of a range (e. g., for the green, yellow, red status), an ongoing generation of actual values, and a comparison to determine the status. This can lead to a tremendous effort and complexity, which is a major pitfall of such an approach. Although the balanced scorecard may be designed to have only 20 metrics, it may actually result in 100 metrics by adding up all the submetrics when using 20 analytics. If it has been designed this way, reporting is usually following a management by exception principle: As long as the aggregated status is green, there is no need to drill down. However, if not, it requires a drill-down to the root cause and an investigation where this red status comes from and what can be done about it. When working with this type of analytics, it is advisable to trigger the aggregated performance warning by one or more red statuses of submetrics (see also fig. 4-82 in chapter 4.3.6.3). For example, a red status of the "fitness index" may only be triggered by a red status on "stress level", whereas all the other four submetrics are green. An alternative is to work with analytics based on weighted averages. This approach needs to be treated with great caution, though, as critical warning information might be averaged out. Anecdotally spoken "if you have your head in the oven and your feet in the freezer you feel well on average". Ultimately, it needs to be decided in which cases the management team wants to talk about certain issues and this needs to be reflected in the status mechanism on the balanced scorecards.

4.3.5 Plan Strategic Actions

Discretionary initiatives contributing to strategic goal achievement

In the previous chapters it has been described how strategic goals and related key performance indicators with corresponding targets can be developed. The logical next step is to define initiatives that help to achieve those goals and targets. Typically, ongoing day-to-day activities are not sufficient to achieve the strategic goals and make the strategy happen. They need to be supplemented with initiatives that are specifically designed for strategy execution. "Strategic initiatives [strategic actions] are the collection of finite-duration discretionary projects and programs, outside the organization's day-to-day operational activities, that are designed to help the organization achieve its targeted performance." (Kaplan/Norton 2008: 103). Thus, initiatives are called strategic actions if they

- are discretionary, that is they are not part of but supplement ongoing day-to-day activities
- are limited in time, that is they have a definite start and completion date
- support the achievement of strategic goals.

Key considerations for the screening, development, prioritization, and documentation of strategic actions are described in the following (Horváth & Partners 2007: 222–230):

4.3.5.1 Developing and Prioritizing Strategic Actions

Inventory of existing "strategic" projects

The development of strategic actions starts by screening initiatives that already exist in the organization. The result is an inventory of running or planned projects that utilize significant resources and are conducted in addition to the day-to-day routine activities. Next, the contribution of these running initiatives to the achievement of strategic goals is validated to see whether they are still relevant from a strategic point of view. This may not only provide a list of strategic actions that are already in place but may also help to stop projects which are not strategically important anymore, and thus, unfreeze resources for other strategic priorities (see also Strategy Practice Example 4.9 in chapter 4.3.2.3).

Ideation on new strategic actions

Existing strategic initiatives may not be sufficient to make the desired strategy happen, particularly, when new or revised strategic goals are defined. In this case, new ideas for strategic actions are developed. Strategy maps with strategic goals as well as key performance indicators and targets are used as orientation. The focus should be on all perspectives of the strategy map *except* the financials perspective. The strategy map reflects how strategic goals in customer/market-, process and potentials-related perspectives are supposed to jointly deliver financial results.

Therefore, defining strategic actions in those three perspectives should be sufficient. There are rarely strategic actions related to financial goals only (e. g., tax optimization) that are not being captured somewhere else in the strategy map. Finding the right strategic actions can be facilitated by asking the following question:

What can we do to achieve the strategic goal?

Ideally, this question is answered by the same team that has developed the corresponding strategic goals and key metrics. To ensure an efficient process, strategic action proposals may be developed individually in advance, and then discussed, refined, supplemented and decided on in a management workshop. Furthermore, there might be a number of strategic action proposals that are strongly related to each other and can be clustered and consolidated to a strategic action with a higher altitude.

Strategic action relevance matrix

Figure 4-62 shows a comparison of strategic actions (existing and new) with strategic goals. The "X" illustrates to which strategic goal achievement an action is supposed to contribute directly. In case of new actions, this is the strategic goal for which the action was developed. Additionally, strategic actions may contribute indirectly to other strategic goals, for example, through the logical linkage in a strategy

Fig. 4-62

Strategic action relevance matrix

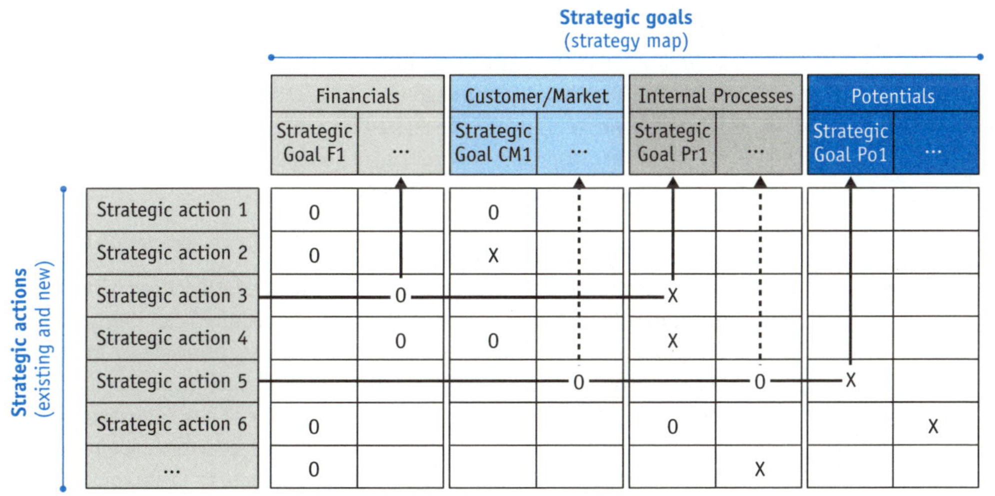

X = Direct impact of action on goal achievement (e.g., action was developed for this goal)

0 = Significant indirect impact of action on goal achievement (e.g., critical enabler)

Source: Horváth & Partners 2007: 225 and Wunder 2001: 138; adapted map. Those indirect contributions are illustrated with an "O". The resulting strategic action relevance matrix shows how strongly the strategy is generally supported with initiatives and if there are any strategic goals that are not supported ("white spots"). When prioritizing and deciding for or against certain strategic actions, this matrix can be used to illustrate what it means for strategy if a strategic action is left out or postponed (Wunder 2001: 138).

Focus on highly relevant strategic actions only.

Strategic action development often results in a fairly high number of initiatives covering both existing strategic projects and new proposals. Those strategic action candidates can be prioritized using the matrix shown in figure 4-63. One criterion for deciding on strategic actions is its contribution to making the strategy happen. For some strategic actions, this may be based on its financial impact such as a rough order of magnitude regarding its revenue or margin contribution. Particularly for strategic actions that relate to intangible elements such as human capital or organization capital, the direct financial contribution cannot be calculated. However, they may still be critical for strategy execution because, for example, they represent crucial enablers for other strategic goals (see fig. 4-62). Those strategic actions would be rated as highly relevant without specifying their financial impact. Companies applying a value-based management approach may be able to investigate the financial impact of non-financial actions through linking them to elements in their value driver tree. The second criterion for prioritization is the resource requirements for strategic actions. For the purpose of the initial prioritization, a rough resource estimate (e. g., total expenditures) is required, which will be refined later on. According to the allocation of strategic actions in the matrix (see fig. 4-63) the following approach can be suggested:

- Quadrant 1: Strategic actions are started as soon as possible and get full funding.
- Quadrant 2: Strategic actions are either started as soon as possible with limited funding or postponed until full funding is available.
- Quadrant 3: Actions are not considered for a strategic action program but may be considered for improvements of operational activities.
- Quadrant 4: Actions are not considered.

4.3.5.2 Refining and Aligning Strategic Actions

Strategic project charter

Once a shortlist of relevant strategic actions is developed, they need to be refined and planned more in detail to ensure that they get the required funding and avoid any cross-functional resource conflicts or time constraints. Information that has to be provided for each strategic action is summarized in a strategic project charter, which is illustrated in figure 4-64. It provides transparency of key information, which is required for effective strategy execution. The project charter helps to understand what the initiative is supposed to deliver and contribute to the strategic goal achievement, in which time frame (incl. milestones) it has to be done, managed by which team, and based on which resource requirements. To ensure effective execution, high level executive sponsors are defined. Furthermore, there is information about involved organizational units that are expected to commit resources. Those expectations need to be aligned with the available resources considering the entire final strategic project portfolio.

Fig. 4-63

Strategic action prioritization scheme

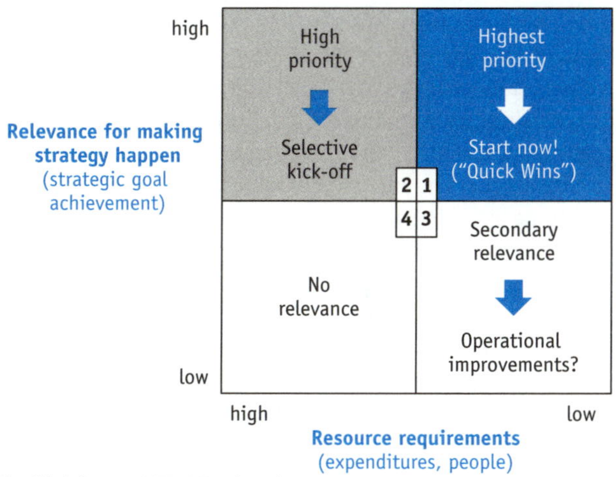

Source: Horváth & Partners 2007: 228; adapted

Fig. 4-64

Illustrative strategic project charter

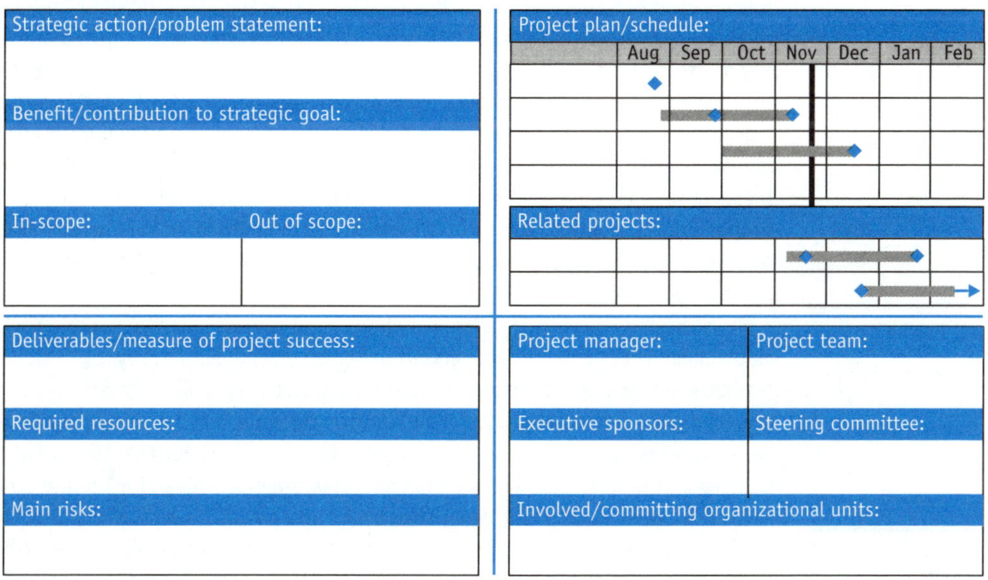

Top-level strategic action programs

Larger organizations in particular consolidate and group important strategic actions of different organizational units that are content-related into a few top level strategic action programs. Additional corporate strategic actions for creating cross-unit synergies and value might supplement those strategic programs at the end of a bottom-up development process (Wunder/Baur 2000: 255). From a company-wide perspective those, for example, 10 or 20 top-level programs are considered most crucial to make the overall strategy come true. They typically require cross-functional commitment, need significant resources (e. g., operational or capital expenditures, internal people capacities, etc.), and thus, require top management approval. Top level strategic action programs are typically subject to a specific governance process that ensures ongoing attention of the management leadership team, for instance in monthly or quarterly strategic performance reviews (see chapter 4.3.7.1). Each strategic program gets a high-level executive sponsor who mitigates cross-functional issues, for example. Their funding gets highest priority and may come from a specifically defined strategic budget bucket (see chapter 4.3.6.1). A regular update of the top level strategic action program can be facilitated by strategic action portfolio management with corresponding processes for strategic project prioritization, approval and formal close (see chapter 4.3.7).

Resource planning

Defining strategic actions with implicit assumptions about how other departments are supposed to contribute can be a major flaw in strategy execution if those units do not have the required resources. Particularly in functional areas, it is of utmost importance that resource requirements for executing own strategic actions (e. g., the strategic action program in the IT department derived from the IT strategy map) are aligned with resource requirements from other organizational units that require support, and thus, resources from the functional team to get their strategic actions executed (e. g., strategic business units that implicitly expect IT support for some of their strategic actions). To avoid this pitfall, a detailed resource planning for key strategic actions or programs is necessary (see fig. 4-65), which is the base for the operational planning and budgeting process (see chapter 4.3.6.1). It captures the required time of staff and executive people in the own unit and the different areas involved as well as corresponding operational expenditures and capital expenditures. If this is done for all key strategic actions in a company, it can be calculated for each support function which resources it is expected to commit for achieving not only their own strategic goals but also the strategic goals that are defined in different other organizational units (e. g., businesses).

Financial summary

In addition to the strategic project charter and resource plan, a preliminary financial summary for each strategic action has to be developed. Hereby, operational and capital expenditures as well as expected financial results directly delivered by the corresponding initiative (if applicable) are populated over the time period of strategy (e. g., five years). This preliminary summary can be used for the alignment of resources and targets during strategic action planning as well as for linking strategic actions to operational planning and the budgeting process (see chapter 4.3.6.1).

Strategic action roadmapping

The alignment of schedules is strongly related to the alignment of resources for strategic actions. This can be supported by developing a roadmap that places all strategic actions on a multi-year project plan. Some companies conduct this exercise in

Fig. 4-65

Strategic action resource requirement plan

Strategic action:	

Resource type	Project		Contributing organizational units					Total
	Sponsor	Manager	IT	Finance	Marketing	Production	R&D...	
Project time [days]								
Staff cost [k$]								
Executive time [days]								
Executive cost [k$]								
Total personnel cost [k$]								
Other cost [k$]								
Oper. expenditures [k$]								
Cap. expenditures [k$]								
Opex + Capex [k$]								

their management teams using a large wallpaper to visualize what the strategic action programs mean from a time perspective. This way it can be validated whether the organization is actually capable of executing the different initiatives according to the project plans defined in the corresponding strategic project charters. In case schedules need to be changed and aligned, it needs to be assessed whether the defined target values for key performance indicators are still realistic.

The resources committed by management need to be integrated into the short-, mid- and long-range planning process. Some companies leverage this process to simplify and speed up their operational planning process (see chapter 4.3.6.1).

4.3.5.3 Benefits and Pitfalls of Strategic Action Planning

Deliver on desired strategic performance

Defining strategic actions increases the likelihood for effective strategy execution if they are embedded in a corresponding management process. Successfully delivering on those initiatives helps the organization to achieve its desired strategic performance. It ensures high-level sponsorship, appropriate funding and resources of different organizational units involved being aligned for effectively executing the strategic action program.

Watch out: micro management versus self-control

One the one hand, a key principle of the suggested strategy execution system is an organization-wide management by objectives (MBO) system through which strategic priorities are cascaded to different management levels and into operational areas (see chapter 4.3.6.2). This type of management process usually relies on self-control mechanisms in the different organizational units and refrains from a mechanistic definition of strategic action programs. On the other hand, defining strategic actions helps to secure required resources and sponsorship for achieving strategic goals. Hereby, the right balance between cross-functional strategic action governance and self-control of the different organizational units need to be found. Strategic action development in various units may be supported by the strategic planning team but ongoing governance at the corporate level is typically related to the most important company-wide strategic action programs only. Unit-specific strategic action reporting and governance is usually left to the discretion of the corresponding organizational leadership teams.

Watch out: top-line or bottom-line growth

When defining prioritization criteria for strategic actions, it needs to be considered carefully whether the company wants to drive top-line growth or bottom-line growth (or both). A top-line growth orientation, for example, may be reflected in defining a hurdle rate of revenue contribution that a strategic action's rough order of magnitude revenue estimate needs to fulfill. This could lead to the situation that ultimately only growth initiatives get top management attention and funding. There is a risk that highly profitable smaller initiatives are neglected in the strategy execution system or cannot be executed because they do not get corresponding resources. For bottom-line growth it is more appropriate to look at profitability contributions of strategic actions for the purpose of prioritization.

4.3.6 Align Organization to Strategy

Organizational alignment

Excellent strategy execution requires an alignment of the organization to strategy. Three key areas for organizational alignment are resources, people, and management information (see fig. 4-66). They are strongly linked to the three key components for strategy execution that have been explained at the beginning of chapter 4.3.1, that is strategy maps, balanced scorecards, and strategic action programs. Those elements articulate and operationalize strategy and how to make it happen. Companies need to make sure that

- necessary **resources** (funds, employees) for making the strategy happen are secured
- **people** are motivated and enabled to help executing the strategy
- **management information** is provided to keep track with strategy execution and facilitate strategic learning.

This requires a calibration and orchestration of key managerial systems such as operational planning and budgeting, strategy communication, target agreement and incentive system, personnel skill development, and management reporting in a way that they are consistent with the strategy. Out of the many managerial systems applied in companies those are considered vital for effectively executing strategy. Organizational alignment is a challenging task as it involves different types of established processes and procedures in different areas of responsibility. The following outlines some key considerations and approaches of how to align those managerial systems to strategy (Horváth & Partners 2007: 271–312; Kaplan/Norton 2001 and 1996: 199–249).

Fig. 4-66

Key components of aligning the organization to strategy

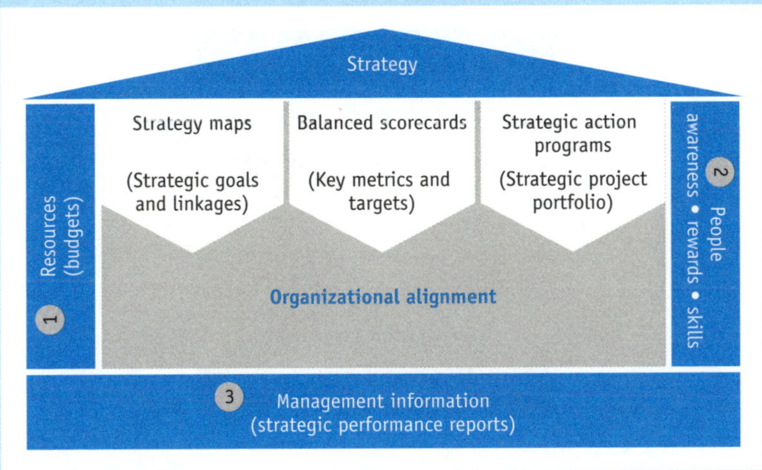

4.3.6.1 Resources

Securing funds for executing strategy

One of the most fundamental requirements for an effective strategy execution is to secure funds for strategic goals and action programs. People in charge of strategy execution need to have the corresponding resources to excel. This means that resource requirements for strategic goals and strategic actions programs need to be considered in the operational planning and budgeting system, which, however, historically might have been following a different logic (see chapter 4.2.2.3). Key questions are:

- What resources do we need in the coming years for current strategic action programs?
- Which resources in our business units do we commit to each strategic action?
- Are there resource gaps and how can we close them?

Funneling limited resources to create the strongest impact

A properly created strategy-based budget funnels funding into strategic activities that promise the highest value for the company. Linking operational planning and budgeting to strategy means that limited funds are allocated to those strategic projects that are supposed to have the strongest overall business impact as articulated in the strategic goals. Hereby, budgets—along with incentives—become a bridge between strategy and operations. Long-term strategy management has to be synchronized with short-term operational management to ensure that resources are in place to deliver the desired results and to reward people for their contributions (see fig. 4-67).

One way of linking strategy to operational planning and budgeting is based on the following three steps that will be explained next (Kaplan/Norton 2001: 288–301; Wunder 2006):

Fig. 4-67

Linking strategy to operations with budgets and rewards

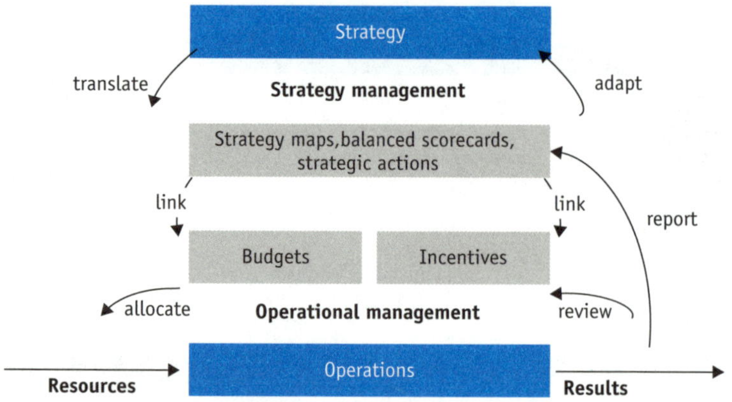

Source: Kaplan/Norton 2001: 275; adapted

1. Separating strategic and operational budgets
2. Linking strategic actions to strategic budgets
3. Simplifying the development of operational budgets

Separating Strategic and Operational Budgets

Base and development business

Separating budgets is based on a management perspective that divides the business into two parts: the base business and the development business. The *base business* reflects the baseline growth that can be achieved with incremental improvement to existing operations ("business as usual"), based on market models and business forecasting within a certain time period (e. g., up to 5 years). The challenge of focusing primarily on base business is that it is likely to hit a growth limit at a certain time (Senge 1990: 95–104). The *development business* goes beyond an optimization of the existing scope of business activities and extrapolated financial outlooks and is typically related to a longer time horizon (e. g., more than five years). It might be related to desired new customer acquisitions, regional market expansion, new product development, etc. caused by, for example, anticipated trends or disruptions in the company's external environment. A company characteristically requires the development business to close the gap between a desired breakthrough performance (e. g., growth aspiration) and what is achievable with continuous improvement of the existing business. Figure 4-68 illustrates the simple logic of a strategy focusing on top-line growth. The existing base business only accounts for a small portion of the growth aspiration. Closing that gap typically requires strategic programs and actions that go beyond operational activities to maintain baseline growth. In the illustrative example, leveraging existing products for new customer segments and new regional markets (market development) and delivering customer value with new products in existing markets (product development) are defined as strategic themes to close the gap. Each of those themes has a set of strategic goals and corresponding strategic actions that need to be funded. How to identify and develop those strategic action programs is captured in step 4 of the applied strategy execution framework (see fig. 4-40) and has been explained in chapter 4.3.5.

Strategy-oriented budgeting

Funding those action programs requires a strategy-oriented budgeting process. Rather than only asking the question of what a company needs to spend on each cost category such as salaries, facilities, material, promotion, training, travel, etc. (budget centric), a strategy-oriented budgeting process is centered on the question of how a budget must look like to support strategic themes and meet strategic objectives such as acquiring new customers, entering new markets, foster innovation, attract talent, etc. (Greiner 2004). This can be achieved, for example, by dividing the total budget into an operational ("business as usual") and a strategic portion following the rationale of base and development business (see fig. 4-69).

Operational budgets

Operational budgets reflect basic costs to operate and are necessary for sustaining and optimizing the business. They consist of forecasted revenue numbers related to the sales of goods and services and the corresponding expenses to produce and deliver them to the customer under efficient operations. Operational budgets are required for achieving financial targets that reflect the base business (e. g., baseline revenue growth at market growth rates), which may also include the launch of new

Fig. 4-68

Mapping revenue-generating strategic programs and actions to growth aspiration

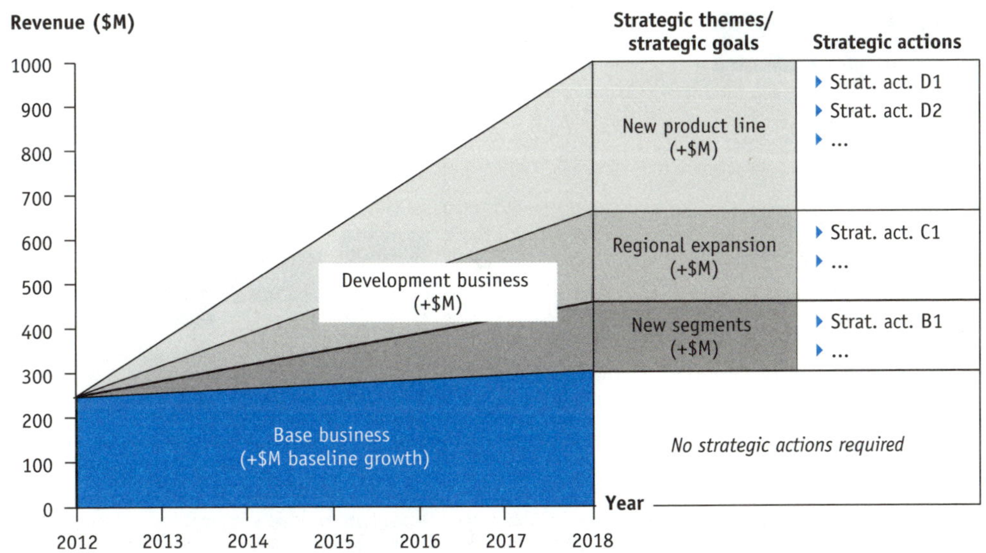

Source: Kaplan/Norton 2008: 91 and Wunder 2006: 21; adapted

products or the acquisition of new customers during the next period. Most annual spending and expenditures are determined by operational budgets.

Strategic budgets

Strategic budgets cover people, capital and financial resources that the organization requires for achieving future growth or other goals beyond the baseline. Compared to operational budgets, they are only a small percentage of the total budget and reflect discretionary expenses that are not driven by day-to-day operations. From a commercial perspective, strategic budgets finance initiatives that make the difference between growing with the market and outpacing it. They are in place to bridge the gap from baseline growth to the desired growth aspiration or the gap between base and breakthrough performance regarding other performance indicators.

Allocating and prioritizing strategic budgets

Operational budgets are generally determined by the volume, product mix or customer structure of the operational business that relate to certain cost and revenue categories. For example, they are allocated to different organizational units such as strategic business units, functional units, regions or countries, etc. based on their requirements for running operations (see chapter 4.2.2.3). Once the allocation is done, the corresponding managers within the organizational units decide how they use those budgets. Contrary to operational budgets that are derived from the operational value creating process, strategic budgets are strongly linked to strategic themes or strategic goals and do not reflect continuous or repetitive expenses. They are additional discretionary funds that are prioritized and allocated across the differ-

Fig. 4-69

Allocating operational and strategic budgets

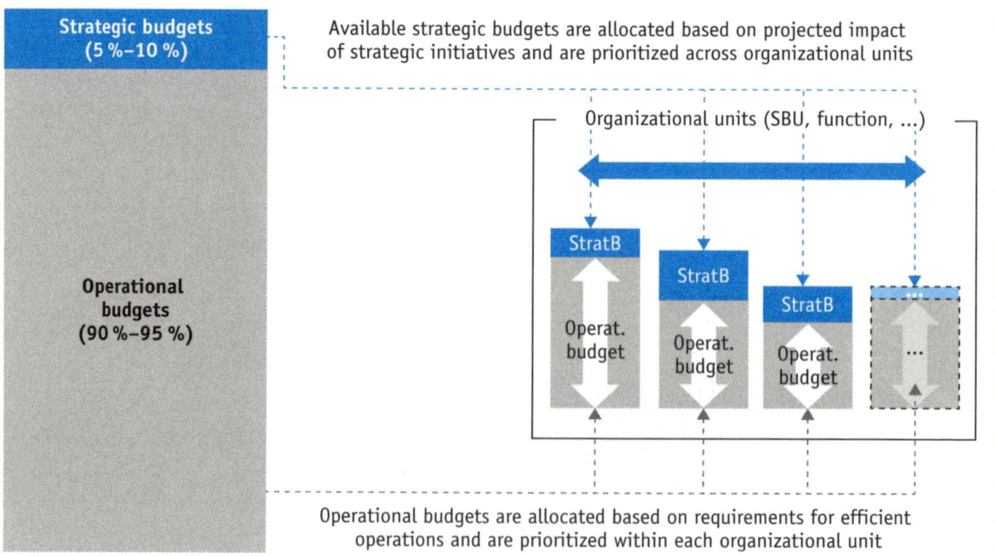

Source: Kaplan/Norton 2001: 289 and Wunder 2006: 17; adapted ent organizational units. Once the available strategic budget is determined by the senior management, the different organizational units are competing for those limited funds based on their individual strategy contribution. To make the case for those strategy contributions, managers may leverage strategy maps, balanced scorecards and strategic action programs of their corresponding unit. The different allocation principles for operating and strategic budgets are illustrated in figure 4-69.

Linking Strategic Actions to Strategic Budgets

What is important versus what is budgeted

The second step in matching operational planning and budgeting with strategy is the link of strategic actions or programs to the strategic budgets. During the annual operational budgeting process, managers typically follow a budget-centric logic by asking the question:

How much will we need to spend for each cost category?

They are requesting an amount of funds for salary and wages, equipment and material, advertisement and promotion, travelling, etc. to run their operational activities and to make investments. In a strategy-oriented budgeting process, the focus shifts from "what's budgeted" to "what's important". The key question is how the budget should look like to achieve the defined strategic goals:

How much will we need to spend to achieve our strategic goals?

This may include resource requirements for acquiring new customer segments, expanding the business into new regional markets, innovating and developing new products, building a certain image or realizing a new business model?

Strategic budgets for high impact strategic actions

The linking pin between strategy in terms of strategic themes—expressed by a strategy map or by strategic goals—and the strategic budget are the strategic actions. As explained earlier, those initiatives drive growth or other business performance indicators above baseline performance and typically cannot be supported with operational budgets. Related strategic actions may be consolidated to strategic programs that are budgeted in order to ensure a significant size and effect. The business impact of each strategic action or action program is planned in terms of financial "effort" such as required internal employee capacity, expenses, investment, and financial "benefit" such as sales increase or cost reduction. This can be facilitated by developing strategic project charters and resource plans for each strategic action (see chapter 4.3.5.2). In the project charter, key information on the desired business impact, resources needed, risk, high level management sponsors, etc. as well as involved (contributing) organizational units and their required resource commitments is summarized. The project charter is finally signed by senior managers to ensure that the required resources are actually committed and integrated in the planning process. One way to prioritize strategic actions as decision support for allocating the limited strategic budgets is by looking at the cost/benefit ratio and the probability of success in terms of risk and feasibility. In the strategic budgeting process, only budget requests above a certain threshold of incremental expenses per year are considered. Smaller amounts might be covered in the operational budgets. Figure 4-70 illustrates the explained budgeting principle: Strategic themes or goals along with the corresponding metrics and targets on different levels in the organization are supposed to be achieved by executing a number of strategic actions or programs. Resource requirements for those strategic actions are budgeted for and expressed in terms of strategic expenditures (stratex). The direct or indirect financial impact of strategic actions is reflected in the revenue and profit plan. To avoid the pitfall of a fixed year-end planning period (see chapter 4.2.2.3), the different line items might be planned with rolling forecasts. Figure 4-70 shows such an approach with an illustration of a five-quarter rolling forecast. The planning period is fixed (five quarters) but plans will be updated each quarter based on actual results and new developments in the competitive arena.

Conventional budgeting of strategic actions

Depending on how a company applies its strategy execution framework (e. g., how it develops strategic goals and strategy maps, defines performance indicators with targets and derives strategic action programs), the resulting strategic actions may

Fig. 4-70

Strategic actions as a link between strategy and operational planning and budgeting

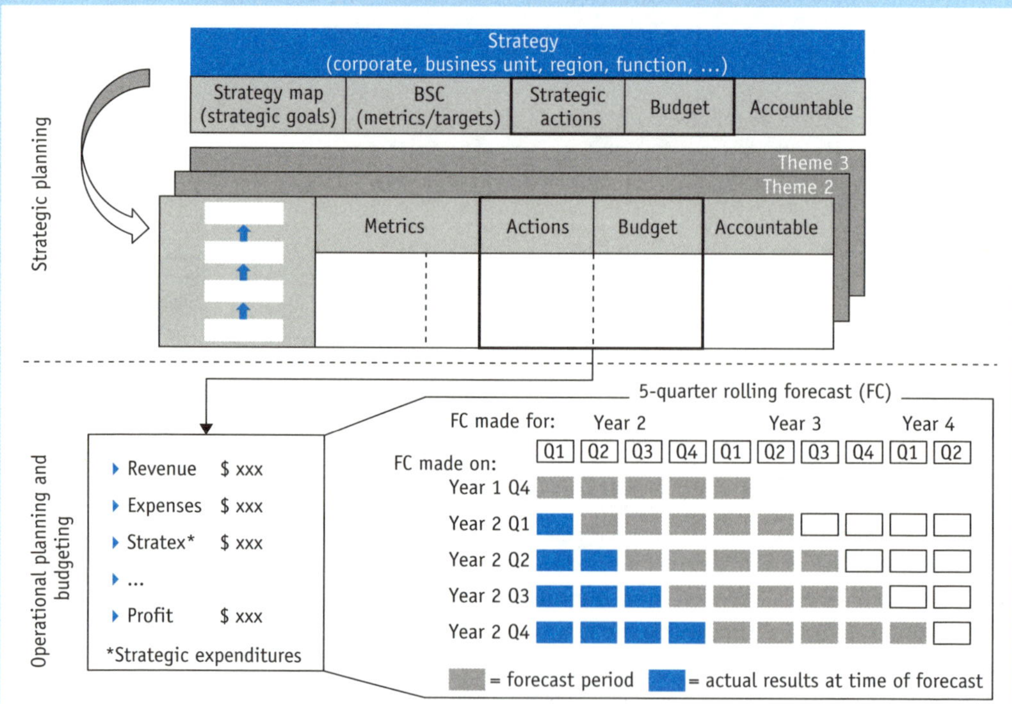

Source: Norton 2006: 9 and Wunder 2006: 9; adapted

not only reflect the previously mentioned development business for closing a revenue gap but may also be related to the base business. In such a case, defining and applying a fixed strategic budget (stratex) for funding all strategic actions and separating it from the operational budget might not be appropriate or feasible. Companies that have to budget their strategic actions in situations like this might be better off avoiding a predefined fixed strategic budget but applying a conventional budgeting process. Hereby, the different cost categories such as salaries and wages, material, marketing, distribution, sales infrastructure, etc. are budgeted with two financial numbers. One reflects the resource requirements for "business as usual" (operational budget), which are related to next year sales volume numbers and other forecasts. The second expresses the expected financial implications of long-term strategic actions that are broken down to annual budgets for each cost category. Thus, the total budget per cost category is composed by an operational and a strategic portion and must ensure that both operational and strategic initiatives are enabled. The risk of this approach is that operational funds are shifted to the strategic portion, and vice versa within the total budget, as those two portions are usually not controlled separately. For example, slack might be built in operational budgets to create "additional" funds for strategic actions. To avoid this risk a firm has to be able to strip out these funds through, for example, a project tracking code in their financial system that allows to report spending on strategic expenditures.

Simplifying the Development of Operational Budgets

Unfreeze time for strategy.

In step 1 and 2 it has been explained how strategic priorities that a company or organizational unit has defined in its strategic planning process can be integrated in the operational planning and budgeting process to ensure sufficient funding. If the budgeting process is disconnected from strategic planning there is a high risk for an organization to fail in strategy execution as the necessary financial resources, capital and people are not provided in the budget. The third step is making the operational planning and budgeting process simpler and more efficient. This unfreezes time and management attention that can be utilized for strategy. Three ways to simplifying operational budgeting are:

- Different levels of planning detail are applied for different areas in the organization. For each organizational unit (e. g., strategic business unit, function, region, country, etc.) the relevant level of detail is determined individually based on effort/benefit considerations.
- Only the major budget positions within a business area are planned. For example, products or regions that account only for a minor portion of the overall business results are consolidated into groups or planned with pre-populated budgets.
- Different planning methods are applied within an organization. The planning effort for a business area is adjusted to its specific environment. Business areas in highly volatile and complex environments may require rolling forecasts and continual monitoring, whereas areas with fairly stable cost positions are budgeted based on past numbers.

Other levers to improve the efficiency and effectiveness of conventional budgeting processes have been elaborated on in chapter 4.2.2.3.

4.3.6.2 People

Awareness, skills and incentives

In addition to providing the necessary resources, the staff members need to be motivated to help to execute the strategy. This covers three areas: First, employees needs to be aware of and inspired by their company or business unit strategy. This could be supported by a formal strategy communication program that is engaging and creates understanding and commitment. Second, individual objectives and incentives of employees need to be linked not only to their operational tasks but to strategy, that is strategic goals, metrics and targets and strategic actions. This is related to the performance management system (target agreement system) as well as reward and incentive system, which needs to reflect and reward strategic priorities. Third, the organization needs to have the corresponding capabilities and its employees the corresponding skills that are necessary to execute strategy. This can be addressed by aligning personal skill development programs with strategy (see also chapter 4.2.2.3).

Creating Awareness through Strategy Communication

Effective strategic communication is meaningful.

For a strategy to be successful, it should not only be formulated collaboratively but communicated to all levels as well. According to Kaplan and Norton "Communicating a mission, values, vision, and strategy is the first step in creating motivation among employees. Executives can use the strategy map and Balanced Scorecard to communicate strategy—both *what* the organization wants to accomplish and *how* it intends to realize its strategic outcomes. Taking all the objectives and measures together provides a comprehensive picture of the organization's value creating activities. This new representation of strategy communicates to everyone what the organization is about: how it intends to create long-term value and how each individual can contribute to organizational objectives. Employees can come to work each day energized about doing their jobs differently and better, helping advance the organization's success and realizing their personal objectives." (Kaplan/Norton 2008: 141) Effective strategy communication can be meaningful to employees as it provides a reasoning and appreciation for their daily activities in the context of the overall strategic direction. Instead of directing people in detail what to do, companies provide strategic context to encourage and guide the right behavior.

The "strategy chart"

The challenge for communicating strategy to the entire organization is to define what key strategic messages everyone in the organization should know and understand and to find simple, clear and compelling ways to communicate those themes. This can be accomplished, for example, by using a strategy chart as a summary sheet that provides the answers to crucial strategic questions on one page (see fig. 4-71). This chart is related to the idea of a strategy map, though it is a more comprehensive model for communicating not only the strategic goals in terms of strategically desired outcomes but also the means how to get there. Strategy maps articulate only strategic goals to highlight which elements the company or department needs to focus its limited resources and management attention on. They are a method for communicating strategic "spotlights" and for aligning them between different areas in the organiza-

Fig. 4-71

The strategy chart—a bird's eye perspective on what matters on one page

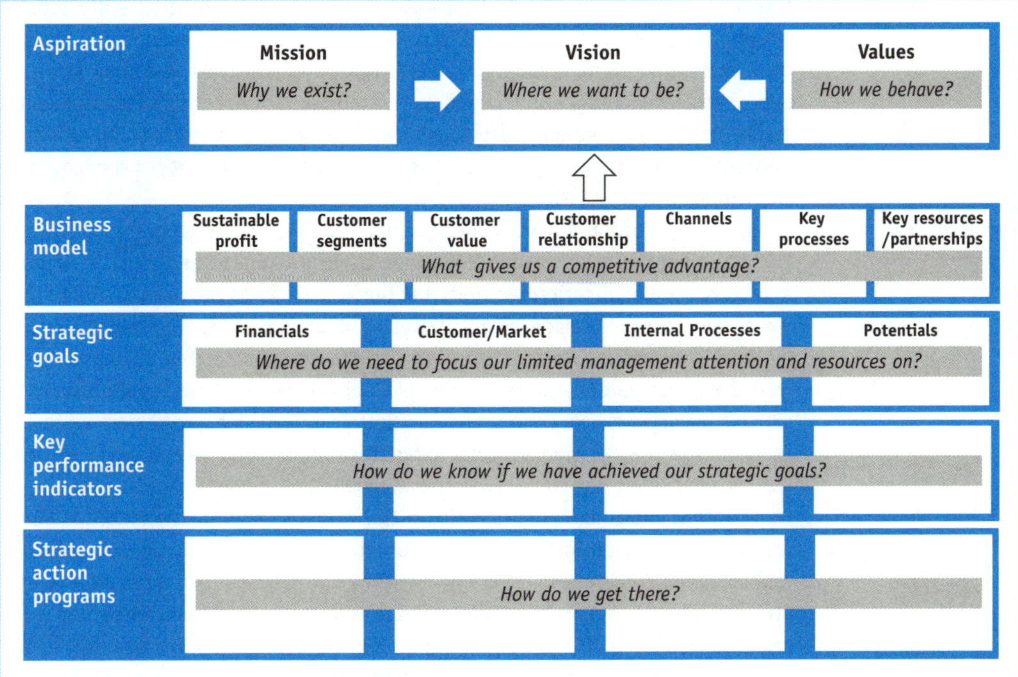

tion. The strategy chart, on the other hand, conveys the entire strategic framework covering a variety of interconnected elements. It summarizes the strategic guideposts ("aspiration"), highlights which pieces of the business model are crucial for competitive advantage, and thus, makes a difference in the marketplace, illustrates the strategic priorities (strategic goals) along the four balanced scorecard perspectives that are important to realize the desired strategic course, presents how success in pursuing those goals is measured (key performance indicators) and what initiatives (strategic action programs) the company drives in order to make the strategy happen.

Rich and lean communication channels

Without effective communication, the employees who are supposed to support the strategy execution are not aware of strategic priorities, and thus, will not understand how they can contribute individually within their area of responsibility. Therefore, the importance of strategy communication cannot be overemphasized. When designing a strategy communication program it is important to consider that communication channels differ in their richness or capacity to convey information (see fig. 4-72). *Rich* communication channels can handle multiple information cues (e.g., words, gestures, facial expressions, etc.) simultaneously, facilitate both verbal and nonverbal rapid feedback, and are very personal. Therefore, face-to-face conversations score highest when it comes to channel richness, followed by video conferences and communication via phone. Lean communication channels score low on the three criteria, which makes impersonal written media such as formal reports or memos and newsletters rate lowest in richness. In general, routine information that is straightforward and has little potential for misunderstanding can be efficiently communicated with channels low in richness. In companies this type of communication can be generally found in, for example, standard reporting processes (see chapter 4.2.2.3). Non-routine communications, such as a strategic reorientation or transformation, can be communicated more effectively by selecting rich channels to foster strategic awareness and create a shared understanding and commitment for strategic priorities (Robbins/Judge 2013: 384 f.).

Written, verbal and drawn communications

Companies typically apply comprehensive communication frameworks to ensure that all employees are captured. Nowadays, there is a huge variety of different written and verbal communication methods and media that firms normally combine to convey their strategy messages (Kaplan/Norton 2008: 141–148). Written strategy communications include strategy booklets, brochures, posters/bulletin boards/racks, wall charts and permanent framed posters, pocket guides, PowerPoint presentations, articles in corporate magazines and newsletters as well as strategy messaging posted in local intranets. Written strategy communication is tangible and physically available for continued reference. It is usually well thought out, logical, clear, easy to understand, and compelling (see fig. 4-73). Verbal communication methods include face-to-face communication, small group staff meetings, large town hall meetings and company conferences, strategy events or exhibits that make people experience strategy, webcasts and video conferences (live or recorded) as well as specific training sessions (Robbins/DeCenzo 2008: 325 f.). More recent means for strategy communication are interactive new and social media, strategy toys, cartoons (see fig. 4-76) or strategy dialog pictures and strategy games (see fig. 4-75) where the strategy is

Fig. 4-72

Information richness and communication channels

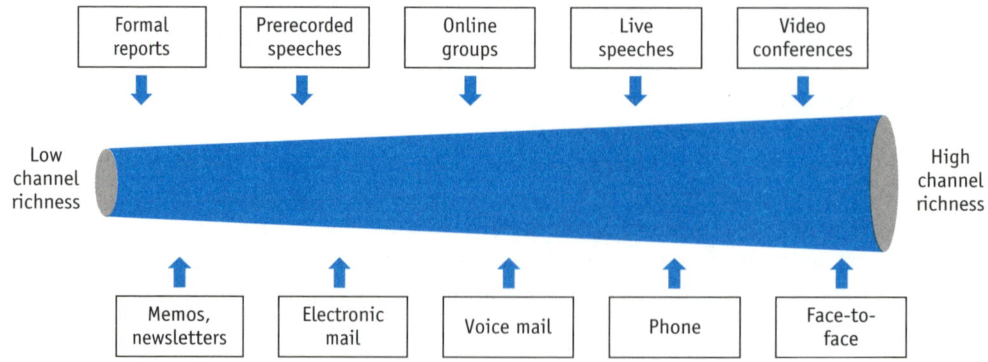

Source: Robbins/Judge 2013: 384; adapted

visualized in a simple and compelling way, ambassador programs, and others (Harringer/Maier 2011). The Strategy Practice Examples 4.13 show different types of strategy communication that are applied in large organizations.

Strategy Practice Examples 4.13

Selected Strategy Communications in Large Organizations: Examples from Nestlé, Toyota, DB Group, and BAYER

One way to communicate its strategic ambition and pillars that Nestlé has been using for many years is the "Nestlé Strategic Performance Framework (4x4)" (see fig. 4-73). It articulates key strategic messages on one page and is used to provide guidance and alignment for people around the world. The framework includes the company's ambition of being the leader in nutrition health and wellness (NHW), an industry reference for financial performance as well as trusted by all stakeholders. It further articulates a set of crucial strategic priorities and pillars for delivering this ambition, which covers hard to copy competitive advantages, key prospective growth areas, operational pillars with four interrelated core competences as well as the "Nestlé Model" of providing profitable growth, improving business results and achieving market leadership year after year. Embedded in the strategic and performance framework are the principles of compliance, sustainability, and creating shared value, that is value for shareholders, value for the communities in which the company operates, value for business partners as well as value

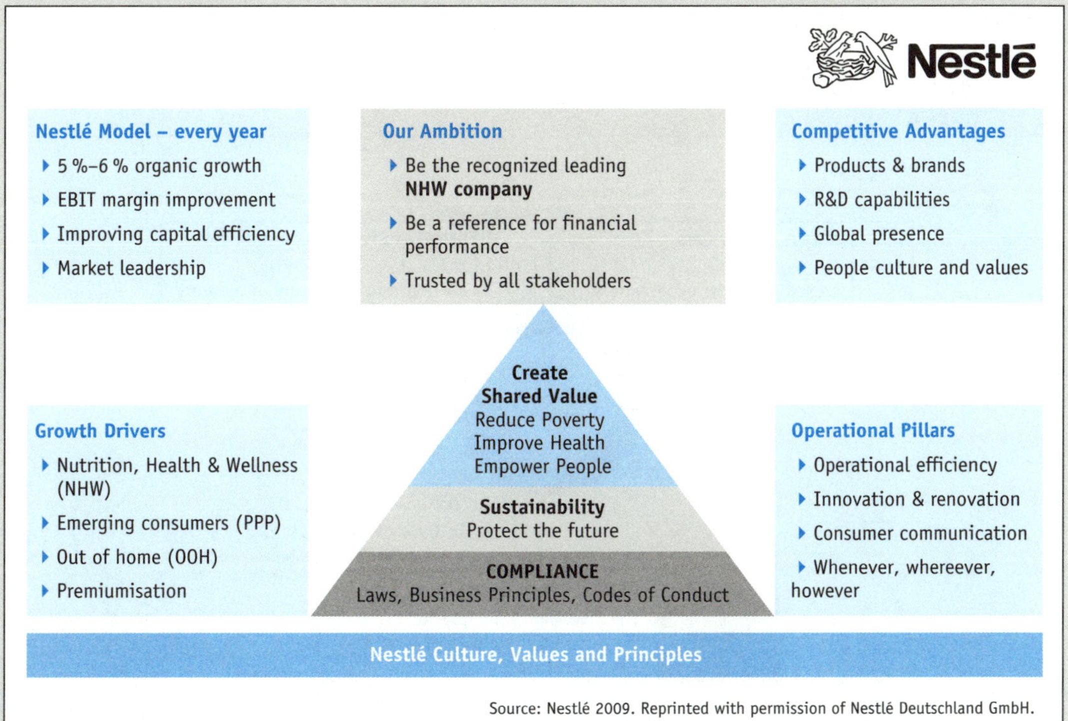

Source: Nestlé 2009. Reprinted with permission of Nestlé Deutschland GmbH.

Fig. 4-73: The Nestlé strategic performance framework (4x4)

for the consumers. Foundational for the strategic and performance framework are the Nestlé culture, values, and corporate business principles.

Toyota describes what kind of company it wants to be with its global vision: "Toyota will lead the way to the future of mobility, enriching lives around the world with the safest and most responsible ways of moving people. Through our commitment to quality, constant innovation and respect for the planet, we aim to exceed expectations and be rewarded with a smile. We will meet our challenging goals by engaging the talent and passion of people, who believe there is always a better way." (Toyota 2015a)

"The image of a tree has been used to symbolize the Toyota Vision from 'roots to fruits'. The tree is a symbol of natural strength. It is useful and long-lasting, something that continues to grow and flourish year after year." (Toyota 2015b) Toyota uses the tree metaphor (see fig. 4-74) to explain and communicate its global vision (see fig. 4-74): "(...) the firm roots stand for Toyota's shared values, the fruit for always better cars and enriching lives of communities, and the trunk for the stable base of business. The firm roots produce fruit and allow the trunk to grow thick and strong, ensuring the next crop of fruit. This virtuous circle reflects Toyota's vision to be a company achieving sustainable growth." (Toyota 2015c: 172)

Another way to create awareness for strategy is the use of specifically designed games and dialog pictures that are intended to bring comprehensive messages to life in a simple and compelling way. Figure 5-44 shows the dialog picture "Rail Station 2020", which has been used within the the DB Group's business business unit "DB Netze" (stations) to convey strategy messages to a large number of employees. In a series of workshops with up to 15 participants each, employees had the opportunity to get inspired by the business unit strategy as well as to experience particular challenges for strategy execution in a playful way. The picture illustrates a rail station of the future that is not only a key element for creating customer value but also a gateway to the company and to rail transportation in general.

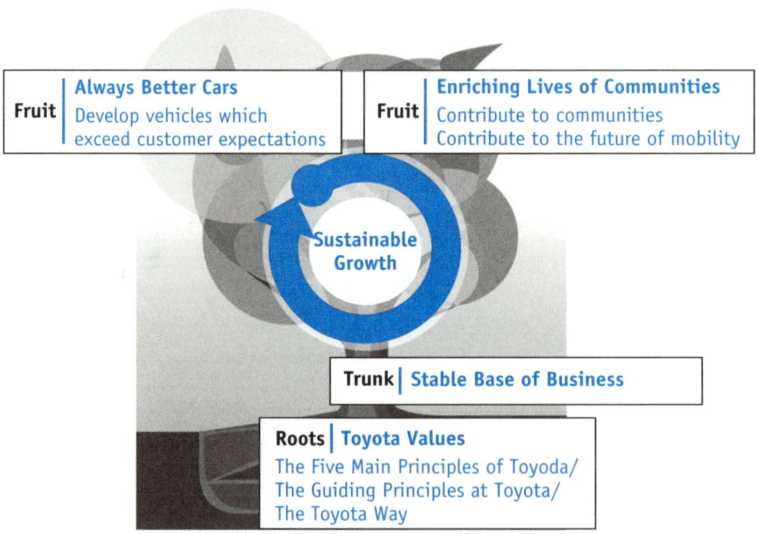

Source: Toyota 2015c: 11 (02-03); adapted

Fig. 4-74: Toyota tree metaphor

Source: © DB Station&Service AG. Reprinted with permission

Fig. 4-75: Aligning people to strategy in a playful way: strategy game at the DB Group

Source: Steiger/Bagel 2012: 34. Reprinted with permission of Bayer MaterialScience.

Fig. 4-76: Strategy communication cartoon at BAYER (2007): "One Business, One Team"

Figure 4-76 shows the cartoon "One Business, One Team" that has been used by Bayer MaterialScience (now Covestro), a business area and subgroup within the Bayer AG, to support an organizational transformation effort toward consolidated and more strategically aligned businesses and functions. With the use of a series of colored pictures, the company was able to communicate crucial strategic messages and milestones within the entire organization. The cartoon from 2007, which is named like the transformation program's strategic slogan, shows four different groups or clusters of employees that—according to Strebel (1998: 6–8)—have a significant impact on the success of an organizational change process: resisters, traditionalists, bystanders, and change agents. The cartoon's intention was to create awareness for the different groups and how to deal with them among the management team as a lever for aligning the different areas and groups as one team around common strategic goals. (Steiger-Bagel 2012: 33–35)

Barriers to an effective strategy communication

From the perspective of the recipients, strategy communication needs to be simple and "digestible", meaningful, clear and understandable as well as compliant to their own national culture. When designing a strategy communication program, companies need to be aware of potential pitfalls (Robbins/DeCenzo: 2008: 327–329), which may include the risk of

- **information overload** in case too many details are communicated regarding the desired strategy
- not considering **emotions**, that is how employees as key receivers of strategic messages feel when receiving the information (e. g., fear of losing the job, resistance to change, feeling of being overwhelmed, etc.)
- missing a commonly shared strategy **language** which means that different people in the same organization might have different interpretations of the words communicated
- disregarding different meanings of words, colors, symbols, etc. in different **national cultures**, which is particularly relevant for strategy communication in multinational enterprises (Homburg/Kuester/Krohmer 2013: 441–443; Kaplan/Norton 2008: 141 f.).

Another barrier to effective strategy execution might occur when it lacks persistence. In case it is just a short-lived onetime communication effort but not followed up, people may not only lose key messages but also credibility in the strategic messages if these are changed over and over again.

Establishing Commitment through Target Agreements (MBO) and Incentives

Linking the strategy execution system with the performance management/MBO system

In addition to fostering intrinsic motivation through leadership and strategy communication frameworks, an effective way of aligning people with strategic priorities is through target agreements along with corresponding rewards and incentives (Kaplan/Norton 2001: 242–271; Horváth & Partners 2007: 304–322). To achieve this type of extrinsic motivation, the strategy execution system needs to be linked to the performance management cycle and the management by objectives (MBO) system along with the corresponding rewards and incentive mechanisms (see chapter 4.2.2.3). This way, the performance management and MBO process become a key element for creating strategy transparency, awareness, and commitment in all areas of the organization. It facilitates the understanding of strategic expectations at an individual level, the definition of individual or group contributions, and the link of rewards and incentives throughout the organization.

Integration of strategic goals in target agreements

Figure 4-77 shows an illustrative target agreement sheet that is used for defining expectations and reviewing achievements in the performance management cycle. For facilitating a highly interactive target agreement meeting, a "strategic" section has been included in the structure. Hereby, the reference for deriving individual goals are key elements from the strategy execution system, that is strategy maps (strategic goals), strategic actions (projects) and key performance indicators or metric with their corresponding defined target values. By outlining those guiding elements from the company or organizational unit strategy, the contribution of individuals or groups can be defined. For example, if cross-functional topics are essential for the company's success, the corresponding company strategic goals are an important point of reference for target agreements. Resulting individual goals are supposed to contribute to their achievement. In case employees are in charge of or significantly

Fig. 4-77

Illustrative example of a target agreement

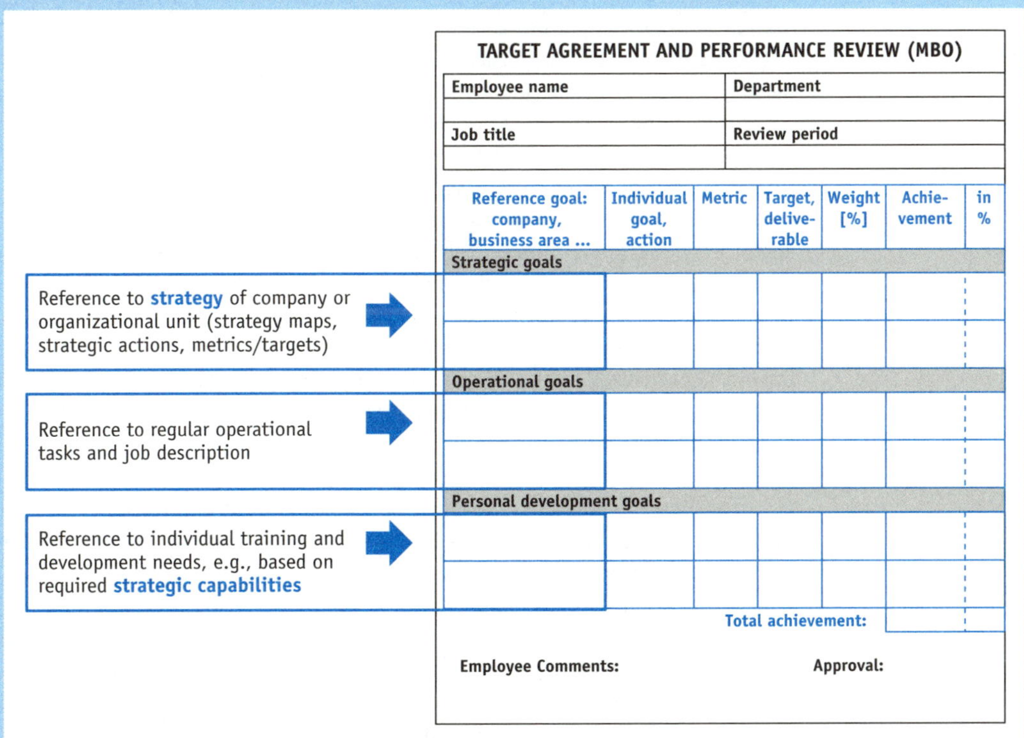

contributing to approved strategic actions, they should include those projects as one of their goals.

Operational goals and personal development goals

The individual strategy contributions, the type of individual goals (e. g., financial versus non-financial) as well as the weighting of goals may well vary depending on the management level and the type of organizational unit. In addition to the strategic component, the target agreement is supposed to also drive greater operational effectiveness and efficiency. Therefore, individual operational goals are derived from the individual's regular operational tasks or job description. Finally, the target agreement can also be used for personal skill development, which is an important stage in the performance management cycle. Personal training and development needs are identified and discussed, which leads to the definition of corresponding personal development goals. Those might be based on strategic capabilities that the organization requires to execute its strategy or on other more operational requirements.

3–7 individual goals/actions with metrics and targets/delivery

For each individual goal or action (project) a clear measurement with a corresponding target value or deliverable that is related to a certain time period is defined. This makes expectations transparent and enables objective performance reviews. As a result, the superior and subordinate will mutually agree on a total of no more than six individual goals and targets for the review period. This method fits to the guideline of defining no more than three to seven elements in a target agreement (see chapter 4.2.2.3). During the performance period (e. g., one year), superiors should provide ongoing feedback and coach their employees to ensure the best results. This culminates in formal—halftime (for possible updates or corrective actions) and year-end—performance review meetings, which represent the last stage in the performance management cycle. In the final performance review, results achieved during the performance period are jointly assessed and summarized. The corresponding achievement rating (here in %) is used for compensation purposes and for linking the reward and incentive system. Figure 4-78 shows an exemplary annual management by objectives (MBO) process of a railroad company that is based on the described principles. In this process, target agreements are flowing directly from the company level strategic goal to individual goals at different consecutive management levels in the organization. At each level, individual goals, metrics and targets values or deliverables that have to be achieved within 12 months are agreed on between the solid line managers and their direct reports.

Ensuring Strategic Readiness through Personnel Skill Development

Different jobs have different impact on executing a strategy.

All jobs are important in a company. People get hired, rewarded and developed because somehow they contribute to the success of an organization. Ongoing programs for continuously developing the skills required by managers and employees to perform their jobs are a standard in most companies. Personnel development is usually explicitly articulated as strategic goals at corporate and subsequent management levels, which frequently reads along the lines of attracting, selecting, developing, rewarding, and retaining top talent. But do all jobs have the same impact on executing strategy? The answer tends to be *no*. Identifying and building the most critical skills, which need to be in place for a successful strategy execution, is one of the most

Fig. 4-78

Consistent target agreements along the chain of command (railroad company)

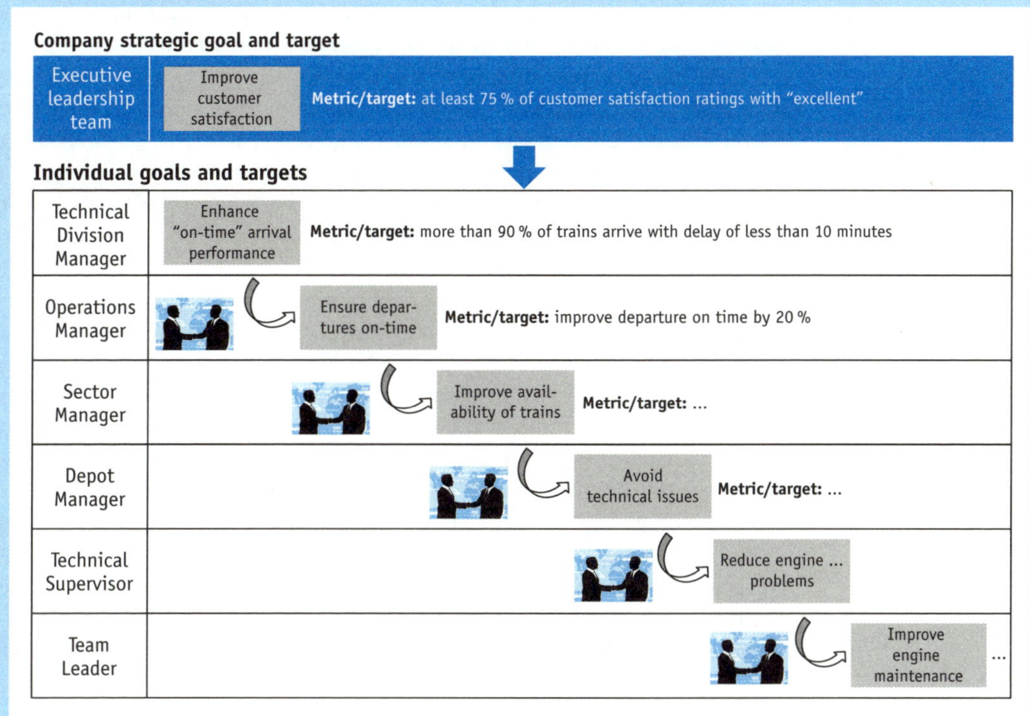

important tasks when aligning the organization and its people to strategy. Aligning the employees is not only about awareness and motivation as outlined in the previous chapters but also about ensuring that the right type and level of skills are in place to make strategies come true.

Identifying the most critical jobs for making a strategy happen

There are different ways of how to systematically identify the type of work that is essential to a company's strategy. One strategic approach distinguishes into A, B, and C positions (Huselid/Beatty/Becker 2005):

- **A positions** have a direct impact on strategy and have to represent a high performance variability among those in the position within the company and compared to other companies. The range in the performance level of the people in the position (i. e. the variability in the quality of their work displayed) represents upside potential that can be captured with strategic workforce management. If everyone's performance is already at a high level there is little upside potential and little opportunity for gaining competitive advantage with systematic workforce management. The idea is to have a lever for improving the firm's competitiveness with positions that make the most difference to the strategy (i. e. A positions). They may relate to less than 20 % of workforce across different areas and levels in the organization. They should be filled with the most talented high performing (or high potential) people with regard to the corresponding level in a company. A positions are not exclusively related to senior executive roles nor are they tied to a firm's hierarchy at all. Jobs that might have a crucial direct impact on a company's strategy can be found throughout the organization and may include relatively "simple" jobs that need to be performed particularly well to make the strategy come true. An example might be point of sales employees such as shop managers at the Spanish firm Zara who are sensing the latest fashion trends based on their intimate customer contacts and who are transmitting those trends every day to the corporate design people in Spain. As a result the designers are able to respond to those trends and launch a new collection with a breakthrough speed to market (Homburg/Kuester/Krohmer 2009: 63).

- **B and C positions** supplement the A positions in an overall portfolio that needs to be managed carefully in order to allocate investments in human capital effectively and have the right people in the right jobs. B positions support the A positions, and thus, have an indirect impact on strategy. They may also include positions with a high strategic impact (potential A positions) but without much performance variability, and thus, little opportunity for competitive advantage. Those positions are unlikely to create value but are often crucial for maintaining it. C positions relate to jobs that characteristically provide little discretion at work, have little economic impact, but are required for a company to function. Whereas hiring the wrong person or missing development needs may represent significant opportunity cost in terms of lost training investments and missed revenue for A positions, it may be fairly easy to remedy those failures through hiring or a replacement with B and C positions.

Strategic human capital readiness

Another approach for identifying strategic jobs and competencies that are required to make a strategy happen is by measuring the strategic readiness of human capital (HC). Originally strongly linked to the internal process perspective of an organization's strategy map (Kaplan/Norton 2004: 225–248), this measure reflects in a more generic approach the degree to which a company has the right kind and level of skills in place for achieving its strategic goals. The human capital readiness index shows what percentage of strategically required employees with a certain competency profile derived from strategic goals are actually in place. It is based on "to be" strategic competency profiles of individual employees, the number of employees that are required with such a profile, and the number of people that the company currently employs with such a profile (see fig. 4-79). The process of estimating HC readiness starts with identifying strategic job families that have the biggest impact on the achievement of strategic goals. Those positions correspond with the A positions that have been explained before. Next, the set of competencies (competency profile) that is best to perform each of those strategic jobs needs to be defined. Finally, the strategic "competency gap" is measured by comparing the company's current capabilities with the required capabilities, and thus, indicates the alignment needs of skills to strategy that need to be bridged by hiring or developing as well as retaining people.

Fig. 4-79

Determining strategic human capital readiness

	Operational excellence		Customer management		Innovation		HR
Strategic goals	Minimize problems	React quickly	Cross-sell/ Product bundling	Shift to relevant channel	Gain customer insights	Develop new products	Diversify workforce
Strategic job families	Quality Manager	Inbound Call Center Agent	Financial Advisor	Outbound Call Center Agent	Consumer Marketer	Joint Venture Manager	Recruiter
Competency profile	Six Sigma and Lean Management; Analytics; Problem management; Coaching	Customer interaction; Service orientation; Problem management; Teambuilding	Solution selling; Relationship management; Product line knowledge; Certification	Phone sales; Product line knowledge; Order management	Market research; Cross-business processes; Market communication	Relationship management; Negotiation skills; E-commerce know-how	Local roots; Social skills; Public relations; Legal know-how
Required #	30	20	100	20	10	30	10
Actual #	30	18	40	10	2	21	8
Strategic job readiness	100 % G	90 % G	40 % R	50 % R	20 % R	70 % Y	80 % G

Human capital readiness: 65 % R

Source: Kaplan/Norton 2004: 216; adapted

The same basic principle for measuring intangible assets is also suggested for information capital, that is the availability of strategically required information systems and infrastructure, and for organization capital, that is the strategically required culture, leadership, alignment of goals and incentives, and teamwork. Those will not be explained further. (Kaplan/Norton 2004: 249–316 and for a critical review of measuring intangibles this way Wunder 2004b)

Information capital and organization capital

4.3.6.3 Management Information

Fewer and more meaningful management reports

The third key area of organizational alignment relates to management information. One of the fundamental questions when introducing a strategy execution system is the link to management reporting. Some practitioners are concerned that it might add another layer to the already existing wealth of information available in companies nowadays and become an addition to the existing reporting system. Its scope will only increase, if the structure and process of current management reporting remains the same. With effective alignment of management information and reporting though, the opposite is true. A new type of strategy reporting can lead to fewer but more meaningful numbers und reports. It can help to overcome some of the key issues frequently stated regarding traditional management reporting systems (Kaplan/Norton 2001: 325 f.; Horváth & Partners 2007: 322–326; see also chapter 4.2.2.3):

- On the one hand, there are too many management reports with too much information and a high level of detail, whereas, on the other hand, some information that is crucial for managerial control is missing.
- Much of the reported information is related to financials. However, those controls are backward oriented and lagging, that is they indicate the impact of right or wrong decisions of the past. Future-oriented and leading controls are often missing (see also Strategic Snapshot 4.23).
- Operational controls that are reported on a monthly or even daily basis have no clear link to strategic goals. They are used for short-term feedback cycles but are not integrated with strategic performance indicators.
- Management reports provide a wealth of descriptive numbers, but it is unclear what really matters. In addition to the missing prioritization of numbers they lack valuable analytical information and commentaries as an important base for management decisions.

Distinguish between available and reported information.

A first step to overcome those issues in strategy reporting is to distinguish between available and reported information. Today's technical possibilities with modern IT systems and the corresponding automated means of data collection, storing and processing have been leading to an ever increasing availability of information. If companies still stick to the principle of what's available will be reported, preferably on paper and on a regular schedule, it can easily lead to information overload. This is a "condition in which information inflow exceeds an individual's processing capacity" (Rob-

Strategic Snapshot 4.23

Types of Controls

Variables or measures that assist managers in steering their company or unit in a certain direction are called steering controls. They are supposed to guide people's behavior based on different mechanisms: **Output controls** seek to influence behavior by specifying the expected end results (outputs). They define what has to be accomplished but leave the means of how to achieve those results open to the individual, team or business unit. Output controls are very popular in management and are frequently tied to employee rewards and incentives. They are the basis for *feedback control*, that is comparing the desired output with what has been achieved and drawing corresponding conclusions. Examples of output controls are predetermined goals such as sales and profitability targets, cost reduction objectives, or customer satisfaction indices. A major drawback of this type of control mechanism is its backward orientation. By the time the manager has the information, the potential "damage" has already been done, that is output controls are lagging.

Input controls are supposed to ensure a predicted outcome by emphasizing its enablers such as resources, skills, motives, maintenance schedules, etc. or by defining corresponding means. A prominent example of input controls in management is the budget as it is set in advance of the actual business activity. Those controls are future-oriented in a way that they are considered before the outcome is accomplished. Input controls are the base for *feedforward control* that allows managers to prevent issues rather than having to fix them later. Specific types of input controls are **behavior controls** that specify how something has to be done through standard operating procedures, policies and rules. Another form of input control is direct supervision, which represents the best-known form of a *concurrent control mechanism*. Managerial control systems in business practice are typically using several types of controls. (Source: Robbins/DeCenzo 2008: 362 f.; Rothaermel 2013: 322 f.; Wheelen/Hunger 2010: 332 f.)

bins/Judge 2013: 383). Some companies report hundreds of performance indicators every month to their senior management and a whole team is necessary to prepare these "performance books". Instead of improving decision support and increasing productivity with information, an overload may actually have negative implications on those items. To avoid this situation, it needs to be determined carefully how information is used within a company so managers do not need to pay unnecessary attention on any variable that is measured. Hereby, it is helpful to distinguish three layers of information as illustrated in figure 4-80. Relevant diagnostic controls and interactive controls (see chapter 4.3.7) are determined, selected and utilized depending on the corresponding information layer (Horváth & Partners 2007: 225 f.):

- **Controls for ad-hoc decisions** are collected, analyzed and distributed on demand to support specific management decisions such as entering a specific new international market, identifying which customers or products are more profitable than others based on a process cost analysis, or selecting the best region for building a new manufacturing site. They are not reported regularly but should ideally be quickly available from the company's data pool or other sources in case they are needed.
- **Operational controls** are important for managing daily operations in different areas efficiently. This detailed diagnostic information is relevant for specialized managers in the company to run their area of responsibility (processes or depart-

Fig. 4-80

Selection of available management information for strategic performance reporting

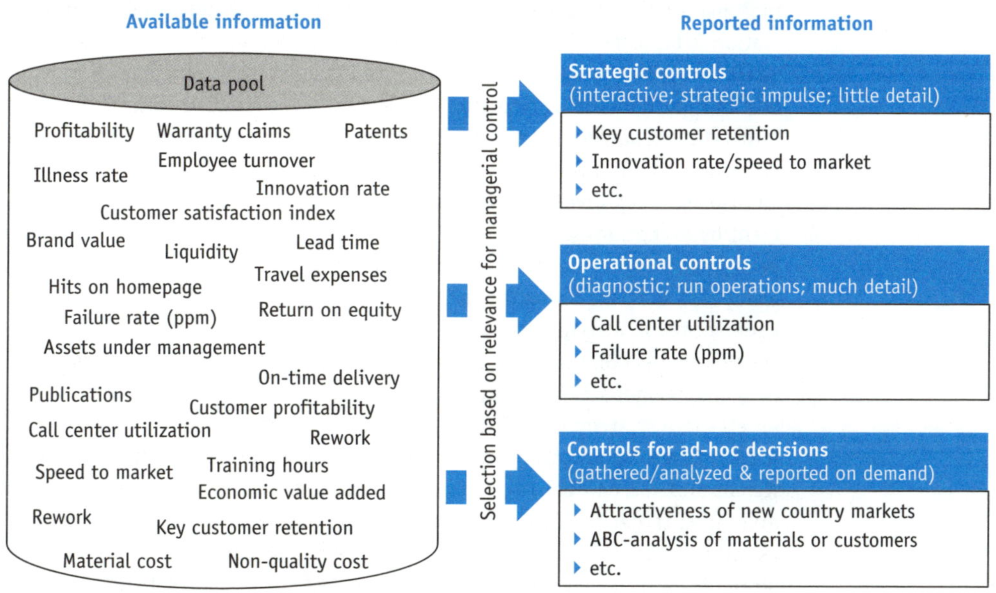

Source: Horváth & Partners 2007: 326; adapted

ments) such as sales, production, finance, etc. Following the concept of management by exception, detailed information typically gets elevated to the corresponding responsible if controls break through a predefined range. In such cases, detailed data is provided for root-cause analysis, and short-term tactical actions are triggered.

- **Strategic controls** are the few critical measures that indicate whether the organization is strategically on course. This requires that they are tightly linked to the strategic goals that articulate what the company wants to accomplish (see chapters 4.3.2 and 4.3.4). Strategic controls provide a "big picture" on the strategic performance and have more of a long-term focus. They are regularly reported to the senior management and typically cover both actual strategic performance results (outcomes, lagging) as well as strategic controls that are supposed to achieve those results (enablers, leading) (see Strategic Snapshot 4.23). The key principle of strategic controls is to facilitate strategic debate in the management team. Strategic impulse is more important than a high level of detail. Additional information is only provided in case it is required to better understand certain strategic issues.

Principles of strategic performance reporting

Strategic controls are the core of strategic performance reporting, which differs significantly from traditional reporting. Key differences and principles are illustrated in figure 4-81 and can be summarized as follows (Horváth & Partners 2007: 322–335): Instead of covering financial and some customer/market-related information in great detail, strategic performance reporting focuses on both financial and non-financial information as well as lagging indicators and leading indicators, both with high materiality and business impact. Compared to traditional reporting, the information is reported with a much lower level of detail but with a high strategic relevance as all indicators are directly derived from strategy. Instead of providing all available performance information, it focuses on the vital few controls that provide a bird's eye perspective on the strategic performance. Strategic issues that require special management attention are identified and characteristically supplemented with additional "deep-dive" information in the strategic performance report (management by exception). One of the most important principles is the fact that strategic performance reporting is not only diagnostic but supposed to be highly interactive in a way that it drives forward-looking strategic discussions and learning in the management team. Although strategic controls are in essence diagnostic in a way that they measure actual results, compare them with predetermined expectations or targets, and trigger action to correct deviations, they are used in an interactive way for an effective strategy execution and control (see Strategic Snapshot 4.24 in chapter 4.3.7). "Some Balanced Scorecard implementations failures occurred", according to Kaplan and Norton, "because organizations used their scorecards only diagnostically, and failed to get the learning and innovation benefit from an interactive system." (Kaplan/Norton 2001: 129) Following the explained principles when designing a strategic performance reporting system may overcome some of its underlying pitfalls explained at the beginning of this chapter.

Elements of strategic performance reports

Strategic performance reporting has two components: The first component is a report that covers the documentation of strategies and business models and is classi-

Fig. 4-81

Principles of strategy-oriented exception reporting

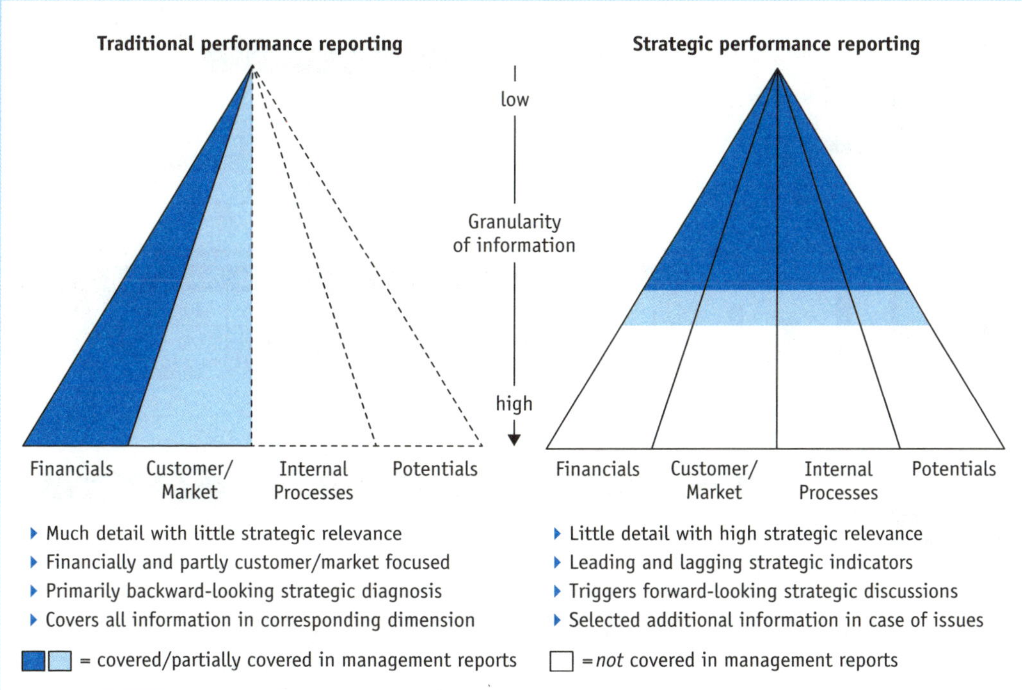

cally updated on an annual basis. In addition to the strategy documentation, the second component is related to management information that is relevant for ongoing strategy execution, and control that is typically prepared on a monthly or quarterly basis. These regular strategic performance reports are related to the state of strategy execution and—with respect to the previously explained strategy execution framework—are supposed to cover the three elements strategy map, balanced scorecard, and strategic actions summarized in figure 4-82 (Horváth & Partners 330–332). Note that strategic performance reports may be provided in a similar format but with different information on various management levels (e. g., strategic performance report for the top management team of the company and another one for the management leadership team of a strategic business unit or region). Key considerations when reporting the three elements to the top management leadership team are as follows.

Strategy map and strategic goals achievement status

High-level summary as entry point

A strategy map with status lights on each strategic goal (e. g., green = well on track; yellow = somewhat off track; red = off track) provides a bird's eye perspective on the strategic performance and the progress of strategy execution. It is the entry point into the strategy report and typically supplemented with qualitative comments on

Fig. 4-82

Elements of strategic performance reports and performance status logic

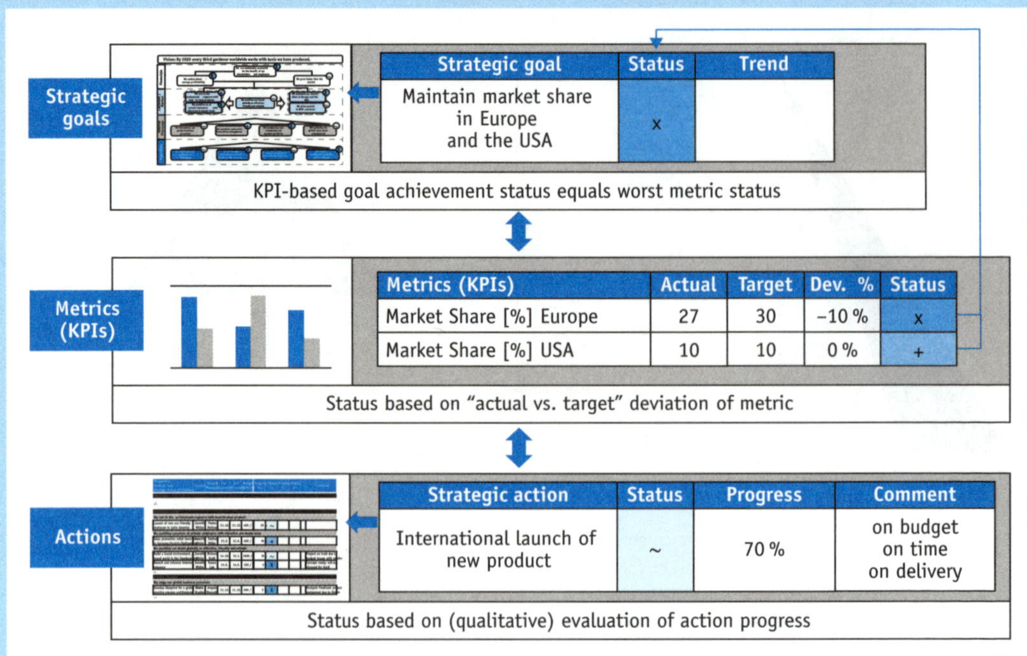

significant developments. The strategy map and strategic goals are frequently used to prepare a management summary—either by perspective or by strategic themes (see chapter 4.3.3.1)—to provide a high-level description on where the company stands strategically and whether there are strategic issues that require special management attention. This is supposed to act as an important trigger for strategic dialogs and learning processes of the corresponding management leadership team. The summary may also include prepared proposals for required management decisions as well as the due date when the decisions need to be made.

The worst metric determines the goal status and triggers the dialog.

Ideally there is only one metric per goal. In case of multiple metrics for one strategic goal the goal status is determined by the worst metric status if the objective is to trigger awareness and strategic dialog. In such a case, a goal achievement status determined by (weighted) averages of the metrics tends to be not effective as it may hide or average out an important strategic issue that the management team is supposed to talk and decide about. These types of issues may be supplemented with additional information to provide deeper insights on what is going on (exception reporting). The trend information indicates whether the goal achievement is improving, stable or deteriorating. In case there are no metrics defined for certain strategic goals, the goal achievement status might be set based on the progress of corresponding strategic actions.

Metrics or key performance indicator status

"As is"-"to be" comparisons with comments on deviations

The overview of financial and non-financial key metrics is based on a comparison of actual values and expectations, which are expressed with target values. The focus is on the vital few aggregated figures. If needed, details are available in deviation reports. The status of metrics is determined by "as is"-"to be" comparisons and the calculation of deviations. The latter is flagged in case they exceed a certain tolerance range. Following the same logic as with strategic goals, the status is typically illustrated with traffic lights. Note that the traffic lights are supplemented with symbols to provide status information also for recipients suffering from color blindness or in case of black-and-white printouts. Responsible managers are usually required to provide explanatory comments on metrics with red or yellow status as they might need corrective action. Those comments along with proposals for corrective action are of utmost importance for focusing management attention and preparing decisions. Comments should not merely repeat in words what is already articulated with numbers and graphs but explain root-causes of the issue and suggest what to do about it particularly in case of red status lights. For example, managers need to understand whether the issue is a one-time exceptional event or will level out over time as in this case there might be no necessity to act. If there is need for action or if it even requires approval by the management leadership team, the corresponding comments should be included in the management summary.

Appropriate value types and comparison logic

The update frequency of values depends on the type of metric. Some metrics are updated on a daily basis, others only once a month or a quarter, and some are once a year such as indicators resulting from annual surveys or other means of annual data collection. Financial metrics are frequently reported year-to-date (YTD) and quarter-to-date (QTD), that is with a period starting from the beginning of the current year or quarter up to the present day. Others may be cumulated over the last 12 months (LTM) independent of the fiscal year. Some target values may relate to a period longer than the reporting cycles (e. g., market share provided only once a year but with quarterly performance reporting cycles). In those cases where values are not due in the current reporting period because of a longer measurement frequency, their status is generally carried forward. As such information is not particularly valuable, it might be manually supplemented by estimated actual end-of-year numbers. By comparing those forecasts with the end-of-year target, the likelihood of achieving the annual target is highlighted, and corrective actions can be initiated at an early stage in case of significant forecasted deviations. In some strategic performance reports, this logic is applied to all metrics. The full year perspective acts as a kind of early warning indicator and requires reliable input of the corresponding responsible. In addition to comparing "as is" with "to be" metrics, reports frequently provide information of how the actual values compare to the preceding periods to highlight the current year's metric and strategic goal evolution as well as how they compare to the actual of the same period in previous years.

Strategic action progress status

Top management governance of most critical strategic actions

The third element of a strategic performance report is an overview of the progress in executing strategic actions. To make strategy happen, approved strategic projects need to get appropriate management attention, funding and governance to be executed effectively. In some companies this is one of the major barriers to strategy execution. The scope of reporting strategic action progress is typically related to the most important strategic action programs that

- have a strong impact on the strategic performance of the company (e. g., significant business impact or enablers for important strategic goals)
- require cross-functional commitment
- need significant resources (e. g., operational or capital expenditures, internal people or FTE)
- require top management attention and approval.

As those top action programs are crucial to make the strategy come true, the management leadership team needs to act as a sponsor and ensure that they get what they need in case of progress issues.

On time, on budget, on quality

As part of the strategic performance report, top strategic action programs are continuously tracked by the management leadership team. It needs to be evaluated whether they are on time, on budget and on quality (delivery) as well as whether they drive the corresponding strategic goal achievement for which they were originally developed. In case there are significant deviations from the desired progress, explanatory comments, proposals for corrective actions as well as additional information for better understanding the issues need to be provided. This typically includes project charters, financial summaries (financial resources and expected business impact) as well as a detailed resource plan highlighting the resource commitments of the different contributing business units (see chapter 4.3.5.2). For assisting the senior leadership team in governing the strategic project portfolio, information about suggested new strategic projects, executed projects (lessons learned) and projects to be put on hold due to resource constraints or other reasons are provided in the strategic performance report (see also chapter 4.3.7).

Strategic actions of business units, regions or functions are typically not included in the top-level strategic performance report but monitored and governed at the corresponding management levels although some of those actions may be linked to or included in the top-level strategic action programs.

Management cockpits and dashboards

Strategic performance reporting can be facilitated by a variety of standardized or customized software applications. Many of them provide access to an information system database and are linked to the company's ERP-system. Management information at the user front end is frequently displayed in form of so-called management cockpits or management dashboards with user-friendly drill-down capabilities. Without doubt, IT-support is important for implementing a strategic control process including a solid strategy execution system, particularly in larger organizations. However, it needs to be emphasized that the business concept and applied strategy framework has to drive IT and not vice versa. Many IT tools have their own business rationale, and it might not be possible to customize them according to the specific company requirements. There are many special texts available on this topic, which will not be further elaborated on here.

4.3.6.4 Benefits and Pitfalls of Organizational Alignment

Ensuring that every resource and activity is linked to strategy

The key benefit of organizational alignment is that it links every resource and activity in the organization to strategy. This requires that traditional established management systems are calibrated and orchestrated in a way that secures funds and time necessary for strategy execution, motivates and enables people to contribute individually, and provides a focused and reliable view on the strategic performance of the company. Key managerial systems that need to be aligned are the operational planning and budgeting system, the human resource system including strategy communication, target agreements and incentives as well as skill development and the management reporting system.

Allocating limited resources to the highest value strategic initiatives

Strategy execution can easily fail if required resources, that is financial funds and employees, are not made available. As a key benefit of organization alignment, the planning and budgeting system is linked to strategy. This way it is ensured that naturally limited resources of the company are allocated to the highest value strategic initiatives to improve the overall business results. Strategy-oriented budgeting helps to provide funds based on the importance and expected performance impact of strategic themes, as well as goals and actions instead of only extrapolating from historical values as it can be frequently found in company practice. Strategy maps and balanced scorecards can be leveraged by organizational units to articulate their "business case" to top management when they compete for scarce company resources with other units. Moreover, a well aligned planning and budgeting process can help to simplify, and thus, reduce the effort and time required for operational budgeting.

People across all levels are aligned with and inspired by strategy.

Effectively aligning the company's performance management system with its corresponding target agreement and incentive mechanisms increases the likelihood of people being motivated or even inspired to individually support strategy execution. Along with an effective strategy communication it endorses the hearts and minds of all employees across the different levels in the organization, not just the few executives at the top management suite. Individual operational and strategic perspectives are aligned with corporate or organizational unit perspectives to optimize the company's overall performance. This, however, can only work if employees have the necessary skills to support strategy execution. Linking the personal skill development process to strategy bridges identified skill gaps and enables employees to deliver what is strategically expected from them.

Focused, balanced, consistent, and reliable view on the strategic performance

Finally, aligning the management reporting system with strategy provides management teams in the organization the opportunity to keep track of strategy execution and drive strategic learning processes. Strategy-oriented reporting ensures a regular focused follow-up on the strategy by the management team, increases transparency about whether the strategy is working as intended or not, supplements the financial reporting with non-financial strategic metrics to provide a more consistent and balanced view on strategy, with both lagging and leading indicators, supports governance of key action programs for all organizational units and, finally, is the basis for issue-focused and action-oriented strategic performance review meetings. Management attention gets directed toward the vital few strategic priorities and issues that are most important to act on.

Watch out: lack of strategic leadership and senior management commitment

Assuming that key components of the strategy execution system are well designed, major pitfalls of organizational alignment are related to poor leadership (Kaplan/Norton 2001: 361–368): Employee inspiration, commitment and buy-in to strategic priorities do, of course, not only come from managerial systems. The key of directing behavior to strategy is leadership. This is the reason for having strategic leadership at the core of the WOS strategy framework underlying this book (see chapter 1.3.3.3). The strategy execution system needs to match the desired strategic leadership culture. Senior management commitment and strategic leadership is not only important for ensuring appropriate strategy alignment but also for organizational alignment. Different traditional and typically well established management systems and processes from different areas of responsibility (Finance, HR and Corporate Communications, IT, etc.) need to be recalibrated and orchestrated toward strategy. High-level executives who are in charge of those corresponding areas need to buy-in to aligning their areas with strategy.

Watch out: long-term transformation

Another pitfall of organizational alignment is its long-term transformational nature. It is neither a one-time event nor a technological systems project. In the early phase of implementing a new strategy execution system, the management may not have the courage to align all those well established management processes as suggested. Key elements such as new metrics may not have been developed yet, there may be a lack of experience regarding certain target values, and management may not be used to a highly focused strategic performance report or strategy-oriented budgeting and skill gap analysis. However, this can lead to a situation where management teams across the organization spend time and effort to define, articulate and align their strategic priorities vertically and horizontally with strategy maps, develop balanced scorecards and define strategic action programs only to realize that the next years budgets, target agreements and incentives, personal skill development initiatives or management reports do not capture those elements but work the same way they used to in the past. In such a situation the strategy execution system will not work, lose acceptance and may ultimately fall apart due to failures in organizational alignment. To avoid this pitfall there needs to be a well defined transformation roadmap of how the organization is gradually aligned with strategy. Quick wins in the sense of short-term results or realized benefits of a new strategy execution process (e.g., identified with the results of a strategy execution excellence survey; see fig. 4-3 in chapter 4.1.2) may help to keep the momentum as part of a professional change management effort.

4.3.7 Execute and Control Strategy

Any company faces two basic demands: It has to execute its current strategic plans effectively and, at the same time, identify needs for adapting those plans. This is addressed in the final element that concludes the described strategy execution system. It covers two key components:

- **execution support and governance of strategic action programs** with the goal to increase the likelihood of successfully realizing individual strategic projects as well as managing the dynamic portfolio of key strategic actions
- **strategic control** based on strategic performance reviews (execution control), examining assumptions as well as monitoring and scanning of environment. Those elements are supposed to continuously test and, if required, adapt strategy.

Strategic action execution support and quick wins

With the development of strategic action programs, the need for initiative has been identified to make strategies come true at different levels in the organization. Although the proposed strategy execution system is primarily based on the principle of self-control by teams and individuals in charge of execution, line staff may not be able to deliver all those strategically required initiatives with the desired pace and passion due to capacity constraints or gaps of experience and knowledge. Some companies establish a pool of project or change management experts that can be leveraged as additional resources to help the various organizational units in executing their strategic actions, and thus, delivering the expected results. In recent years, more and more companies have been following such an approach of involving their operational project management resources—particularly their project management office (PMO)—also in the strategic management process (Hobbs 2007: 25 f.). This type of support typically increases the likelihood for short-term success stories (quick wins), which are important to show employees that the applied strategy execution system works and pays pays off. Furthermore, it increases the urge for delivering results at other strategic actions and keeps or even accelerates the strategy execution momentum in the organization.

Strategic action portfolio management

The total of top-level strategic action programs represents a portfolio of strategic projects that are particularly crucial for the company. Typically, it includes strategic actions that have a high strategic importance because they have a significant direct business impact or enable important strategic goals. The portfolio may also include resource-intense projects (e. g., capital or operational expenditures) from various other areas in the company such as research & development or mergers & acquisitions that are not part of the defined strategic action programs. The company needs to ensure that resources and management attention is continuously allocated to the highest value strategic projects for improving overall business impact. This can be supported by a systematic strategic action portfolio management, which includes effective prioritization and approval mechanisms as well as a dynamic governance approach. Goals of such a process are to

- continuously provide transparency about the status of top-level strategic projects
- decide on and commit resources to new strategic projects
- ensure that committed resources are actually used correspondingly

- formally close executed strategic projects (incl. review of lessons learned)
- put projects on hold due to unforeseen changes in the environment or a lack of resources.

Types of strategic control

In addition to supporting and governing the actual execution of strategic projects, the effectiveness of strategy execution and the validity of the strategy itself need to be controlled continuously. Strategy maps, key metrics, strategic action programs as well as budgets and incentives have been established to put the organization strategically in motion. In the final stage, processes must be established to evaluate continuously whether all those elements are actually leading to the desired performance and if there is any need for altering them. This is addressed with strategic control, that is formal internal steering mechanisms that are supposed to provide answers to the following three questions and will be explained in the following sections (Lombriser/Abplanalp 2004: 385–393; Steinmann/Schreyögg 2000: 245–250):

1. Are we executing our intended strategy well and is the strategy working?
2. Are our explicit strategic planning assumptions still valid?
3. Are there new developments in the business environment that we need to consider?
4. As a result of question 1–3: Is our strategy still valid or do we need to adapt it?

To provide answers to those questions, strategic control covers three different mechanisms that are strongly interrelated (Steinmann/Schreyögg 2000: 245 f.; see also Strategy Practice Example 4.15 in chapter 4.4.7.2): providing performance feedback (**execution control**), control of strategic planning assumptions (**assumptions control**), and monitoring of external developments (**strategic monitoring and scanning**). The types of strategic control are illustrated in figure 4-83 and used as a frame for the following chapters. Based on the insights of those three components, it needs to be determined whether the strategy is still valid or has to be adapted, which ultimately closes the strategic management process and feeds back to the initial strategic analysis phase.

Levers of control for implementing strategy

Another practical approach to strategic control is based on four critical levers for the control of business strategy (see Strategic Snapshot 4.24). They are supposed to help managers to deal with tensions such as rigid control versus organizational learning, planned versus emergent strategy formation, rewarding people versus punishment and holding them accountable, etc. The levers are supposed to support both strategy execution as well as strategy formation. Key principles of the two levers diagnostic and interactive control systems will be referred to in the following chapter when discussing performance reviews. Key ideas of the belief system have been addressed in chapter 3 as strategic guideposts and elements of the boundary system have been applied when defining the competitive arena and deciding which strategies to pursue and which ones explicitly not to.

4.3.7.1 Control Execution: Strategic Performance Reviews

Determining the state of strategy execution

Execution control is dealing with the implementation of defined strategic plans that typically cover goals, metrics, and action programs as well as corresponding mile-

Fig. 4-83

The strategic control process with three control mechanisms

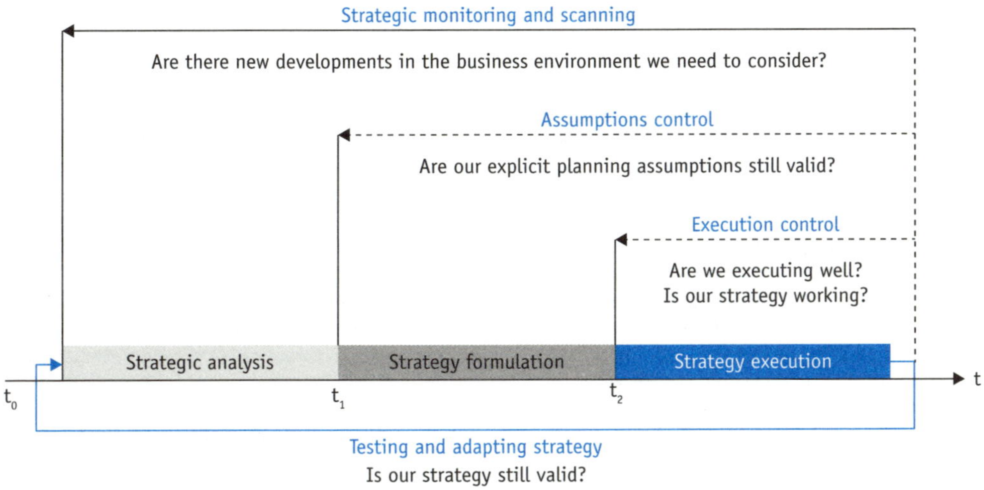

Source: Steinmann/Schreyögg 2000: 246; adapted

stones regarding time and deliverables (Steinmann/Schreyögg 2000: 246 f.). Based on an examination of goal achievements and deviations as well as the realization of strategic action programs and projects it needs to be anticipated whether strategy execution is working as intended or not. In addition to those elements, a regular employee survey about the state of strategy execution (execution excellence survey; see chapter 4.1.2) can help to identify potential issues and abolish barriers to the effective realization of the intended strategies. Based on the evaluation of the state of strategy execution and root-cause analyses of issues, corrective actions are derived. Effective execution control provides insights not only about whether the company is executing well but, through strategic dialog and learning, also whether the planned strategy is working as intended. If not, the reasons need to be identified and may be related to planning assumptions that become invalid (see chapter 4.3.7.2). However, even in case the planning assumptions hold true, elements of the defined strategy may not be effective. Those need to be identified (e. g., through "learning by doing") and elaborated on during execution control, which may actually lead to a strategy adaption (Lombriser/Abplanalp 2004: 388). Methodologically, execution control can be supported with well designed strategic performance reviews, which will be explained next.

Two components of strategic performance reviews

Strategic performance reviews are fundamental to govern the strategy execution process and drive strategic learning in the management leadership team. They ensure a regular follow-up by the corresponding executives and consist of two components: the strategic performance report (see chapter 4.3.6.3) and the strategic performance

Strategic Snapshot 4.24

Levers of Control Framework

According to Robert Simons, strategic control can be achieved by leveraging and integrating four levers: diagnostic control systems, interactive control systems, beliefs systems, and boundary systems. Effectively applying these four levers for strategy execution (including emergent strategy formation) requires that they are used in an integrated way so that the different mechanisms complement each others. Figure 4-84 illustrates the levers of control, their basic mechanisms and their relation to strategy. With reference to the differentiation between intended and emergent strategies as well as the different "Ps" of strategy (see chapter 1.1.1.2), "**diagnostic control systems** coordinate and monitor the implementation of intended strategies. The targets and goals embedded in formal plans are the embodiment of management's intended strategies. Diagnostic control systems, then, relate to strategy as a plan. These systems are essential management tools for transforming intended strategies into realized strategies because they focus attention on goal achievement for the business and the individual. (...) **Interactive control systems** give managers tools to influence the experimentation and opportunity-seeking that may result in emergent strategies. Thus interactive control systems facilitate and shape the emergence of new strategies. These systems relate to *strategy as patterns of action*. (...) Tactical day-to-day actions and creative experiments can be welded into a cohesive pattern that respond to strategic uncertainties and may, over time, become realized strategy. (...) The **beliefs system** of the organization inspires both intended and emergent strategies. Management's vision (...) motivates organizational participants to search for and create opportunities to accomplish the overall mission of the firm. Beliefs systems appeal to the innate desires of organizational participants to belong and contribute to purposive organizations. These systems relate to strategy as perspective. (...) **Boundary systems** ensure that realized strategies fall within the acceptable domain of activities. Boundary systems control strategy as 'position', ensuring that business activities occur in defined product markets and at acceptable levels of risk." (Source: Simons 1995: 155 f.)

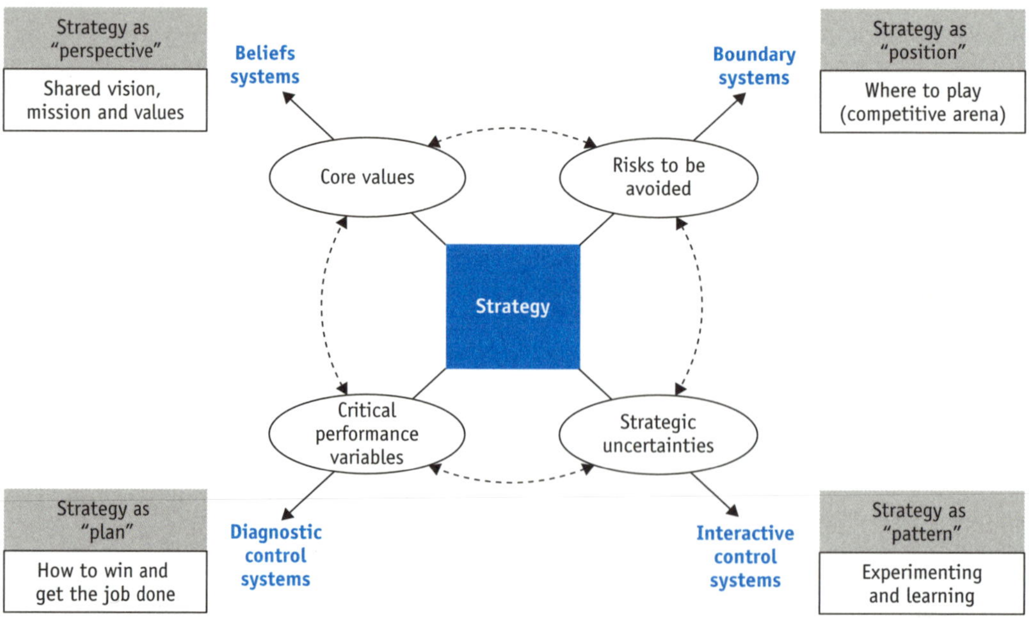

Source: Simons 1995: 159; adapted

Fig. 4-84: Levers of control

review meeting (see fig. 4-85). The report provides transparency about the state of strategy execution. It is distributed prior to the management meeting to provide everybody the opportunity of reviewing the strategic performance in advance and getting prepared for the meeting. The primary purpose of the strategic performance review meeting is to elaborate collaboratively on strategic issues, that is to understand root-causes and to initiate corresponding actions for tackling the issues. It typically follows a standardized agenda that is supplemented by specific actual strategic topics that need to be discussed.

Variance analysis drives strategic learning.

Strategic performance reviews are a specific type of feedback process that is supposed to facilitate strategic dialog and learning instead of focusing only on reporting and control. Hereby, diagnostic information provided in the strategic performance report is used by the corresponding management teams in an interactive way (Kaplan/Norton 1996: 163 and Strategic Snapshot 4.24). A key trigger of strategic dialog and learning in the strategic performance review is a variance analysis of both key metrics determining corresponding strategic goal achievements as well as progress of strategic action programs. As explained in chapter 4.3.6.3,, the calculation of deviations from set target values of metrics trigger status lights on the strategy map. In case of undesired deviations, a red (or yellow) goal status is shown that is supposed to attract management attention and trigger root-cause analysis and ulti-

Fig. 4-85

Two components for strategic performance reviews

Strategic performance review

Strategic performance report

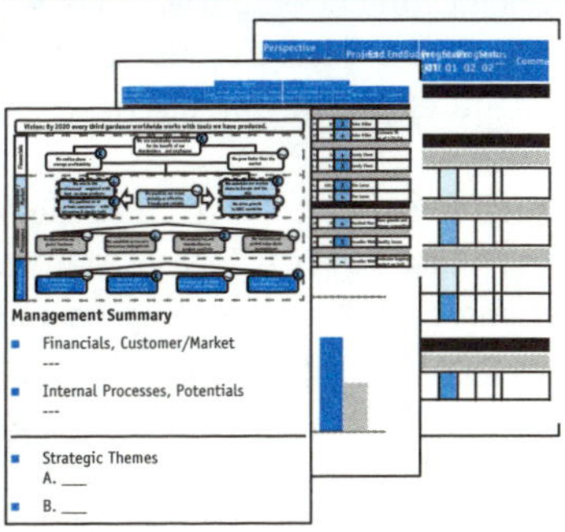

- Strategic goal achievement status and strategic theme summaries
- Metrics/KPIs (actual vs. target)
- Strategic action progress

Strategic performance review meeting

Agenda
- exemplary -

Time	Topic (... responsible)
09:00	Review last meeting follow up items: action log
09:15	Performance review corporate and SBUs
09:45	Performance review regions and functions
10:15	Review of top-level strategic action programs
10:30	Break
10:45	Deep dives: discussion of selected strategic issues and decisions on corrective actions
12:45	Recap and new follow up items: action log
13:00	Meeting close

mately constructive discussions in the management team. Instead of looking for "responsible" to blame for the performance deviation, the focus should be on gaining insights what is going on and on action. Reasons for deviations can be manifold (Simons 2000: 133). With respect to the explained three components of a strategic performance report such insights may include answers to the following questions:

- Is our performance standard (target) realistic and appropriate?
- Are the assumed cause-and-effect relationships between strategic goals on the map true?
- Is our strategic intention expressed in the strategic goal working in the marketplace?
- Do new and unforeseen external developments interfere with our strategic intent?
- Have the defined strategic actions for achieving the goals been executed?
- Do executed strategic actions actually contribute to goal achievements?
- Is the time frame for expecting performance improvements appropriate?
- Are there any enablers for achieving goals that are not in place (e. g., resources, skills, etc.)?

Early warning indicators

In addition to understanding the reasons for deviations, execution control based on strategic performance reviews can also function as an early warning mechanism. For example, significant deviations and gaps in the achievement of strategic goals from the potentials or process perspective (leading indicators) are an early warning that desired future customer/market-related or financial goal achievements (lagging indicators) may be at risk. Furthermore, strategic goals and key metrics may be specifically tied to early warning indicators. An example is a software company who wants to differentiate in the marketplace with the highest quality software products and services and articulates this aspiration as strategic goal related to customer/market. In order to achieve this goal, the company has additionally defined "attract and retain top IT talent" in the potentials perspective. The number of freshmen in computer science could act as an early warning indicator showing that there will be a significant gap of highly qualified IT experts in 3–4 years (Lombriser/Abplanalp 2004: 390 f.). This can be considered as a strategic challenge that has to be solved in order to make the strategy happen.

Shifting time to valuable strategic dialogs

Gaining valuable insights through discussing those questions requires the agenda of strategic performance reviews to be designed in a way that focuses management attention on strategic issues and its implications. Instead of spending most of their time listening to report presentations of information that has been distributed in advance, management team members should discuss root-causes and solve problems (see fig. 4-85 and fig. 4-86). Whereas participants in traditional management meetings tend to spend about 60 % of the time on reviewing performance, 30 % on the discussion of consequences and only 10 % on developing action plans for dealing with the issues identified, those time allocations are reversed in a strategy review meeting (Kaplan/Norton 2008: 236). Key benefits of this type of strategic performance review meetings are:

- Management resources are focused on relevant strategic topics.
- Strategic decision making is supported with a regularly updated analytical basis.

Fig. 4-86

Key questions, information sources and principles of strategic performance review meetings

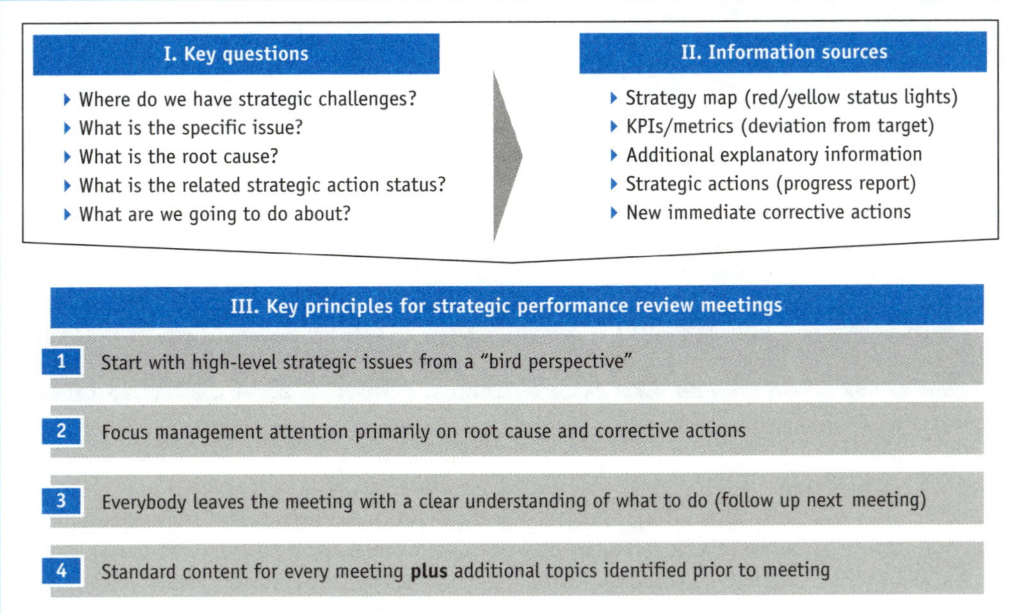

- Strategic issues are solved from an integrated and holistic company perspective.
- Clear strategic guidance for the top management team and the organization is provided.
- Strategic dialog and learning is encouraged which may also capture emergent strategies.
- The likelihood of planned strategies to happen is increased.
- A performance- and execution-oriented culture is developed.

Effective performance reviews require solid preparation.

Figure 4-87 provides an overview of an integrated strategic performance reporting and review process. It starts with a collection and update of data that is required for the report. This includes an update of actual values for key metrics to determine deviations from the set targets, and thus, the status of strategic goals on the strategy map. Furthermore, the progress in executing strategic action programs (e.g., time and milestones, use of budgets, deliverables) needs to be determined. In the second step, deviations are analyzed and trends are identified. Those are supplemented with explanatory commentaries and, in case there are issues that require management attention, with additional detailed information that provides deeper insights for the management. After that, the strategic performance report can be generated and is distributed to the relevant executives and leadership teams prior to the strategic performance review meeting. Some sections of the performance report might be discussed individually

Fig. 4-87

Strategic performance reporting and review process

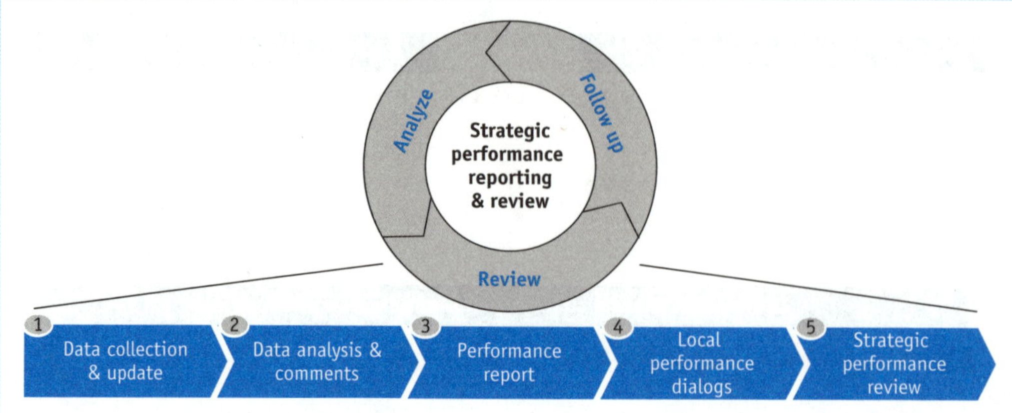

with responsible managers or teams to gain more insights in current issues and develop proposals for corrective actions that are presented to the top-level management team for decision making. The final step is a well prepared and well structured strategic performance review that follows a standardized agenda and focuses on team-based strategic issue-solving, learning and action with a professional follow-up.

Linking strategy and operational performance

Structured, issue-focused, and action-oriented strategic performance review meetings with a professional follow-up can be established at different levels and areas in the organization. Essentially, there are three types of performance-related review meetings (Kaplan/Norton 2008: 277). First, operational review meetings typically happen on a weekly or daily basis and focus on solving immediate problems that may be indicated by data and metrics from operational management dashboards. Second, strategy review meetings are conducted in a quarterly or monthly cycle and are facilitated with strategy maps, balanced scorecards and strategic action programs as explained before. Third, strategy testing and adaption meetings are happening on an annual or quarterly basis and are strongly linked to the strategic performance reviews (see chapter 4.3.7.3). One of the challenges when establishing those meetings tends to be the link of operations back to strategy. The Strategy Practice Example 4.14 shows how strategic performance that is reviewed quarterly or monthly at the first and second management levels can be connected to operational performance which is reviewed weekly and daily at the third and fourth management level.

Strategy maps provide orientation for operational meetings.

Typically, the scope (e. g., strategic goals, KPIs, strategic actions to be reviewed) and depth (e. g., management levels and organizational units to be reviewed) of performance reviews evolve over time depending on the maturity level of information as well as the entire strategy execution system. Some managers include elements of a strategic performance report in their regular operational management meetings at an early stage even if key metrics are not established yet. For example, they start every

Strategy Practice Example 4.14

Linking Operational Performance at a Food Production Site to Strategy

Figure 4-88 provides an anonymized example how operational performance at a production site for producing food and beverages can be linked to strategy across four management levels. At the top level, the company has defined cost leadership as a strategic goal to win in the marketplace of the highly competitive food industry. Food producers in particular are facing strong price pressures from multiple sides, that is the few large trade chains (food retailers and discounters), the consumers who want high quality products at low prices, and the suppliers of raw materials asking for higher prices due to increasing global price developments. Progress in strategic goal achievement is reviewed regularly by the top management as part of a **quarterly or monthly strategic performance review meeting.** To make the strategy happen, the strategic company goal is translated into "reduce production cost" for the global production/operations unit and, subsequently, into "reduce non-quality cost from 4.3 % to less than 2 %" for a specific production site XY in a certain country (third management level). The factory management team has installed a **weekly meeting to review operational performance** including non-quality cost. Different sources of non-quality cost have been identified in the three operational departments "semi-finished product", "powder" and "aseptic". Corresponding goals and metrics related to different types of non-quality cost in each specific department

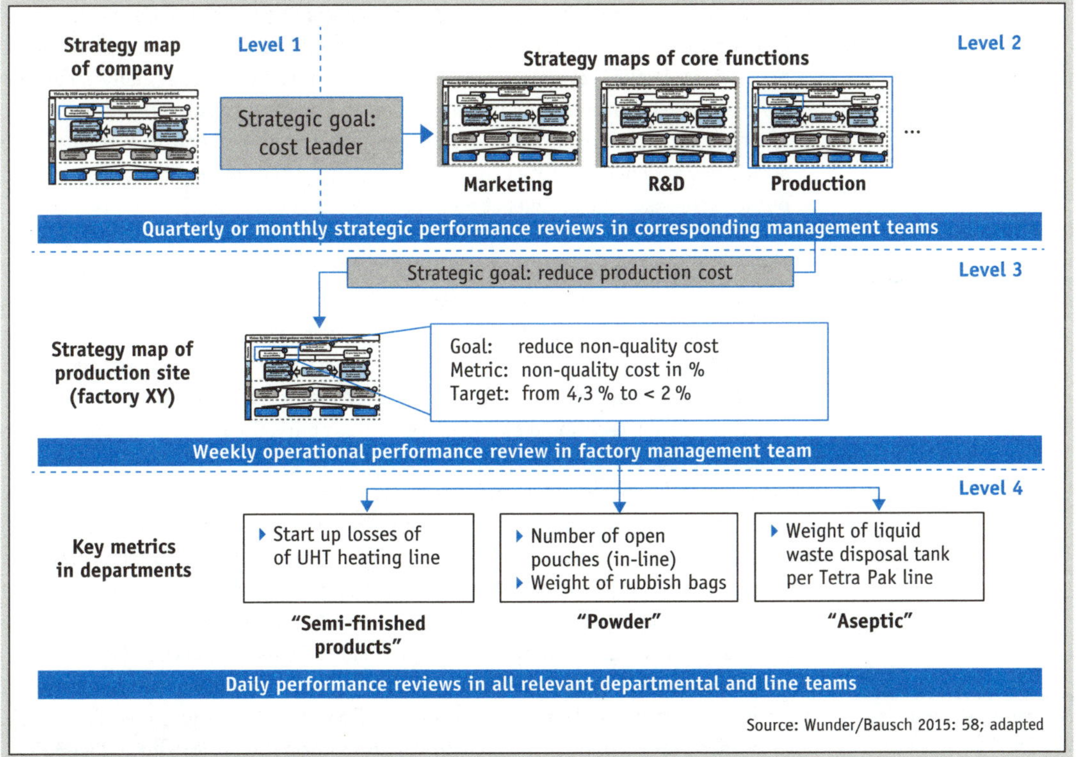

Fig. 4-88: Linking strategic to operational performance

(e. g., start-up losses, failures, scrap and waste) are assessed by the departmental teams and corresponding production line workers in **departmental and line performance reviews on a daily basis.** The linking pins between strategy and operations are strategic goals and key metrics that are strongly related across the different management levels and teams.

meeting by presenting the strategy map and highlighting the corresponding strategic goals to which the topical operational issue to be discussed that day logically relates to. This way they keep strategy "alive" not only on the executive suite but also across the entire organization. Strategy contributions of more tactical and short-term activities become clear, and thus, operational performance is linked to strategic performance.

4.3.7.2 Control Assumptions and Observe Environment

Whereas execution control is focused on *what* is executed and *how*, the control of assumptions and the continuous observation of the environment relates to *why* the company pursues those strategies. Key questions to be answered are (Lombriser/Abplanalp 2004: 387–389):

- Are the assumptions about the company's environment and its internal situation that influenced the decision for or against certain strategy options or strategic goals are still valid? Have anticipated developments and events actually occurred? Do available insights about customer wants and preferences hold true?
- Are there any new opportunities or threats that were not considered when the strategy was developed? Was something considered to be negligible that turned out to have a significant impact on the success of the strategy?

Assumptions control

The formulation of strategy options (see chapter 3) is classically based on a number of explicit or implicit assumptions. This way the highly complex and dynamic environment is simplified to support the making of management decisions. Many of those strategic planning premises relate to external developments that have been addressed during strategic analysis (see chapter 2) such as anticipated demand patterns, assumptions about technological advancements, political developments, exchange rate fluctuations, etc. Additionally, assumptions that relate to the likelihood of strategy options to be successful may also relate to internal factors such as expected new product development times or the availability of resources and capabilities (Bea/Haas 2009: 244). For the purpose of strategic control, those strategic planning assumptions need to be stated and recorded explicitly. It needs to be assessed continuously whether they are still true as their validity may change over time due to changes in the company's environment (Steinmann/Schreyögg 2000: 246 f.).

Monitoring and scanning of environment

Assumptions control and environmental observation requires two different types of activities: environmental monitoring and environmental scanning (see fig. 4-90). Both methods are related to the discovery of new information (Pillkahn 2008: 95–99). The control of assumptions is typically done by focused observation, that is monitoring developments that directly relate to the corresponding items for which premises

Strategy Practice Example 4.15

Strategic Control at BOSCH

The Bosch Group is a leading international technology and service company focusing on automotive and industrial technologies as well as products and services for professional and private use. Founded in 1886, Bosch generated € 47.3 billion revenues with its 285,000 employees in 2010. With respect to increasing volatility in the business environment, strategic control (original: "Strategic Controlling") is considered a key management process for both keeping the company on its defined strategic course as well as ensuring strategic adaptability regarding technological developments and changes in the competitive arena. Strategic control at Bosch covers three core elements (see fig. 4-89). **Environment and assumptions control** takes a close look at the market and technological environment of the company and its business sectors and monitors megatrends and their potential impact on the Bosch Group. Underlying assumptions of anticipated developments need to be made transparent and comprehensively recorded. Bosch is applying scenario- as well as sensitivity analyses methods for this type of control, which is considered crucial, particularly in case of developments with a high level of uncertainty. The second pillar is **consistency control**, which is related to the evaluation of strategy options as a base for management decisions. Hereby, it needs to be evaluated whether suggested strategic initiatives are in line with the defined strategic orientation of the Bosch Group such as strategically focused regions, business sectors, or technologies. In addition to the assessed "strategic fit", the probability of success of strategy options in the corresponding markets, technology fields and value chain structures has to be evaluated. **Execution control** is the third element of the strategic control framework at Bosch. Hereby, defined milestones and actions are focused including the comparison with previously defined criteria for stopping and terminating projects. (Source: Asenkerschbaumer 2012)

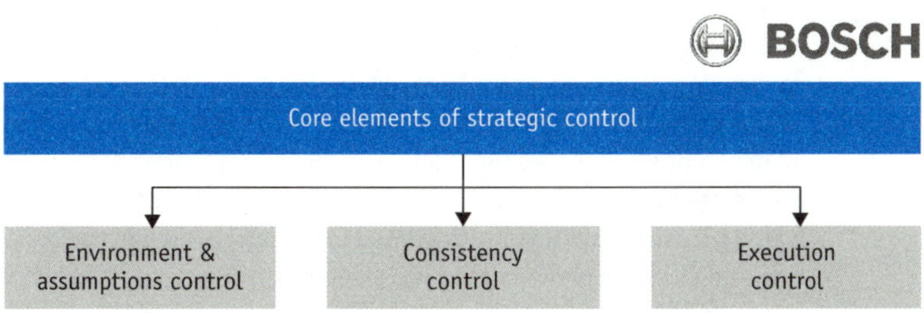

Source: Asenkerschbaumer 2012: 337; adapted. Reprinted with permission of Robert Bosch GmbH.

Fig. 4-89: Core elements of strategic control at the BOSCH Group

have been made. Hereby, the field of control has already been defined, and selected internal and external information sources are skimmed to identify comprehensive information about those items. In addition, a more general or unfocused observation (scanning) of the environment is necessary as there are a number of developments and events in the environment that either have been overlooked or ignored during assumption recording or they have evolved afterwards. Hereby, generic scanning activities can be directed toward certain external or internal areas of interest such as macroeconomical developments, key stakeholders, competitive forces within an

Fig. 4-90

Information gathering methods based on the scope and degree of maturity

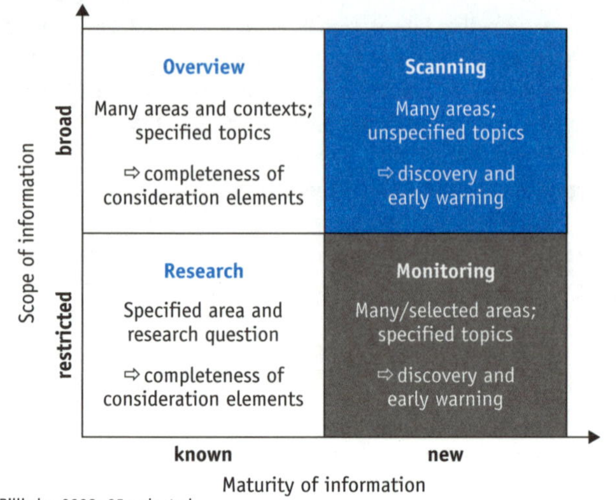

Source: Pillkahn 2008: 95; adapted

industry, or specific developments at key customers or competitors as well as internal resources and capabilities. All of those areas have been addressed during external strategic analysis (see chapter 2.2).

Early warning and weak signals

The goal of environmental monitoring and scanning is to sense and identify critical developments or events at an elementary stage. Those discoveries or weak signals can be important early warning indicators allowing managers to consider a pro-active adaption of strategy before changes in the environment have a negative impact on the company. It may be more effective to act quickly based on weak signals instead of waiting for more mature and validated information but with only little room for reaction and strategic maneuvering left (Lombriser/Abplanalp 2004: 125). To accomplish the task of ongoing environmental observation, there is a wealth of information sources available including new forms of social media such as internet discussion forums and blogs (Pillkahn 2008: 404–414). Of particular relevance nowadays are online sources that are easily accessible and continuously updated. To make the most out of environmental monitoring and scanning, companies have begun to apply systematic approaches for gathering, selecting, compressing and compiling, presenting, reporting, and recording information. Recommendations for an effective environmental observation are summarized in Strategic Snapshot 4.25.

4.3.7.3 Test and Adapt Strategy

Periodically assess whether the strategy is still valid.

The insights gained from execution control and assumptions control as well as from environmental monitoring and scanning are crucial for testing the validity of a strategy over time (see fig. 4-91). It may ultimately lead to an adaption of the strategic

Strategic Snapshot 4.25

Recommendations for an Effective Environmental Observation

- The implementation of an institutional, systematic and continuous process helps to ensure that future orientation is not forgotten because of operational day-to-day requirements.
- Work from general to specific. To identify weak signals, it is particularly important to avoid research efforts that are too focused. In case of internet-based search activities, search strategies such as pattern-, semantic- or selective methods need to be designed carefully as internet sources do not only come with manifold opportunities but also with significant risk (Theisen 2011: 69–75).
- Watch out for confirmation bias (see chapter 1.2.2.3). The question should be "What is changing in our company environment?" and not "What will support the assumptions and conclusions we already have in mind?"
- No single source or state of information regarding a particular topic qualifies as complete. Moreover, it is important to watch out for contradictions, divergent statements and uncertainties and to deep-dive into those items if necessary.
- Management decisions have to be based on valid information, that means the reliability of data sources needs to be ensured, particularly when it comes to manifold sources in the internet. Ambiguous information needs to be recognized and labeled correspondingly.
- To make information utilizable for strategic decision making, it needs to be filtered and compressed to manageable units. Special cognitive skills and experience are crucial for this process to be effective, which must not be underestimated when designing and particularly staffing this process.
- As the quality of strategic decisions strongly depends on the care with which environmental scanning and monitoring is carried out, an arbitrary budget might be the wrong criterion for setting the scope. The right cost-benefit ratio to reach a reasonable state of information is related to, as a rule of thumb, as soon as observation and research efforts turn up repeats of the same information.

(Source: Pillkahn 2008: 100–104)

direction. As explained in chapter 2 and 3, a company and its businesses formulate strategies based on information available for the external environment and competitive arena. It decides on strategy options and defines strategic goals based on assumed correlations and cause-and-effect relationships. The strategy execution system needs to capture a process of periodically assessing whether those strategic assumptions are still valid. Hereby, disappointing business results arising from both bad execution or executing a bad strategy may be an indicator for the validity of the strategy. Furthermore, results may still be satisfactory but scanning the environment might reveal some signals for upcoming major changes that are likely to influence the strategic performance if not addressed early and adequately by the company.

Capturing emergent strategies

In chapter 1.1.2.5 it has been emphasized that successful strategies are not only an outcome of a rational planning approach (i. e. intended strategies) but may emerge from within the organization (i. e. emergent strategies). There are examples of great ideas for new strategies that were not found and formulated by senior management teams and then deployed top-down in the organization but a result of strategic initiatives that evolved bottom-up (Kaplan/Norton 2008: 275 f.; Rothaermel 2013: 43 f.). One key element of the described strategy execution system is to create shared understanding and commitment among all employees in the organization about the strategic direction and priorities of the company. Many of those individuals have firsthand knowledge about key components of the strategy such as customers, processes, products or services, etc. The strategic context provided by, for example, stra-

Fig. 4-91

Strategic control and organizational learning

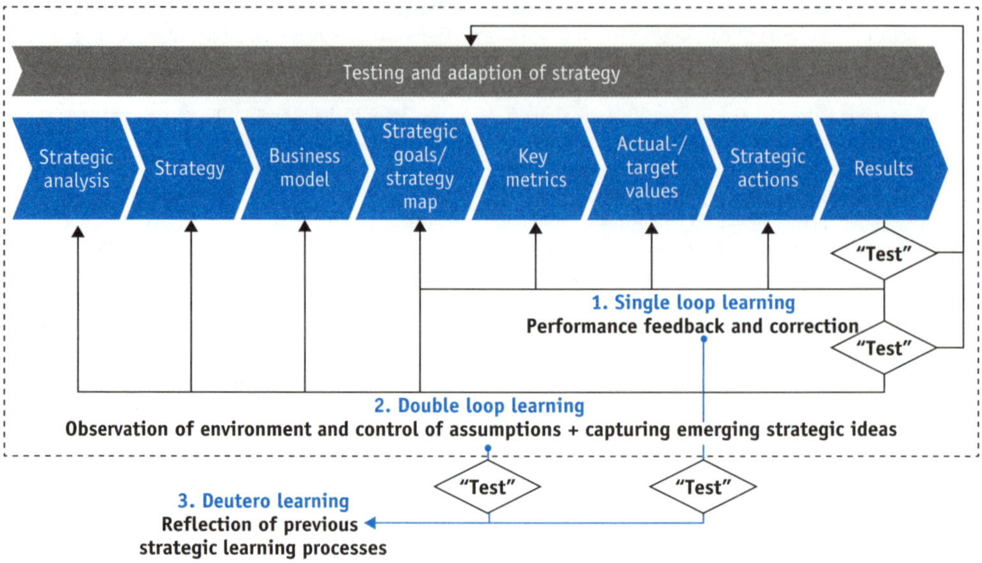

Source: Wunder 2004a: 169; adapted

tegic themes and strategy maps, can trigger strategic experimentation and ideation that may result in strategic suggestions originating outside the formal strategy development and review process. This can be referred to as "planned emergence" (Rothaermel 2013: 45 f.). "Employees who understand the strategy—and are motivated to help the organization succeed with the strategy—are in an excellent position to see where gaps exist either in strategy implementation or in an underlying hypothesis. They may even identify where a new strategic approach might yield better performance." (Kaplan/Norton 2008: 275) As part of strategy testing and strategy adaption, members of senior management are advised to be alert, solicit actively and assess those strategic ideas that have emerged from within the organization.

Three types of outcomes from testing strategy

The executive leadership team has to cycle back and determine regularly whether there is a need to update the strategy in the light of strategic lessons learned since the previous meeting that reviewed and updated strategy. Depending on the industry-related speed of change in the macroenvironment and competitive arena, those fundamental strategy reviews and updates should be conducted at least annually if not quarterly. Strategic learnings may result from strategic performance reviews, from significant changes or emerging strategic ideas inside the company as well as from changes in the company's environment that have occurred since the strategy was updated last. Strategy testing can essentially lead to three types of outcome (Kaplan/Norton 2008: 253):

- **The existing strategy is reaffirmed.** In this case, the executive leadership team may only update some target values, reprioritize strategic action programs, adjust time frames, budgets and responsibilities, and realign the organization to those new expectations and refinements.
- **Incremental changes to strategy are necessary.** This type of strategy adaption may be reflected by erasing, adding or changing one or more strategic goals in the strategy maps while keeping the fundamental strategic direction. Furthermore, key metrics may be replaced with new ones, target values are recalibrated and strategic action programs are revised.
- **Fundamental elements of the strategic direction become obsolete.** In this case, the company cycles back to the comprehensive strategic analysis and strategy formulation processes described in chapters 2 and 3 and develops a new strategy. Based on this, the six steps for making a strategy come true are reapplied (see fig. 4-40 in chapter 4.3.1).

Strategic control and organizational learning

For an effective testing and adaption of a strategy, a company ultimately needs to find the right mix of strategic control and learning (see Strategic Snapshot 4.26). The question is not whether to follow a top-down *or* bottom-up process but how to best combine top-level strategic intent with continual learning and emergent strategy formation. Figure 4-91 shows how testing and adapting strategy relates to different learning types based on a common classification of organizational learning (Argyris/Schön 2002: 35–44; Kaplan/Norton 1996: 267–269; Steinmann/Schreyögg 2000: 466–468). The comparison of desired and actual performance, which leads managers to review the cause-and-effect relationships of strategic actions and the corresponding metrics and targets, can be seen as a **single-loop learning** process. It is based on a closed-loop feedback mechanism within a pre-specified set of strategic goals. An analysis of deviations from defined targets may lead to corrective action so that expectations can still be met. Insights from such a process as well as from environmental observation, assumptions control and from soliciting emerging strategic ideas are the base for assessing whether the defined strategic goals and hypotheses are still valid. This can be seen as **double-loop learning** where the underlying strategic intent

Strategic Snapshot 4.26

Henry Mintzberg on Integrating Strategic Control and Strategic Learning

"As implied earlier, few, if any, strategies can be purely deliberate, and few can be purely emergent. One suggests no learning, the other, no control. All real-world strategies need to mix these in some way—to attempt to control without stopping the learning process. Organizations, for example, often pursue what may be called *umbrella* strategies: the broad outlines are deliberate while the details are allowed to emerge within them. Thus emergent strategies are not necessarily bad and deliberate ones good; effective strategies mix these characteristics in ways that reflect the conditions at hand, notably the ability to predict as well as the need to react to unexpected events." (Source: Mintzberg 2000: 25).

and its premises are questioned fundamentally. In addition, those two learning processes themselves and their effectiveness in the context of strategic management need to be continuously assessed as well. This can be referred to as **deutero-learning** or learning how to learn best. Thus, testing and adaption is not only related to the strategy but also to the corresponding strategic learning processes. This concludes the final step of the described strategy execution and control system.

4.3.7.4 Benefits and Pitfalls of Strategic Control

Effectively execute in the present and adapt to the future.

The execution and control of strategy is the final step of the described strategy execution system. For an organization to survive, it must execute effectively in the present and at the same time be alert and willing to adapt to the future. Although the three key components of strategic control—execution control, assumptions control as well as monitoring and scanning of environment—fulfill different roles, they all help to test current strategies and determine whether they need refinement or adaption. The continuous assessment of execution effectiveness and validation of the strategic direction is one of the key benefits of strategic control. This way, strategy becomes a continuous process as it supports both driving short- and mid-term strategic performance as well as sustaining it in the long run. Companies with effective strategic control mechanism know that past performance is no guarantee for future success (see chapter 1.1). Strategic control supports them in adapting their strategies to the changing environment, internal dynamics as well as lessons learned from their ongoing strategic performance reviews.

Driving effectiveness of strategy execution

Strategic control is driving feedback and different types of strategic learning. The benefit of strategic performance reviews (execution control) and the corresponding single-loop learning processes is a systematic and ongoing evaluation of the effectiveness of strategy execution. Issues are identified and corrective actions are initiated to get back on course, if required. This feedback, though, is not only limited to "as is" versus "to be" comparisons. In addition to looking at deviations from the defined aspiration, feedback is used to test the validity of expectations and the effectiveness of means to fulfill them. Those strategic hypotheses are questioned in strategic performance reviews and may be adjusted. Furthermore, effective execution in the present is not only driven by control mechanisms but also by supporting line managers to deliver on their defined strategic actions at the desired pace with, for example, the help of experienced project or change management experts. Furthermore, strategic action portfolio management helps to ensure that resources and management attention is continuously allocated to the highest value strategic projects.

Identification of opportunities and threats at an early stage

The focus of assumptions control and ongoing environmental observation is not on how well strategies are executed but on whether the right strategies are pursued (double-loop learning). The benefit is that internal or external signals can be sensed and discovered at an early stage, which may require an adjustment of the strategic course. This can have a significant advantage in competition. On the one hand, it allows companies to sense upcoming opportunities prior to their competitors and such provide them the chance for realizing first mover advantages. This may include internally emerging strategic ideas and initiatives that are captured and leveraged for competitive advantage. On the other hand, environmental observation helps to investigate weak signals of potential threats that may require action and preparation before they could have a negative impact on the company's performance. In both cases, strategic control supports the flexibility and strategic adaptability of a company.

Watch out: strategic myopia

A strategy execution and control system can only be as good as the skills of the managers who use it. A key pitfall is the lack of strategic thinking, which can be labeled as strategic myopia (Lombriser/Abplanalp 2004: 135–138): Managers may ignore or even reject reported signals that significantly deviate from their current strategic assumptions and business logic or unconsciously only use information sources that confirm their premises (see also "confirmation bias" in chapter 1.2.2.3). Many companies that were "suddenly" hit by a strategic crisis had ignored corresponding signals at an earlier stage. Strategic adaptability tends to be particularly difficult in cases where the existing strategy has been successful for a long time. Instead of selected individuals, heterogeneous teams with diverse backgrounds tend to be better positioned for identifying particularly weak signals. Additionally, fostering free expression of opinions as well as critical views within the management leadership team may help to reduce the risk of strategic myopia. Hereby it needs to be accepted that early signals of opportunities and threats are not always hard facts but also intuitive subjective opinions that need to be considered and evaluated.

Watch out: dealing with unfavorable results

Another pitfall particularly to the effectiveness of strategic performance reviews is the way how senior leaders and management leadership teams react to reported unfavorable numbers. Strategic decision makers typically embrace messages reflecting that they are part of success stories and not of failures (see also "self-serving bias" in chapter 1.2.2.3). This may lead to a reaction that leaves people or organizational units that are reporting unfavorable results feeling isolated or punished. Executives may also question the effectiveness of the strategy execution system. Though, poor performance based on a failure to achieve competitive advantage may not be a reflection of a poor strategy execution system but of a strategy failure that needs to be investigated and solved. This type of dealing with unfavorable results does not foster a culture of information sharing, strategic learning, teamwork, and joint problem solving. Moreover, it may even lead people to withhold important information that is relevant for improving the effectiveness of strategy execution or for adapting the strategy itself.

4.4 Summary

Strategy execution is not just something adjacent to the "rigor" of strategic analysis and the "art" of strategy formulation. In fact, companies cannot win in the market only by crafting a worthwhile strategy. They need to make sure that the organization has or gets what is required to make the strategy happen. Excellence in execution has become one of the top priorities of many senior leaders knowing that bad performance in this area is one of the major reasons for flawed strategic management. Several barriers, which have evolved over time in organizations today, need to be overcome for effective strategy execution. Generally speaking, companies need to master the challenging task of establishing the right

goals, actions, skills, resources, incentives, attitudes, and **information**

for making their strategies come true. However, they naturally find it difficult to calibrate and orchestrate the interrelated areas of organizational design in a way that will most likely enable a strategy to come true, that is

- strategy
- culture
- structure (managerial infrastructure).

Key elements of managerial infrastructure to be considered for strategy execution are organizational structure, business processes and managerial systems such as planning and budgeting, human resources, and management reporting. Companies need to determine how all those components of an organization need to be designed in order to match strategy, knowing that the relationship is twofold: Existing organizational design is influencing strategy formation and a new strategy typically imparts the organizational design. Assuming that the strategic direction is right, a company's ability to master the challenges of strategy-oriented organizational design is a key success factor for effective execution.

Developing a strategy execution system is the last step in the design of an effective strategic management process. It is based on key principles and advancements primarily derived from the balanced scorecard (BSC) approach, which is the most widely known and used management concept for strategy execution today. Hereby, the balanced scorecard is not meant as a specific "tool" but as a conceptual framework and method of key components, principles and processes that drive strategy execution throughout the organization. Six integrated steps are necessary to develop a corresponding management system for strategy execution:

- describe strategy
- cascade and align strategy
- make strategy measurable
- plan strategic actions
- align organization to strategy
- execute and control strategy.

Each of those steps fulfills a crucial role for making a strategy happen and is conducted following various guidelines and principles. It starts with clarifying and refining the fundamental strategic direction and priorities by developing strategic goals and a strategy map. Hereby, selected elements or changes in the company's business models and other strategic choices are key linking pins to strategy formulation, which are translated into strategic goals. In the next step, strategic goals and strategy maps are used to cascade higher-level strategies to organizational units (strategic business units, functional units, regional and country units) and to vertically and horizontally align strategic priorities throughout the organization. This way, forces are joined toward the realization of a common strategic orientation. Making the strategy measurable by developing key metrics and targets clarifies what exactly should be achieved. Strategic action programs are derived as means for achieving those defined aspirations. Goals, metrics/targets as well as actions are the foundation for organizational alignment (i. e. resources, people, and management information). The last step in the framework is the actual execution and control of strategy that not only includes feedback on the effectiveness of strategy execution but also whether the strategy is still valid as well as, if required, an adaption of strategy. This is closing the loop of the strategic management process and the WOS strategy framework by cycling back to strategic analysis and strategy formulation. All those described elements are the foundations for strategy to become a continuous and highly iterative strategic learning and adaption process. This requires both sophisticated governance and effective strategic leadership to deliver results in the short as well as in the long run.

Appendix: Strategy Workout and Review Questions

Strategy Workout

Kick-off

Firm selection

Please select a company to study for this strategy workout. The goal is to use the same company for each section of the workout, and thus, ensure a seamless integration of the different strategic management tools and methods you will apply—from strategic analysis to strategy formulation and execution. You may identify *any company* in an industry you find interesting and would like to know more about. For example, select a potential future employer, an industry you would consider joining, or a firm just out of curiosity. You may also select your *own company* if you are a practitioner who has been assigned to develop a strategy, solve a particular strategic management issue in your firm such as execution, or just want to enhance your knowledge and skills in this field. In this case you may consider having the strategy workout conducted by different people in your organization either jointly or individually.

An important criterion for your firm selection is to choose a company that has data available that you can obtain and analyze. Publicly held large or medium-sized companies might be best suited for this purpose as they are typically widely covered in popular media and business literature. Information about these companies can be obtained freely using a variety of web services, annual reports, the firm's own websites, magazines from industry trade associations, and databases of articles at most universities and public libraries. You may also select a smaller firm, but you need to make sure that you will have access to a broad range of data and information. If feasible you might want to supplement the data gathered by conducting strategy interviews with key managers at the firm and study internal company material. In these cases, please ensure that the firm is willing to share its information with you.

If your firm is a large multi-business corporation, you may need to choose only one of its major strategic business units (SBU) for this workout. In this case you can apply most of the tasks in the workout to analyze the SBU more deeply while some questions may be related to the parent company. In case you want to analyze the entire multi-business firm, you need to apply the different tasks and questions to the entire firm and all its various businesses.

Company background

1. Please provide a brief background and history of the company you selected.
2. Who are the current top executives of your company and what experience do they have? Can you determine the major mode of decision making in the executive leadership team (i. e., primarily analytical and number-driven or intuitive with less emphasis on numbers)?

Brief strategy review

1. What does the company view as the reasons for its current or expected future successes?
2. Does your firm have a competitive advantage? How sustainable is it? Justify your answer with reference to the firm's dominant perspective on performance.
3. Trace any major changes in the company's strategy that you can identify over time. Did the firm face strong dynamics in its external environment in the past and could it effectively adapt its resources and capabilities to those changes? Was the firm able to set own trends in its external environment?
4. Would you say the firm's strategy is different than its competitors? Briefly describe the principle business model the firm applies to make most of its profits? (Note: There is no comprehensive business model analysis required at this stage, only a brief description). How is the firms positioning regarding strategy and operational excellence?
5. What is the articulated long-term aspiration of the company that it wants to achieve with its strategy? Is there a time horizon articulated? Where does the aspiration come from and how inspirational and challenging does it appear to you?
6. Do you see consistency in the strategic behavior of your company? How would you describe this behavior? If possible, explain whether the behavior of your selected firm is a result of intended strategies, emergent strategies or a combination of both.
7. If you selected a large multi-business corporation, what are the divisions and strategic business units of the company? What functional units are particularly important for the firm's strategy and why? What is its geographic scope? Does the organization appear to be aligned around the firm's strategy?

Strategic analysis

Stakeholders and strategic analysis framework

1. What are the most relevant stakeholders for your company and why? Have there been any particular issues with certain stakeholder groups? To which degree have stakeholders been involved in strategy formulation in the past?

2. Based on the results of stakeholder analysis, what areas do you consider most relevant for determining the initial strategic situation of the company? Illustrate a framework that you suggest for strategic analysis. Make sure that you document the most important strengths and weaknesses as well as opportunities and threats you identified in each of the different analysis parts when completing the strategic analysis.

External strategic analysis

1. What are major developments in the firm's external environment that it cannot affect or change? Apply the PESTEL framework to identify relevant factors in the macroenvironment that may affect the company's performance. Set a time scope and develop scenarios of what the external environment of your company may look like in the future.
2. Define your relevant market (i. e. industry boundaries) and conduct an industry structure analysis by applying the five forces model to your relevant market. Analyze selected additional forces if you believe this is valuable. What key insights about the nature of competition in the industry can you derive based on this model?
3. Identify your key customer groups in your competitive arena and determine their needs, wants, demands, and customer requirements. How does your company currently fulfill these customer requirements? If necessary, make a segmentation of your relevant market into submarkets by choosing an appropriate segmentation approach. In this case apply the customer analysis to your specific competitive arena in the submarkets.
4. Is your company's market approach rather broad or narrow (i. e. niche strategy)? Select some key dimensions such as market size and market growth and make an analysis of your relevant market and submarkets.
5. Identify strategic groups that exist in the industry. How do the intensity of competition and profitability differ across the strategic groups you have identified? How would you characterize the nature of competition within the firm's strategic group? Which current direct or indirect competitors affect the company most? Are there any potential competitors that may become a threat in the future? Develop profiles for current and potential key competitors that cover relevant information.
6. Are there any other external stakeholders that strongly affect the firm's performance, and thus, need a more comprehensive analysis? Identify these stakeholders and explain how they may impact the company's strategy.

Internal strategic analysis

1. Develop a catalog of areas and evaluation criteria you want to consider for internal strategic analysis based on internal success factors for your company.

2. Analyze these areas and criteria and develop a strengths-and-weaknesses profile of the firm in comparison to its most important competitors. Hereby, make sure that you consider both the firm's tangible and intangible resources as well as its capabilities.
3. Try to identify additional strengths and weaknesses by conducting a value chain analysis. Put a special emphasis on analyzing linkages between activities within one value chain as well as on synergies between different value chains you may find within your company.
4. Apply the 7-S-model to identify strengths and weaknesses with a special emphasis on soft factors that impact company performance as well as the "fit" of the different elements in the model. Would you say the firm's current strategy, structure, and culture are aligned? If not, what are areas for improvement.
5. Now extend beyond the resource and capabilities (i. e. competence) base and use the VRIO framework to identify core competencies in your firm. Which competency, if any, is a core competency, and thus, likely to help gaining and sustaining competitive advantage?

Consolidation of the initial strategic situation

1. Prioritize all the strategic factors you determined in your strategic analysis and consolidate the most important strengths and weaknesses from internal analysis and the most relevant opportunities and threats from external analysis by using a SWOT framework.
2. Based on the results from your strategic analysis, illustrate the initial strategic situation of one or more of your product lines or strategic business units by developing a market attractiveness–business strength matrix (GE/McK-matrix). Where do you see these organizational unit(s) positioned in the future?
3. Based on your SWOT framework derive initial focal points for strategy formulation by developing a TOWS matrix. Where do you see the most promising strategic areas to consider for crafting strategy?

Strategy formulation

Strategic guideposts

1. Search for statements about the company's vision, mission as well as its articulated values. Do you see any need to adapt these strategic guideposts based on your insights from strategic analysis? Do any of these guideposts refer in any way to sustainability?
2. Which type of vision has the company defined? Would you say this vision provides a clear orientation and is inspiring to employees in the organization?

3. Is the purpose of the company (i. e., its mission) defined product-based or customer-need based? Does it address a broader role of the company in society? Identify the major brand claim of the company and evaluate whether it also reflects the company's mission? Would you say that the current mission is appropriate given the results from strategic analysis? What about the claim?
4. Do the articulated core values reflect the actual behavior of members in the organization and what the company stands for? If strategic circumstances changed, would the executive leadership team still keep the proclaimed values even though it might be penalized for holding them?
5. Make suggestions in case strategic guideposts have not been defined yet or should be adapted or refined based on your evaluation.

Corporate strategy

1. What corporate parenting approach does your company follow for creating value in the interaction of its different business units? How effective is this strategy? Does the company have a parenting advantage? Do you suggest any changes to corporate parenting given the insights from strategic analysis?
2. What is the articulated growth aspiration of the company? When looking at the projected baseline growth in its relevant markets, what is the growth gap (if any) that needs to be filled with strategy?
3. When analyzing the different businesses within your company would you say it is diversified or rather focused? What diversification strategy, if any, do you suggest for expanding the scope of its business in the future?
4. What role have mergers or acquisitions played in the growth of your company? Did it participate in any merger or acquisition in the past three to five years, and did this result in an improved competitiveness? Given the articulated growth aspiration, do you suggest to focus on external growth in the future? To what degree (key driver or only opportunity based)?
5. Draw out the vertical value chain for your firm's industry. Is your firm highly vertically integrated along the chain? If yes, where within the industry value chain has the firm chosen to compete? Based on your strategic analysis results, would you suggest any changes in the degree of vertical integration that might expand or focus the company's business scope?

International strategy

1. Does your company have a track record of successful international expansion? What are the most important motives for your firm, if any, to further expand the scope of its international activities? What basic geographic areas and country markets would you suggest for such an international expansion?
2. Does your company strongly integrate its global activities or apply a more multinational approach with strongly varying go-to-market strategies for different geo-

graphic markets? Which of the four strategy postures in multinational corporations does your firm pursue with its various businesses? How would you evaluate this strategic approach given your strategic analysis insights?

3. Are there layers of worldwide competitive advantage the company has not yet exploited? Have a particular focus on how the company manages its various international subsidiaries and their interrelations. Can the firm improve its overall competitive position by better leveraging the diversity of international environments in which it operates? How?

Business strategy

1. Which intensive growth strategies has the company been primarily pursuing to improve the competitive position of a selected *existing* business? Which approach would you suggest for further growing this business given the growth aspiration or identified growth gap?
2. How important is innovation to your firm and how do you evaluate its innovativeness? Is the company primarily focused on radical or incremental product innovation? Outline and evaluate the company's current innovation process. How can it be improved?
3. Does your selected business follow best-product, lock-in, or complete customer solution strategies? If so, what specific strategic approaches does it pursue in the different generic strategy categories?
4. Would you say that your selected business has a clear strategic positioning or is it in the zone of ambiguity? What can the business do to hold, improve, or establish an effective strategic positioning?
5. Using the answers to the preceding questions, what suggestions do you have to improve the firm's generic business strategies and strategic position? Do you see opportunities for supplementing current business strategies with other generic strategy types?
6. Aside from traditional competition, have a look at cooperative strategies such as strategic alliances your business has pursued in the past three to five years. Were they successful and can these approaches be leveraged to improve the performance of your selected business?
7. Do you see any opportunities for creating uncontested market space? Answer the question by applying the strategy canvas framework for a selected business leveraging the insights you gained earlier in strategic analysis such as the identified strategic groups.
8. Analyze the principal business model the firm applies to make most of its profits by describing the seven business model building blocks. Where do you see opportunities for business model innovation? Leverage generic business strategies and common business model patterns for ideation on how to further develop your business model.
9. How important is sustainability for your current business strategies? Does your firm pursue or create a business case for sustainability? Do you see any opportuni-

ties for developing sustainable business models and achieve economic value creation through value creation for society?

Strategy option evaluation and choice

1. Summarize the various strategy options you have identified in the previous sections by briefly explaining each option with a few sentences.
2. Evaluate the suitability of each strategy option: Is it plausible for dealing with the strategic situation identified in strategic analysis? How do the strategy options relate to the combination of strengths and weaknesses with opportunities and threats identified with the TOWS-matrix? Is the strategy option consistent with the strategic guideposts and with other strategies pursued by the company?
3. For suitable strategy options, evaluate whether they are also acceptable in terms of their expected contribution to the company's performance and the related business risk to the organization.
4. Finally, evaluate whether strategy options that are suitable and acceptable are also feasible. Are they compatible with expectations of key stakeholders that you have identified earlier, and are there internal constraints such as resources, skills, and cultural or structural issues that currently cannot be overcome for making the strategy option happen?
5. Consolidate all strategy options that are suitable, acceptable, and feasible. Choose which of those options you want to execute. These strategic directions or strategic themes are the input for the next section of the strategy workout.

Strategy execution

1. Evaluate the firm's current capability to execute strategies by applying the strategy execution excellence radar.
2. What have been the biggest challenges or barriers, if any, to effective strategy execution in your company? Which of the drivers for strategy execution have been identified as being ineffective in your firm?

Organizational design

1. What words or phrases summarize the organizational culture of your company? Where does it not fit with the strategy you want to execute? Identify which specific culture dimensions are most relevant for your strategy and evaluate their degree of alignment with your strategy by applying a strategic culture alignment (SCA) matrix. What are your suggestions to overcome a potential misalignment?

2. What is your basic organizational structure (product, geography, function, matrix or tensor)? Does it generally support the strategy you want to execute? If not, what is your suggestion?
3. Identify the processes that are most relevant for your strategy and evaluate their level of alignment with your strategy by applying a strategic process alignment (SPA) matrix. How do key processes that are not aligned need to be addressed to make the strategy happen effectively?
4. Do the company's key managerial systems such as operational planning and budgeting, human resources, and management reporting generally support your strategy? Identify areas that need to be linked stronger with strategy. You will have to consider these areas later on when designing your strategy execution system and defining corresponding organizational alignment initiatives.

Describe the strategy

1. Translate the strategy you want to execute into a maximum of 20 strategic goals along the four perspectives financials, customer/market, internal processes, and potentials by answering the following questions and applying the eight principles for developing strategic goals:
   a. Which financial goals reflect the desired economic consequences (i. e. success) of your strategy?
   b. Which goals regarding targeted customers, markets, and value proposition need to be accomplished for achieving the financials?
   c. Which most critical processes have to be performed excellently in order to achieve the customer/market and financial goals?
   d. Which goals regarding human, information, and organizational capital are required to achieve all other goals and ensure long-term success?
2. In case you are following a triple bottom line philosophy you may want to replace the financials perspective with a corporate sustainability perspective and supplement the desired economic impact with social and ecological consequences of your strategy.
3. Develop a strategy map that clearly articulates the strategy based on the strategic goals derived. What key linkages between the strategic goals best reflect the specific strategic intention of the company or business?
4. Consolidate the overall strategic themes articulated in the strategy map by highlighting logical vertical flows between strategic goals and describe them in a short essay-type format (i. e. story of strategy).

Cascade and align the strategy

1. Identify the specific strategy alignment needs in the organization for an effective company-wide execution of the overall strategic themes and goals defined earlier.

2. What organizational units, sequence, and methods for cascading strategy do you recommend based on the company's steering logic and the identified alignment needs?
3. Cascade strategic goals and derive strategy maps in the involved organizational unit, or by applying other methods if appropriate.
4. Vertically and horizontally align strategies. Are the overall strategic goals of the company sufficiently supported by the various organizational units? Are there any conflicts of strategic priorities between organizational units? If so, refine the strategic goals correspondingly.

Make strategy measurable

1. Once the strategic goals in various areas of the organization are aligned, identify one or two key performance indicators (metrics) for each strategic goal. These indicators must be suitable for determining whether you will be successful in achieving the corresponding strategic goal. Is the resulting scorecard "balanced"? Make sure you follow the six principles for developing key performance indicators.
2. What are appropriate targets for the key performance indicators you identified? Identify reference values and define corresponding target values that reflect the company's desired aspiration level along the time horizon of the strategy for each key performance indicator. Make sure that target values in different organizational units fit together.
3. What are acceptable deviation ranges for each target value? You will need these ranges for establishing a strategic control process later on.

Plan strategic actions

1. What are strategic projects the company is currently running? Make an inventory of these projects and validate the relevance of these projects for achieving strategic goals. How do you proceed with current strategic projects that do not seem to be relevant?
2. What new strategic actions are required for achieving the defined strategic goals? Develop and prioritize corresponding proposals of strategic actions.
3. Refine relevant strategic actions by adding start and end dates, resource requirements, expected deliverables, and responsible project managers and sponsors.
4. Is your strategic action program sufficient for executing the defined strategy? Answer this question by developing a strategic action relevance matrix.
5. Analyze whether there are any resource or time constraints in the strategic action program by developing a financial summary, strategic action resource requirement plan, and a roadmap.

Align organization to strategy

1. Based on insights from organizational design earlier, identify alignment needs regarding the firm's key managerial systems such as operational planning and budgeting, strategy communication, target agreement (MBO) and incentive system, as well as personal skill development. In particular, how does the company currently ensure that
   a. resources (funds, employees) for making the strategy happen are secured?
   b. people are motivated and enabled to help with executing the strategy?
   c. relevant management information is provided to track strategy execution and learn?
2. In case you identify any misalignments by answering the preceding questions, what suggestions do you have for improving the firm's organizational alignment to strategy?

Execute and control strategy

1. How does the firm drive and support the effective execution of strategic actions? How is the dynamic portfolio of strategic action programs governed (e. g., closing of finalized strategic actions, initiation of new strategic actions)?
2. What strategic control mechanisms such as strategic performance reviews (i. e. execution control), examining assumptions (i. e. assumptions control), as well as monitoring and scanning of the environment need to be installed to make strategy a continuous process?
3. How does the company test and adapt its strategy to changing conditions?
4. What suggestions do you have for institutionalizing the management of strategy execution and control?

Review Questions

Chapter 1: Foundations of Strategic Management

1. What are triggering events when companies typically rethink their strategic orientation and engage in a (re-)definition of their strategy? Provide at least three triggers.
2. What are the 5 P's of strategy and how do they relate to a strategic planning logic?
3. What is the major difference between prescriptive and descriptive schools of strategy? Provide one example for each category.
4. What is competitive advantage? Describe the concept and what types of competitive advantage can be distinguished regarding its duration.
5. What is the relationship between external and internal strategic factors of a firm? What are the corresponding views of strategy? Would you rather follow an outside-in or inside-out approach to ensure satisfactory firm performance in the long run?
6. What is meant by the statement "strategy is about being different"? Explain your answer with reference to the productivity frontier. Based on this, describe the two key principles of strategic positioning.
7. What is more likely to drive superior performance: operational excellence or strategy? Explain your answer.
8. Please comment on the statement: "Our strategy is to become the biggest and most profitable firm in our industry in five years." Is this a strategy? Explain your answer.
9. How can a realized strategy arise? In this context, explain the concept of strategy as planned emergence.
10. In which distinct organizational levels in a company would you typically find strategies? Name the strategy and provide a typical strategic theme for each level.
11. What are the four different approaches of how competitive advantage can be measured?
12. What is producer surplus? What are the two general levers to influence it? Give a brief example for each of these levers of how to increase the producer surplus.
13. When comparing Toyota and Tesla on the absolute measure of profit, does having a higher absolute profit mean that the firm has a competitive advantage? Explain your answer.
14. What is the advantage and what are two important limitations of using accounting data to measure competitive advantage?
15. What is the difference between the Shareholder Value and the Shared Value approach?
16. What are the three components of the triple bottom line? Why is it referred to as "bottom line"?

17. Why would a firm apply a triple bottom line approach for evaluating its corporate performance?
18. What are typical characteristics regarding the ambiguity of strategic decision situations? Explain which type of decision making you would recommend in these situations.
19. What are cognitive biases? Explain which impact they can have on decision making and provide three examples. Suggest one way to counter them.
20. What are the building blocks of the wheel of strategy framework and how are they interrelated?
21. What are the three areas that have to be clarified for designing strategy process governance? Provide at least two important questions for each area.
22. What are two potential strategic leadership issues in the interaction between different levels in an organization that could prevent a strategy process from being effective?
23. Which guidelines would you recommend to follow when practicing strategic management? Provide at least three guidelines.
24. Which benefits can a firm expect from strategic management assuming that it applies the concept effectively?
25. What might be reasons why a company would not engage in a formal strategic management process? Provide at least three reasons.

Chapter 2: Strategic Analysis

1. What are the three distinct stakeholder categories a firm needs to consider in strategic analysis? Provide one example of a stakeholder group for each category and how it might affect the long-term performance of a firm.
2. Explain the four steps of conducting a stakeholder analysis.
3. Explain the four layers that need to be considered when developing a strategic analysis framework?
4. What are the six macro level areas of analysis that make up the PESTEL model? Provide an example for each area.
5. What is strategic foresight and how does it relate to the phenomenon of future myopia?
6. What is the main intention of firms when they use scenario planning? When would you suggest not to link scenarios to any kind of probabilities?
7. Define the relevant market for an automobile manufacturer: Provide three exemplary dimensions and suggest corresponding categories for each dimension to illustrate the entire market scope. How could the company define its relevant market?
8. What is the primary relationship between the forces and the profitability level in an industry according to the industry structure analysis model (Porter's five forces model)?

9. Will each of the following exemplary industry conditions make that industry more or less attractive? Explain your answer with reference to the corresponding force in Porter's five forces model for analyzing the industry structure.
   a. Government policies in favor of existing industry participants (e. g., trade restrictions or laws)
   b. Existence of many entirely different/alternative products that fill the same need as the focal firm's product but in a different way
   c. Buyers have high switching costs
   d. A large number of competitors in a market with slow or declining growth
   e. A large number of buyers for the firm's output
   f. Highly differentiated supplier products
10. Suppose a firm is not yet in an industry, but it is considering entering. Is it better to have high or low entry barriers with regard to the attractiveness of that industry in general?
11. What are the two additional forces that can be added to the five forces model? Give an example of each and how it affects industry profitability.
12. How do a consolidated and a fragmented industry structure generally differ in profitability?
13. What are the key areas of analysis within the competitive arena? Provide at least four areas.
14. What are different criteria or approaches for market segmentation within a competitive arena? Provide at least three.
15. What are key dimensions that are of interest for an analysis of a market segment or submarket? Provide at least three.
16. Explain the difference between needs, wants, demands, and customer requirements. Evaluate the two statements below to verify your understanding:
   a. "Marketers create needs"
   b. "Marketers get people to buy things they don't want"
17. List the three types of customer requirements according to the Kano-Model (customer enthusiasm model) and provide an example for each of them using a product of your choice.
18. Select two firms in the automobile industry that would be in different strategic groups. Explain what is different about their groups. What could be a mobility barrier for a member in one of your identified strategic groups to move into the other?
19. What are important elements that need to be considered for evaluating competitors or strategic groups? Provide at least five.
20. What tools or methods can be applied to determine strengths and weaknesses of an organization? Provide at least three.
21. What is the principle of the resource based view of strategy? Explain with reference to tangible and intangible resources, capabilities, and core competencies.
22. What is a core competency? Explain with reference to the VRIO framework and give a practical example. Do you see any pitfalls for applying this approach?
23. What is meant with the step "investigate risk position" when applying the market attractiveness–business strength matrix (GE/McK-matrix)?

24. What are the two dimensions of the market growth–market share matrix (BCG-matrix) and their conceptual foundations? What does a firm want to ensure by applying this portfolio?
25. How can insights from external and internal strategic analysis be combined to identify initial focal points for strategy formulation? Name and explain the four alternative strategic combinations.

Chapter 3: Strategy Formulation

1. Which three questions do managers need to ask as a first step in strategy formulation to guide strategic ideation and decision making?
2. Which benefits can a firm expect from articulating a compelling vision? Please provide four benefits and explain their role with regard to dynamic and volatile environments.
3. How does a mission differ from a firm's proclaimed values? Explain and provide an example for each.
4. What is the difference between a customer-oriented and a product-oriented mission? Explain and provide an example for each. Which one would you prefer and why?
5. What is the difference between corporate and business strategy? For which three main areas should corporate strategy provide an answer?
6. What is parenting advantage? Explain the concept and provide an example of a value-adding parental strategy that is directed toward the utilization of economies of scale and economies of scope?
7. Why would a firm choose to diversify and which types of corporate diversification can it use? Explain the two types and provide a company example for each.
8. What is vertical integration and what types can be distinguished? Provide an example of a firm using vertical integration for both types and the benefit it gets from it.
9. Which three major questions do the owners of a company that is considering going international for the first time have to answer? Give two recommendations of what the owners should consider for answering each of these three questions.
10. Why do firms go international and what are related pitfalls? Provide three motives and three items a company should understand before making the decision to go international.
11. Give an example of a regional knowledge cluster and explain which role it could play in a firm's international expansion strategy?
12. What are areas you would analyze for evaluating the attractiveness of a country market that a company is considering to enter? Provide six areas.
13. Provide six examples of potential entry barriers that a company may need to overcome when entering a foreign country market. Are high entry barriers preferable or not when analyzing the sustained profit potential of an international country market?

14. Which two distinct strategic options does a company have regarding the timing of an international market entry for multiple countries? What are the advantages and disadvantages of each option?
15. Why do some companies pursue a global strategy? Provide three reasons.
16. Which sources of competitive advantage does a company try to leverage when pursuing a multinational strategy? Provide three sources.
17. What is a transnational strategy and which three layers of competitive advantage is it pursuing?
18. What are the three intensive growth strategies for a business and under which conditions would they be favored? Provide one favorable condition for each strategy.
19. What are the three generic competitive strategies on a business level? Provide two distinct strategic approaches and company examples for each of them and explain what the "zone of ambiguity" refers to.
20. How can a business identify and exploit uncontested market space? What would be another approach of breaking out of traditional competition? Explain your answer.
21. What is the difference between a strategy and business model and how are these two concepts related?
22. What are the seven building blocks of a business model? Name the building blocks and provide an example of a real-life business model of a company of your choice by explaining at least one element in each of the seven building blocks.
23. What is business model innovation and which types can be distinguished? Why do business model innovations tend to be more important than product innovations in a given industry over time?
24. What is the business case for sustainability, what are drivers to create it, and what is a sustainable business model? Provide two patterns of sustainable business models and corresponding company examples.
25. How can strategy options be evaluated? Please provide the three categories and their corresponding evaluation criteria. How is the real options perspective different from using the net present value for evaluating the performance impact of a strategy option?

Chapter 4: Strategy Execution

1. What are reasons why an effectively planned strategy might fail and what are the barriers to strategy execution?
2. What are the seven drivers for strategy-oriented behavior and why should a company have all of them in place?
3. Which elements have to be considered and integrated in organizational design? Name the different elements and explain the relationship of organizational design and strategy.

4. What are the four key elements of organizational culture and their visibility? Name the elements and provide at least three dimensions that can be used when conducting a culture audit.
5. How does the development of corporate culture relate to competitive advantage?
6. What are the three elements of managerial infrastructure? Should structure follow strategy? What does company practice look like? Explain you answer.
7. Describe three major types of organizational structure and list one advantage and disadvantage of each.
8. What are the three types of managerial systems which are particularly important for executing strategies?
9. What are the key principles of the better or advanced budgeting approach? Provide at least three principles.
10. Which organizational structure would you consider most appropriate for effectively executing
    a. a strategy with a high degree of related or unrelated diversification?
    b. a cost leadership strategy with only a few products or services?
11. What is the relationship between process and strategy and what is strategic process alignment?
12. What are the four perspectives and their underlying logic in the classic "balanced-scorecard" approach for a private company?
13. What are three key components of a strategy execution system and the key questions for strategy execution they address? Explain the three components and the corresponding management process with its benefits for strategy execution. Name at least two benefits.
14. What are the six steps for developing a strategy execution system after a firm has outlined its strategy and business model and what are the expected results from applying these six steps?
15. Which principles should a company apply when developing strategic goals? Provide at least four principles.
16. How can a firm ensure that the strategic goals it selects for strategy execution are relevant? Explain with reference to the two criteria that can be applied for this purpose.
17. A company can apply a "balanced" approach to defining strategic goals and metrics. Which items are being "balanced" in this approach?
18. What is a strategy map and what benefits can a firm expect from applying this concept? Provide at least three benefits.
19. What is the difference between strategy alignment and organizational alignment? Explain the two approaches and the sequence in the strategy execution system.
20. What are key questions that need to be clarified for effective strategy alignment? Explain at least three questions.
21. Which methods would you use for cascading strategic goals to international country subsidiaries within a multinational enterprise? Explain your answer.
22. What is the difference between vertical and horizontal alignment? Explain your answer.

23. What are the four benefits of defining metrics (key performance indicators) for strategic goals to operationalize a strategy?
24. How would you link strategy to operational planning and budgeting and why is this critical for effective strategy execution?
25. What are the three types or mechanisms of strategic control and how do they relate to organizational learning processes?

References

Aaker, D. A. (2011): Strategic Market Management, 9th ed., Hoboken, NJ: John Wiley & Sons.

Abdelkafi, N./Makhotin, S./Posselt, T. (2013): Business Model Innovation for Electric Mobility—What Can be Learned from Existing Business Model Patterns? In: International Journal of Innovation Management, vol. 17 (2013), no. 1, pp. 1–41.

Adelberger, W. (2013): Produktportfoliomanagement – Im Spannungsfeld von Absatzpotenzialen, Renditen und Kapitaleinsatz [Product portfolio management—between the conflicting priorities of sales potentials, return and capital expenditure]. In: Horváth, P./Michel, U. (eds.): Controlling integriert und global. Erfolgreiche Steuerung von komplexen Organisationen, Stuttgart: Schäffer-Poeschel, pp. 27–39.

Adelberger, W./Haft-Zboril, N. (2013): Portfoliomanagement als Aufgabe der Optimierung von Rendite, Marktanteil und Ressourceneinsatz [Portfolio management to optimize return, market share and use of resources]. In: Controlling, vol. 25 (2013), no. 1, pp. 41–48.

Adjouri, N. (2014): Alles was Sie über Marken wissen müssen. Leitfaden für das erfolgreiche Management von Marken [All you need to know about trademarks. Guideline for a successful management of brands], 2nd ed., Wiesbaden: Springer Gabler.

Al-Laham, A. (1997): Strategieprozesse in deutschen Unternehmen. Verlauf, Struktur und Effizienz [Strategy processes in German companies. Sequence, structure and efficiency], Wiesbaden: Gabler Verlag.

Alexander, C. (1979): The Timeless Way of Building, Center for Environmental Structure, (Berkeley, California)/New York: Oxford University Press.

Alexander, C./Ishikawa, S./Silverstein, M. (1978): A Pattern Language. Towns. Buildings. Construction, Oxford: Oxford University Press.

Alexander, L. D. (1985): Successfully Implementing Strategic Decisions. In: Long Range Planning, vol. 18 (1985), no. 3, pp. 91–97.

Amit, R./Zott, C. (2010): Business Model Innovation: Creating Value in Times of Change, IESE Working Paper, No. WP-870, Barcelona: IESE Business School/University of Navarra.

Andrews, K. R. (1980): The Concept of Corporate Strategy, revised edition, New York: Irwin IL.

Ansoff, H. I. (1976): Managing Surprise and Discontinuity—Strategic Response to Weak Signals. In: Zfbf – Schmalenbachs Zeitschrift für betriebswirtschaftliche Forschung, vol. 28 (1976), pp. 129–152.

Ansoff, H. I. (1965): Corporate strategy. An analytical approach to business policy for growth and expansion, New York: McGraw-Hill.

Ansoff, H. I. (1957): Strategies for Diversification. In: Harvard Business Review, vol. 35 (1957), no. 5, Sep-Oct, pp. 113–124.

Argyris, C./Schön, D. A. (2002): Die Lernende Organisation: Grundlagen, Methode, Praxis. [The Learning Organization: Foundations, Methods, Practice], 2nd ed., Stuttgart: Klett-Cotta.

Asenkerschbaumer, S. (2012): Strategisches Controlling bei Bosch: Volatilität ist die neue Normalität [Strategic controlling at Bosch: Volatility is the new normality]. In: ZfCM – Zeitschrift für Controlling & Management, vol. 56 (2012), no. 5, pp. 336–340.

Axson, D. A.J. (2010): Best Practices in Planning and Performance Management: Radically Rethinking Management for a Volatile World, 3rd ed., Hoboken, NJ: John Wiley & Sons.

Backhaus, K./Büschken, J./Voeth, M. (2003): Internationales Marketing [International Marketing], 5th ed. Stuttgart: Schäffer-Poeschel.

Baldrige Performance Excellence Program (2015): Why Baldrige? National Institute of Standards and Technology NIST (eds.), retrieved from http://www.nist.gov/baldrige/enter/index.cfm (accessed: 8/30/2015).

Barney, J. B. (1991): Firm Resources and Sustained Competitive Advantage. In: Journal of Management, vol. 17 (1991), no. 1, pp. 99–120.

Barney, J. B. (2001): Is the Resource-based "View" a Useful Perspective for Strategic Management Research? Yes. In: Academy of Management Review, vol. 26 (2001), no. 1, pp. 41–56.

Barney, J. B./Hesterly, W. S. (2010): Strategic Management and Competitive Advantage. Concepts and Cases, 3rd edition, Upper Saddle River, NJ: Prentice Hall.

Baron, R. S. (2008): Thinking and Deciding, 4th ed., New York: Cambridge University Press.

Barry, K./ Domb, E./Slocum, M. S. (2014): TRIZ – What is TRIZ? In: http://www.triz-journal.com/triz-what-is-triz/ (accessed: 12/30/2014).

Bartlett, C. A./Beamish, P. W. (2011): Transnational Management. Text, Cases, and Readings in Cross-Border Management, 6th ed., New York: McGraw-Hill/Irwin.

Bartlett, C. A./Goshal, S. (2002): Managing Across Borders. The Transnational Solution, 2nd ed., Boston, MA: Harvard Business School Press.

Bayer AG (2015): Mission & Values, retrieved from: http://www.bayer.com/en/mission---values.aspx (accessed: 07/29/2015).

Bazerman, M. H./Moore, D. A. (2013): Judgment in Managerial Decision Making, 8th ed., Hoboken, NJ: Wiley.

Bea, F. X./Haas, J. (2009): Strategisches Management [Strategic Management], 6th ed., Konstanz/München: UVK Verlagsgesellschaft mbH.

Beck, H. (2009): Wirtschaftspolitik und Psychologie: Zum Forschungsprogramm der Behavioral Economics [Economic policy and psychology: the research program of behavioral economics]. In: ORDO. Jahrbuch für Ordnung von Wirtschaft und Gesellschaft, vol. 60 (2009), pp. 119–151.

Becker, W. M./Freeman, V. M. (2006): Going from Global Trends to Corporate Strategy. In: McKinsey Quarterly, no. 2 (2006), pp. 17–27.

Beiersdorf AG (2015): Our Profile, retrieved from: http://www.beiersdorf.com/about-us/our-profile/at-a-glance (accessed: 6/1/2015).

Beinhocker, E. D. (2006): The Adaptable Corporation. In: The McKinsey Quarterly, no. 2 (2006), pp. 77–87.

Beinhocker, E. D. (1999): Robust Adaptive Strategies. In: Sloan Management Review, vol. 40 (1999), no. 2, pp. 95–106.

Bergsmann, S. (2012): End-to-End-Geschäftsprozessmangement. Organisationselement – Integrationsinstrument – Managementansatz [End-to-end business process management. Organizational elements—integration tools—management approach], Wien: Springer.

BERI S. A. (2015): Business Environment Risk Intelligence. Risk Rating, Analysis, and Forecasts for over 140 Countries, retrieved from: http://www.beri.com (accessed: 6/19/2015).

Bertelsmann Stiftung (editor, 2006): Messen, werten, optimieren: Erfolg durch Unternehmenskultur. Ein Leitfaden für die Praxis [Measuring, evaluating, optimizing: success through corporate culture. A guideline for practical application], Gütersloh: Bertelsmann Stiftung.

Bertelsmann Stiftung (2003): Car Bertelsmann-Preis 2003: Unternehmenskultur [Corporate Culture].

Berlet-Maiweg, K./Oppermann, O. (2014): Begeisterung! Strategiearbeit ohne Weisungshierarchien [Enthusiasm! Strategy work without hierarchical command and control]. In: Horváth & Partners, 14. Jahreskonferenz Strategisches Management, Stuttgart 8. Mai 2014.

Besanko, D./Dranove, M./Shanley, M./Schaefer, S. (2010): Economics of Strategy, 5th ed., Hoboken, NJ: Wiley.

Bhimani, A./Horngren, C. T./Datar, S. M./Foster, G. (2008): Management and Cost Accounting, 4th edition, Harlow: Pearson Education Limited.

Bieger, T./Reinhold, S. (2011): Das wertbasierte Geschäftsmodell – Ein aktualisierter Strukturierungsansatz [The value-based business model—an updated structurizing approach]. In: Bieger, T./zu Knyphausen-Aufseß, D./Krys, C. (2011): Innovative Geschäftsmodelle. Konzeptionelle Grundlagen, Gestaltungsfelder und unternehmerische Praxis, Berlin/Heidelberg: Springer, pp. 13–70.

Bieger, T./Knyphausen-Aufseß, D. zu/Krys, C. (2011): Innovative Geschäftsmodelle. Konzeptionelle Grundlagen, Gestaltungsfelder und unternehmerische Praxis [Innovative business models. Conceptual foundations, design areas and company practice], Berlin/Heidelberg: Springer.

Birkenmeier, B./Brodbeck, H. (2010): Wunderwaffe Innovation. Was Unternehmen unschlagbar macht – ein Ratgeber für Praktiker, [Silver bullet innovation. What makes companies unbeatable], Zürich: Orell Füssli Verlag.

Blanke, M./Godemann, J./Herzig, C. (2007): Internetgestützte Nachhaltigkeitsberichterstattung. Eine empirische Untersuchung der Unternehmen des DAX 30 [Internet-based sustainability reporting. An empirical study of DAX 30 companies], Lüneburg: CSM.

Bleicher, K. (2011): Das Konzept integriertes Management: Visionen – Missionen – Programme [The concept of integrated management: visions—missions—programs], 8th ed., Frankfurt/Main: Campus.

Bleicher, K. (1991): Organisation. Strategien – Strukturen – Kulturen [Organization. Strategies—structures—cultures], 2nd ed., Wiesbaden: Gabler.

Bocken, N. M.P./Short, S. W./Rana, P./Evants, S. (2014): A literature and practice review to develop sustainable business model archetypes. In: Journal of Cleaner Production, vol. 65 (2014), pp. 42–56.

Boesen, T. M. (2005): Case Study: Transforming the Planning and Budgeting Process. In: Balanced Scorecard Collaborative/Palladium Corporate Performance Management Conference (May 18–20, 2005), Cambridge, MA (USA).

Boons, F./Lüdeke-Freund, F. (2013): Business Models for Sustainable Innovation: State of the Art and Steps Towards a Research Agenda. In: Journal of Cleaner Production, vol. 45 (2013), pp. 9–19.

Booz Allen Hamilton (2003): Carl Bertelsmann-Preis 2003: Unternehmensprofile [Company profiles], unpublished study, Gütersloh: Bertelsmann Stiftung.

Bossidy, L./Charan, R. (2002): Execution. The Discipline of Getting Things Done, New York: Crown Business.

Bragg, S. M. (2002): Accounting Reference Desktop, New York: John Wiley & Sons.

Brandenburger, A. M./Nalebuff, B. J. (1996): Co-opetition, New York, NY: Doubleday.

Brown, M. G. (2007): Beyond the Balanced Scorecard. Improving Business Intelligence with Analytics, New York: Productivity Press.

Bruhn, M. (2014): Marketing. Grundlagen für Studium und Praxis [Marketing. Fundamental principles for study and practice], 12th ed., Wiesbaden: Springer Gabler.

Burmeister, K./Neef, A./Beyers, B. (2004): Corporate Foresight – Unternehmen gestalten Zukunft [Corporate Foresight—Companies shape the future], Hamburg: Murmann.

Burns, T./Stalker, G. M. (1961): The management of innovation, London: Tavistock Publications.

Burr, W. (2004): Innovationen in Organisationen [Innovation in organizations], Stuttgart: Kohlhammer.

Buschmann, F./Meunier, R./Rohner, H./Sommerlad, P./Stal, M. (1996): A system of patterns: Pattern-oriented software architecture, vol. 1, Chichester: John Wiley & Sons.

Bruwer, J./Li, E./Reid, M. (2002): Segmentation of the Australian wine market using a wine-related lifestyle approach. In: Journal of Wine Research, vol. 13 (2002), no. 3, pp. 217–242.

Calabro, L. (2001): On Balance: Almost ten years after developing the balanced scorecard, authors Robert Kaplan and David Norton share what they've learned. In: CFO Magazine for Senior Finance Executives, vol. 17 (2001), no. 2, pp. 72–78.

Camerino, O. F./Chaverri, J./Anguera, M. T./Jonsson, G. K. (2012): Dynamics of the game in soccer: Detection of T-patterns. In: European Journal of Sport Science, vol. 12 (2012), no. 3, pp: 216–224.

Cameron, K. S./Quinn, R. E. (1999): Diagnosing and Changing Organizational Culture: Based on the Competing Values Framework, Reading, MA: Addison-Wesley.

Campbell, A./Goold, M. (1997): Building core skills. In: Campbell, A./Sommers Luchs, K. (eds., 1997): Core Competency-Based Strategy, London/Boston, MA: International Thomson Business Press.

Casadesus-Masanell, R./Ricart, J. E. (2010): From Strategy to Business Models and onto Tactics. In: Long Range Planning, vol. 43 (2010), no. 2–3, pp. 195–215.

Cateora, P. R./Gilly, M. C./Graham, J. L. (2009): International Marketing, 14th ed., New York, NY: McGraw-Hill.

Center for Strategic Business Wargaming (2015): Strategic Wargaming, retrieved from http://wargaming.ibu.uni-karlsruhe.de/en/strategic-wargaming (accessed: 8/29/2015).

Chandler, A. D. (1962): Strategy and structure. Chapters in the history of American industrial enterprise, Cambridge, MA: MIT Press.

Chandler, D./Werther, W. B. (2014): Strategic Corporate Social Responsibility. Stakeholders, Globalization, and Sustainable Value Creation, 3rd ed., Los Angeles, CA et al.: Sage.

Charan, R./Colvin, G. (1999): Why CEO's Fail. In: Fortune, 1999, June 21, pp 68–78.

Chesbrough, H. (2010): Business model innovation: opportunities and barriers. In: Long Range Planning, vol. 43 (2010), no. 2–3, pp: 354–363.

Choi, J./Lovallo, D./Tarasova, A. (2007): Better Strategy For Business Units: A McKinsey Global Survey. In: McKinsey Quarterly Online (July 2007).

Christensen, C. R./Andrews, K. R./Bower, J. L./Hamermesh, P./Porter, M. (1982): Business Policy: Text and Cases, 5th ed., Homewood, Il: Irwin.

Clinton, L./Whisnant, R. (2014): Model Behavior—20 Business Model Innovations for Sustainability. London: SustainAbility.

Collins, J./Porras, J. I. (2002): Built to Last. Successful Habits of Visionary Companies, New York, NY et al.: Harper Business.

Collis, D. J./Rukstad, M. G. (2008): Can You Say What Your Strategy Is? In: Harvard Business Review, vol. 86 (2008), no. 4, pp. 82–90.

Cordence Worldwide (2014): retrieved from http://www.cordenceworldwide.com/about-us.html (accessed: 9/29/2014).

Covey, S. M.R. (2008): The Speed of Trust. The One Thing that Changes Everything, New York, NY: Free Press.

Cullen, J. B./Parboteeah, K. P. (2008): Multinational Management. A Strategic Approach, 5th ed., Mason, OH: South-Western Cengage Learning.

Dammann, G. (2007): Narzissten, Egomanen, Psychopathen in der Führungsetage. Fallbeispiele und Lösungswege für ein wirksames Management [Narcissts, egomaniacs, psychopaths in the executive suite. Cases and solutions for an effective management] Bern: Haupt.

D'Aveni, R. A. (1994): Hypercompetition, New York: Free Press.

Davenport, T. H. (1993): Process Innovation—Reengineering Work through Information Technology, Boston, MA: Harvard Business Review Press.

David, F. R. (2013): Strategic Management. Concepts and Cases, global edition, 14th ed., Harlow: Pearson.

Davis, S. M. (1984): Managing Corporate Culture, Cambridge, MA: Ballinger Publishing.

DB Group (2015a): At a Glance. Facts & Figures, retrieved from http://www.deutschebahn.com/en/group/ataglance/facts_figures.html (accessed: 8/11/2015).

DB Group (2015b): Sustainability at Deutsche Bahn. Vision for 2020, retrieved from http://www.deutschebahn.com/en/sustainability/overview/vision.html (accessed: 8/11/2015).

Deal T. E./Kennedy, A. A. (1982): Corporate Cultures: The Rites and Rituals of Corporate Life, Harmondsworth: Penguin.

De Geus, A. (1988): Planning as Learning. In: Harvard Business Review, vol. 66 (1988), no. 2, pp: 70–74.

Deming, E. W. (1982): Quality, Productivity, and Competitive Position, Cambridge, MA: MIT, Center of Advance Engineering Study.

Denison, D. R. (2014): Culture and Leadership Diagnostics, retrieved from http://www.denisonconsulting.com/diagnostics (accessed: 12/11/2014).

Denison, D. R./Haaland, S./Goelzer, P. (2003): Corporate Culture and Organizational Effectiveness: Is There a Similar Pattern Around the World? In: Dorfman, P. W./Mobley, W. H. (eds.): Advances in Global Leadership, vol. 3, Oxford: Elsevier Science Ltd., pp. 205–227.

Deresky, H. (2014): International Management. Managing Across Borders and Cultures. Text and Cases, Global Edition, 8th ed., Harlow: Pearson.

Deutsche Bank Group (2012): Sustainable investing. Establishing long-term value and performance, New York, NY: Deutsche Bank Climate Change Advisors.

Dobelli, R. (2011): Die Kunst des klaren Denkens. 52 Denkfehler, die Sie besser anderen überlassen [The art of thinking clearly. 52 flaws you better leave to others], München: Carl Hanser Verlag.

Dörner, D. (1992): Die Logik des Mißlingens. Strategisches Denken in komplexen Situationen [The logic of failure. Strategic thinking in complex situations], Hamburg: Rowohlt.

Doole, I./Lowe, R. (2008): International Marketing Strategy. Analysis, development and implementation, 5th ed., Andower: Cengage Learning EMEA.

Drosdek, A. (1998): Die Visionen der deutschen Unternehmen. Unternehmensphilosophien im Zeitalter der Globalisierung. Eine Studie zum Entwicklungsstand in den großen deutschen

Unternehmen [Visions of the German companies. Company philosophies in an era of globalization. A study of state of the art in large German companies], Freiburg/Berlin/München: Haufe.
Drucker, P. F. (2008): Management. Revised and Updated by Joseph A. Maciariello, New York, NY: HarperCollins.
Drucker, P. F. (1954): The Practice of Management, New York, NY: Harper.
Dye, R./Sibony, O. (2007): How to Improve Strategic Planning. In: McKinsey Quarterly, no. 3 (2007), pp. 40–48.
Dyllick ,T./Hockerts K. (2002): Beyond the Business Case for Corporate Sustainability. In: Business Strategy and the Environment, vol. 11 (2002), no. 2, pp. 130–141.
Eccles, R. G./Krzus, M. P. (2010): One report. Integrated reporting for a sustainable strategy, Hoboken, NJ: Wiley.
EFFAS CESG (2015): http://www.effas-esg.com (accessed: 3/6/2015).
EFQM (2015): Model Criteria, retrieved from http://www.efqm.org/efqm-model/model-criteria (accessed 8/30/2015).
Elkington, J. (1997): Cannibals With Forks: The Triple Bottom Line of 21st Century Business. Oxford: Capstone.
Elkington, J./Zeitz, J. (2014): The Breakthrough Challenge. 10 Ways to Connect Today's Profits with Tomorrow's Bottom Line, San Francisco, CA: Jossey-Bass.
Ernst & Young GmbH (2013): Die größten Automobilhersteller weltweit. Eine Analyse wichtiger Bilanzkennzahlen Januar bis Dezember 2012 [The biggest automobile manufacturers globally. A ratio analysis of key performance indicators January to December 2012].
Esty D. C./Winston A. S. (2006): Green to Gold: How Smart Companies Use Environmental Strategy to Innovate, Create Value, and Build a Competitive Advantage, New Haven, CT: Yale University Press.
Eyer, E./Haussmann, T. (2014): Zielvereinbarung und variable Vergütung. Ein praktischer Leitfaden – nicht nur für Führungskräfte. Mit neun ausführlichen Fallbeispielen [Target agreements and variable payroll. A practical guide – not only for executives. With nine comprehensive case studies], Wiesbaden: Springer Gabler.
Figge, F./Hahn, T./Schaltegger, S./Wagner, M (2002): The sustainability balanced scorecard—linking sustainability management to business strategy. In: Business strategy and the Environment, vol. 11 (2002), no. 5, pp. 269–284.
Figge, F./Schaltegger, S. (2000): Was ist „Stakeholder Value"? Vom Schlagwort zur Messung [What is "stakeholder value"? From buzzword to measurement], Lüneburg: Centre for Sustainability Management e. V. and Bank Pictet in collaboration with UNEP.
Fink, C. A. (2003): Prozessorientierte Unternehmensplanung. Analyse, Konzeption und Praxisbeispiele [Process-oriented corporate planning. Analysis, conception and practical examples], PhD Thesis, Stuttgart: Universität Stuttgart.
Fink, A./Siebe, A. (2011): Handbuch Zukunftsmanagement. Werkzeuge der strategischen Planung und Früherkennung, [Handbook future management. Tools of strategic planning and early warning] Frankfurt, New York, NY: Campus.
Finkelstein, S./Hambrick, D. C./Canella Jr., A. A. (2009): Strategic Leadership. Theory and Research on Executives, Top Management Teams, and Boards, New York, NY: Oxford University Press.

Fischermanns, G. (2012): Praxishandbuch Prozessmanagement [Practical handbook process management], ibo Schriftenreihe, vol. 9, 10th ed., Gießen: Verlag Dr. Götz Schmidt.

Fleisher, C. S./Bensoussan, B. E. (2015): Business and Competitive Analysis: Effective Application of New and Classic Methods, 2nd ed., Upper Saddle River, NJ: Pearson Education.

Foster, R./Kaplan, S. (2001): Creative Destruction. Why Companies That Are Built to Last Underperform the Market—and How to Successfully Transform Them, New York, NY et al.: Currency published by Doubleday.

Frank, R. (2014): Integrierte Berichte: Gehört dem Integrated Reporting die Zukunft? [Integrated reports: Does integrated reporting have a future?]. In: Schulz, T./Bergius, S. (eds.): CSR und Finance – Beitrag und Rolle des CFO für eine Nachhaltige Unternehmensführung, Berlin/Heidelberg: Springer Gabler, pp. 237–250.

Freeman, R. E. (2010): Strategic Management. A Stakeholder Approach Cambridge, UK: Cambridge University Press.

Freeman, R. E. (1984): Strategic Management. A Stakeholder Approach, Boston, MA: Pitman.

Gälweiler, A. (2005): Strategische Unternehmensführung [Strategic management], 3rd ed., Frankfurt/New York, NY: Campus.

Gänßlen, S./Kraus, U./Ette, D. (2011): Green Controlling – Green Profit – Nachhaltigkeitscontrolling bei Hansgrohe [Green controlling—green profit—sustainability controlling at Hansgrohe]. In: Controlling, vol. 23 (2011), no. 8/9, pp. 461 f.

Gaiser, B. (2006): Strategiekonformes Verhalten in Organisationen – zur Deckung von persönlich Erwünschtem mit strategisch Gewolltem. [Strategy conform behavior in organizations—aligning individual desires to strategic intention]. In: Horváth, P. (ed.): Wertschöpfung braucht Werte. Wie Sinngebung zur Leistung motiviert, Stuttgart: Schäffer-Poeschel, pp. 155–168.

Gaiser, B./Wunder, T. (2004): Strategy Maps und Strategieprozess. Einsatzmöglichkeiten, Nutzen, Erfahrungen [Strategy maps and strategy process. Application areas, benefits, experiences]. In: Controlling, vol. 16 (2004), no. 8/9, pp. 457–463.

Gaitanides, M. (2012): Prozessorganisation. Entwicklung, Ansätze und Programme des Managements von Geschäftsprozessen [Process organization. Development, approaches and programs of business processes management]. 3rd ed., München: Vahlen.

Gassmann, O./Frankenberger, K./Csik, M. (2014): The Business Model Navigator. 55 Models That Will Revolutionise Your Business, London/New York: FT Publishing, Pearson.

Ghauri, P. N./Cateora, P. (2010): International Marketing, 3rd ed., Maidenhead: McGraw Hill.

Ghemawat, P. (2001): Distance Still Matters. The Hard Reality of Global Expansion. In: Harvard Business Review, vol. 79 (2001), no. 5, pp. 137–147.

Gigerenzer, G. (2010): Bauchentscheidungen. Die Intelligenz des Unbewussten [Gut Feelings. The intelligence of the unconscious], original lecture, Jokers Hörsaal, Bernd Ulrich (ed.), Müllheim/Baden: Auditorium Netzwerk.

Gigerenzer, G. (2007): Gut Feelings. The Intelligence of the Unconscious, New York, NY: Viking.

Gigerenzer, G./Brighton, H. (2009): Homo heuristicus: Why biased minds make better interferences. In: Topics in Cognitive Science, no. 1 (2009), pp. 107–143.

Gigerenzer, G./Todd, P. M./ABC Group (1999): Simple Heuristics That Make Us Smart, New York, NY: Oxford University Press.

Gilbert, X./Strebel, P. (1987): Outpacing Strategies. In: Journal of Business Strategy, vol. 8 (summer 1987), pp. 28–36.

Glauner, R. (2013): CSR und Wertecockpits. Mess- und Steuerungssysteme der Unternehmenskultur [CSR and value cockpits. Measurement and control systems for corporate culture], Berlin/Heidelberg: Springer-Gabler.

Gleich, R./Kopp, J. (2001): Ansätze zur Neugestaltung der Planung und Budgetierung [Approaches for a redesign of planning and budgeting]. In: Controlling, vol. 13 (2001), no. 8/9, pp. 429–436.

Gleich, R./Kopp, J./Leyk, J. (2003): Advanced Budgeting: better and beyond. In: Horváth, P./Gleich, R. (eds.): Neugestaltung der Unternehmensplanung. Innovative Konzepte und erfolgreiche Praxislösungen, Stuttgart: Schäffer-Poeschel.

Gleich, R./Temmel, P. (2008): Die Rolle und Organisation des Controllings im Management Reporting [Role and organization of controlling in management reporting]. In: Gleich, R./Horváth, P./Michel, U. (eds.): Management Reporting: Grundlagen, Praxis und Perspektiven, München: Haufe, pp. 63–91.

Global Reporting Initiative (2015a): G4 Sustainability Reporting Guidelines. Reporting Principles and Standard Disclosures, Amsterdam: Global Reporting Initiative, retrieved from: https://www.globalreporting.org/standards/g4/Pages/default.aspx (accessed: 7/12/2015).

Global Reporting Initiative (2015b): G4 Sustainability Reporting Guidelines. Implementation Manual, Amsterdam: Global Reporting Initiative, retrieved from: https://www.globalreporting.org/standards/g4/Pages/default.aspx (accessed: 7/12/2015).

Gomez, P./Probst, G. (1999): Die Praxis des ganzheitlichen Problemlösens [The practice of holistic problem solving]. 3rd ed., Bern: Haupt.

Gomez, P./Probst, G. (1987): Vernetztes Denken im Management. Eine Methodik des ganzheitlichen Problemlösens [Network thinking in management. A method of holistic problem solving]. In: Die Orientierung, Nr. 89, Bern: Schweizer Volksbank.

Google (2015): Company, retrieved from: http://www.google.com/about/company/ (accessed: 5/22/2015).

Goold, M./Campbell, A./Alexander, M. (1994): Corporate-Level Strategy. Creating Value in the Multibusiness Company, New York: John Wiley & Sons.

Gräf, J./Nase, D. (2008): Objekte des Management Reportings im Überblick [Objects of management reporting at a glance]. In: Gleich, R./Horváth, P./Michel, U. (eds.): Management Reporting: Grundlagen, Praxis und Perspektiven, München: Haufe, pp. 43–61.

Grant, R.M (2013): Contemporary Strategy Analysis. Text and Cases, 8th ed., Chichester (UK): Wiley.

Grant, R.M (2010): Contemporary Strategy Analysis. Text and Cases, 7th ed., Chichester (UK): Wiley.

Grant, R. M. (1997): The knowledge-based view of the firm: implications for management practice. In: Long Range Planning, vol. 30 (1997), no. 3, pp. 450–454.

Greiner, O. (2006): Musterbasierte Strategieentwicklung [Pattern-based strategy development]. In: Controlling: Zeitschrift für erfolgsorientierte Unternehmenssteuerung, vol. 18 (2006), no. 11, pp. 611 f.

Greiner, O. (2004): Strategiegerechte Budgetierung [Strategy-oriented budgeting]. München: Vahlen.

Greiner. O./Wolf, T. (2010): Das 7-K-Prinzip. Geschäftsmodelle gestalten – Strategien entwickeln [The 7-C-principle. Designing business models—developing strategies], Horváth & Partner GmbH (eds.), Stuttgart: Whitepaper.

GRI G4 (2014): https://www.globalreporting.org (accessed: 3/6/2015).

Grimm, U. (1996): Evolutorische Unternehmensführung. Eine programmatische Skizze [Evolutionary corporate management. A programmatic outline]. In: European Business School (ed.): Erfahrung-Bewegung-Strategie, Wiesbaden: DUV, pp. 317–343.

Hahn, D./Kaufmann, L. (2002): Praktische Probleme bei der Einführung und Nutzung von Balanced Scorecards [Practical challenges when introducing and using balanced scorecards]. In: Albach, H./Kaluza, B./Kersten, W. (eds.): Wertschöpfungsmanagement als Kernkompetenz, Wiesbaden: Festschrift für Horst Wildemann, pp. 527–542.

Hamel, G. (1996): Strategy as Revolution. In: Harvard Business Review, vol. 74 (1996), no. 4, pp. 69–82.

Hamel, G./Prahalad, C. K. (1994): Competing for the Future, Boston: Harvard Business School Press.

Hammer, M. (2001): The Agenda. What Every Business Must Do to Dominate the Decade, London: Random House.

Hammer, M./Champy, J. (1993): Reengineering the Corporation. A Manifesto for Business Revolution, New York: HarperCollins.

Handy, C. B. (1976): Understanding Organizations, Oxford: University Press.

Hansen, E. G./Schaltegger, S. (2014). The Sustainability Balanced Scorecard: A Systematic Review of Architectures. Journal of Business Ethics. 10.1007/s10551-014-2340-3.

Hanser, Peter (2009): Keine Kompromisse bei der Kaffeequalität, Marken-Vitalisierung: Interview mit Holger Feldmann [No compromise on the coffee quality, brand vitalization: Interview with Holger Feldmann] In: Absatzwirtschaft, spezial edition 2009, pp. 56–60.

Harringer, C./Maier, H. (eds., 2011): Change Communications Jahrbuch 2011, [Change Communications Yearbook] Berlin/Heidelberg: Springer.

Harrison, R. (1972): Understanding your Organizational Character. In: Harvard Business Review, vol. 50 (1972), no. 3, pp. 119–128.

Harvard Business Review (2014): http://hbr.org/authors/kaplan-norton (accessed: 7/31/2014).

Harvard Business School (2015): Robert S. Kaplan, retrieved from http://www.hbs.edu/faculty/Pages/profile.aspx?facId=6487&click=bestbet (accessed: 11/14/2015).

Haßler, R./Häßler, R. D. (2014): Kapitalbeschaffung: Die Bedeutung von Nachhaltigkeitsratings, -indizes und -fonds für die Emission von Aktion und Anleihen [Capital acquisition: the importance of sustainability ratings, indices and fonds for emission of stocks]. In: Schulz, T./Bergius, S. (eds.): CSR und Finance – Beitrag und Rolle des CFO für eine Nachhaltige Unternehmensführung, Berlin/Heidelberg: Springer Gabler, pp. 253–262.

Hauschildt, J./Salomo, S. (2007): Innovationsmanagement [Innovation management], 4th ed., München: Vahlen.

Hax, A. C. (2009): The Delta Model. Reinventing Your Business Strategy, New York: Springer Science+Business Media.

Hax, A. C./Wilde, D.L. (2001): The Delta Project: Discovering New Sources of Profitability in a Networked Economy, New York: Palgrave.

Hax, A. C./Majluf, N. S. (1996): The Strategy Concept and Process. A Pragmatic Approach, 2nd ed., Upper Saddle River, NJ: Prenctice Hall.

Hax, A. C./Majluf, N. S. (1991): Strategisches Management. Ein integratives Konzept aus dem MIT [Strategic management. An integrative concept from the MIT], überarbeitete Studienausgabe, Frankfurt: Campus Verlag.

Helm, R./Meiler, R. C. (2003): Unternehmensvision, Interne Kommunikation und Effizienz des Wissensmanagements [Company vision, internal communication and efficiency of knowledge management]. In: Controlling, vol. 15 (2003), no. 3/4, pp. 201–207.

Hengstmann, R./Seidel, S. D. (2014): Berechnung externer Umweltkosten: Die Ökologische Gewinn- und Verlustrechnung von PUMA [Calculation of external environmental cost: The environmental profit and loss calculation at Puma]. In: Schulz, T./Bergius, S. (eds.): CSR und Finance – Beitrag und Rolle des CFO für eine Nachhaltige Unternehmensführung, Berlin/Heidelberg: Springer Gabler, pp. 179–191.

Henkel (2015a): Company profile, retrieved from: http://www.henkel.com/press-and-media/specials/company-profile (accessed: 5/22/2015).

Henkel (2015b): Vision & Values, retrieved from: http://www.henkel.com/company/corporate-culture (accessed: 10/31/2015).

Henkel (2015c): Vision & Values, retrieved from: http://www.henkel.com/company/corporate-culture/vision-and-values (accessed: 5/22/2015).

Henzler, H. (1988): Vision und Führung [Vision and leadership]. In: Henzler, H. A.: Handbuch Strategische Führung, Wiesbaden: Gabler, pp. 17–33.

Hess, T./Schuller, D. (2005): Business Process Reengineering als nachhaltiger Trend? Eine Analyse der Praxis in deutschen Großunternehmen nach einer Dekade [Business process reengineering as sustained trend? An analysis of company practice in large German corporations after one decade]. In: Zeitschrift für betriebswirtschaftliche Forschung, vol. 57 (2005), no. 5, pp. 355–373.

Hill, C. W.L./Jones, G. R. (2010): Strategic Management Theory: An Integrated Approach, 9th ed., Mason, OH: South-Western Cengage Learning.

Hinterhuber, H. H. (2015): Strategische Unternehmensführung. Das Gesamtmodell für nachhaltige Wertsteigerung [Strategic management. The complete model for sustained value increase], 9th completely new and revised ed., Berlin: Erich Schmidt Verlag.

Hinterhuber, H. H. (2004): Strategische Unternehmensführung: I. Strategisches Denken [Strategic management. I. Strategic Thinking], 7th completely new and revised ed., Berlin: De-Gruyter.

Hinterhuber, H. H. (1997): Strategische Unternehmungsführung. II. Strategisches Handeln Strategic management. 1. Strategic action], 6th completely new and revised ed., Berlin, New York: de Gruyter.

Hitt, M. A./Ireland, R. D./Hoskisson, R. E. (2011): Strategic Management: Competitiveness and Globalization: Concepts, 9th ed., Mason, Ohio: South-Western Cengage Learning.

Hobbs, B. (2007): The Multi-Project PMO: a global analysis of the current state of practice, Montreal, Canada: White Paper prepared for the Project Management Institute, University of Quebec.

Holtbrügge, D./Welge, M. K. (2010): Internationales Management. Theorien, Funktionen, Fallstudien [International management. Theories, functions, case studies], 5th ed., Stuttgart: Schäffer-Poeschel.

Homburg, C./Kuester, S./Krohmer, H. (2013): Marketing Management. A Contemporary Perspective, 2nd ed., Maidenhead: McGraw-Hill Education.

Homburg, C./Kuester, S./Krohmer, H. (2009): Marketing Management. A Contemporary Perspective, Maidenhead: McGraw-Hill Education.

Hope, J./Frazer, R. (2003): Beyond Budgeting: How Managers Can Break Free from the Annual Performance Trap, Boston, MA: Harvard Business Review Press.

Horngren, C. T./Sundem, G. L./Selto, F. H. (1993): Introduction to Management Accounting, 9th edition, Englewood Cliffs, NJ: Prentice-Hall.

Horváth, P. (2011): Controlling, 12th edition, München: Vahlen.

Horváth, P./Sauter, R. (2004): Why Budgeting Fails: One Management System Is Not Enough. In: Balanced Scorecard Report, vol. 6 (2004), no. 5, pp. 3–6.

Horváth & Partner GmbH (eds., 2014): Strategiestudie 2013/2014. 12 Schlüssel erfolgreicher Strategiearbeit. Studienbericht [Strategy Study 2013/14. 12 key factors for successful strategy work. Evaluation report], Competence Center Strategy, Innovation & Sales, Stuttgart.

Horváth & Partner GmbH (eds., 2008): Balanced Scorecard Studie 2008. Ergebnisbericht [Balanced Scorecard Study 2008. Evaluation report], Competence Center Strategic Management & Innovation, Stuttgart.

Horváth & Partners (2008): "Strategic Process Alignment": Strategie ist ein Prozessthema ["Strategic Process Alignment": Strategy is a process topic]. In: Horváth AG: The Performance Architect. Das Kundenmagazin für Deutschland, Österreich und die Schweiz, 12th ed., Februar 2008, pp. 4–6.

Horváth & Partners (eds., 2007): Balanced Scorecard umsetzen [Implementing balanced scorecard], 4th ed., Stuttgart.

Huber, A. (2006): Strategische Planung in deutschen Unternehmen: Empirische Untersuchung von über 100 Unternehmen [Strategic planning in German companies: Empirical study of over 100 companies], Berlin: Phius.

Humphrey, A. S. (2005): SWOT Analysis for Management Consulting. In: SRI International, December 2005 Newsletter, pp. 7 f.

Hungenberg, H. (2011): Strategisches Management in Unternehmen. Ziele – Prozesse – Verfahren [Strategic management in companies. Objectives—processes—methods], 6th ed., Wiesbaden: Gabler.

Hunger, J. D./Wheelen, T. L. (2007): Essentials of Strategic Management, 4th ed., Upper Saddle River, NJ: Pearson Prentice Hall.

Huselid, M. A./Beatty, R. W./Becker, B. E. (2005): "A Players" or "A Positions"? The Strategic Logic of Workforce Management. In: Harvard Business Review, vol. 83 (2005), no. 12, pp. 110–117.

ICV (ed., 2011): Green Controlling – eine (neue) Herausforderung für das Controlling? Relevanz und Herausforderungen der Integration ökologischer Aspekte in das Controlling aus Sicht der Controllingpraxis [Green Controlling—a (new) challenge for Controlling? Relevance and challenges of integrating ecological aspects into managerial control from a Controlling practice perspective], Ergebnisse einer Studie im Internationalen Controller Verein (ICV) durch die ICV-Ideenwerkstatt, Gauting/Stuttgart.

IGC—International Group of Controlling (eds., 2011): Controlling Process Model. A Guideline for Describing Designing Controlling Processes, Freiburg: Haufe.

ISO 26000 (2015): http://www.iso.org/iso/home/standards/iso26000.htm (accessed: 3/6/2015).

ITIL—IT Infrastructure Library (2014): http://www.itil-officialsite.com/ AboutITIL/ WhatisITIL.aspx (accessed 10/20/2014).

Jensen, M. C. (1989): Eclipse of the Public Corporation. In: Harvard Business Review, vol. 67 (1989), no. 5, pp. 61–74.

Jochem, R./Landgraf, K. (2010): Die Prozessorganisation [The process organization]. In: Jochem, R./Mertins, K./Knothe, T. (eds.): Prozessmanagement. Strategien, Methoden, Umsetzung, Düsseldorf: Symposion Publishing.

Johnson, M. W. (2010): Seizing the white space. Business model innovation for growth and renewal, Boston, MA: Harvard Business Press.

Johnson, M. W./Christensen, C. M./Kagermann, H. (2008): Reinventing your business model. In: Harvard Business Review, vol. 86 (2008), no. 12, pp. 50–59.

Johnson, G./Whittington, R./Scholes, K. (2012): Fundamentals of Strategy, 2nd edition, Harlow, Essex: Pearson Education Limited.

Kagermann, H./Österle, H. (2007): Geschäftsmodelle 2010. Wie CEOs Unternehmen transformieren [Business models 2010. How CEOs transform companies], 2nd ed., Frankfurt am Main: Frankfurter Allgemeine Buch.

Kahneman, D. (2011): Thinking, Fast and Slow, New York, NY: Farrar, Straus and Giroux.

Kahneman, D./Lovallo, D./Sibony, O. (2011): Before you make that big decision. In: Harvard Business Review, vol. 89 (2011), no. 6, pp. 50–60.

Kano, N./Seraku, N./Taxahashi, F./Tsjuji, S. (1984): Attractive quality and must-be quality. Hinshitsu (Quality: The Journal of the Japanese Society for Quality Control), vol. 14 (1984), no. 2 (April), pp. 39–48.

Kaplan, R. S. (1994): Devising a Balanced Scorecard Matched to Business Strategy. In: Planning Review (September-October 1994), no. 5, pp. 15–48.

Kaplan, R. S./Norton, D. P. (2008): The Execution Premium, Boston, Massachusetts: Harvard Business Review Press.

Kaplan, R. S./Norton, D. P. (2006): Alignment, Boston, Massachusetts: Harvard Business Review Press.

Kaplan, R. S./Norton, D. P. (2004): Strategy Maps. Converting Intangible Assets Into Tangible Outcomes, Boston, Massachusetts: Harvard Business Review Press.

Kaplan, R. S./Norton, D. P. (2001): The Strategy-Focused Organization. How Balanced Scorecard Companies Thrive in the New Business Environment, Boston, Massachusetts: Harvard Business Review Press.

Kaplan, R. S./Norton, D. P. (1996): The Balanced Scorecard. Translating Strategy Into Action, Boston, Massachusetts: Harvard Business Review Press.

Kaplan, R. S./Norton, D. P. (1993): Putting the Balanced Scorecard to Work. In: Harvard Business Review, vol. 71 (1993), no. 5, pp. 134–147.

Kaplan, R. S./Norton, D. P. (1992): The Balanced Scorecard—Measures That Drive Performance. In: Harvard Business Review, vol. 70 (1992), no. 1, pp. 71–79.

Kellinghusen, G. (2012): RECARO Group – Modernes Unternehmertum und Steuerung eines familiengeführten Markenunternehmens [RECARO Group—modern entrepreneurship and controlling of a family managed branded company]. In: Horváth, P./Michel, U. (eds.): Controlling und Finance. Steuerung im volatilen Umfeld, Stuttgart: Schäffer-Poeschel, pp. 113–124.

Kerth, K./Asum, H./Stich, V. (2009): Die besten Strategietools in der Praxis. Welche Werkzeuge brauche ich wann? Wie wende ich sie an? Wo liegen die Grenzen? [The best strategy tools for practice. Which tools do I need and when? How do I use them? Where are the limits?], 4th extended ed., München: Hanser.

Kets de Vries, M. F.R. (2004): Führer, Narren und Hochstapler: Die Psychologie der Führung [Leaders, fools and frauds: The psychology of leadership], 2nd ed., Stuttgart: Klett-Cotta.

Kieninger, M. (1998): Reengineering und Prozeßoptimierung [Reengineering and process optimization]. In: Horváth & Partner GmbH (eds.): Prozesskostenmanagement. Methodik und Anwendungsfelder, 2nd ed., München: Vahlen, pp. 29–45.

Kim, W. C./Mauborgne, R. (2005): Blue Ocean Strategy. How to Create Uncontested Market Space and Make the Competition Irrelevant, Boston, MA: Harvard Business School Press.

Kirsch, W. (1997): Strategisches Management: die geplante Evolution von Unternehmen [Strategic management: The planned evolution of companies], Münchener Schriften zur angewandten Führungslehre 88, Herrsching: Kirsch.

Kirsch, W. (1991): Unternehmenspolitik und strategische Unternehmensführung [Corporate policy and strategic management], 2nd ed., Münchener Schriften zur angewandten Führungslehre 60, Herrsching: Kirsch.

Kleinfeld, A./Martens, A. (2014): Transparenz: Berichterstattung über Nachhaltigkeitsleistungen [Transparency: Reporting of sustainability performance]. In: Schulz, T./Bergius, S. (eds.): CSR und Finance – Beitrag und Rolle des CFO für eine Nachhaltige Unternehmensführung, Berlin/Heidelberg: Springer Gabler, pp. 219–235.

Knorr, A./Arndt, A. (2003): Why did Wal-Mart fail in Germany? In: Knorr, A./Lemper, A./Sell, A./Wohlmuth, K: (eds.): Materialien des Wissenschaftsschwerpunktes „Globalisierung der Weltwirtschaft“, vol. 24 (June 2003), Universität Bremen: IWIM – Institut für Weltwirtschaft und Internationales Management.

Kotabe, M./Helsen, K. (2008): Global Marketing Management, 4th ed., Hoboken, NJ: Wiley.

Kotler, P./Armstrong, G. (2012): Principles of Marketing, 14th ed., Upper Saddle River, NJ: Pearson Education.

Kotler, P./Berger, R./Bickhoff, N. (2010): The Quintessentials of Strategic Management. What You Really Need to Know to Survive in Business, Berlin/Heidelberg: Springer.

Kotler, P./Keller, K. L. (2009): Marketing Management, 13th ed., Upper Saddle River, NJ: Pearson Prentice Hall.

Kuhn, T. S. (1970): The Structure of Scientific Revolutions, 2nd ed., Chicago, IL: University of Chicago Press.

Kumar, V./Stam, A./Joachimsthaler, E. A. (1994): An Interactive Multicriteria Approach to Identifying Potential Foreign Markets. In: Journal of International Marketing, vol. 2 (1994), no. 1, pp. 29–52.

Lafley, A. G./Martin, R. L. (2013): Playing to Win. How Strategy Really Works, Boston, MA: Harvard Business Review Press.

Latham, G. P./Locke, E. A. (1995): Zielsetzung als Führungsaufgabe [Goal setting as leadership task]. In: Kieser, A. (eds.): Handwörterbuch der Führung, 2nd ed., Stuttgart: Schäffer-Poeschel, pp. 2222–2234.

Latham, G. P./Locke, E. A. (1991): Self-Regulation through Goal Setting. In: Organizational Behavior and Human Decision Processes, vol. 50 (1991), pp. 212–247.

Learned, E. P./Christensen, C. R./Andrews, K. R./Guth, W. D. (1965): Business Policy: Text and Cases, Homewood, IL: Irwin.

Levicki, C. (1996): The Strategy Workout. A Journey to the Heart of Your Business, London: Pitman Pub.

Likert, R. (1967): The Human Organization: Its Management and Value, New York: McGraw-Hill.

Lindgardt, Z./Reeves, M./Stalk, G./Deimler, M. S. (2009): Business Model Innovation. When the Game Gets Tough, Change the Game, Whitepaper, The Boston Consulting Group, December 2009.

Lombriser, R./Abplanalp, P. A. (2004): Strategisches Management. Visionen entwickeln. Strategien umsetzen. Erfolgspotenziale aufbauen [Strategic management. Developing visions. Implementing strategies], 3rd ed., Zürich: Versus.

Loew, T./Clausen, J. (2010): Wettbewerbsvorteile durch CSR. Eine Metastudie zu den Wettbewerbsvorteilen von CSR und Empfehlungen zur Kommunikation an Unternehmen [Competitive advantages through CSR. A meta study of competitive advantages of CSR and recommendations to companies for communication], Berlin/Hannover: Institute 4 Sustainability.

Löbbert, S. (2014): Bonitätsanalyse: Bedeutung von Nachhaltigkeit in den Kreditanforderungen der Banken. [Analysis of creditworthiness: Relevance of sustainability in credit requirements of banks]. In: Schulz, T./Bergius, S. (eds.): CSR und Finance, CSR und Finance – Beitrag und Rolle des CFO für eine Nachhaltige Unternehmensführung, Berlin/Heidelberg: Springer Gabler, pp. 263–274.

Lovallo, D./Sibony, O. (2010): The case for behavioral strategy. In: McKinsey Quarterly, March (2010), pp. 1–16.

Lüdeke-Freund, F. (2013): Business Models for Sustainability Innovation: Conceptual Foundations and the Case of Solar Energy, PhD Thesis, Lüneburg: Leuphana University.

Lüdeke-Freund, F. (2010): Towards a Conceptual Framework of 'Business Models for Sustainability' (September 19, 2010). Knowledge Collaboration & Learning for Sustainable Innovation. Wever, R./Quist, J./Tukker, A./Woudstra, J./Boons, J./Beute, N. (eds.), Delft, 2010 ERSCP-EMSU Conference, The Netherlands, October 25–29, 2010. Available at SSRN: http://ssrn.com/abstract=2189922.

Lynch, R. (2012): Strategic Management, 6th ed., Harlow, Essex: Pearson.

Macharzina, K. /Wolf, J. (2010): Unternehmensführung: Das internationale Managementwissen. Konzepte – Methoden – Praxis [Corporate management. The international management knowledge. Concepts–Methods–Practice], 7th ed., Wiesbaden: Gabler.

Malik, F. (2008): Strategie des Managements komplexer Systeme: Ein Beitrag zur Management-Kybernetik evolutionärer Systeme [Strategy of management of complex systems: A contribution for management cybernetics of evolutionary systems], 10th ed., Bern: Haupt.

March, J. G./Simon, H. A. (1958): Organizations. New York, NY: Wiley.

Maucher, H. (2007): Management breviary. A guideline to corporate success, Frankfurt/New York: Campus.

Mayer, R. (2005): Prozessmanagement. Erfolg durch Steigerung der Prozessperformance [Process management. Success through increasing process performance]. In: Horváth & Partners (eds.): Prozessmanagement umsetzen. Durch nachhaltige Prozessperformance Umsatz steigern und Kosten senken, Stuttgart: Schäffer-Poeschel, pp. 1–6.

Mayer, R. (1998): Kapazitätskostenrechnung. Neukonzeption einer kapazitäts- und prozeßorientierten Kostenrechnung [Capacity costing. New design of capacity- and process-oriented cost accounting], München: Vahlen.

Mayer, R./Brenner, M. (2005): Das Horváth & Partners-Konzept zur Umsetzung des Prozessmanagements [The Horváth & Partners concept for implementing process management]. In: Horváth & Partners (eds.): Prozessmanagement umsetzen, a. a. O., pp. 17–23.

McDermott, B./Snabe, J. H. (2013): Letter from the Co-CEOs: SAP 2012 Annual Report: Helping the World Run Better, Walldorf, p. 8.

McKinsey, J. O. (1922): Budgetary Control, New York: The Ronald Press Company.

Meffert, H./Bolz, J. (1998): Internationales Marketing-Management [International Marketing Management], 3rd revised and supplemented ed., Stuttgart: Kohlhammer.

Merriam-Webster (2014): "pattern". In: Merriam-Webster.com; http://www.merriam-webster.com/dictionary/pattern (accessed 12/28/2014).

Mikel, H. J./Schroeder, R. (2005): Six sigma: the breakthrough management strategy revolutionizing the world's top corporations, New York, NY: Random House.

Mintzberg, H. (2000): The Rise and Fall of Strategic Planning, Harlow: Pearson Education Limited.

Mintzberg, H. (1987): The Strategy Concept I: Five Ps For Strategy. In: California Management Review, vol. 30 (fall 1987), pp. 11–24.

Mintzberg, H. (1978): Patterns in Strategy Formation. In: Management Science, 24/9 (1978), pp. 934–948.

Mintzberg, H./Ahlstrand, B./Lampel, J. (1998): Strategy Safari: a guided tour through the wilds of strategic management, New York, NY: Free Press.

Mitchell, D./Coles, C. (2003): The ultimate competitive advantage of continuing business model innovation. In: Journal of Business Strategy, vol. 24 (2003), no. 5, pp. 15–21.

Mitchell, D./Coles, C. (2004): Business model innovation breakthrough moves. In: Journal of Business Strategy, vol. 25 (2004), no. 1, pp. 16–26.

Möller, K./Menninger, J./Robers, D. (2011): Innovationscontrolling. Erfolgreiche Steuerung und Bewertung von Innovationen [Innovation Controlling. Successful control and evaluation of innovations], Stuttgart: Schäffer-Poeschel.

Montier, J. (2007): Behavioural investing. A practicioner's guide to applying behavioural finance, Chichester: John Wiley & Sons.

Morschett, D./Schramm-Klein, H./Zentes, J. (2010): Strategic International Management. Text and Cases, 2nd ed., Wiesbaden: Gabler.

Müller, A. W. (2008): Strategic Foresight – Prozesse strategischer Trend- und Zukunftsforschung in Unternehmen [Strategic foresight—processes of strategic trend and future research], PhD Thesis, St. Gallen: Universität St. Gallen (HSG).

Müller, A. W./Müller-Stewens, G. (2009): Strategic Foresight: Trend- und Zukunftsforschung in Unternehmen – Instrumente, Prozesse, Fallstudien [Strategic foresight: Trend and future research in companies—tools, processes, case studies], Stuttgart: Schäffer-Poeschel.

Müller-Stewens, G./Lechner, C. (2011): Strategisches Management. Wie strategische Initiativen zum Wandel führen [Strategic management. How strategic initiatives lead to change], 4th ed., Stuttgart: Schäffer-Poeschel.

Murray, K. (2014): Communicate to inspire. A guide for leaders. London: Kogan Page Limited.

Nagel, P. (1992): Techniken der Zielformulierung [Techniques of goal formulation]. In: Frese, E. (ed.): Handwörterbuch der Organisation, 3rd ed., Stuttgart: Schäffer-Poeschel.

Nagji, B./Tuff, G.: Managing Your Innovation Portfolio. In: Harvard Business Review, vol. 90 (2012), no. 5, pp. 66–74.

Naujoks, O./Gogoll, A. (2006): Geschäftsprozessmanagement in der Siemens AG. Konzept–Umsetzung–Assessment [Business process management in the Siemens AG. Concept– Implementation–Assessment], edited by H. Ellringmann, München: Carl Hanser Verlag.

Nelson, T. (2003): The persistence of founder influence: Management, ownership, and performance effects at initial public offering. In: Strategic Management Journal, vol. 24 (2003), no. 8, pp. 707–724.

Nespresso (2014): https://www.nespresso.com (accessed December 2014).

Nespresso/CIMS Sustainable Markets Intelligence Center/INCAE Business School (2014): Background on Nestlé Nespresso S. A., Nespresso Sustainability MBA Challenge 2014. Pushing the Boundaries of Sustainable Quality Coffee Sourcing (retrieved 2/8/2015 from http://sustainabilitymbachallenge.com/wp-content/uploads/2012/09/Backgrounder-2014.pdf).

Nestlé Nespresso (2014): http://www.nestle-nespresso.com (accessed December 2014).

Nicolai, C. (2009): Betriebliche Organisation [Corporate organization], Stuttgart: Lucius & Lucius.

Nidumolu, R./Prahalad, C. K./Rangaswami, M. R. (2009): Why Sustainability is Now the Key Driver of Innovation. In: Harvard Business Review, vol. 87 (2009), no. 9, pp. 56–64.

Niven, P. R. (2002): Balanced Scorecard Step-by-Step. Maximizing Performance and Maintaining Results, New York: John Wiley & Sons.

Noe, R. A./Hollenbeck, J. R./Gerhart, B./Wright, P. M. (2009): Fundamentals of Human Resource Management, 3rd edition, New York, NY: McGraw-Hill /Irwin.

Nørreklit, H. (2000), "The balance on the balanced scorecard: a critical analysis of its assumptions", Management Accounting Research, vol. 11 (2000), no. 1, pp. 65–88.

Norton, D. P. (2006): Linking Planning and Budgeting to Strategy, Balanced Scorecard Collaborative/Palladium. In: Balanced Scorecard Collaborative/Palladium Executive Conference, Miami, FL (USA), January 25–27, 2006.

O'Riordan, L./Fairbrass, J. (2008): Corporate Social Responsibility (CSR): Models and Theories in Stakeholder Dialogue. In: Journal of Business Ethics, vol. 83 (2008), no. 4, pp. 745–758.

Ohmae, K. (1982): The Mind of the Strategist. The Art of Japanese Business, New York, NY et al.: McGraw-Hill.

Osterloh, M./Frost, J. (2006): Prozessmanagement als Kernkompetenz. Wie Sie Business Reengineering strategisch nutzen können [Process management as core competence. How you can use business reengineering strategically], 5th ed., Wiesbaden: Gabler.

Osterwalder, A./Pigneur, Y. (2010): Business Model Generation, Hoboken, NJ: self published.

Osterwalder, A/Pigneuer, Y./Tucci, C. L. (2005): Clarifying business models: origins, present, and future of the concept. In: Communications of the Association of Information Systems, vol. 15 (May 2005).

Pascale, R./Athos, A. (1981): The Art of Japanese Management, London: Penguin Books.

Perlitz, M./Schrank, R. (2013): Internationales Management [International management], 6th ed., Konstanz, München: UVK.

Peteraf, M. A. (1993): The Cornerstones of Competitive Advantage. A Resource-Based View. In: Strategic Management Journal, vol. 14 (1993), no. 3, pp. 179–191.

Peters, T./Waterman, R. (1982): In Search of Excellence. Lessons from America's Best-Run Companies, New York, London: Harper & Row.

Pillkahn, U. (2008): Using Trends and Scenarios as Tools for Strategy Development. Shaping the Future of Your Enterprise, Erlangen: Publicis Corporate Publishing.

Pohl, R. F. (ed., 2012): Cognitive illusions. A handbook on fallacies and biases in thinking, judgement and memory, Hove and New York: Psychology Press.

Porter, M. E. (2008): The Five Competitive Forces That Shape Strategy. In: Harvard Business Review, vol. 86 (2008), no. 1, pp. 79–93.

Porter, M. E. (1996): What is Strategy? In: Harvard Business Review, vol. 74 (Nov./Dec. 1996), pp. 61–78.

Porter, M. E. (1985): Competitive Advantage. Creating and Sustaining Superior Performance, New York, NY: The Free Press.

Porter, M. E. (1980): Competitive Strategy. Techniques for analyzing industries and competitors, New York, NY: The Free Press.

Porter, M. E./Kramer, M. R. (2011): Creating Shared Value. In: Harvard Business Review, vol. 89 (2011), no. 1, pp. 62–77.

Powell, T. C./Lovallo, D./Fox, C. R. (2011): Behavioral strategy. In: Strategic Management Journal, vol. 32 (2011), no. 13, pp. 1369–1386.

Prahalad, C. K. (2005): The Fortune at the Bottom of the Pyramid. In: Eradicating Poverty through Profits, Upper Saddle River, NJ: Wharton School Publishing.

Prahalad, C. K./Doz, Y. (1987): The Multinational Mission, Balancing Local Demands and Global Vision, New York, NY: The Free Press.

Prahalad, C. K./Hamel, G. (1990): The core competence of the corporation. In: Harvard Business Review, vol. 68 (1990), no. 3, pp. 79–91.

Prahalad, C. K./Hart, S. L. (2002): The fortune at the bottom of the pyramid. In: Strategy and Business (Reprint), vol. 26 (2002), no. 1, pp. 1–14.

Puma (2015): History, retrieved from http://about.puma.com/en/this-is-puma/history (accessed: 5/21/2015).

PWC (2015): Total Impact Measurement and Management, retrieved from http://www.pwc.com/gx/en/services/sustainability/publications/total-impact-measurement-management.html (accessed: 10/27/2015).

Qiagen (2015): Customers, retrieved from https://www.qiagen.com/us/about-us/who-we-are/customers/ (accessed: 8/20/2015)

Rainey, D. L. (2010): Enterprise-wide Strategic Management. Achieving Sustainable Success through Leadership, Strategies, and Value Creation, Cambridge: Cambridge University Press.

Rappaport, A. (1986): Creating Shareholder Value: The New Standard for Business Performance, New York, NY: Free Press.

Rausch, A. (2013): Analyse ist gut, Intuition ist besser – Oder umgekehrt? [Analysis is god, intuition is better—or vice versa?]. In: Controlling & Management Review, vol. 57 (2013), no. 3, pp. 14–22.

RECARO 2014: http://www.recaro.com/en/the-world-of-recaro/our-world.html and /tradition.html; http://www.recaro.com/en/business-areas/automotive-seat ing.html and /special-fields-of-operation.html (all accessed 8/14/2014).

Reinecke, S. (2010): Best Practice in Marketing: Erfolgsbeispiele zu den vier Kernaufgaben im Marketing [Best practice in marketing: success stories regarding the four core tasks in marketing], Zürich: Compendio.

Richardson, J. (2008): The business model: an integrative framework for strategy execution. In: Strategic Change, vol. 17 (2008), no. 5–6, pp. 133–144.

Rieg, R./Esslinger, A. S. (2012): Die Wirksamkeit der Balanced Scorecard. Eine Analyse empirischer Studien [The effectiveness of the balanced scorecard. An analysis of empirical studies], in: Controlling – Zeitschrift für erfolgsorientierte Unternehmenssteuerung, vol. 24 (2012), no. 10, pp. 568–574.

Riesenhuber, M (2006): Die Fehlentscheidung. Ursache und Eskalation [The wrong decision. Causes and escalation], Wiesbaden: DUV/Gabler.

Rigby, D./Bilodeau, B. (2015): Management Tools & Trends 2015. Bain and Company [Electronic Version], retrieved from http://www.bain.com/publications/articles/management-tools-and-trends-2015.aspx (accessed: 11/14/2015).

Robbins, S. P. (2004): Decide & Conquer: Making Winning Decisions and Taking Control of Your Life, Upper Saddle River, NJ: Financial Times/Prentice Hall.

Robbins, S. P./DeCenzo, D. A. (2008): Fundamentals of Management. Essential Concepts and Applications, 6th edition, Uppers Saddle River, NJ: Pearson Prentice Hall.

Robbins, S. P./Judge, T. A. (2013): Organizational Behavior, 15th ed., global edition, Harlow, England: Pearson.

Roland Berger Strategy Consultants (2015): Roland Berger Trend Compendium 2030. Munich: Roland Berger Strategy Consultants, retrieved from: http://www.rolandberger.com/expertise/trend_compendium_2030/ (accessed: 8/20/2015).

Ross, S. A./Westerfield, R. W./Jordan, B. D. (2010): Fundamentals of Corporate Finance, 9th ed., New York, NY: McGraw-Hill/Irwin.

Rothaermel, F. T. (2013): Strategic Management: Concepts, New York, NY: McGraw-Hill/Irwin.

Roxburgh, C. (2003): Hidden flaws in strategy. Can insights from behavioral economics explain why good executives back bad strategies? In: The McKinsey Quarterly, no. 2 (2003), pp. 27–39.

Rühli, E. (1992): Koordination [Coordination]. In: Frese, E. (ed.): Handwörterbuch der Organisation, 3rd ed., Stuttgart: Schäffer-Poeschel, pp. 1164–1175.

Rühli, E. (1990): Zeitgemäße Konzernführung und -gestaltung. Anforderungen aus der Trilogie Strategie-Struktur-Kultur [Contemporary corporate management and design. Requirements derived from the strategy-structure-culture trilogy]. In: Zeitschrift Führung + Organisation (zfo), vol. 59 (1990), no. 5, pp. 310–314.

Rummler, G. A./Ramias, A. J./Rummler, R. A. (2010): White Space Revisited. Creating Value Through Process, San Francisco, CA: John Wiley & Sons.

Russow, L. C./Okoroafo, S. C. (1996): On the Way Towards Developing a Global Screening Model. In: International Marketing Review, vol. 13 (1996), no. 1, pp. 46–64.

Sackmann, S. A. (2006a): Betriebsvergleich Unternehmenskultur. Welche kulturellen Faktoren beeinflussen den Unternehmenserfolg? [Intercompany comparison of corporate culture. Which cultural factors impact company success?]. Neubiberg: Institut für Personal- und Organisationsforschung, Universität der Bundeswehr München.

Sackmann, S. A. (2006b): Erfassung von Unternehmenskultur: Eine Auswahl geeigneter Vorgehensweisen [Capturing corporate culture: a selection of effective methods]. In: Bertelsmann Stiftung (eds.): Messen, werten, optimieren: Erfolg durch Unternehmenskultur. Ein Leitfaden für die Praxis. Gütersloh: Bertelsmann Stiftung, pp. 36–39.

Sackmann, S. A. (2002): Unternehmenskultur: Erkennen, Entwickeln, Verändern [Corporate culture: identifying, developing, changing], Neuwied/Kriftel: Luchterhand.

Sackmann, S. A./Bertelsmann Stiftung (2006): Success Factor: Corporate Culture. Developing a Corporate Culture for High Performance and Long-term Competitiveness. Six Best Practices, Gütersloh: Verlag Bertelsmann Stiftung.

SAP (2015): SAP at a glance: company information, retrieved from: http://www.sap.com/corporate-en/about/our-company/index.html (accessed: 5/2/2015).

SAP (2013): New Ride-Sharing App from SAP Helps Organizations Reduce Costs, Save Resources and Better Engage Employees, April 29, 2013, Press Release, by SAP News. http://www.news-sap.com/new-ride-sharing-app-from-sap-helps-organizations-reduce-costs-save-resources-and-better-engage-employees/ (accessed: 9/12/2014).

Sawczyn, A./Fink, H./Werner, M. (2014): Leistungssteuerung (III): Managementsysteme für die Steuerung einzelner sozialer und gesellschaftlicher Leistungen [Performance management (III): managerial systems for control of single social and societal performance elements]. In: Schulz, T./Bergius, S. (eds.): CSR und Finance – Beitrag und Rolle des CFO für eine Nachhaltige Unternehmensführung, Berlin/Heidelberg: Springer Gabler, pp. 141–155.

SCC—Supply Chain Council (2012): SCOR® Quick Reference Guide, version 11.0, Cypress, TX.

Schallmo, D. (2013): Geschäftsmodell-Innovation. Grundlagen, bestehende Ansätze, methodisches Vorgehen und B2B-Geschäftsmodelle [Business model innovation. Foundations, existing concepts, methodological approach, and B2B business models], Wiesbaden: Springer.

Schaltegger, S. (2012): Die Beziehung zwischen CSR und Corporate Sustainability [The relationship between CSR and corporate sustainability]. In: Schneider, A./Schmidpeter, R. (Eds.): Corporate Social Responsibility. Verantwortungsvolle Unternehmensführung in Theorie und Praxis, Heidelberg: Springer, pp. 165–174.

Schaltegger, S./Burritt, R. (2005): Corporate Sustainability. In: Folmer, H./Tietenberg, T. (eds.): International Yearbook of Environmental and Resource Economics 2005/2006, Cheltenham: Edward Elgar, pp. 185–222.

Schaltegger, S./Lüdeke-Freund, F./Hansen, E. (2012): Business cases for sustainability: the role of business model innovation for corporate sustainability. In: International Journal of Innovation and Sustainable Development, vol. 6 (2012), no. 2, pp. 95–119.

Schaltegger, S./Synnestvedt, T. (2002): The Link between "Green" and Economic Success. Environmental Management as the Crucial Trigger between Environmental and Economic Performance. In: Journal of Environmental Management, vol. 65 (2002), no. 2, pp. 339–346.

Schaltegger, S./Lüdeke-Freund, F./Hansen, E. G. (2011): Business Cases for Sustainability and the Role of Business Model Innovation Developing a Conceptual Framework, Lüneburg: Centre for Sustainability Management (CSM).

Schaltegger, S./Wagner, M. (eds., 2006): Managing the Business Case of Sustainability, Sheffield: Greenleaf.

Schaltegger, S./Wagner, M. (2006): Managing and measuring the business case for sustainability. Capturing the relationship between sustainability performance, business competitiveness and economic performance. In: Schaltegger, S./Wagner, M. (eds.): Managing the business case for sustainability, Sheffield: Greenleaf, pp. 1–27.

Scheier, C./Held, D. (2012): Was Marken erfolgreich macht. Neuropsychologie in der Markenführung [What makes brands successful. Neuropsychology in brand management], 3rd ed., Freiburg: Haufe.

Schein, E. H. (2009): The Corporte Culture Survival Guide, new and revised edition, San Francisco, CA: Jossey-Bass.

Scheuss, R. (2012): Handbuch der Strategien. 220 Konzepte der weltbesten Vordenker, [Handbook of strategies. 220 concepts of world-class pioneers], 2nd ed., Frankfurt am Main: Campus.

Schiemann, W. A./Lingle, J. H. (2000): Strategieziele erreichen. Erfolgsfaktoren identifizieren und messen, München/Wien: Hanser.

Schmelzer, H. J./Sesselmann, W. (2010): Geschäftsprozessmanagement in der Praxis. Kunden zufrieden stellen, Produktivität steigern, Wert erhöhen [Business process management in practice. Satisfying customers, improving productivity, increasing value], 7th ed., München: Hanser.

Schmidpeter, R. (2015a): Globale gesellschaftliche Probleme als unternehmerische Chance. Neues Managementparadigma [Global societal problems as enterpreneurial opportunity. A new management paradigm]. In: Verantwortung Zukunft. Das Magazin, no. 1–2015, pp. 38–41.

Schmidpeter (2015b): CSR als betriebswirtschaftlicher Ansatz [CSR as management approach]. In: Schneider, A./Schmidpeter, R. (eds.): Corporate Social Responsibility. Verantwortungsvolle Unternehmensführung in Theorie und Praxis, 2nd ed., Berlin/Heidelberg: Springer Gabler, pp. 1229–1238.

Schrager, J. E./Madansky, A. (2013): Behavioral strategy: a foundational view. In: Journal of Strategy and Management, vol. 6 (2013), no. 1, pp. 81–95.

Scheyögg, G./Klisch-Eberl, M. (2007): How Dynamic Can Organizational Capabilities Be? Toward a Dual-Process Model of Capability Dynamization. In: Strategic Management Journal, vol. 28 (2007), no. 9, pp. 913–933.

Schürmann, M. (2011): Marketing. In vier Schritten zum eigenen Marketingkonzept [Marketing. In four steps to your own marketing concept], 2nd ed., Zürich: vdf Hochschulverlag AG an der ETH Zürich.

Schulz, T. (2014): CFO-Agenda: Gute Gründe, Nachhaltigkeit auf die Tagesordnung zu setzen [CFO-agenda: Good reasons for putting sustainability on the agenda]. In: Schulz, T./Bergius, S. (eds.): CSR und Finance, CSR und Finance – Beitrag und Rolle des CFO für eine Nachhaltige Unternehmensführung, Berlin/Heidelberg: Springer Gabler, pp. 3–34.

Schwartz, H./Davis, S. M. (1981): Matching Corporate Culture and Business Strategy. In: Organizational Dynamics 10, pp. 30–48.

Seidenschwarz, W. (ed., 2008): Marktorientiertes Prozessmanagement. Wie Process Mass Customization Kundenorientierung und Prozessstandardisierung integriert [Market oriented process management. How process mass customization integrates customer orientation and process standardization], 2nd ed., Munich: Vahlen.

Seidenschwarz, W. (1997): Nie wieder zu teuer! – 10 Schritte zum marktorientierten Kostenmanagement [Never again overselling!—10 steps to market oriented cost management], Stuttgart: Schäffer-Poeschel.

Senge, P. M. (1990): The Fifth Discipline. The Art and Practice of the Learning Organization, New York, NY: Currency.

Servatius, H.-G. (2002): Integration des Performance Measurement in die Führungsprozesse [Integration of performance measurement into the management processes]. In: Horváth, P. (ed.): Performance Controlling. Strategie, Leistung und Anreizsystem effektiv verbinden, Stuttgart: Schäffer-Poeschel, pp. 179–204.

Shingo, S. (1986): Zero Quality Control: Source Inspection and the Poka-Yoke System, Cambridge, MA: Productivity Press.

Shingo, S. (1985): A Revolution in Manufacturing—The SMED System, Cambridge, MA: Productivity Press.

Sibbet, D. (1997): 75 Years of Management Ideas and Practice. 1922–1997, Harvard Business Review, special supplement to the 75th anniversary issue of the Harvard Business Review.

Sidler, P. (2004): Branding Nespresso. Unterwegs dazugelernt [Branding Nespresso. Enhanced learnings on the run], (retrieved 2/8/2015 from http://www.mckini van.com/fileadmin/download/de/Wissen/Archiv/Branding_Nespresso.pdf).

Siemens AG (2015): Pictures of the Future, retrieved from http://www.siemens.com/innovation/en/home/pictures-of-the-future.html (accessed: 9/21/2015)

Siemens AG/Pictures of the Future (2015): Smart Cities Scenario 2060: From Parking to Parkland, retrieved from http://www.siemens.com/innovation/en/home/pictures-of-the-future/infrastructure-and-finance/smart-cities-scenario-2060.html (accessed: 9/21/2015).

Simon, H. A. (1957): Models of man. New York, NY: Wiley.

Simons, R. (2000): Performance Measurement & Control Systems for Implementing Strategy: Text and Cases, Upper Saddle River, NJ: Prentice Hall.

Simons, R. (1995): Levers of Control. How Managers Use Innovative Control Systems to Drive Strategic Renewal, Boston, MA: Harvard Business School Press.

Slaughter, R. A. (2002): Developing and applying strategic foresight, retrieved from http://www.forschungsnetzwerk.at/downloadpub/2002slaughter_strategic_foresight.pdf (accessed: 7/27/2015).

Slywotzky, A. J./Morrison, D. J./Moser, T./Mundt, K. A./Quella, J. A. (1999): Profit Patterns. 30 Ways to Anticipate and Profit from Strategic Forces Reshaping Your Business, New York, NY: Times Books/Random House.

Slywotzky, A. J./Morrison, D. J./Andelman, B. (1998): The Profit Zone. How Strategic Business Design Will Lead You to Tomorrow's Profits, Chichester: John Wiley & Sons.

Social Value UK (2015): What is Social Value? retrieved from http://socialvalueuk.org/what-is-sroi (accessed: 10/27/2015).

Sommer, A. (2012): Managing green business model transformation, Berlin/Heidelberg: Springer.

Speckbacher, G./Bischof, J./Pfeiffer, T. (2003): A descriptive analysis on the implementation of Balanced Scorecards in German-speaking countries. In: Management Accounting Research, vol. 14 (2003), pp. 361–387.

Spitzenpfeil, T./Schneider, S. (2012): Volatilität als strategische Herausforderung in Hightech-Unternehmen [Volatility as strategic challenge in high-tech companies]. In: Horváth, P./Michel, U. (eds.): Controlling und Finance. Steuerung im volatilen Umfeld, Stuttgart: Schäffer-Poeschel, pp. 93–111.

Spitzenpfeil, T./Schneider, S. (2011): Von der Unternehmensvision zu zielorientierten KPIs [From company vision to goal-oriented KPIs]. In: Horváth, P. (ed.): Exzellentes Controlling, exzellente Unternehmensleistung. Best Practice und Trends im Controlling, Stuttgart: Schäffer-Poeschel.

Stadler, C./Wältermann, P. (2012): Die Jahrhundert-Champions: Fünf Prinzipien für dauerhaften Unternehmenserfolg oder Was wir aus der Geschichte europäischer Top-Unternehmen lernen können [The champions of a century: Five principles for sustained company success or What we can learn from the history of European top companies], Stuttgart: Schäffer-Poeschel.

Stanovich, K. E./West, R. F. (2000): Individual differences in reasoning: Implications for the rationality debate. In: Behavioral and Brain Sciences, vol. 23 (2000), no. 5, 645–665.

Stewart, G. B. (1991): The Quest for Value, New York, NY: Harper Business.

Steiger-Bagel, A. (2012): Gegen den Strom – unser Weg zu einer High-Performance-/Low-Cost-Kultur [Against the tide—our journey to a high-performance/low-cost culture]. In: Horváth, P./Michel, U. (eds.): Controlling und Finance. Steuerung im volatilen Umfeld, Stuttgart: Schäffer-Poeschel, pp. 31–41.

Steinmann, H./Schreyögg, G. (2000): Management. Grundlagen der Unternehmensführung. Konzepte – Funktionen – Fallstudien [Management. Foundations of Corporate Management. Concepts – Functions – Case Studies], 5th ed., Wiesbaden: Gabler.

Stöger, R. (2013): Die Toolbox für Manager [The toolbox for managers], Stuttgart: Schäffer-Poeschel.

Stöger, R. (2011): Prozessmanagement. Qualität, Produktivität, Konkurrenzfähigkeit [Process management. Quality, productivity, competitiveness], 3rd ed., Stuttgart: Schäffer-Poeschel.

Stockmeyer, B./Weindlmaier, H.: Produktentwicklung in der Ernährungsindustrie – Ausgestaltung und Erfolg [Product development in the Food industry—design and success], Arbeitsbericht 99/2, Institut für Betriebswirtschaftslehre, FML Weihenstephan 1999.

Strebel, P. (1998): The Change Pact. Building Commitment to Ongoing Change, London: Financial Times/Pitman.

Sull, D./Homkes, R./Sull, C. (2015): Why Strategy Execution Unravels—and What to Do About It. In: Harvard Business Review, vol. 93 (2015), no. 3, pp. 58–66.

Teece, D. J. (2010): Business models, business strategy and innovation. In: Long Range Planning, vol. 43 (2010), no. 2/3, pp. 172–194.

Teece, D. J./Pisano, G./Shuen, A. (1997): Dynamic capabilities and strategic management. In: Strategic Management Journal, vol. 18 (1997), no. 7, pp. 509–533.

The Economist Intelligence Unit (2013): Why good strategies fail. Lessons for the C-suite, London et al.: White Paper sponsored by Project Management Institute (PMI).

The North Highland Company (2014): Case Study. Getting Beyond Strategy Definition to Successful Execution. Retrieved from http://www.northhighland.com/insights/case-studies (accessed: 6/28/2014).

The Palladium Group (2014a): http://www.thepalladiumgroup.com/Results/hof/Pages/Overview.aspx (accessed: 7/31/2014).

The Palladium Group (2014b): http://www.thepalladiumgroup.com/Knowledge Object Repository/HOF_2013/Palladium_Hall_of_Fame_for_Executing_Strategy_2013.pdf (accessed: 7/31/2014).

Theisen, M. R. (2011): Wissenschaftliches Arbeiten [Scientific work], 15th ed., München: Vahlen.

Thompson, A. A./Peteraf, M. A./Gamble, J. E./Strickland III, A. J. (2014): Crafting and Executing Strategy. The Quest for Competitive Advantage. Concepts and Cases, 19th ed., New York, NY: McGraw-Hill/Irwin.

Toyota (2015a): Toyota Global Vision, retrieved from http://www.toyota-global.com/company/vision_philosophy/toyota_global_vision_2020.html (accessed: 11/10/2015)

Toyota (2015b): Toyota Global Vision, Learn about The Tree, retrieved from http://www.toyota-global.com/company/vision_philosophy/ toyota_global_vision_2020.html (accessed: 11/10/2015)

Toyota (2015c): Sustainability Report 2015. Corporate Planning Div., Toyota Motor Corporation retrieved from http://www.toyota-global.com/sustainability/report/sr/pdf/sustainability_report15_fe.pdf (accessed: 11/10/2015)

Trapp, M. (2014): Realizing Business Model Innovation. A Strategic Approach for Business Unit Managers, Wiesbaden: Springer Gabler.

Treacy, M./Wiersema, F. (1995): The Discipline of Market Leaders. Choose Your Customers, Narrow Your Focus, Dominate Your Market, Cambridge, MA: Perseus Books.

Tversky, A./Kahneman, D. (1974): Judgment under uncertainty: Heuristics and biases. In: Science, vol. 185 (1974), pp. 1124–1130.

Ulrich, H. (2001): Systemorientiertes Management. Das Werk von Hans Ulrich [System-oriented management. The work of Hans Ulrich], Studienausgabe, Bern: Haupt.

Unerman, J. (2007): Stakeholder engagement and dialogue. In: Unerman, J./Bebbington, J./O'Dwyer, B. (eds): Sustainability Accounting and Accountability, Abingdon, New York, NY: Routledge.

Ungericht, B. (2012): Strategiebewusstes Management. Konzepte und Instrumente für nachhaltiges Handeln [Strategy-aware management. Concepts and tools for sustainable practices], München: Pearson.

Vahs, D. (2007): Organisation: Einführung in die Organisationstheorie und -praxis [Organization: Introduction to organizational theory and practice], 6th ed., Stuttgart: Schäffer-Poeschel.

Volkswagen AG (2015): Annual Report 2014, moving progress, Wolfsburg: Volkswagen AG Group Communications.

Volkswagen 2014: http://www.volkswagen.de/de/markenwelt/verantwortung/Standorte/amerika/Chattanooga.html (accessed: 8/13/2014).

Watson, R. (2012): Future files. A brief history of the next 50 years. Reprint with revisions, London/Boston: Nicholas Brealey Publishing.

WCED—World Commission on Environment and Development (1987): Our Common Future. Oxford.

Weber, J./Schäffer, U. (2011): Einführung in das Controlling [Introduction to Controlling], 13th ed., Stuttgart: Schäffer-Poeschel.

Weber, D./Wunder, T. (2002): Was kommt nach Zielkosten? – Strukturiertes Kostenkneten in der Wehrtechnikbranche [What will follow after determining market-oriented costs?—Structured cost shaping in the defence technology sector]. In: krp-Kostenrechnungspraxis, vol. 46 (2002), no. 4, pp. 240–248.

Wehking, J. (2014): DB2020 – Wie transformiert man die DB? [DB2020—how to transform the German Rail?]. In: 14. Jahreskonferenz Strategisches Management, Horváth & Partners, Stuttgart May 8, 2014.

Weihrich, H. (1982): The TOWS-Matrix: A Tool for Situational Analysis. In: Long Range Planning, vol. 15 (1982), no. 2, pp. 54–66.

Welch, J./Welch, S. (2006): Stop the B. S. Budget. In: Business Week, June 26, 2006, The Welch Way by Jack and Suzy Welch.

Werle, K.: Strategen – Die Rückkehr der Kampfdenker [Strategists—return of the warlords]. In: Manager Magazin, 10/2005, pp. 190–199.

Welge, M.K./Al-Laham, A. (2012): Strategisches Management: Grundlagen – Prozess – Implementierung [Strategic Management: Foundations—Process—Implementation], vol. 6, Wiesbaden: SpringerGabler.

Wernerfeld, B. (1984): A Resource-based View of the Firm. Ten Years After. In: Strategic Management Journal, vol. 5 (1984), no. 2, pp. 171–180.

Wheelen, T. L./Hunger, J. D. (2010): Strategic Management and Business Policy. Achieving Sustainability, 12th ed., Upper Saddle River, NJ: Prentice Hall.

Wiggins, R. R./Ruefli, T. W. (2005): Schumpeter's ghost: Is hypercompetition making best of times shorter. In: Strategic Management Journal, vol. 26 (2005), no. 10, pp. 887–911.

Williamson, P. J./Zeng, M. (2009): Value-for-money strategies for recessionary times. In: Harvard Business Review, vol. 87 (2009), no. 3, pp. 66–74.

Williamson, P. J. (2010): Cost innovation: preparing for a 'value-for-money' revolution. In: Long Range Planning, vol. 43 (2010), no. 2–3, pp. 343–353.

Wirtz, B. (2011): Business Model Management. Design, Instrumente, Erfolgsfaktoren von Geschäftsmodellen [Business model management. Design, tools, success factors of business models], 2nd ed., Wiesbaden: Gabler.

Wirtz, B. (2010): Eletronic Business, 3rd ed., Wiesbaden: Gabler.

Womack, J. P./Jones, D. T. (2003): Lean Thinking. Banish Waste and Create Wealth in Your Corporation, 2nd edition, New York, NY: Free Press.

World Economic Forum (2015): Strategic Foresight, retrieved from http://www.weforum.org/community/strategic-foresight (accessed 8/29/2015).

World Economic Forum (2009): retrieved from The Future of the Global Financial System. A Near-Term Outlook and Long-Term Scenarios, World Scenarios Series, Cologny/Geneva: World Economic Forum, http://www3.weforum.org/docs/WEF_Scenario_FutureGlobalFinancialSystem_Report_2010.pdf (accessed 8/29/2015).

Wunder, T. (2014a): Strategisches Management: Integration ökologischer Nachhaltigkeit in den Strategieprozess [Strategic management: Integration of ecological sustainability into the strategy process]. In: Schulz, T./Bergius, S. (eds.): CSR und Finance – Beitrag und Rolle des CFO für eine Nachhaltige Unternehmensführung, Berlin/Heidelberg: Springer Gabler, pp. 65–81.

Wunder, T. (2014b): Strategie-Alignment. Eine gemeinsame strategische Orientierung schaffen [Strategy alignment. Creating a shared strategic orientation]. In: Zeitschrift für Führung und Organisation (zfo), vol. 83 (2015), no. 6, pp. 408–412.

Wunder, T. (2013): Geschäftsmodelle. Die Erfolgslogik des Geschäfts verstehen und gestalten [Business models. Understanding and designing the business success logic]. In: Zeitschrift für Führung und Organisation (zfo), vol. 82 (2013), no. 5, pp. 354–360.

Wunder, T. (2012): Strategy Maps. Strategien klären, beschreiben und abgleichen [Strategy maps. Clarifying, describing and aligning strategies]. In: Zeitschrift für Führung und Organisation (zfo), vol. 81 (2012), no. 4, pp. 273–277.

Wunder, T. (2006): Advanced Budgeting—Making Planning More Relevant, Péter Horváth & Partners Inc. In: ASMI 2006 Budgeting & Forecasting Conference, American Strategic Management Institute, Chicago, IL (USA), September 20–21, 2006.

Wunder, T. (2005): New Strategy Alignment in Multinational Corporations. In: Strategic Finance, vol. 87 (2005), no. 5, pp. 35–41.

Wunder, T. (2004a): Transnationale Strategien. Anwendungsorientierte Realisierung mit Balanced Scorecards [Transnational strategies. Actionable implementation with balanced scorecards], Wiesbaden: DUV/Gabler, PhD thesis at European Business School, Oestrich-Winkel 2003.

Wunder, T. (2004b): Wichtiges Fundament. Immaterielle Werte: Grünes Licht für Ihre Strategie [Important fundament. Intangibles: green light for your strategies]. In: Harvard Business Manager, vol. 26 (2004), no. 7, Forum, p. 111.

Wunder, T. (2001): Wie konkret muss eine Balanced Scorecard sein? [How precise does a balanced scorecard have to be?]. In: Controller Magazin, vol. 26 (2001), no. 2, pp. 133–139.

Wunder, T./Baur, M. (2000): Die Balanced Scorecard unternehmensweit einführen. Der BSC Roll-out an einem Beispiel aus der Energiewirtschaft [Company-wide implementation of balanced scorecard. The BSC rollout illustrated with a utility company]. In: Controlling, vol. 12 (2000), no. 11, pp. 549–556.

Wunder, T./Bausch, J. (2015): Auf dem Weg zu operativer Exzellenz. Erfolgsfaktoren und Handlungsempfehlungen [On the way to operational excellence. Success factors and recommendations]. In: Controller Magazin, vol. 40 (2015), no. 3, pp. 53–61.

Wunder, T./Bausch, J. (2014a): Unternehmerische Kreativität zielgerichtet entfalten [Goal-oriented unfolding of entrepreneurial creativity]. In: Controlling & Management Review, vol. 58 (2014), no. 4., pp. 46–55.

Wunder, T./Bausch, J. (2014b): Vier Erfolgsfaktoren für einen effektiven Strategieprozess [Four success factors of an effective strategy process]. In: Controlling & Management Review, vol. 58 (2014), no. 1, pp. 54–62.

Wunder, T./Müller, M. (2009): Navigating through the Downturn & how to end up amongst the Winners. In: German American Trade, vol. 20 (2009), no. 4, pp. 6–10.

Wunder, T./Müller, M. (2008): In Search of Finance Excellence. In: The Journal of Corporate Accounting & Finance, vol. 19 (2008), no. 3, pp. 41–49.

Wunder, T./Stemmermann, K. (2013): Strategisches Management und Führungskultur. Warum auch die besten Strategieprozesse manchmal scheitern [Strategic management and leadership culture. Why even the best strategy processes sometimes fail]. In: SEM | Radar Zeitschrift für Systemdenken und Entscheidungsfindung im Management, vol. 12 (2013), no. 2, pp. 105–143.

Wunderer, R. (2001): Führung und Zusammenarbeit. Eine unternehmerische Führungslehre [Leadership and collaboration. An entrepreneurial management science], 4th ed., Neuwied/Kriftel: Luchterhand.

Zeitschrift Führung + Organisation (ed., 2015): Toolkit – Managementinstrumente für die Praxis. Strategie und Leadership, Organisation und Change Management, Projekt- und Prozessmanagement [Toolkit—management methods for practice. Strategy and leadership, organisation and change management, project and process management], Stuttgart: Schäffer-Poeschel.

Zeitz, J. (2014): The Courage to Redefine Business Purpose, Talk by Jochen Zeitz at Zermatt Summit 2014, published on Jul 10, 2014, retrieved from https://www.youtube.com/watch?v= 5NvRL5v-c2k (accessed: 5/21/2015).

Zenie, F. H. (2011): StreetSmart Entrepreneuring, 3rd ed., Lulu.com.

Zink, K. J. (2004): TQM als integratives Managementkonzept. Das EFQM Excellence Modell und seine Umsetzung [TQM as an integrative management concept. The EFQM Excellence Model and its implementation], 2nd edition, München/Wien: Carl Hanser.

Zollenkop, M. (2006): Geschäftsmodellinnovation. Initiierung eines systematischen Innovationsmanagements für Geschäftsmodelle auf Basis lebenszyklusorientierter Frühaufklärung [Business model innovation. Initiating a systematic innovation management for business models based on lifecycle-oriented early warning], Wiesbaden: DUV/Gabler.

Zott, C./Amit, R. (2010): Business model design: an activity system perspective. In: Long Range Planning, vol. 43 (2010), no. 2–3, pp. 216–226.

Zott, C./Amit, R./Massa, L. (2011): The Business Model: Recent Developments and Future Research. In: Journal of Management, vol. 37 (2011), no. 4, pp. 1019–1042.

Index